Cell Technology

CELL TECHNOLOGY

Khushboo Chaudhary

Delve
PUBLISHING

www.delvepublishing.com

Cell Technology

Khushboo Chaudhary

Delve Publishing
224 Shoreacres Road
Burlington, ON L7L 2H2
Canada
www.delvepublishing.com
Email: orders@arclereducation.com

ISBN: 978-1-77469-516-6 (Hardcover)

ABOUT THE AUTHOR

Dr. Khushboo Chaudhary is presently working as a Research Associate in NRCE, Hisar Haryana, India and has one year of teaching experience and seven years of research experience. Previously, she worked on "Improvement of Phytoremediation efficiency of Fluoride". She has published several research papers in international and national journals. She has published five international textbooks. She has got the seven best paper and poster presentation awards from the Indian Society of Genetics and Biotechnology Research and Development and received the president appreciation awards also in the International conference. She has got the best poster award from ISSGPU Central Institute Research on Goats, Makhdoom. She has published several gene banks in NCBI Pubmed. She has also published a research article in the virology journal. She is likely to be a co-author in several of the publications and coauthor in J. Virological Methods.

TABLE OF CONTENTS

LIST OF GLOSSARY

Agarose Gel Electrophoresis: It is a process in which a matrix composed of a highly purified form of agar is used to separate large DNA and RNA molecules.

Allele: One of the pair of variant forms of a gene occurring at a given locus of a gene.

Amplification: There are formations of many copies of the DNA segment of PCR.

Annealing: The process of healing and slowly cooling of double-stranded DNA to allow the formation of hybrid DNA or DNA-RNA molecules.

Antibiotic: A biological substance produced by one organism inhibiting the growth of another organism.

Antiparallel orientation: Arrangement of two strands of DNA molecule oriented in opposite directions so that the 5' PO_4 end of one strand is aligned with the 3'-OH end of the complementary strand.

Antisense DNA: The sequence of Chromosomal DNA that is transcribed.

Antisense RNA: An RNA sequence complementary to all or part of a functional RNA.

Antisense therapy: *In vitro* treatment of genetic disease by blocking translation of a protein with DNA or RNA sequences that is complementary to a specific mRNA.

Artificial seeds: A synthetic seed consisting of a somatic embryo surrounded by a nutrient medium which is protected by a thin membrane of chemicals.

Autoclave: An instrument used for sterilization of glassware and culture media.

Auxins: A class of plant growth regulators that stimulate cell division, cell elongation, apical dominance and root initiation. They are used in plant cell and tissue culture e.g. 2,4-D, IAA, NAA, etc.

Axenic culture: A pure culture of a single propagule of an organism not contaminated by any other microorganism.

Axillary bud: A bud produced in the axils of leaves.

Bioaccumulation: The concentration of a chemical agent e.g DDT in the increasing amount in the organisms of a food chain.

Biodegradation: The gradual breakdown of a compound to its constituents by a living organism.

Biodiversity: The variability among the living organism from all sources, soil, water, air, extreme habitat or associated with organisms.

Biofertilizers: The commercial preparation of microorganisms by which nitrogen and phosphorus levels and growth of plants are increased.

Bioremediation: The process of using living organisms to remove contaminants, pollutants or unwanted substances from soil or water.

Biosensor: An electronic device that uses biological molecules or cells to detect specific compounds.

Biotic stress: Stress resulting from living organisms that can harm other plants such as viruses, bacteria, fungi, weeds, harmful insects.

Callus: An organized mass of plant cells capable of cell division and growth *in vitro*.

cDNA clone: A double-stranded DNA complement of mRNA synthesized *in vitro* by using reverse transcriptase and DNA polymerase.

cDNA library: A collection of cDNA clones generated *in vitro* from the mRNA sequences of a single tissue or cell population.

Cell culture: The cell culture of growing cells *in vitro* in a liquid medium.

Cell line: The cell line is derived from a primary culture and selected at the time of the first subculture or derived from a cell population with particular attributes.

Chimera: Recombinant DNA molecules containing sequences from different organisms.

Clone: A collection of genetically identical cells or organisms derived from a common ancestor where all members have a similar genetic composition.

Consensus sequences: The nucleotide sequences that are present in the majority of genetic signals that perform a specific function.

Cosmid: A plasmid vector that consists of the cos site of phage lambda (λ) and one or more selectable markers such as an antibiotic resistance gene.

Crown gall: A tumor occurring at the base of certain plants due to infection of the plant by *Agrobacterium tumifaciens*.

Cytokinins: The class of plant growth regulators which causes cell division, cell differentiation, shoot differentiation and breaking of apical dominance e.g BAP, Kinetin and zeatin.

Denaturation: The separation of DNA double-stranded molecules into single strands.

Differentiation: The process of biochemical and structural changes by which cells become specialized in form and function.

DNA construct: A suitable DNA that has been prepared for cloning purposes.

DNA polymerase: An enzyme that catalyzes the phosphodiester bond in the formation of DNA.

DNA probe: A radiolabelled (^{32}P) DNA segment used to hybridize the base between the probe and complementary base sequences in a DNA sample.

Electrophoresis: The method of separation of a molecule such as DNA, RNA or protein based on their relative migration applying a strong electric field.

Embryo transfer: Implantation of embryos from donor animals or generated by *in vitro* fertilization into the uterus of the recipient animals.

Excision: Enzymatic removal of a DNA segment from a chromosome.

Exonucleases: An enzyme that digests DNA or RNA from the ends of the strand; 5' exonuclease requires a free 5' end and degrades the molecule in the 5'-3 direction; 3' exonuclease requires a free 3' end and degrades the DNA IN 3'-5 direction.

Explant: The tissue that is taken from a plant or seed and transferred to a culture medium to establish a tissue culture system or regenerate the plant.

Fermentation: The process by which microorganisms turn raw materials such as glucose into products such as alcohol.

Fusion product: A single protein molecule encoded by parts of two or more genes combined together as one unit.

Gene: A hereditary unit or the segment of DNA coding a specific protein.

Genetic fingerprinting: A technique of analyzing the DNA of an individual to reveal the pattern of repetition of particular nucleotide sequences through the genome.

***In vitro*:** Any process carried out in sterile cultures or measurement of biological processes outside the intact organism such as enzyme reaction.

***In vivo*:** The natural condition in which organisms live.

Leader sequences: It is the non-translated sequence at the 5' end of mRNA that precedes the initiation codon.

Lysogeny: The phenomenon in which prophage survives within a host bacterium as a part of the host genome or as an extrachromosomal element and does not cause lysis.

Micropropagation: The small piece of tissue such as meristem grown in culture to produce a large number of plants.

Nod genes: Genens encode nodulins.

Oncogene: A gene that causes cancer.

Operon: A cluster of genes that are coordinately regulated.

Palindromic sequences: The complementary DNA sequences that are the same when each strand is read in the same direction e.g (5'-3). These sequences act as recognition sites for type II restriction endonuclease.

Patent: A legal document issued by a government that permits the holder the exclusive right to manufacture, use of, to sell an invention for a defined period, which varies from country to country.

Plasmid: Extrachromosomal, self-replicating, circular, double-stranded DNA molecules containing some non-essential genes.

Primer: A short oligonucleotide that hybridizes the template strand and gives a 3' OH end for the initiation of nucleic acid synthesis.

Recombinant: An individual whose genes on chromosomes result from one or more cross-over events.

Secondary metabolites: The metabolites produced as end products of primary metabolism and not involved directly in metabolic activity.

Splicing: The gene manipulation where the one DNA molecule is attached to one another.

Totipotency: A property of normal cells that they have the genetic potential to give rise to complete individuals except terminally specialized cells.

Transfection: DNA transfer into a eukaryotic cell.

Transposon: A DNA segment that can jump from a locus and join a new locus on the DNA molecule by involving the enzyme transposase.

Vectors: The vehicles for transferring DNA from one cell to another or DNA molecule that can carry inserted DNA and be perpetuated in a host cell.

Western blot: A technique in which protein is transferred from an electrophoresis gel to a cellulose or nylon support membrane following electrophoresis.

Yeast artificial chromosomes: A vector is used for hundreds to kilobases long used for the cloning of a DNA fragment.

PREFACE

"The finding of the double helix thus brought us not only joy but great relief. It was unbelievably interesting and immediately allowed us to make a serious proposal for the mechanism of gene duplication".

- James Watson

The goal of this guidebook termed "Cell Technology" is to assist students understand the concepts of this vast field. It has been our experience as biology faculty that many students lack confidence to find notes so this book has been written in such a way that students gain the basic knowledge to make notes for subjective examinations.

Thus the information has been updated but keeping the basic structure and simplicity of the presentation. The topics covered are inclusive but not limited to Cell Biology, Cell Biochemistry, Cell Enzymology, Cell Immunology, Cell Molecular, Cell Genetics, Cell Microbiology, Cell Metabolic, Biotechnology, Plant Biotechnology, Animal Biotechnology, Bioinstrumentation and Techniques. All chapters have well-explained figures some of which have been drawn by the authors that can easily be understood by students. While preparing this book, the authors had in mind difficulties faced by our students so to enhance the utility of this book numerous examination-oriented problems and their solutions have been added.

This book is written in a very simple and easy style. Scientists and authors who have been referred to write the book have been quoted or fully acknowledged in the bibliography at end of the book. We claim no originality for the matter or text presented in this book but the method of presentation and all illustrations are our own.

"Biology is also more important than physics, as measured by its economic consequences, by its ethical implications, or by its effects on human welfare."

- Freeman Dyson

Khushboo Chaudhary

CHAPTER 1

CELL BIOLOGY

1. CELLULAR BASIS OF LIFE, EVOLUTION OF CELL THEORY AND ORGANIZATION OF PLANT AND ANIMAL CELLS

Cellular Basis of Life and Application of Microscope

Organisms are made of one or more cells. Multicellular organisms have a hierarchical organization of cells, tissues, organs and organ systems that work together to keep the cells alive. "Cells need a supply of energy and molecules to carry out life processes. Living organisms depend on their environment to survive. Animals, including humans, need a plentiful supply of air, water, nutrients from food, and a suitable temperature to survive. Plants need air, light, water, a suitable temperature, nutrients from soil, and room to grow. All living organisms carry out some characteristic processes that distinguish them from non-living things. Living organisms can move, can respond to the world around them, use food as a source of energy and materials, get rid of waste, grow and can reproduce – though something that is alive will not show all of these life processes all of the time, and in some cases may not show them in an obvious way. All living organisms eventually die. Tissues, organs and systems The bodies of humans and other animals are divided into different parts with different functions, including senses. Humans and some other animals have skeletons and muscles for support, protection and movement. The human digestive system includes the mouth, stomach and intestines. In the human circulatory system, the heart pumps blood around the body through blood vessels. Flowering plants are also divided into parts with different functions. The roots absorb water, which is transported up the stem or trunk to the leaves and flowers.

Cells Organisms, living and dead, are made up of cells. Cells are made of molecules organized into membranes and other structures. Most cells are too small to be seen with the naked eye but can be seen using a light microscope. There are many different types of cells with different shapes and sizes, but all cells are made up of common parts: all cells have a genome and cytoplasm contained by a cell membrane; all animal and plant cells store their genome within a nucleus, and they also have mitochondria; plant cells additionally have a cell wall and can have chloroplasts and a vacuole. These parts have common functions in all cells. Molecules move through the cytoplasm by diffusion, and some molecules can enter and leave a cell by diffusing through the cell membrane. A single cell can carry out all the processes of life. An organism may be made up of a single cell or many cells working together. This is why scientists think of cells as the basic units of life". Tissues, organs and systems to stay alive, cells need a constant supply of energy and molecules for chemical reactions, and they need to get rid of waste. In a multicellular organism, the cells are organized into tissues, organs and organ systems that work together to support the life processes of cells to keep the organism alive. In humans, the circulatory system transports useful molecules and waste around the body. The blood transports useful molecules to cells from food that has been broken down by the digestive system. The blood also transports oxygen to cells from the gas exchange system. Biochemistry Chemical reactions that make and break down substances take place all the time in living cells. Photosynthesis and cellular respiration are important chemical processes in living cells. Plants do not take in food. Instead, they make carbohydrates inside their cells, which is their only source of food. Plants are called producers because they make their own food. A chemical process called photosynthesis in plant cells makes the carbohydrate food, and also makes oxygen as a waste product. Photosynthesis requires light, water and carbon dioxide. Photosynthesis can be modeled very simply using a summary of the inputs and outputs. Photosynthesis takes place in chloroplasts inside cells in parts of a plant such as the leaves. Plants use the carbohydrate food made by photosynthesis to make other substances for growth and they also use it as a fuel. Although plants do not take in food, they do take in non-food substances, such as water and carbon dioxide for photosynthesis, and minerals that the plant cannot make. Plants are the source of the food that is transferred through food chains. Plant and animal cells break down glucose in a process called cellular respiration, which makes energy available for life processes". Aerobic cellular respiration uses oxygen in the breakdown of glucose, and produces carbon dioxide and water. Aerobic cellular respiration takes place in the cytoplasm and mitochondria inside cells. Aerobic cellular respiration can be modeled very simply using a word summary of the inputs and outputs of the process.

Microscope

Microorganisms are very small to be seen with the unaided eye. The study of microorganisms is greatly enhanced by the use of microscopes. The word microscope, first coined by Giovanni in 1625, derived from the Latin word micro (small) and the Greek word scopos (to look at), is an important tool for microbiologists and the technology of making very small things visible to the human eye (Figure 1.1). The simple microscope used by Anton van Leuwenhoek in the 17th century in 1673 had only one lens and was similar to a magnifying glass that could magnify a microbe 300X.

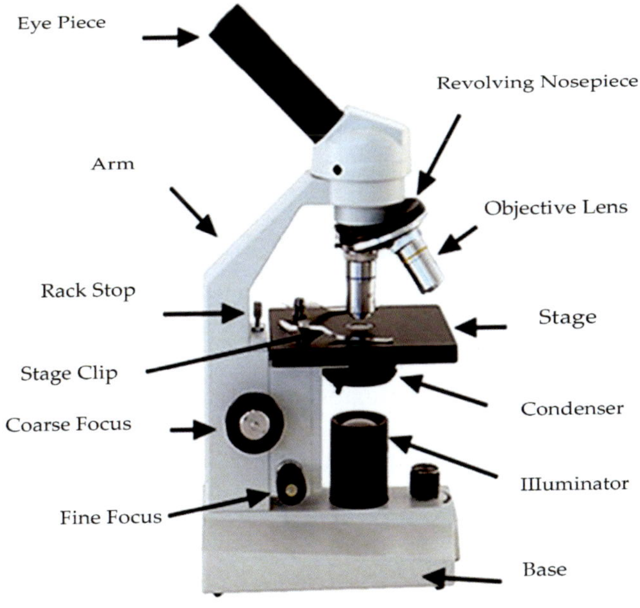

Figure 1.1: Microscope.

The simple microscopes enabled us see bacteria and protozoa. Robert Hooke, a contemporary of van Leeuwenhoek built compound microscopes that have multiple lenses. The microscopes used today are of two categories: light and electron. Light microscopes are sub-divided into bright-field and fluorescent microscopes, while the electron microscope is subdivided into scanning and transmission electron microscopes. The bright-field microscope or compound light microscope that allows light rays to pass directly through to the eye without being deflected by an intervening opaque plate in the condenser is routinely used in all the microbiology laboratories to study the size, shape, arrangement of microbial cells and sporulating structures. Dark-field microscopy is a kind of light microscopy that uses a special dark field condenser, which contains an opaque disk that causes the light to reflect off the specimen at an angle rather than pass directly through it. This microscopy is used to examine live microorganisms. Fluorescent microscopy uses ultraviolet light instead of white light to excite molecules within the specimen or fluorescent dye. Fluorescence microscopy must be intrinsically fluorescent or stained with a dye such as auramine O, acridine orange, fluorescein and DAPI (diamidino-2-phenylinodole).

Bright-field Light Microscopy

A microscope is a compound optical system, a compound magnifying glass. The real meaning of the functioning of a microscope is a test object positioned between the single and double focus points of the objective. The light coming from the object and passing through the lens creates a magnified, inverted and real image of the subject on the other side of the objective, behind the double focus point. The eyepiece is at a distance from the objective lens so that the image formed by the objective is generated within the focus of the eye piece. Thus, looking throughout the eyepieces, one can see a further enlarged, direct but virtual image of this real, inverted and magnified image. The magnification power of a bright-field microscope can be calculated by multiplying the magnification of the objective and of the eyepiece, respectively. The objective lens system consists of multiple lenses. The first is the front lens facing the object. This determines the magnification and resolution of the microscope. Other elements are responsible for the exclusion of lens errors. The features of an image formed by the objective depend on the optical characteristics of the objective lens. The quality of a lens represents its property of how sharp the object image is drawn. The image defect, i.e. blurring is caused by lens errors (aberrations). The spherical aberration (spherical divergence) is caused when the lens away from the optical axis increasingly breaks the light rays passing through it, therefore a point-like object in the image will be blurred at the edge. The reason for chromatic aberration is that the focal points of different wavelengths of light do not coincide on the optical axis.

- The shorter wavelength rays unite closer, while longer wavelength rays unite farther.
- Thus, a sharp image cannot be obtained using white light, and the image has a colored (rainbow) border.
- A concave flint glass lens with lead content and high refractive index is fitted to the convex lens to correct for chromatic aberration (the achromatic lens has a two-color correction, while the apochromatic lens has three-color correction).
- The resolution of the objective lens is the ability of how detailed the image of a subject can be drawn.
- The resolving power is quantified with the minimum distance between two points that are just distinguishable.
- The resolution (d) depends on the illumination wavelength of light used (λ), the half-angular aperture of the objective lens (α) and the refractive index of the material between the front lens and the cover slip (n).

$d = 1.22\ \lambda\ /2\ n \times \sin\alpha$

where $n \times \sin\alpha$ = numerical aperture (NA)

- The greater the resolution the smaller the values of d are.
- This can be achieved by reducing the wavelength of light used, increasing the angular aperture of the objective, or increasing the refractive index of the material between the front lens and the cover slip.

For this purpose, cedar oil (oil immersion) is the most suitable material, because it has almost the same refractive index as that of the glass, so the light passes through a virtually homogeneous medium. The denominator of the above formula and $2\ nX\ \sin\alpha$ is the value of numerical aperture (NA), which may vary between 0.20 and 1.4 (always indicated on the lens). The eyepiece draws a direct image of the test object. The fine structure of an image observed in the microscope depends on the details of the real picture, which in turn is determined by the resolution of the objective. This image, however, is not visible to the naked eye; it can only be visualized in the magnification of the eyepiece.

Microorganisms Examined from Natural Waters by Bright-field Light Microscopy

Principle

A microscope is a series of finely ground lenses and visible light as a source of illumination are used. The microscope field or area observed is brightly lighted and the objects being studied appear dark because they absorb some of the light. However, microbes absorb light after staining them with a dye greatly increases their light absorbing ability resulting in greater contrast and color differentiation.

Requirements

- Environmental water sample (e.g. from a lake, a stream, a creek or an aquarium)
- Glass slides
- Cover slips
- Glass dropper dispenser alcohol (for sterilization)
- Bunsen burner
- Light microscope

Procedure

1. Dry the surface of a glass slide with alcohol over a Bunsen burner and then label the slide.
2. Place one drop of the sample on the slide with a glass dropper.
3. Put the edge of a cover slip on the slide so that it touches the edge of the water drop.
4. Slowly lower the cover slip to prevent the formation and trapping of air bubbles.
5. Put the sample under the microscope, and locate the focal plane.
6. Apply first an objective lens of 16X, and then 40X magnification.
7. Observe the shape and movement of microbes, in the case of protists and eukaryotic algae, try to identify the different cell organelles (e.g. chloroplasts and contractile vacuoles)
8. Make drawings of the observations.

Fluorescence Microscopy

Specimens that absorb light of color and subsequently emit light of another color (fluoresce) can be visualized by using the fluorescence microscope. The basis of fluorescence microscopy is the principle of the removal of incident illumination by selective absorption, where light absorbed by the specimen and reemitted at another wavelength is transmitted. The light source must produce light of appropriate wavelength and ineffective wavelengths that are unable to excite the fluorochrome used are removed by an excitation filter. The phenomenon when a sample (e.g. cell organelle of a microbe) shows fluorescence without prior staining is called auto-fluorescence. Otherwise, bacterial cells can be stained with a fluorescent dye. Using a fluorescent dye, e.g. DAPI (4',6-diamidino-2-phenylindole), cells are stained bright blue because the dye forms a complex with the cell's DNA. Fluorescent dyes are therefore widely applied to visualize and enumerate bacteria in different natural habitats or clinical samples. Microbes are examined with DAPI staining (microbes from water sample) DAPI molecules are able to penetrate cell membranes and bind to the double helix of the DNA.

- Cells can be easily counted, if a known volume of a fixed water sample is filtered through a membrane and, after staining the filter, the surface is investigated by a fluorescence microscope (Figure 1.2).
- A microscope is a series of finely ground lenses and visible light is the source of illumination.
- The microscope field or area observed is brightly lighted and the objects being studied appear dark because they absorb some of the light.
- However, microbes do not absorb much light but after staining them with a dye greatly increases their light absorbing ability resulting in greater contrast and color differentiation.
- The effective magnification of a microscope depends upon its resolving power.

Typical magnification values are:

Scanning 10X × 4X= 40X magnification

Low power 10X × 10X = 100X magnification

High power 10X × 40X = 400X magnification

Oil immersion 10X ×100X = 1000X magnification

Total magnification = Magnifying power of the objective lens* magnifying power of the ocular lens.

Requirements

- Disposable gloves laboratory scales paraformaldehyde (PFA)
- Beaker
- Magnetic stirrer
- Phosphate buffered saline (PBS)
- NaOH solution
- Pasteur pipette
- Membrane filtration apparatus
- Polycarbonate and cellulose nitrate membrane filters (0.22 or 0.45 μm pore size)
- 50 ml Falcon tube
- Plastic Petri dishes
- Scalpel
- Pipette with pipette tips
- DAPI solution (1μg/ml)
- 80% ethanol
- Double distilled water
- Glass slide
- Cover slip
- Vectashield Mounting Medium (H-1000, Vector Laboratories Ltd) immersion oil (non-fluorescent)
- Fluorescence microscope with a mercury lamp digital camera computer with software

Procedure

1. For the preparation of the fixative solution, dissolve 1 gm PFA in 50 ml PBS. (PFA causes irritation when inhaled; therefore, the use of a fume hood is recommended.)
2. Dissolution can be aided with heating (60°C), permanent stirring and adding some drops of NaOH solution.
3. Adjust the pH to 7.0.

4. Filter the solution through a 0.22 µm pore size membrane filter. (The prepared 2% PFA solution can be stored in the fridge for one week).
5. Filter the water sample (2-50 ml, depending on the type of sample) using a polycarbonate membrane filter (slowly with occasional stirring).
6. To help uniform cell distribution, place a 0.45 µm pore size cellulose nitrate membrane filter between the sieve of the filtration unit and the polycarbonate filter.
7. Fill up the Falcon tube with fixative (PFA solution) and immerse the filter in it with sterile forceps (PFA solution must cover the entire membrane filter).
8. Incubate the filter overnight at 4°C.
9. Fill PBS into an empty Petri dish, then transfer the filter into the PBS solution for 1-2 minutes (liquid must cover the entire membrane filter).
10. Transfer the filter to another empty Petri dish and let it dry.
11. Cut a 0.5 by 0.5 cm piece from the filter with a scalpel or scissors, and pipette 30 µl DAPI solutions onto its surface. From this step onwards, work in a dark place.
12. The filter piece can be marked with a soft pencil.
13. After 2 minutes, transfer the filter into 80% ethanol for a few seconds.
14. Dip the filter paper into double distilled water for a few seconds.
15. Dry the filter.
16. Place the filter onto the surface of a glass slide, put a drop of Vectashield Mounting Medium onto the filter and then cover with a cover slip.
17. Cover with a paper towel and gently press the cover slip to remove any excess mounting medium.
18. Examine the slide with an epifluorescence microscope using a 100X objective and immersion oil under UV excitation (the absorption maximum of DAPI is at 358 nm, and the emission maximum is at 461 nm).
19. Record images from at least 20 different microscopic fields with a digital camera.
20. Count the cells on each picture and determine the mean values.
21. Determine the cell count for one ml of the water sample based on the amount of filtered water and the size of the membrane filter field.

Phase-contrast Light Microscope

Most of the microorganisms are colorless and so are their organelles, which often are too difficult to see by dark field microscopy and common bright field microscopy. The observations are clear by phase-contrast microscope such as the structure of endospores, unstained bacteria and living organisms. The species are often arranged in pairs or chains, rod shaped and rounded or square ends known as *Bacillus* species and their endospores are mostly oval shape or sometimes cylindrical. *Clostridium* species are repeatedly set in short chains or pairs, mostly with rounded or sometimes pointed ends. The endospores regularly swell up the cell. Consequently, by means of ready slides of *Clostridium* and *Bacillus,* the students gain knowledge of a phase-contrast microscope and observe precise bacterial structures such as diverse endospores are present in bacteria. Pond water is generally crowded with protists and bacteria, the internal structure can be seen by using a phase-contrast microscope. Apart from this we can also see the internal structure of paramecium (protist) and slowing down the different microorganisms present in pond water.

Principle

- The colorless and transparent living microorganisms having organelles are generally impossible to see by bright field microscope due to their absorption, refract and reflect limits.
- Another reason is sufficient amount of light diffract from the nearby environment for the rest of the microbes.
- The microbes are visible when they absorb much more light or to refract and reflect from the environment.
- Thoroughly, we can see microbe organelles by this process.
- The observations of living and invisible unstained microbes are by using only phase-contrast microscope.
- The condenser has an annular diaphragm and hollow cone which produces light, glass disk (phase plat) has an objective with a thin film transparent material deposited on it for phase changes produced in the specimen.
- The phase change of images is observed due to the difference of intensity in light.
- The positive phase plate is diffracted due to undiffracted light in dark phase-contrast light microscopy or negative phase plate

same as above given light relative to the directed light producing by phase-contrast light microscopy.

Requirements

- Pond waste water
- Phase-contrast light microscope
- New slides and coverslips
- Pipette
- Microorganisms
- Methyl cellulose "Protoslo, Carolina Biological Supply"
- Tweezers
- Lens paper
- Lens cleaner
- Prepared slides of "*Bacillus* or *Clostridium*"

Procedure

1. Collect waste pond water.
2. Add one drop of methyl cellulose to slow down the swimming of the microorganisms.
3. Prepare the slides of selected microbes given above in the introduction part.
4. The slide is placed on the stage of phase-contrast microscope so that specimen is under the light hole.
5. The lenses are set into place and rotate the position (annular diaphragm) of the specimen if it is required.
6. It is necessary that the cone of light produced by the annular diaphragm below the condenser be centered.
7. Center with the 10X objective and observe the specimen that is under over light.
8. Rotate the annular diaphragm into the correct position for the observation with a 40X objective if the image isn't clear.
9. Do the same with the oil immersion lens.
10. In the report for exercise 4, sketch several of the microorganisms that you have observed.
11. If you examine pond water (collection of microbes), use the pictorial guides provided by your teacher to help you in identifying microorganisms present in pond water.

Scanning Electron Microscope

A scanning electron microscope (SEM) is a very specific type of microscope; it uses scanned electron beams to produce an image of a specimen. A complete labeled picture of the SEM instrument is shown in figure 1.2. The scanned electrons when falling on a specimen interact with the atoms of the specimen and start producing various signals that are captured by the SEM software to produce an image. SEM is used to know the structure of specimens with their composition as well as topography. By using SEM we can achieve much better resolution (better than 1 nm). Mostly the image formation and detection of the specimen are done with the help of a secondary electron beam of SEM. Different states of the sample can be imaged like in their dry, wet, hot or cool states.

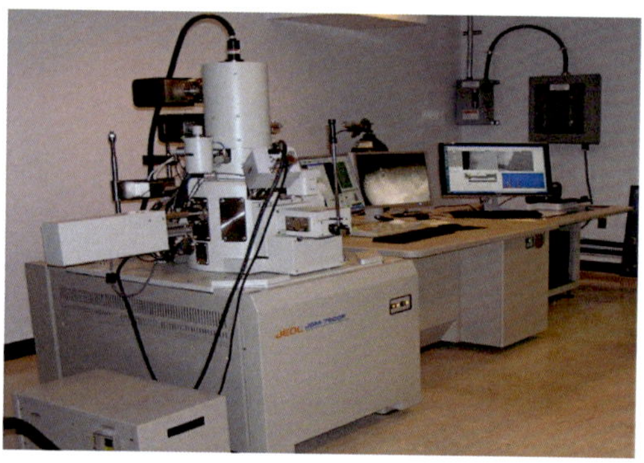

Figure 1.2: A complete labeled picture of an SEM instrument.

In 1935, a well-known scientist named M. Knoll first described the concept of a scanning electron microscope, he was working in Germany. Another scientist named M. Von Ardenne constructed a scanning transmission electron microscope (STEM) by adding scan coils to a transmission electron microscope (TEM) after 3 years in 1938. ZnO crystals were imaged for the first time by using STEM, the operating voltage was 23 kV and magnification 8000 times with spatial resolution between 50-100 nm. But this instrument had no system to store the image. The first SEM was used to examine the surface of a solid specimen, which was described by Zworykin in 1942 after STEM. A scientist of Cambridge University, England named C. W. Oatley start reinvestigating the SEM with Dennis McMullan and they built their first SEM in 1948. In 1952, SEM had achieved various advancements like a resolution of 50 nm and three-dimensional imaging capacity. A research student named O.C. Wells in 1953 under Oatley's supervision started building a second SEM, this instrument had the gun at the bottom of the column, a configuration considered better for experimental work. In 1955, Oatley's another research student working on SEM improved the secondary electron (SE) detector by using a scintillator to convert electrons to photons, which were then transmitted by a light pipe directly interfaced to the photomultiplier tube.

Principle

- Scanning electron microscope is a kind of specific microscope (Figure 1.3).
- In this microscope, the image is produced by using scanned electron beams of the sample.
- The image is produced on both top-down sides.
- There are different ways by which the primary electron beam interacts with the sample.
- The sample of primary electrons generates low energy secondary electrons that fall on the sample and produce an image.
- The primary electrons can also produce images.
- The X-rays produced by relaxed ionized atoms also help in producing the image of the sample.

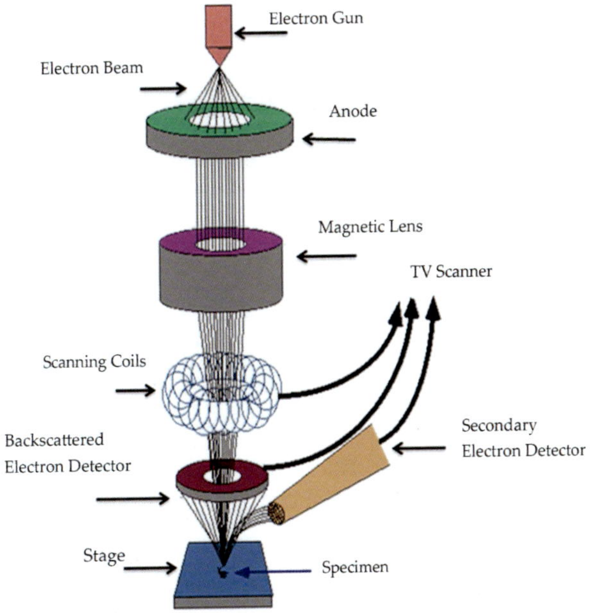

Figure 1.3: A complete schematic diagram of the working principle of SEM.

How to Work

Sample Preparation

- The sample needs to be electrically connected to the sample holder to prevent the electron beam from "charging" the sample and distorting the image: called sample grounding.
- The samples for SEM are sputtered with gold and platinum coating: called sample coating.
- After sample preparation, the SEM software is opened.
- The sample is placed on the SEM sample mount.
- When the sample is properly placed on the SEM stage, close the door of the chamber gently.
- Now click OK in the box on the screen asking "Press OK to pump."
- Click on the "Vacuum" tab to view the vacuum level in the chamber and the gun.
- The voltage and resolution are saved according to the nature of the sample.
- Activate the SCM (sample current monitor) window to check whether it is on or off.
- Remove the sample and collect your results from the system

Usage

- There are multiple usages of SEM, some are mentioned below:

Life Sciences

Biomedical engineering: For designing a variety of technologies such as support matrices for cell growth, artificial tissue, and implantable biomedical device. Cell and tissue morphology: For investigating cellular and tissue structure with high resolution (Figure 1.4). Typical applications involve observing shape changes of grooves, pores, blebs or microvilli on the cellular in response to the changes in the extracellular environment. Microbiology: High-resolution imaging of the microbial surface using SEM helps to have a better understanding of the morphology of microbial populations, bacteria communication, and biofilm formation. It helps researchers visualize microbial populations with great focal depth and high resolution. Pharmaceuticals: SEM offers very high-resolution imaging and superb analytical performance, which make them perfect and reliable tools for quality control of new preparations.

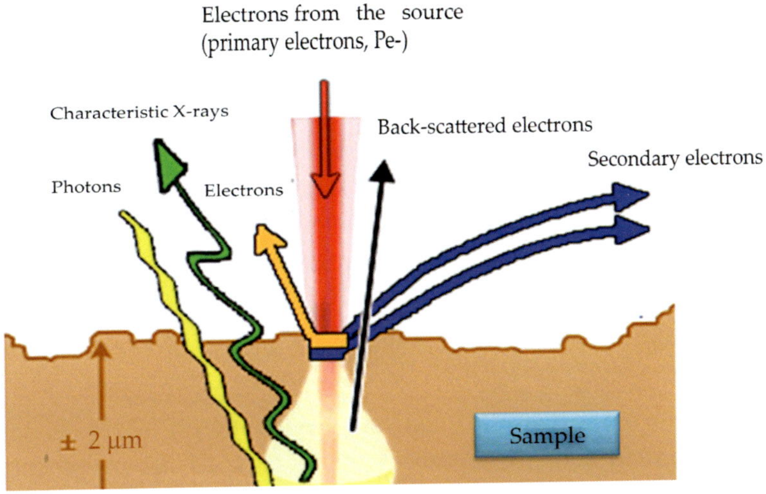

Figure 1.4: An image of electrons scatter during sample inhaled.

Earth Sciences

Magnetic and metamorphic petrology and mineralogy: Petrology and mineralogy benefit from the electron microscopy technique because of several reasons. The high magnification factor, as well as the phase-contrast provided by the back, scattered electron imaging play a crucial role in these disciplines. The analytical capabilities of the technique make SEM an ultimate solution in this field. Ore Processing: Scanning electron microscopy helps to characterize the mineral of interest as well as the surrounding gangue. Such a technique is capable of both phase identification and evaluation of textural relationships of individual minerals. Oil and Gas: Scanning electron microscopy is an integral part of reservoir rock characterization workflow. The elementary imaging function of SEM is used when searching for basic rock properties such as porosity or grain shape. Paleontology: Scanning electron microscopy is a key tool in microfossil characterization helping to distinguish features that are otherwise irresolvable by optical microscopy. Sedimentary Petrology: Electron microscopy is helpful in tackling a wide range of petrological problems. Rock characterization mostly relies on optical microscopy, however, there is a growing demand to characterize the rocks with greater detail nowadays. Recycling: Automated mineralogy is currently finding its way to other areas than just ore processing. This technique can be applied e.g. in the field of electronic scrap recycling where it is used to characterize the composition and structure of comminuted e-scrap (www.scribd.com., www.slideshare.com., www.coursehero.com., www. biocyclopedia.com., www. molecular-plant-biotechnology.info., www.slideplayer.com., www.documents.mx., www.docplayer.net., www. bmb.psu.edu.,www.archive.org.,www.mafiadoc.com., www.powershow. com., www.bioscience.jbpub.com., www.outhorstream.com., www. docslide.com., www.msluay.com., www.scienceprofonlue.com., www.csus. edu., www.krishikosh.egrouth.ac.in., www.tankonyvtor.hu., www.silo.pub., www.studyblue.com., www.Preservearticles.com., www.biotechuniverse. blogspot.com).

Cell Theory

The cell theory states that all organisms are composed of similar units of organization called cells. There are **3** major parts of the cell theory: All organisms are made of cells. All existing cells are produced by other living cells. The cell is the most basic unit of life.

There are **5** contributors to the cell theory

1. Robert Hooke
2. Anton van Leeuwenhoek
3. Matthias Schleiden
4. Theodor Schwann
5. Rudolf Virchow

Each of the above contributors produced evidence through investigations and experiments that led to today's cell theory. Before the cell theory people believed in spontaneous generation figure 1.5 which means living things come from non-living things. Two scientists disproved spontaneous generation. Francisco Redi proved that microorganisms do not spontaneously generate. People thought that maggots came from meat. Redi disproved this in a controlled experiment.

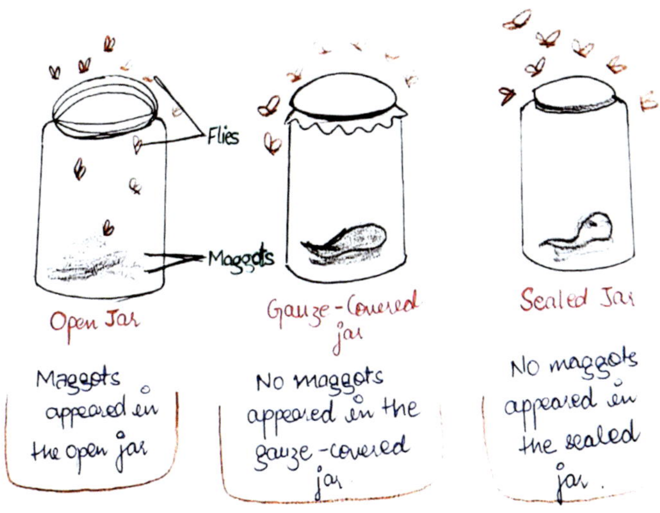

Figure 1.5: Spontaneous generation.

Pasture proved that microorganisms do not spontaneously generate figure 1.6.

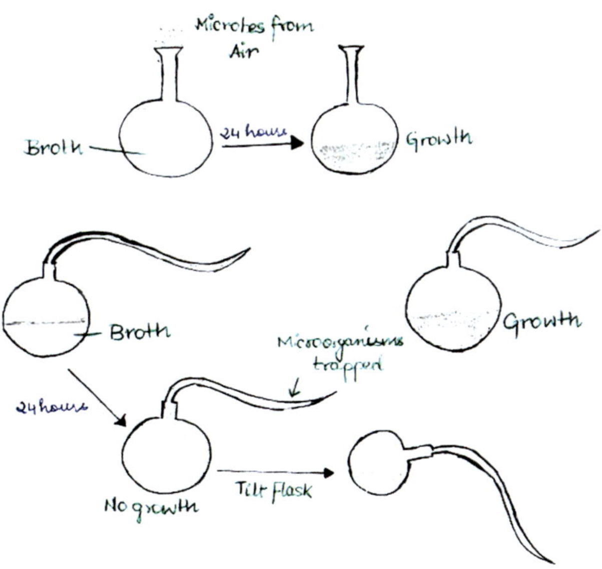

Figure 1.6: Microorganisms do not spontaneously generate.

Theodor Schwann and Matthias Schleiden are credited with coming up with the cell theory.

Schwann

- He discovered in animal cells.

Schleiden

- He discovered in plant cells.

Janssen

- He invented the first compound microscope.

Robert Hooke

In 1663 an English scientist, Robert Hook improved Janssen's invention. He used the compound microscope to observe cork. Hooke was observed that cork is composed of small and hollow compartments. The parts prompted Hooke to think of small rooms (cells) in a monastery, so he gave them the same name:

Cells: It was investigated in cork through experimenting with the compound microscope and came up with the name cells.

Key Note: Hooke discovered the cell in 1665, which started formulating the cell theory.

Anton van Leeuwenhoek

Dutch Tradesman

He studied new approaches for creating lenses to observe cloth. Leeuwenhoek's microscope was more powerful than Hooke's compound microscope. From investigating and experimenting with his microscope, **Leeuwenhoek** became one of the first scientists to refer to living cells when he observed an abundant number of single-celled organisms, which he called animalcules (plant & animal), swimming in a drop of pond water.

Matthias Schleiden (German Scientist)

He was fascinated with plant cells, **Schleiden** used the compound microscope and studied plant cells from investigating and experimenting with plants, projected plant parts were made of cells. It was discussed what he observed with his dear friend, German scientist **Theodor Schwann**.

Theodor Schwann

German Scientist

He studied plant & animal cells, and was intrigued by the similarities between the two from investigating and experimenting with plant & animal cells, Schwann was able to determine that all animals are made of cells. Schwann published the 1st statement of the cell theory: all living things are made of cells and cell products.

Rudolf Virchow

Based on his investigations and experiments, he stated that all cells come from preexisting cells, which is the 2nd part of the cell theory: all existing cells are produced by other living cells. **Key Note:** After Virchow's completed his work, the cell theory was finally formulated in 1838.

Cell structure, organelles and their function

Prokaryotic cells are single-celled entities that are primitive in structure and function as they lack a membrane-bound nucleus and other organelles. The term "prokaryote" is derived from two Greek words, 'pro' meaning 'before' and 'karyon' meaning 'nucleus'. Prokaryotes are considered to be the first living organisms of the earth as they are the simplest form of life. Most organisms that we can see, such as trees, grass, worms, flies, mice, humans, mushrooms and yeast are eukaryotes. It can either be single-celled or multi-celled. It can reproduce in one of several ways (Ex. meiosis and mitosis). It has a cell nucleus within containing its DNA. The nucleus is the most evident distinction between these cell types.

- A plasma membrane: an outer covering that separates the cell's interior from its surrounding environment.
- Cytoplasm: a jelly-like cytosol within the cell in which other cellular components are found
- DNA: the genetic material of the cell.
- Ribosomes: where protein synthesis occurs.

However, "prokaryotes differ from eukaryotic cells in several ways. A prokaryote is a simple, single-celled (unicellular) organism that lacks an organized nucleus or any other membrane-bound organelle **figure 1.7**. We will shortly come to see that this is significantly different in eukaryotes. Prokaryotic DNA is found in a central part of the cell: the nucleoid. Most prokaryotes have a peptidoglycan cell wall and many have a polysaccharide capsule. The cell wall acts as an extra layer of protection, helps the cell maintain its shape, and prevents dehydration. The capsule enables the cell to attach to surfaces in its environment. Some prokaryotes have flagella, pili, or fimbriae. Flagella are used for locomotion. Pili are used to exchange genetic material during a type of reproduction called conjugation". Fimbriae are used by bacteria to attach to a host cell figure 1.8.

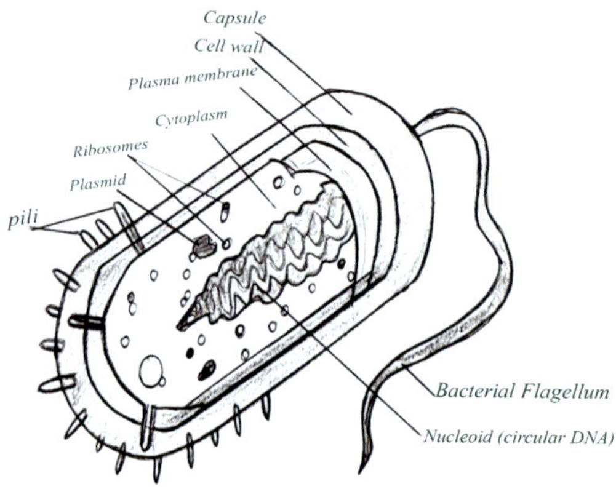

Figure 1.7: Prokaryote.

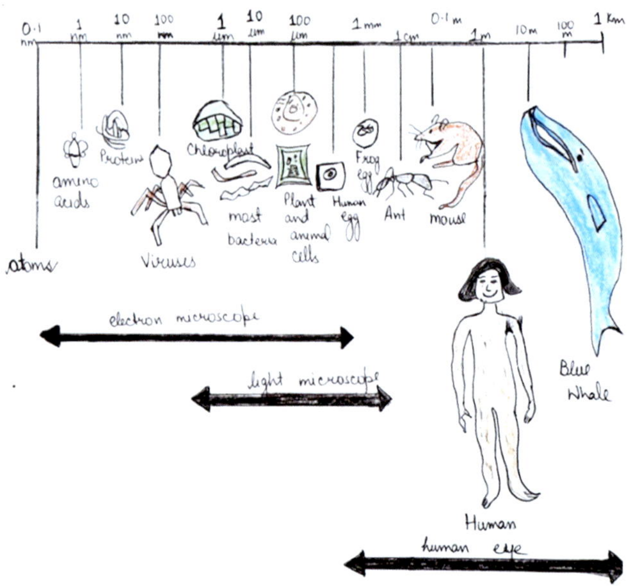

Figure 1.8: Stage of life.

Nucleoid

The region of cytoplasm where prokaryote's genome (DNA) is located and it usually a singular, circular chromosome. "The nucleoid, which means nucleus-like, is an irregularly shaped area containing the genetic material of the prokaryotic cell. It is different from the nucleus of a eukaryotic cell in a way that the genetic material is not enclosed in a membrane to separate it from the cytoplasm".

Plasmid

The small extra piece of chromosome/genetic material is called a plasmid. It has 5 - 100 genes. It can provide genetic information to promote:

- Antibiotic resistance
- Virulence factors

The molecules produced by pathogens specifically influence the host's function to allow the pathogen to thrive. It promotes conjugation (transfer of genetic material between bacteria through cell-to-cell contact).

Other Definition

"A plasmid is a small, circular, double-stranded DNA molecule that is distinct from a cell›s chromosomal DNA. Plasmids naturally exist in bacterial cells, and they also occur in some eukaryotes. Bacteria can also transfer plasmids to one another through a process called conjugation».

Cytoplasm

It is also known as protoplasm. Gel-like matrix of water, enzymes, nutrients, wastes, and gases and contains cell structures. The site of growth, metabolism, and replication.

"Cytoplasm is a thick solution that fills each cell and is enclosed by the cell membrane. It is mainly composed of water, salts, and proteins. All of the organelles in eukaryotic cells, such as the nucleus, endoplasmic reticulum, and mitochondria, are located in the cytoplasm".

Granules

This is the way of storing nutrients and staining of some granules aids in identification.

"In cell biology, a granule is a small particle. It can be any structure barely visible by light microscopy. The term is most often used to describe a secretory vesicle".

Cytoskeleton

The cellular "scaffolding" or "skeleton" is present within the cytoplasm. The major advance in prokaryotic cell biology in the last decade has been the discovery of the prokaryotic cytoskeleton.

"The cytoskeleton is a structure that helps cells maintain their shape and internal organization, and it also provides mechanical support that enables cells to carry out essential functions like division and movement. Rather, several different components work together to form the cytoskeleton".

Ribosomes

It is found within the cytoplasm or attached to the plasma membrane. It is made of protein & rRNA. It is composed of two subunits. The cell may contain thousands of ribosomes.

"Ribosomes are macromolecular machines, found within all living cells, that perform biological protein synthesis. Ribosomes link amino acids together in the order specified by the codons of messenger RNA molecules to form polypeptide chains".

Prokaryotic Plasma Membrane

It separates the cell from its environment. Phospholipid molecules are oriented so that hydrophilic water-loving heads are directed outward and hydrophobic water-hating tails are directed inward. Proteins are embedded in two layers of lipids (lipid bilayer). The membrane is semi-permeable.

"The plasma membrane is a lipid bilayer that surrounds the cytoplasm of a prokaryotic cell. It physically separates the cytoplasm from the outside environment. ... The plasma membrane is also the site of many metabolic reactions for the prokaryotic organism, such as respiration, fermentation, and photosynthesis".

Prokaryotes: Plasma Membrane as a Barrier

Osmosis

The diffusion of water across a semi-permeable membrane is called osmosis. The environment surrounding cells may contain amounts of dissolved substances (solutes).

"Osmosis is the spontaneous net movement or diffusion of solvent molecules through a selectively permeable membrane from a region of high water potential to a region of low water potential, in the direction that tends to equalize the solute concentrations on the two sides".

Tonicity and Osmosis

Isotonic: equal concentration of a solute inside and outside of the cell.

Hypertonic: a higher concentration of solute.

Hypotonic: a lower concentration of solute.

Water will always move towards a hypertonic environment.

Plasma Membrane as a Barrier

Active Transport

The molecules move across the plasma membrane. It is a pump moving water uphill. Types of active transport are classified by the type of energy used to drive molecules across membranes. In ATP-driven active transport, energy from adenosine triphosphate (ATP) drives substances across the plasma membrane with the aid of carrier molecules.

Prokaryotes: Cell Wall

All bacteria are very similar inside the peptidoglycan wall, the bacterial world is divided into two major classes gram-positive and gram-negative figure 1.9.

"A cell wall is a structural layer surrounding some types of cells, just outside the cell membrane. It can be tough, flexible, and sometimes rigid. It provides the cell with both structural support and protection, and also acts as a filtering mechanism".

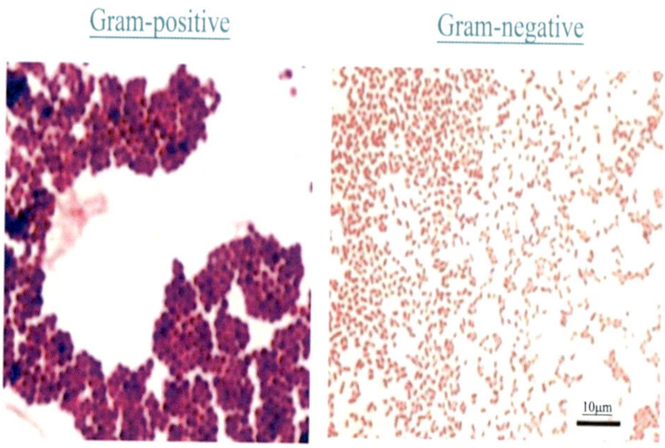

Figure 1.9: Gram-positive and gram-negative.

Bacterial Cell Wall

Peptidoglycan is a huge polymer of interlocking chains of alternating monomers. It provides a rigid support while freely permeable to solutes. The backbone of the peptidoglycan molecule is composed of two amino sugar derivatives of glucose. The "glycan" part of peptidoglycan.

- N-acetylglucosamine (NAG)
- N-acetlymuramic acid (NAM)

NAG / NAM strands are connected by interlocking peptide bridges. The "peptide" part of peptidoglycan is shown in figure 1.10.

"The bacterial cell wall consists of peptidoglycan, an essential protective barrier for bacterial cells that encapsulates the cytoplasmic membrane of both Gram-positive and Gram-negative bacterial cells. Peptidoglycan is a rigid, highly conserved, complex structure of polymeric carbohydrates and amino acids".

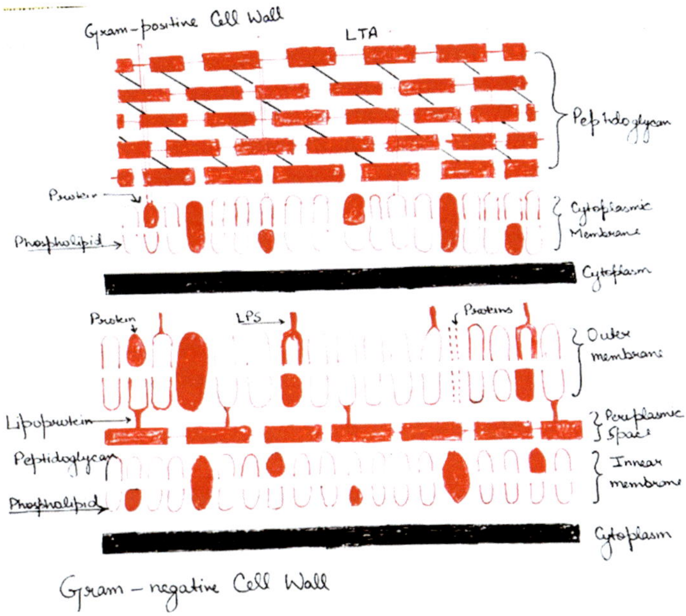

Figure 1.10: Prokaryotes - Cell Wall.

Prokaryotes – Glycocalyx

Some bacteria have an additional layer outside of the cell wall called the glycocalyx. This additional layer can come in one of two forms:

1. Slime Layer

Glycoproteins loosely associated with the cell wall. Slime layer causes bacteria to adhere to solid surfaces and helps prevent the cell from drying out. The slime layer of Gram+ *Streptococcus mutans* allows it to accumulate on tooth enamel (yuck mouth and one of the causes of cavities). Other bacteria in the mouth become trapped in the slime and form a biofilm & eventually a buildup of plaque.

2. Capsule

Polysaccharides are firmly attached to the cell wall. The capsules adhere to solid surfaces and to nutrients in the environment. The adhesive power of capsules is a major factor in the initiation of some bacterial diseases. The capsule also protects bacteria from being phagocytized by cells of the host's immune system.

Prokaryotes: Endospores

It is a dormant, tough, non-reproductive structure produced by a small number of bacteria. It is resistant to radiation, desiccation, lysozyme, temperature, starvation, and chemical disinfectants. Endospores are commonly found in soil and water, where they may survive for very long periods of time.

Prokaryotes – Surface Appendages

Some prokaryotes have distinct appendages that allow them to move about or adhere to solid surfaces. They consist of delicate stands of proteins. Flagella: They are long, thin extensions that allow some bacteria to move about freely in aqueous environments. Axial filament (endoflagella): Wind around bacteria, causing movement in waves. Fimbriae: Most Gram-negative bacteria have these short, fine appendages surrounding the cell. Gram-positive bacteria don't have. It has no role in motility. It helps bacteria adhere to solid surfaces. They have a major factor in virulence. **Pili:** Tubes that is longer than fimbriae, usually shorter than flagella. Use for movement, like grappling hooks, and also use conjugation pili to transfer plasmids.

Prokaryotes – Cell Shapes

Most bacteria are classifying according to shape:

1. Bacillus *(bacilli)* = rod-shaped
2. Coccus *(cocci ... sounds like cox-eye)* = spherical
3. Spiral shaped
   - Spirillum *(spirilla)* = spiral with rigid cell wall, flagella.
   - Spirochete *(spirochetes)* = spiral with flexible cell wall, axial filament.

There are many more shapes beyond these basic ones. A few examples

- Coccobacilli = elongated coccal form
- Filamentous = bacilli that occur in long threads
- Vibrios = short, slightly curved rods
- Fusiform = bacilli with tapered ends

Prokaryotes: Arrangements of Cells

Bacteria sometimes occurr in groups, rather than singly. Bacilli divide along a single axis, seen in pairs or chains. Cocci divide on one or more planes, producing cells in:

- Pairs (diplococci)
- Chains (streptococci)
- Packets (sarcinae)
- Clusters (staphylococci).

Size, shape and arrangement of cells often are the first clues in the identification of a bacterium. Many "look-alikes", so shape and arrangement are not enough for identifying the genus and species.

Eukaryotic Cells

"Eukaryotes are organisms whose cells have a nucleus enclosed within a nuclear envelope. Eukaryotic cells typically contain other membrane-bound organelles such as mitochondria and Golgi apparatus; and chloroplasts can be found in plants and algae".

Like prokaryotes, and most living things, the genome is made of DNA. It may include several to many linear chromosomes within a membrane-bound nucleus.

Cytoplasm

The cytoplasm is present inside the plasma membrane; the nucleus is surrounded by cytoplasm. A water-like substance is that fills inside the cells. The cytosol is made up of water, salts, organic molecules and many enzymes that catalyze reactions.

Membrane-bound Organelles

Eukaryotic cells have many organelles. Prokaryotes only have ribosomes, which are not bound by a membrane. Membrane-bound eukaryotic organelles organize functions within the cell. The system is present of internal membranes within eukaryotic cells that divide the cell into compartments, or organelles. The transport system, for moving molecules through the interior of the cell, as well as interactive surfaces for lipid and protein synthesis. The membranes of the endomembrane system are made of a lipid bilayer, with proteins.

Endomembrane System Organelles: Nucleus

It is a control center and separates the genetic material (DNA) from the rest of the cell. DNA, the genetic material, is a blueprint, or code for making proteins. It allows nutrients to enter the cell and waste to be removed. This is referred to as selective permeability (Selective=Chooses, Permeability=filter through). Keeping a healthy balance of nutrients and water within the cell is called homeostasis

Cell Types

Prokaryotic cells are mainly single-celled organisms. Eukaryotic cells are mainly multicellular organisms, although some are single-celled (amoeba, algae, yeast).

Prokaryotic Cell

According to the morphological point of view, prokaryotic cells are the most primitive cells. "They do not have a definite nucleus which includes bacteria and cyanobacteria (blue-green algae). The chromatin bodies remain scattered inside the cytoplasm. In prokaryotes, asexual division occurs basically binary fission. Prokaryotes are smaller than eukaryotes". Example – Bacteria cell

Eukaryotic Cells

It is believed that eukaryotes have been evolved from the prokaryotes. "They have been characterised by their membrane nucleus. They contain organelles like mitochondria bounded by membranes and are located in the cytoplasm. That is, they contain a definite nucleus. The chromatin bodies are enclosed by a nuclear membrane. Both asexual and sexual division occurs in eukaryotes. They are larger than prokaryotes and show better structural organization and increased functional efficiency than prokaryotes". **Example – Plant and animal cells** "Animal cells have centrosomes (or a pair of centrioles), and lysosomes, whereas **plant cells do not**. Plant cells have a cell wall, chloroplasts, plasmodesmata, and plastids used for storage, and a large central vacuole, whereas animal cells do not".

2. CELL ORGANELLE, CELL DIVISION, CELL CYCLE INTERS AND INTRACELLULAR COMMUNICATION

Cell Organelles

The cellular components are called cell organelles. "These cell organelles include both membrane and non-membrane bound organelles, present within the cells and are distinct in their structures and functions. They coordinate and function efficiently for the normal functioning of the cell. A few of them function by providing shape and support, whereas some are involved in the locomotion and reproduction of a cell. There are various organelles present within the cell and are classified into three categories based on the presence or absence of membrane".

Organelles without membrane: The Cell wall, Ribosomes, and Cytoskeleton are non-membrane-bound cell organelles. They are present both in prokaryotic and the eukaryotic cells.

Single membrane-bound organelles: Vacuole, Lysosome, Golgi Apparatus, Endoplasmic Reticulum are single membrane-bound organelles present only in a eukaryotic cell.

Double membrane-bound organelles: Nucleus, mitochondria and chloroplast are double membrane-bound organelles present only in a eukaryotic cell.

Plasma Membrane

The plasma membrane is also termed as a cell membrane or cytoplasmic membrane. "It is a selectively permeable membrane of the cell, which is composed of a lipid bilayer and proteins. The plasma membrane is present both in plant and animal cell, which functions as the selectively permeable membrane, by permitting the entry of selective materials in and out of the cell according to the requirement. In an animal cell, the cell membrane functions by providing shape and protects the inner contents of the cell". Based on the structure of the plasma membrane, it is regarded as the fluid mosaic model. According to the fluid mosaic model, the plasma membranes are subcellular structures, made of a lipid bilayer in which the protein molecules are embedded.

Cytoplasm

The cytoplasm is present both in plant and animal cells. They are jelly-like substances, found between the cell membrane and nucleus. They are mainly composed of water, organic and inorganic compounds. The cytoplasm is one of the essential components of the cell, where all the cell organelles are embedded. These cell organelles contain enzymes, mainly responsible for controlling all metabolic activity taking place within the cell and are the site for most of the chemical reactions within a cell.

Nucleus

Largest organelle in the cell and it is the most inner compartment of the cell "which contains chromatin (DNA); genetic information on strands called chromosomes "control center" for cell metabolism and reproduction. The nucleus is a double-membraned organelle found in all eukaryotic cells. It is the largest organelle, which functions as the control center of the cellular activities and is the storehouse of the cell's DNA. By structure, the nucleus is dark, round, surrounded by a nuclear membrane. It is a porous membrane (like a cell membrane) and forms a wall between the cytoplasm and nucleus. Within the nucleus, there are tiny spherical bodies called the nucleolus. It also carries another essential structure called chromosomes. Chromosomes are thin and thread-like structures that carry another important structure called a gene. Genes are a hereditary unit in organisms i.e., it helps in the inheritance of traits from one generation (parents) to another (offspring). Hence, the nucleus controls the characters and functions of cells in our body. The primary function of the nucleus is to monitor cellular activities including **metabolism** and growth by making use of DNA's genetic information. Nucleoli in the nucleus are responsible for the synthesis of protein and RNA».

Centrosome and Centrioles

The centrosome organelle is made up of two mutually perpendicular structures known as centrioles. Each centriole is composed of 9 equally spaced peripheral fibrils of tubulin protein, and the fibril is a set of interlinked triplets. The core part of the centriole is known as a hub and is proteinaceous. The hub connects the peripheral fibrils via radial spoke, which is made up of proteins. The centrioles from the basal bodies of the cilia and flagella give rise to spindle fibers during cell division.

Chromatin

The directions on how to make proteins.

Nucleolus

It is found inside nucleus and ribosomes are made here.

Ribosomes

It makes proteins (made up of RNA and protein); thought of as "factories"

Cytoplasm

It is a clear gel-like fluid inside the cell, which suspends all organelles.

Endoplasmic Reticulum

The Endoplasmic Reticulum is a network of membranous canals filled with fluid. They are the transport system of the cell, involved in transporting materials throughout the cell.

There are two different types of Endoplasmic Reticulum:

- **Rough Endoplasmic Reticulum** – They are composed of cisternae, tubules, and vesicles, which are found throughout the cell and are involved with protein manufacture.
- **Smooth Endoplasmic Reticulum** – They are the storage organelle, associated with the production of lipids, steroids, and also responsible for detoxifying the cell.

Golgi apparatus: Sorts proteins made by the ribosomes and sends them to needed places in the cell.

Lysosomes- Organelles that are filled with digestive enzymes to remove waste and invading bacteria.

Mitochondria

Mitochondria are called the powerhouses of the cell as they produce energy-rich molecules for the cell. The mitochondrial genome is inherited maternally in several organisms. It is a double membrane-bound, sausage-shaped organelle, found in almost all eukaryotic cells. The double membranes divide its lumen into two distinct aqueous compartments. The inner compartment is called 'matrix' which is folded into cristae whereas the outer membrane forms a continuous boundary with the cytoplasm. They usually vary in their size and are found either round or oval in shape. Mitochondria are the sites of aerobic respiration in the cell, produces energy in the form of ATP and helps in the transformation of the molecules. For instance, glucose is converted into adenosine triphosphate – ATP. Mitochondria have their own circular DNA, RNA molecules, ribosomes (the 70S), and a few other molecules that help in protein synthesis. It is referred to as the "powerhouse" of the cell that releases energy for the cell. It converts the energy stored in glucose into ATP for the cell.

Vacuoles are fluid which filled organelles enclosed by a membrane. It stores materials such as food, sugar, water, and waste products. Vacuoles are mostly defined as storage bubbles of irregular shapes which are found in cells. They are fluid-filled organelles enclosed by a membrane. The vacuole stores the food or a variety of nutrients that a cell might need to survive. In addition to this, it also stores waste products. The waste products are eventually thrown out by vacuoles. Thus, the rest of the cell is protected from contamination. The animal and plant cells have different size and number of vacuoles. Compared to animals, a plant cell has larger vacuoles.

Plastids

Plastids are large, membrane-bound organelles that contain pigments. Based on the type of pigments, plastids are of three types:

Chloroplasts – Chloroplasts are double membrane-bound organelles, «which usually vary in their shape – from a disc shape to spherical, discoid, oval and ribbon. They are present in mesophyll cells of leaves, which store chloroplasts and other carotenoid pigments. These pigments are responsible for trapping light energy for photosynthesis. The inner membrane encloses a space called the stroma. Flattened disc-like chlorophyll-containing structures known as thylakoids are arranged in a stacked manner like a pile of coins. Each pile is called as granum (plural: grana) and the thylakoids of different grana are connected by flat membranous tubules known as stromal lamella. Just like the mitochondrial matrix, the stroma of chloroplast also contains a double-stranded circular DNA, 70S ribosomes, and enzymes required for the synthesis of carbohydrates and proteins.

Chromoplasts – The chromoplasts include fat-soluble, carotenoid pigments like xanthophylls, carotene, etc. which provide the plants with their characteristic color – yellow, orange, red, etc.

Leucoplasts – Leucoplasts are colorless plastids that store nutrients. Amyloplasts store carbohydrates (like starch in potatoes), aleuroplasts store proteins, and elaioplasts store oils and fats.

Ribosomes

Ribosomes are nonmembrane-bound and important cytoplasmic organelles found in close association with the endoplasmic reticulum. Ribosomes are found in the form of tiny particles in a large number of cells and are mainly composed of 2/3rd of RNA and 1/3rd of protein. They are named as the 70s (found in prokaryotes) or 80s (found in eukaryotes) The letter S refers to the density and the size, known as Svedberg's Unit. Both 70S and 80S ribosomes are composed of two sub-units. Ribosomes are either encompassed within the endoplasmic reticulum or are freely traced in the cell's cytoplasm. Ribosomal RNA and Ribosomal proteins are the two components that together constitute ribosomes. The primary function of the ribosomes includes protein synthesis in all living cells that ensure the survival of the cell.

Golgi Apparatus

Golgi apparatus is also termed as Golgi complex. It is a membrane-bound organelle, which is mainly composed of a series of flattened, stacked pouches called cisternae. This cell organelle is primarily responsible for transporting, modifying, and packaging proteins and lipids to targeted destinations. Golgi apparatus is found within the cytoplasm of a cell and is present in both plant and animal cells.

Cytoskeleton

It is a continuous network of filamentous proteinaceous structures that run throughout the cytoplasm, from the nucleus to the plasma membrane. "It is found in all living cells, notably in the eukaryotes. The cytoskeleton matrix is composed of different types of proteins that can divide rapidly or disassemble depending on the requirement of the cells. The primary functions include providing the shape and mechanical resistance to the cell against deformation, the contractile nature of the filaments helps in motility and during cytokinesis".

Cilia and Flagella

Cilia are hair-like projections, small structures, present outside the cell wall and work like oars to either move the cell or the extracellular fluid. "Flagella are slightly bigger and are responsible for the cell movements. The eukaryotic flagellum structurally differs from its prokaryotic counterpart. The core of the cilium and flagellum is called an axoneme, which contains nine pairs of gradually arranged peripheral **microtubules** and a set of central microtubules running parallel to the axis. The central tubules are interconnected by a bridge and are embedded by a central sheath. One of the peripheral microtubular pairs is also interconnected to the central sheath by a radial spoke". Hence there is a total of 9 radial spokes. The cilia and flagella emerge from centriole-like structures called basal bodies.

Mitosis

The division of **somatic** cells (non-reproductive cells) in eukaryotic organisms is called mitosis. A single cell divides into two identical daughter cells. Daughter cells have the same numbers of chromosomes as parent cells figure 2.1 and 2.2.

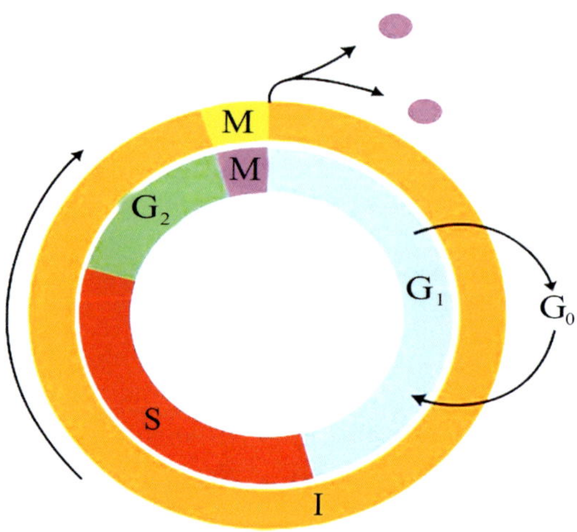

Figure 2.1: Cell cycle.

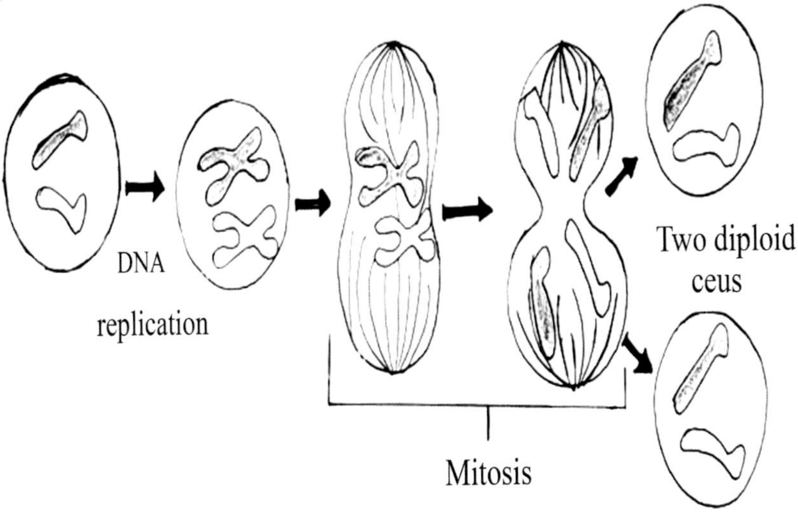

Figure 2.2: Stages of mitosis.

The DNA molecules are in extended form. The uncondensed form is called **chromatin**. A cell can only replicate and transcribe DNA when in an extended state. DNA molecules condense to form **chromosomes** prior to division. Each chromosome is a single molecule of DNA. Easier to sort and organize the replicated DNA into the daughter cells.

There are 4 sub-phases

- 1st – Prophase
- 2nd – Metaphase
- 3rd – Anaphase
- 4th – Telophase
- Cytokinesis

Prophase

It contains chromosomes condense. Spindle fibers form (Spindle fibers are specialized microtubules radiating out from centrioles) figure 2.3. Chromosomes are captured by the spindle.

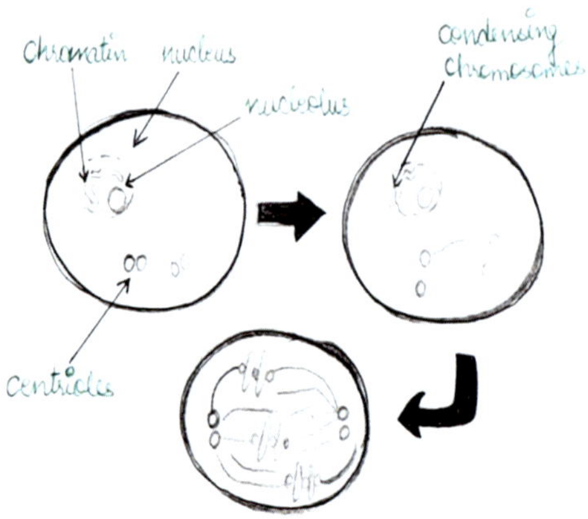

Figure 2.3 Prophase.

Metaphase

Chromosomes align along the equator of the cell, with one kinetochore facing each pole **figure 2.4**.

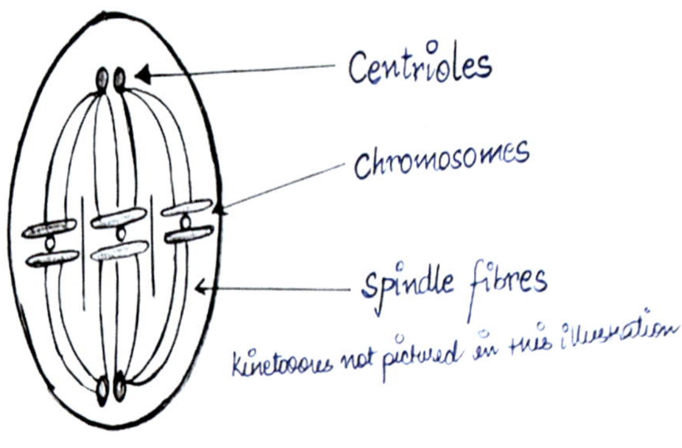

Figure 2.4: Metaphase.

Anaphase

Spindle fibers attached to kinetochores shorten and pull the chromatids towards the poles. The free spindle fibers lengthen and push the poles of the cell apart figure 2.5.

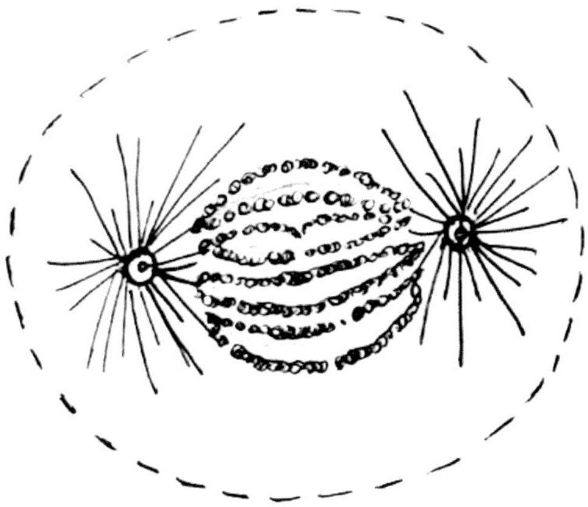

Figure 2.5: Anaphase.

Telophase

Spindle fibers disintegrate and nuclear envelopes form around both groups of chromosomes. Chromosomes revert to their extended state. Cytokinesis occurs, enclosing each daughter nucleus into a separate cell **figure 2.6.**

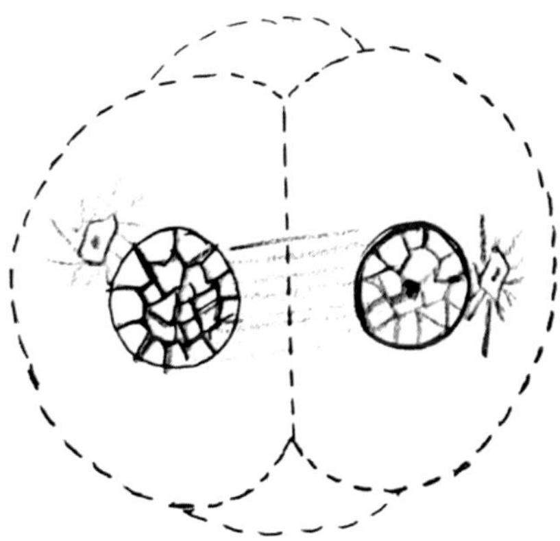

Figure 2.6: Telophase.

Cytokinesis

It undergoes cytokinesis by forming a cell plate between the two daughter nuclei in plant cells. They undergo cytokinesis through the formation of a cleavage furrow. A ring of microtubules contracts, pinching the cell in half in animal cells figure 2.7 and 2.8.

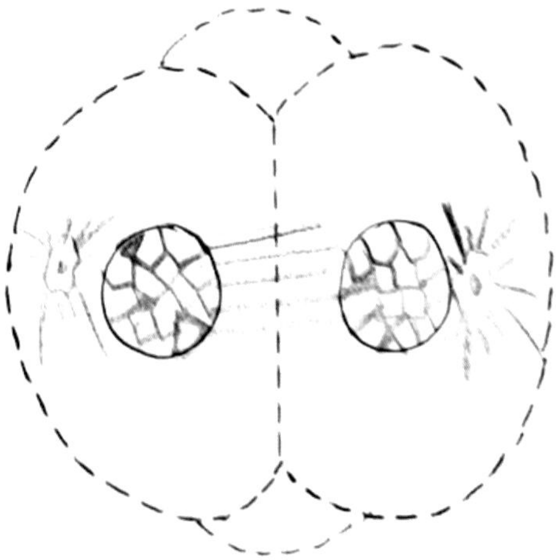

Figure 2.7: Cytokines.

Stages of Mitosis

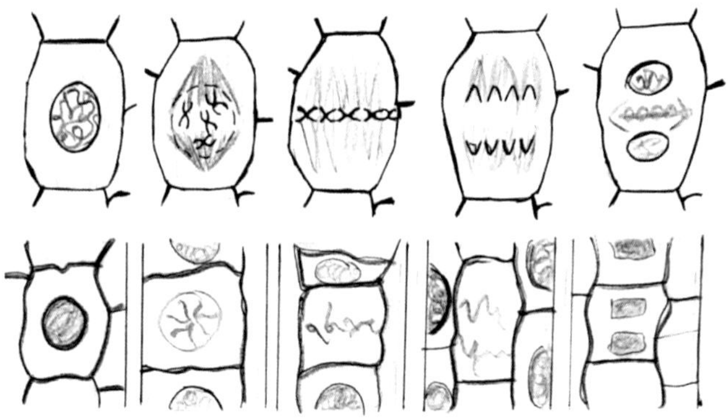

Figure 2.8: Stages of Mitosis.

Meiosis

0The cell division of gametes is called meiosis. A single germ cell divides into four unique daughter cells. Daughter cells have half of the chromosomes as a parent cell, so they are considered haploid **figure 2.9**.

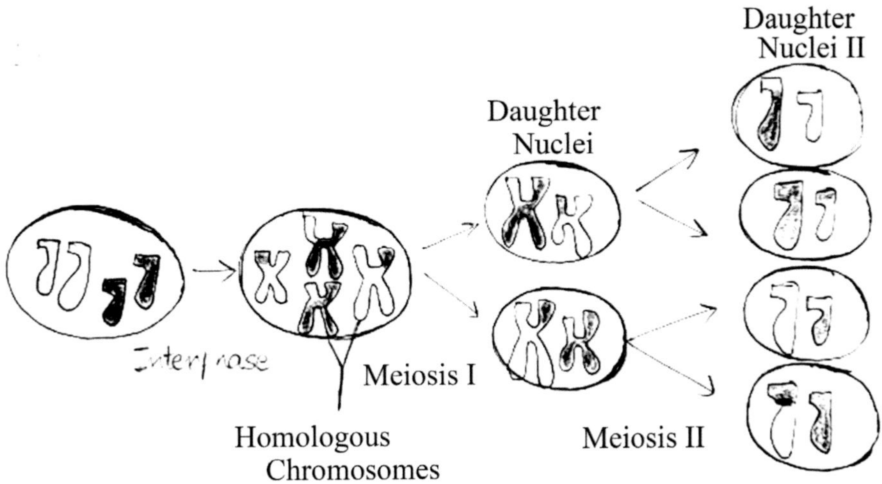

Figure 2.9: Stages of Meiosis.

Haploid

One copy of each chromosome. It was designated as "n", the number of chromosomes in one "set" gametes.

Diploid

Two sets of chromosomes and it is designated as "2n".

Somatic Cells

Diploid organisms receive one of each type of chromosome from the female parent (maternal chromosomes) and one of each type of chromosome from the male parent (paternal chromosomes).

Genetics: Homologs

The chromosomes exist in homologous pairs in diploid (2n) cells figure 2.10.

Figure 2.10: Homologous Chromosomes.

Exception: Sex Chromosomes (X and Y)

Other chromosomes, known as autosomes have homologs. "A sex chromosome is a chromosome that differs from an ordinary autosome in form, size, and behavior. The human sex chromosomes, a typical pair of mammal allosomes, determine the sex of an individual created in sexual reproduction".

Asexual Reproduction

Single-celled organisms reproduce by splitting and budding. Some multicellular organisms can reproduce asexually and produce clones (offspring genetically identical to parent).

Sexual Reproduction

The fusion of two gametes to produce a single zygote is called sexual reproduction. It presents greater genetic variation and allows genetic recombination. With the exception of self-fertilizing organisms, the zygote has gametes from two different parents.

Sexual Reproduction in Humans

Human reproduction naturally takes place as internal fertilization by sexual intercourse. This process is also known as "mating". The sperm and the ovum are known as gametes (each containing half the genetic information of the parent, created through meiosis). It is donated by each parent *(*total = 46 or 23 pairs*)*. Gametes (sperm/ova): Contain 22 autosomes and 1 sex chromosome. Are haploid (haploid number "n" = 23 in humans). Fertilization results in the diploid zygote. Diploid cell; 2n = 46. (n = 23 in humans)

Meiosis

In meiosis, a sex cell (Gamete) is formed. In meiosis, there are 2 divisions of the nucleus: meiosis I and meiosis II figure 2.11. "Meiosis is a type of cell division that reduces the number of chromosomes in the parent cell by half and produces four gamete cells. This process is required to produce egg and sperm cells for sexual reproduction. Meiosis begins with a parent cell that is diploid, meaning it has two copies of each chromosome».

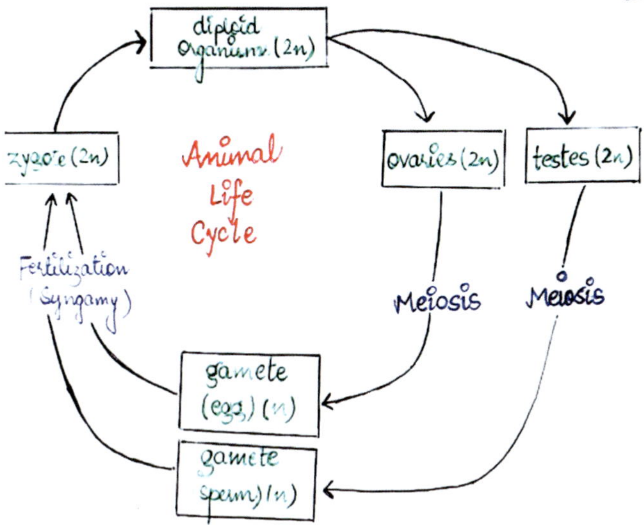

Figure 2.11: Meiosis & sexual reproduction life cycle.

Genetic Variation in Diploid Organisms

Fusion of the sperm and egg results in a unique offspring. So, the young are a product of two individuals with different genetic makeup. Meiosis "shuffles" the genes so that the individual's gametes are genetically different from one another.

Genetic Shuffling of Meiosis I

In addition to a new combination of chromosomes resulting from **fertilization**, there are also events in Meiosis I that shuffle the genes.

- **Crossing over** in Prophase I.
- **Independent assortment** in Metaphase I.

Crossing Over

Homologs break at identical locations, and then rejoin opposite partners. This creates new combinations of the alleles on each chromosome. It occurs randomly several times on every chromosome. So, the results in the mixing of the genes you inherited from your parents are called crossing over.

Spermatogenesis

Males produce sperm throughout life, after the onset of puberty, about 1,500 sperm per second figure 2.12.

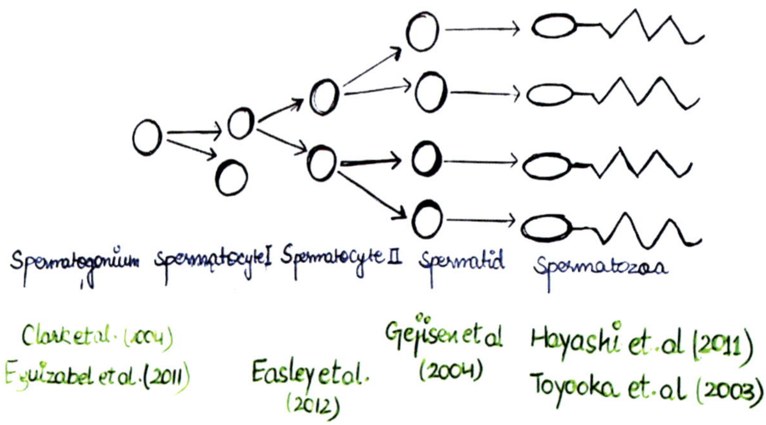

Figure 2.12: Structure of spermatogenesis.

Oogenesis

Oogenesis in females is probably complete either before or shortly after birth. During oogenesis, three polar bodies develop as the mature ovum is generated. Polar bodies contain little cytoplasm and eventually degenerate figure 2.13.

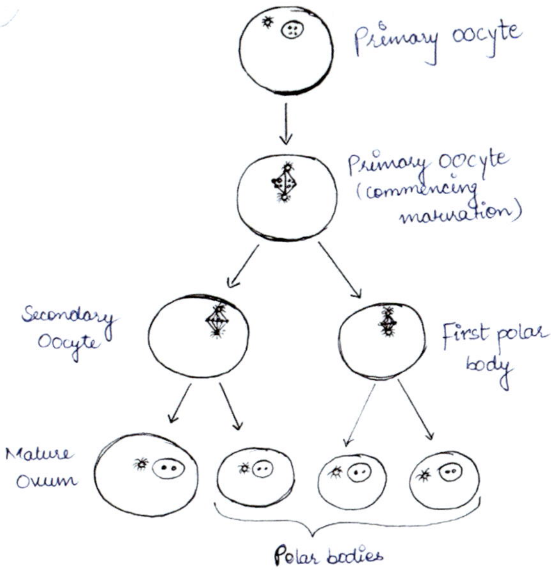

Figure 2.13: Stages of Oogenesis.

3. CELL GROWTH, GROWTH FACTORS ACTIVE AND PASSIVE TRANSPORT

Cell Growth and Proliferation and Molecular Constituents of Cells

In unicellular organisms, "cell division is the means of reproduction; in multicellular organisms, it is the means of tissue growth and maintenance. Survival of the eukaryotes depends upon interactions between many cell types, and it is essential that a balanced distribution of types be maintained. This is achieved by the highly regulated process of cell proliferation. The growth and division of different cell populations are regulated in different ways" but the basic mechanisms are similar throughout multicellular organisms.

Regulation of Cellular Function by Growth Factors and Hormones

Cell growth factors are the term for a class of gene products that play important roles in the regulation of cell division and tissue proliferation. "Each growth factor has a specific cell-surface receptor. The binding of the growth factor to the receptor initiates or, in some cases, blocks cell division. Most growth factors regulate only certain types of cells and tissues. Control of cell division is an active process involving both stimulation and inhibition (by antagonists). Mitosis, for example, will not occur in the presence of active inhibition and the absence of stimulation. Growth factors regulate defined phases of the cell cycle». The two most important phases of the cell cycle are the transition from:

The transition from G_0 to G_1 is controlled by growth factors for:

- Skin cells (epidermal growth factor, EGF)
- Nerve cells (nerve growth factor, NGF)
- Connective tissue or mesenchymal cells (fibroblast growth factor, FGF)
- Thrombus-forming cells that line blood vessels (platelet-derived growth factor, PDGF)

An important growth factor regulating the transition from G1 to S is insulin-like growth factor, IGF-1. The stimulating effect of these growth factors can be opposed by such antagonists as:

- transforming growth factor-beta (TGF-beta)
- tumor necrosis factor (TNF).

Activation of a Growth Factor Receptor

A growth factor receptor is activated by binding to a specific growth at the cell's surface. "The activated receptor, in turn, activates an intracellular protein (i.e., substrate protein). A receptor such as platelet-derived growth factor (PDGF) can stimulate a number of substrates including *Ras* protein, the *Src* protein (the name is derived from the tumor, a sarcoma, in which it was first found), or phospholipase C, a signal transmitter.

Cell Pigments

The principal types of pigment cells are melanocytes and melanophores, which contain melanin; xanthophores, which contain carotenoids, flavins, and pterins; and iridocytes, or guanophores, which contain guanidine crystals.

Cell Viability

Cell viability is a measure of the proportion of live, healthy cells within a population. Cell viability assays are used to determine the overall health of cells, optimize culture or experimental conditions, and to measure cell survival following treatment with compounds, such as during a drug screen.

Active and Passive Transport

Active transport *is the movement of molecules across a membrane from a region of lower concentration to a region of higher concentration against the concentration gradient, often assisted by enzymes and requires energy"*

"**Passive transport** is the movement of ions and molecules across the cell membrane without requiring energy." Active and passive transport are the two main biological processes that play a crucial role in supplying nutrients, oxygen, water and other essential molecules to the cells along with the elimination of waste products. In essence, active and passive transport work for the same goals/ purposes, but with different movement".

Active Transport

This is the biological process in which molecules move against the concentration gradient and require chemical energy to move biochemical compounds from a lower region to the high region. "Therefore, this process uses ATP – Adenosine triphosphate to pump molecules through a concentration gradient. Complex sugar, ions, large cells, proteins and other particles are transported in this process. There are two types of Active transport:

- Primary Active transport
- Secondary Active transport

Exocytosis, endocytosis and sodium-potassium pump are a few examples of active transport. The process of endocytosis and exocytosis are utilized by all the cells for the transportation of molecules that cannot passively permeate via the membrane. Endocytosis is the process of active transportation of molecules into the cells by the action of engulfing them along with their membrane. Exocytosis produces a counter function thereby forcing molecules out of the cell. The process of homeostasis facilitates an equal flow of molecules in and out of a cell which confers that the number of molecules that enter the cell through endocytosis equates to the number of molecules that exit a cell through the process of exocytosis. Both the processes assure that nutrients and wastes are balanced for the smooth functioning of the cells".

Passive Transport

In this biological process, energy is not required for transporting the molecules, as the biochemicals move from a region of higher concentration to a region of lower concentration. "All particles which are easily soluble are transported through passive transport. This process is carried out to maintain the balance and the equilibrium level in a cell. All the waste molecules including, water and carbon dioxide are separated and moved out of the cell using passive transport. Meanwhile, nutrients like oxygen that are functional for the cell are diffused in this process. Osmosis, diffusion and facilitated diffusion are some of the examples of passive transport".

Key Points on Active and Passive Transport

- Active transport requires energy for the movement of molecules whereas passive transport does not require energy for the movement of molecules.
- In active transport, the molecules move against the concentration gradient whereas in passive transport, the molecules move along the concentration gradient.
- Uptake of glucose in the human intestine works on the principle of active transport.
- Simple diffusion, facilitated diffusion, osmosis and filtration are examples of passive transport.

Movement of Molecules into and out of Cells

The cell membrane is one of the great multi-taskers of biology. "It provides the structure for the cell, protects cytosolic contents from the environment, and allows cells to act as specialized units. A membrane is the cell's interface with the rest of the world - it's gatekeeper, if you will. This phospholipid bilayer determines what molecules can move into or out of the cell, and so is in large part responsible for maintaining the delicate homeostasis of each cell".

Semi-Permeability

Some cells function best at a pH of 5, while others are better at pH 7. "The steroid hormone aldosterone is made in the adrenal gland but affects mostly the kidney. Sodium is more than ten times more concentrated outside of cells rather than inside. If our cells couldn't control what crossed their membranes, either no molecules would make it across, or they'd be traveling willy-nilly and the internal environment would always be in flux. It'd be like taking every item on a menu and blending it together before serving (not the tastiest idea). So how do cells maintain different concentrations of proteins and molecules despite the pressures on them to be homogenous? Cell membranes are semipermeable, meaning they have control over what molecules can or cannot pass through. Some molecules can just drift in and out, others require special structures to get in and out of a cell, while some molecules even need an energy boost to get across a cell membrane. Each cell's membrane contains the right mix of these structures to help that cell keep its internal environment just right. Cartoon representing the cell membrane as a shield preventing things from entering the cell which is represented as a castle".

Movement across a Membrane and Energy

There are two major ways that molecules can be moved across a membrane, and the distinction has to do with whether or not cell energy is used. "Passive mechanisms like diffusion use no energy, while active transport requires energy to get done. Diffusion is the movement of particles down their gradient. A gradient is any imbalance in concentration, and moving down a gradient just means that the particle is trying to be evenly distributed everywhere, like dropping food coloring in water. This is what happened when we made our granola - a bunch of separate ingredients came together and spread out across the whole mixture. We call this evening-out moving "downhill", and it doesn't require energy. The molecule most likely to be involved in simple diffusion is water - it can easily pass through cell membranes. When water undergoes simple diffusion, it is known as osmosis. Simple diffusion is pretty much exactly what it sounds like – molecules move down their gradients through the membrane. Molecules that practice simple diffusion must be small and nonpolar*, in order to pass through the membrane. Simple diffusion can be disrupted if the diffusion distance is increased. If the alveoli in our lungs fill with fluid (pulmonary edema), the distance the gases must travel increases, and their transport decreases. Facilitated diffusion is diffusion that is helped along (facilitated by) a membrane transport channel. These channels are glycoproteins (proteins with carbohydrates attached) that allow molecules to pass through the membrane. These channels are almost always specific for either a certain molecule or a certain type of molecule (i.e. an ion channel), and so they are tightly linked to certain physiologic functions. For example, one such transporter channel, GLUT4, is incredibly important in diabetes. GLUT4 is a glucose transporter found in fat and skeletal muscle. Insulin triggers GLUT4 to insert into the membranes of these cells so that glucose can be taken in from the blood. Since this is a passive mechanism, the amount of sugar entering our cells is proportional to how much sugar we consume, up to the point that all our channels are being used (saturation). In type II diabetes mellitus, cells do not respond as well to the presence of insulin, and so do not insert GLUT4 into their membranes. This can lead to soaring blood glucose levels which can cause heart disease, stroke, and kidney failure».

Subjective Questions

1. What is the importance of a vacuole in a plant cell?
2. What is a satellite chromosome?
3. State the characteristics of prokaryotic cells?
4. Multicellular organisms exhibit division of labor. Comment?
5. Why is the cell the basic unit of life?
6. What do you mean by plasmids?
7. What role do they play in bacteria?
8. Describe the cell theory in brief?
9. What is the biochemical composition of the plasma membrane?
10. Define and explain the cell theory?
11. Describe the cell organelle?
12. Differentiate between the plant cell and animal cell?
13. Explain the mitochondrial role in a cell?
14. Differentiate between mitosis or meiosis cycle?
15. Explain the check points?
16. Differentiate plant and animal cells?
17. Differentiate prokaryotic and eukaryotic cells?
18. Describe cell growth and cell proliferation?
19. Describe the active and passive transport movement and molecules across the membrane?
20. Explain the eukaryotic cell organelle?
21. Define the cell theory with explanations?

22. What is a sex chromosome and explain about the autosome with examples?
23. Explain microscope and its application in life sciences?
24. Differentiate about the phase-contrast microscope and bright field microscope?
25. Explain cell communication?
26. What are chromosomes and their types?
27. Explain the chloroplast and plastids?
28. Define spontaneous theory with examples?
29. Explain the plasma membrane?
30. Explain the cell growth receptor?

Objective Questions

1. Cell was discovered by

a. Swanson

b. Leeuwenhoek

c. Robert Hooke

d. Robert Brown

Answer: C

2. Cells were observed prior to Robert Hooke by

a. Aristotle

b. Malpighi

c. Bauhin

d. Eicher

Answer: B

3. Cell was discovered in

a. 18th Century

b. 19th Century

c. First half of 17th Century

d. Second half of 17th Century

Answer: D

4. Cell, as basic unit of plants, was discovered by

a. Robert Brown

b. Robert Hooke

c. Virchow
d. Schleiden

Answer: D

5. Cell theory states that

a. All cells are living
b. All cells have nucleus
c. Cells are fundamental structural units of living organisms
d. Cells reproduce by mitosis and meiosis

Answer: C

6. Robert Brown Is known for his discovery of

a. Chloroplasts
b. Respirometer
c. Nucleus
d. Mitochondria

Answer: C

7. Who applied cell theory to plants

a. Schwann
b. Schleiden
c. Swanson
d. Jenssen

Answer: B

8. The living substance of cell was named sarcode by

a. Corti
b. Dujardin
c. Lamarck
d. Dutrochet

Answer: B

9. The modern cell theory is called

a. Protoplasmic Theory
b. Cell Principle
c. Cell Doctrine
d. Both B and C

Answer: D

10. Which ones do not have cellular structure

a. PPLO

b. Rickettsia

c. Viruses

d. Archaebacteria

Answer: C

11. Robert Hooke discovered cell in

a. 1665

b. 1725

c. 1545

d. 1595

Answer: A

12. 'Micrographia' was written by

a. Grew

b. Hooke

c. Brown

d. Lamarck

Answer: B

13. Cell theory was modified by

a. Schleiden

b. Brown

c. Schwann

d. Grew

Answer: C

14. Omnis cellula e cellula is a generalization given by

a. Lamarck

b. Dutrochet

c. Leeuwenhoek

d. Virchow

Answer: D

15. Cells of Robert Hooke were actually

a. Cell walls

b. Protoplasts

c. Wall-less cells

d. Walled cells

Answer: A

16. Cell principle is not applicable to

a. Bacteria

b. Viruses

c. Algae

d. Fungi

Answer: B

17. Distinction of individual cells is absent in coenocytic organism

a. Ulothrix

b. Volvox

c. Escherichia

d. Rhizopus

Answer: D

18. Who believed in the individuality of cells?

a. Lamarck

b. Leeuwenhoek

c. Dutrochet

d. Malpighi

Answer: C

19. Who saw the living matter for the first time?

a. Leeuwenhoek

b. Hooke

c. Grew

d. Corti

Answer: D

20. A nucleus is absent in the mature

a. Sieve tube cells

b. Mammalian erythrocytes

c. Monocytes

d. Both A and B

Answer: D

21. A tissue having more nonliving material than the living matter is

a. Epithelial tissue

b. Parenchyma

c. Connective tissue

d. Nervous system

Answer: C

22. Who proposed protoplasmic theory as opposed to cell theory?

a. Virchow

b. Schultze

c. Sachs

d. Strasburger

Answer: B

23. The theory proposing that body an organism consists of incompletely divided cells is

a. Organismal theory

b. Protoplasmic theory

c. Cell theory

d. Theory of cell Lineage

Answer: A

24. Organismal theory was proposed by

a. Van Mohl

b. Sachs

c. Virchow

d. Haberlandt

Answer: B

25. Cells are autonomous because

a. They synthesise components of living protoplasm from nonliving materials

b. They are able to grow and divide

c. Each cell has its own life span

d. All the above

Answer: D

26. "Each cell leads a double Life" was first proposed by

a. Schleiden

b. Grew

c. Von Mohl

d. Malpighi

Answer: D

27. Aging is slow or absent in

a. Planta

b. Parrot

c. Hydra

d. Unicells

Answer: D

28. Which are less efficient?

a. Multicellular animals

b. Multicellular plants

c. Colonial organisms

d. Unicellular organisms

Answer: D

29. A multicellular organism possesses

a. Differentiated cells

b. Undifferentiated cells

c. Dedifferentiated cells

d. All the above.

Answer: D

30. Number of type of cells found in the human body is

a. 20

b. 30

c. 200

d. 300

Answer: C

31. Cells that lose their nucleus during differentiation are

a. Nerve cells

b. Muscle cells

c. Erythrocytes

d. Leucocytes

Answer: C

32. A nucleated differentiated cell that has lost the power to dedifferentiate is

a. Nerve cell

b. Kidney cell

c. Liver cell

d. All the above

Answer: D

33. First successful culture was obtained by

a. Haberlandt

b. White

c. Skoog and Miller

d. Steward et al

Answer: D

34. Callus was grown successfully for the first time by

a. White

b. Gautheret

c. Nobecourt

d. All the above

Answer: D

35. Morphogenesis in tissue culture was discovered by

a. Gautheret

b. Skoog and Miller

c. Muir et al

d. Steward et al

Answer: B

36. Who proposed for the first time that cells are totipotent

a. White

b. Haberlandt

c. Steward

d. Halperin and Wetherell

Answer: B

37. Steward et al performed an experiment to prove cellular totipotency on

a. Tomato

b. Carrot

c. Tobacco

d. Potato

Answer: B

38. Explant is

a. Propagule

b. Callus used for subculturing

c. Part of plant used in tissue culture

d. Part of tissue culture used for planting

Answer: C

39. What was done by Steward et al in order to separate individual cells of carrot root?

a. Shaking in liquid medium

b. Homogenization

c. Pressure sieving

d. Microsurgery

Answer: D

40. Single cells in Steward's culture formed

a. Cellular clumps

b. Embryoids

c. Plantlets

d. All the above

Answer: A

41. Embryoids are

a. Somatic embryo-like structures

b. Small embryos through fertilization in culture

c. Early embryo stages used for propagation in tissue culture

d. All the above

Answer: A

42. In animals, cellular totipotency has been restricted only to

a. Germinal cells

b. Epithelial cells

c. Zygote

d. Zygote and early blastomeres

Answer: D

43. Animal cloning is carried out by

a. Artificial fertilization of ovum

b. Direct growth of ovum

c. Ovum with somatic nucleus

d. All the above

Answer: C

44. The first successful animal cloning, Dolly, was accomplished by

a. Fisher and Velton

b. Wilmut and Campbell

c. Morgan

d. Bernstein

Answer: B

45. Cells capable of division are

a. Stem cells
b. Meristematic cells
c. Undifferentiated cells
d. All the above

Answer: D

46. Differentiated cells are

a. Premitotic specialized
b. Post-mitotic specialized
c. Premeiotic specialized
d. Post-meiotic specialized

Answer: B

47. RBCs are

a. Differentiated cells
b. Undifferentiated cells
c. Dedifferentiated cells
d. Dead cells

Answer: A

48. During differentiation, RBCs lose

a. Aerobic respiration
b. DNA replication
c. RNA synthesis
d. All the above

Answer: D

49. Dedifferentiated cells are formed in the region of

a. Injury
b. Regeneration
c. Secondary growth
d. All the above

Answer: D

50. Functionally important dead cells are

a. Cork cells

b. Tracheary elements

c. Both A and B

d. Endothelial cells

Answer: D

CHAPTER 2

CELL BIOCHEMISTRY

1. BIOCHEMISTRY: DEFINITION, SCOPE AND IMPORTANCE

Introduction

The chemistry of life that's what biochemistry is all about, "the chemical changes that happen between plants, animals, and other organisms when they undergo their cycle of life, that's what Biochemistry is all about. Today we need to know more about biochemistry than ever as the world is heating up from global warming, and our crops are dying, causing lots of farmers to go without money and food for days. Biochemistry is the field of science that can save the future of our world from all the horrific things humankind is doing to nature. Today we will be talking about What is biochemistry and the importance of biochemistry, and it is one of those topics which are fun to learn while still too important to know about. The study of chemical processes inside the organisms, including the human body and both living and dead, is called biochemistry. Metabolism is the branch of biochemistry that governs all the flow of energy through metabolism. In recent years this field of science was the center of breakthroughs in finding the explanation of the living process.

As a result, biochemistry is being used in research related to botany, medicine, and gene enhancement. Biochemistry is now working on finding the secret of life and how biological molecules boost the processes that occur within living cells. This study will help in understanding the whole working organism. Well, in terms of biochemistry definition and marks point of view, biochemistry is the study of structures and the interactions of biological macromolecules. These macromolecules include protein, nucleic acids, lipids, and carbohydrates present in your body. There are various scenarios where you can easily find the scope of biochemistry. We are going to talk about some of them in the next section. We might not think about it, but biochemistry is playing a vital role in keeping you safe all the time from infection and diseases. Farmers using chemicals to keep their crop safe from animals, and microorganisms is one of the examples of biochemistry in nature. When packing food, companies check their nutrition value to see if the food satisfies all the nutritional demands set by the government". With the help of biochemical tests, scientists make sure food is good to eat and holds all the healthy nutrients inside. Medications have a date of expiration, and to know if they are suitable for use, we can use a biochemical test to find its condition and whether it's safe to take it or not. It is one of the questions only curious minds ask, the father of biochemistry was Carl Neuberg, he was the one that pioneered the study of chemistry and made it to the study of science. Before his research, people can't digest that biology and chemistry both are working hand in hand in our lives to help us and other organisms and animals grow, live, and die (Neuberg was born on July 29th, 1877, in Hanover, Germany, and took his last breath in New York, United States, on May 30th, 1956)". Living organisms are composed of lifeless molecules. When these molecules are isolated and examined individually, they conform to all the physical and chemical laws that describe the behavior of inanimate matter. Yet living organisms possess extraordinary attributes not shown by any random collection of molecules. In this chapter, we first consider the properties of living organisms that distinguish them from other collections of matter. After arriving at a broad definition of life, we can describe a set of principles that characterize all living organisms. These principles underlie the organization of organisms and the cells that make them up, and they provide the framework for this book. They will help you to keep the larger picture in mind while exploring the illustrative examples presented in the text. Second, living organisms extract, transform, and use energy from their environment usually in the form of either chemical nutrients or the radiant energy of sunlight. This energy enables living organisms to build and maintain their own intricate structures and to do mechanical, chemical, osmotic, and other types of work. By contrast, inanimate matter does not use energy in a systematic way to maintain structure or to do work. Inanimate matter tends to decay toward a more disordered state, to come to equilibrium with its surroundings. The third and most characteristic attribute of living organisms is the capacity for precise selfreplication and self-assembly a property that can be regarded as the quintessence of the living state. A single bacterial cell placed in a sterile nutrient medium can give rise to a billion identical "daughter" cells in 24 hours. Each of the cells contains thousands of different molecules, some extremely complex; yet each bacterium is a faithful copy of the original, constructed entirely from information contained within the genetic material of the original cell". By contrast, mixtures of inanimate matter show no capacity to grow and reproduce in forms identical in mass, shape, and internal structure, generation after generation.

The ability to self-replicate has no true analog in the nonliving world, but there is an instructive analogy in the growth of crystals in saturated solutions. "Crystallization produces more material identical in lattice structure with the original "seed" crystal. Crystals are much less complex than the simplest living organisms, and their structure is static, not dynamic as are living cells. Nonetheless, the ability of crystals to "reproduce" themselves led the physicist Erwin Schrodinger to propose in his famous essay "What Is Life?" that the genetic material of cells must have some of the properties of a crystal. Schrodinger's 1944 notion (years before the modern understanding of gene structure was achieved) describes rather accurately some of the properties of deoxyribonucleic acid, the material of genes. Each component of a living organism has a specific function. This is true not only of macroscopic structures such as leaves and stems or hearts and lungs, but also of microscopic intracellular structures such as the nucleus or chloroplast. Even individual chemical compounds in cells have specific functions". The interplay among the chemical components of a living organism is dynamic; changes in one component cause coordinating or compensating changes in another, with the result that the whole ensemble displays a character beyond that of the individual constituents. The collection of molecules carries out a program, the end result of which is the reproduction of the program and the self-perpetuation of that collection of molecules.

2. BIOMOLECULES: CARBON AND ITS BRIEF CHEMISTRY AND GENERAL PROPERTIES OF BIOMOLECULES

Introduction

They are polyhydroxy aldehydes or polyhydroxy ketones or compounds that can be hydrolyzed to them. In the majority of carbohydrates, H and O are present in the same ratio as in water hence also called as hydrates of carbon. Carbohydrates are polyhydroxy aldehydes or polyhydroxy-ketones of formula $(CH_2O)_n$, or compounds that can be hydrolyzed to them (aldehyde sugars or saccharides). It is classified into the following classes depending upon whether these undergo hydrolysis and if so on the number of products formed.

Monosaccharides

It consists of a single polyhydroxy aldehyde or ketone unit. It is simple sugars and they have the general formula $C_nH_{2n}O_n$. It is freely soluble in water but insoluble in non-polar solvents. Carbohydrates cannot be hydrolyzed to simpler carbohydrates such as glucose or fructose.

Disaccharides

It is the simplest and most common oligosaccharide. A disaccharide consists of two monosaccharides joined by an O-glycosidic bond. Carbohydrates can be hydrolyzed into two monosaccharide units such as sucrose, which is hydrolyzed into glucose and fructose.

Oligosaccharides

Oligosaccharides containing three or more residues are relatively rare and occurring almost entirely in plants. These carbohydrates can be hydrolyzed into a few monosaccharide units.

Polysaccharides

It is ubiquitous in nature. It is also called glycans. They are two types: structural and storage polysaccharides. Structural polysaccharide provides mechanical stability to cells, organs and organisms. Storage polysaccharides serve as carbohydrate stores that release monosaccharides as required.

Carbohydrates are polymeric sugars; eg starch or cellulose such as Triose, tetrose, pentose, hexose, etc. carbohydrates that contain three, four, five, six, etc figure 2.1 and 2.2. Hydrates of carbon hve a general formula is $C_n(H_2O)_n$. In plants: photosynthesis $6\ CO_2\ +\ 6\ H_2O\quad C_6H_{12}O_6\ +\ 6\ O_2$.

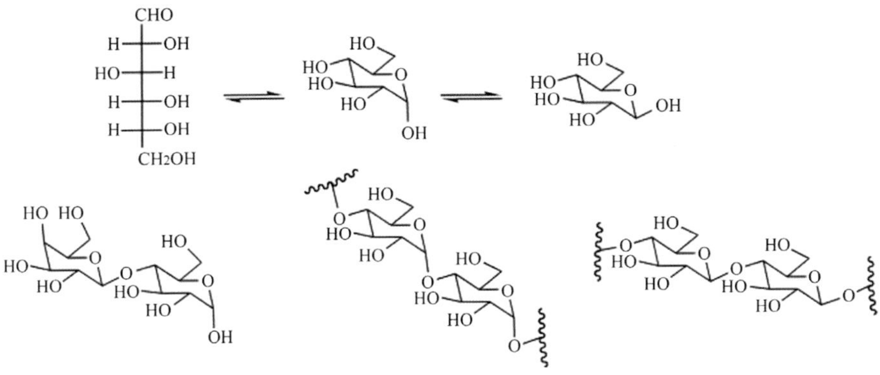

Figure 2.1: Structure of Carbohydrate.

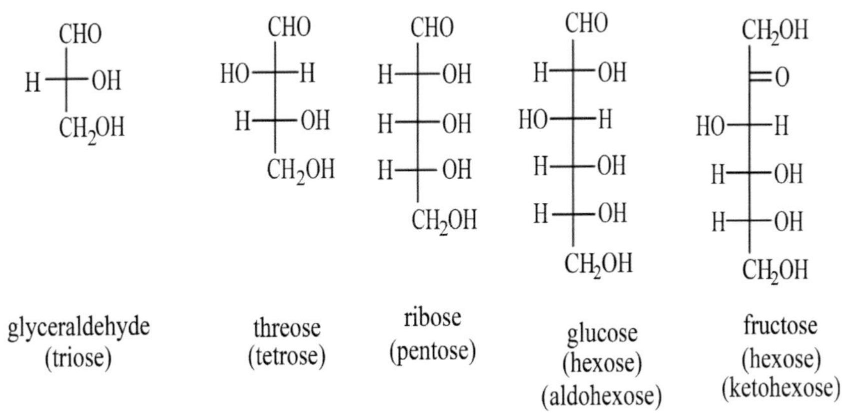

Figure 2.2: Structure of polysaccharides.

- **Polymers:** Large molecules are made up of repeating smaller units (monomer) is called polymers.
- **Aldose:** Polyhydroxyaldehyde **example** glucose.
- **Ketose:** Polyhydroxyketone **example** fructose.
- Carbons per molecule (usually five or six); eg. Aldohexose and ketopentose, etc.

- **Reducing sugar:** It is a carbohydrate that is oxidized by Tollen's, Fehling's or Benedict's solution is called reducing sugar.

Tollen's: $Ag^+ \rightarrow Ag$ (silver mirror)

Fehling's or Benedict's: Cu^{3+} (blue) $\rightarrow$ Cu^{2+} (red ppt).

These are reactions of aldehydes and alpha-hydroxyketones. All monosaccharides (both aldoses and ketoses) and most disaccharides are reducing sugars. Sucrose (table sugar) is a disaccharide, is not a reducing sugar figure 2.3 and 2.4.

Glucose (a monosaccharide)

Plants:

photosynthesis

$$6\ CO_2 + 6\ H_2O \xrightarrow[\text{sunlight}]{\text{chlorophyll}} \underset{(+)\text{-glucose}}{C_6H_{12}O_6} + 6\ O_2$$

$$(+)\text{-glucose} \rightleftharpoons \text{starch or cellulose}$$

respiration

$$C_6H_{12}O_6 + 6\ O_2 \longrightarrow 6\ CO_2 + 6\ H_2O + \text{energy}$$

Figure 2.3: Glucose in plants.

Animals

$$\text{plant starch} \longrightarrow (+)\text{-glucose}$$

$$(+)\text{-glucose} \longrightarrow \text{glycogen}$$

$$\text{glycogen} \longrightarrow (+)\text{-glucose}$$

$$(+)\text{-glucose} \longrightarrow \text{fats or aminoacids}$$

respiration

$$(+)\text{-glucose} + 6\ O_2 \longrightarrow 6\ CO_2 + 6\ H_2O + \text{energy}$$

Figure 2.4 Glucose in animals

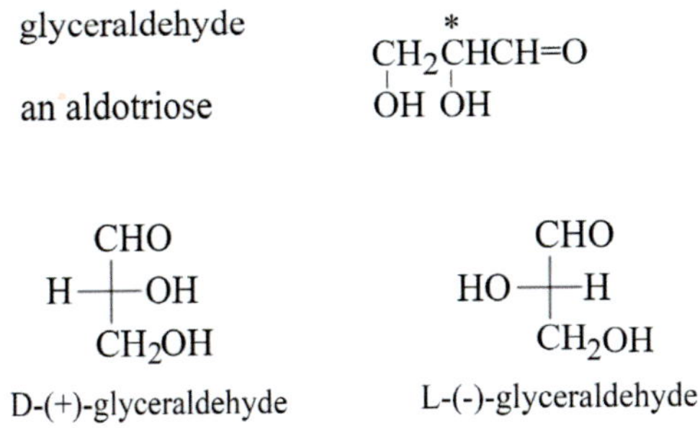

D & L are used to relate configuration of the chiral center most removed from the reducing group (C=O). If the -OH is on the right in the Fischer projection, then it is D, if the -OH is on the left, then it is L

Figure 2.5: Fischer projections and the D, L Notation.

- Representation of a three-dimensional molecule as a flat structure figure 2.5.
- Tetrahedral carbon is represented by two crossed lines figure 2.6.

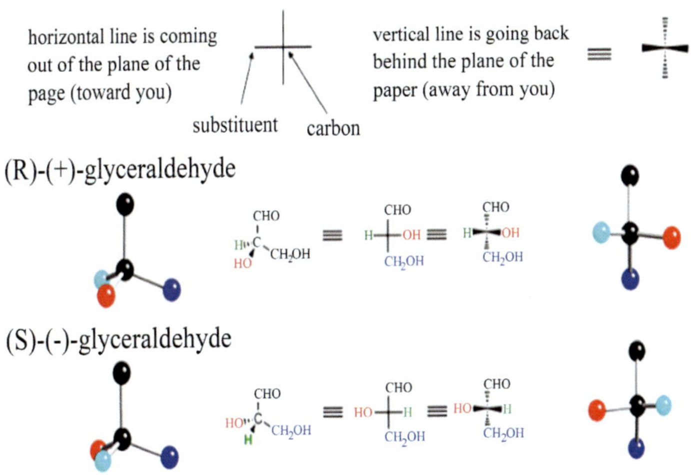

Figure 2.6: Absolute configuration of Glyceraldehyde is specified by the RS system in Fisher projection formulas**.** Before the *R*/*S* convention, stereochemistry was related to (+)-glyceraldehyde.

D-glyceraldehyde *R*-(+)-glyceraldehyde (+)-rotation = dextrorotatory = d	L-glyceraldehyde *S*-(-)-glyceraldehyde (-)-rotation = levorotatory = l

D-carbohydrates have the -OH group of the highest numbered chiral carbon pointing to the right in the Fischer projection as in R (+)-glyceraldehyde figure 2.4, 2.5. For carbohydrates, the convention is to arrange the Fischer projection with the carbonyl group at the top for aldoses and closest to the top for ketoses. The carbons are numbered from top to bottom. Carbohydrates are designated as D- or L- according to the stereochemistry of the highest numbered chiral carbon of the Fischer projection. If the hydroxyl group of the highest numbered chiral carbon is pointing to the right, the sugar is designated as D (*Dextro*: Latin for *on the right side*). If the hydroxyl group is pointing to the left, the sugar is designated as L (*Levo*: Latin for on *the left side*) figure 2.6. Most naturally occurring carbohydrates are of the D-configuration. Chiral molecules such as glucose and arabinose can exit in two configurations that are non-superimposable mirror images of each other. These two configurations are called enantiomers figure 2.7.

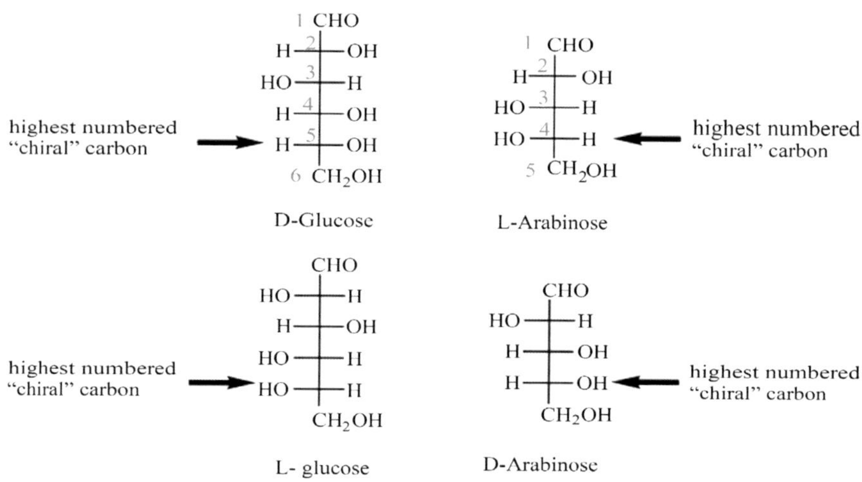

Figure 2.7: Enantiomers of glucose and arabinose.

Aldotetroses: Glyceraldehyde is the simplest carbohydrate (C_3, aldotriose, 2,3-dihydroxypropanal). The next carbohydrates are aldotetroses (C_4, 2,3,4-trihydroxybutanal) figure 2.8.

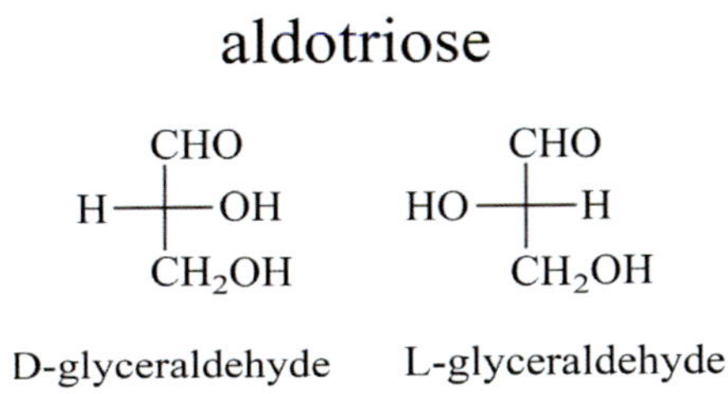

Figure 2.8: Structure of DL aldoses with three carbons.

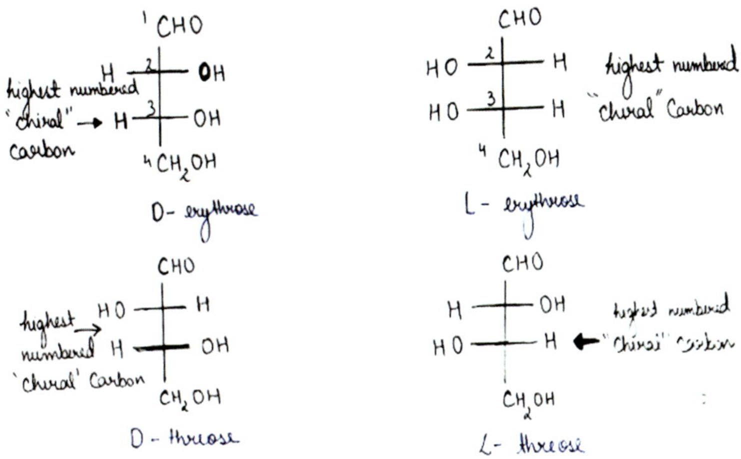

Figure 2.9: Chiral carbon.

Aldopentoses and Aldohexoses

- **Aldopentoses:** It contains five carbon **C_5**, three chiral carbons and eight stereoisomers figure 2.10.

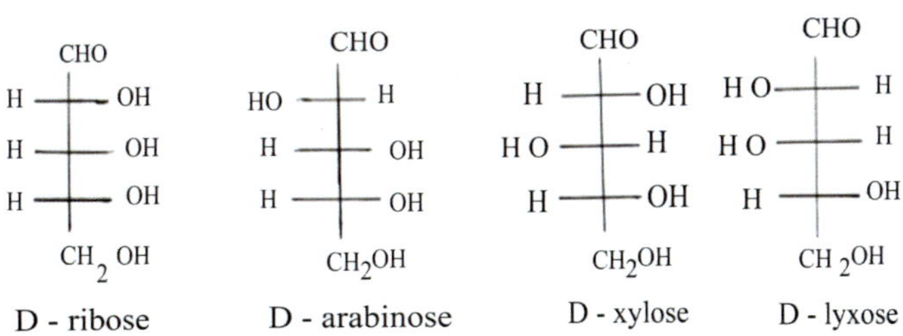

Figure 2.10: Aldopentoses.

Aldohexoses: It contains six carbon **C_6**, four chiral carbons and sixteen stereoisomers figure 2.11 and 2.12.

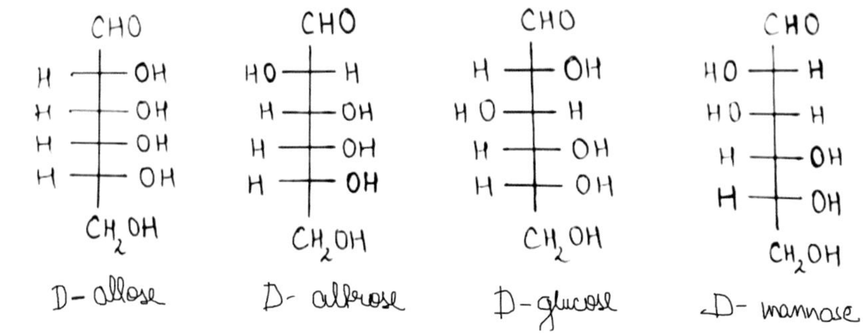

Figure 2.11: Aldohexoses.

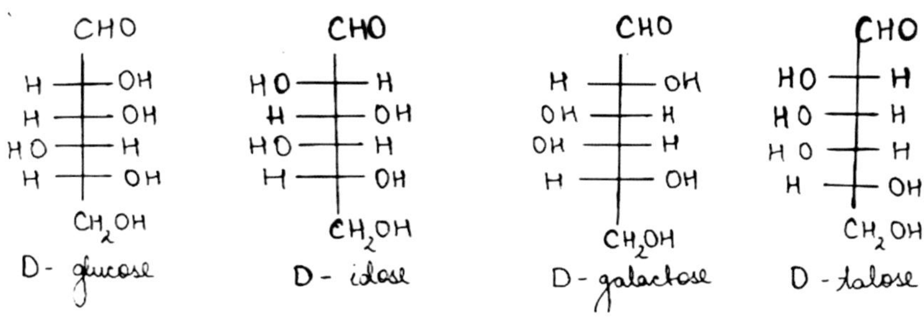

Figure 2.12: Aldohexoses.

Fischer Projections

Fischer projections can be rotated by 180° figure 2.13.

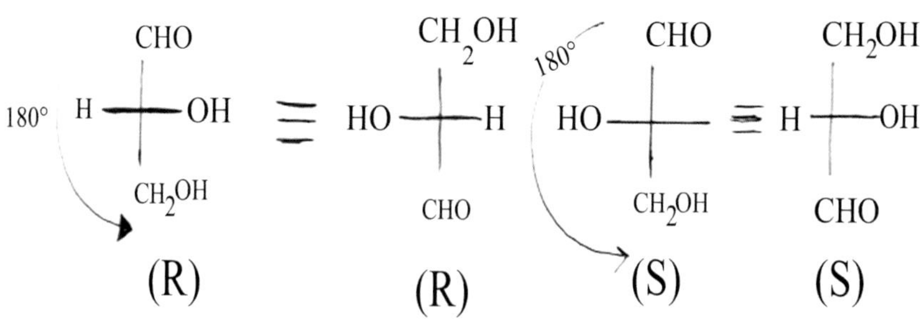

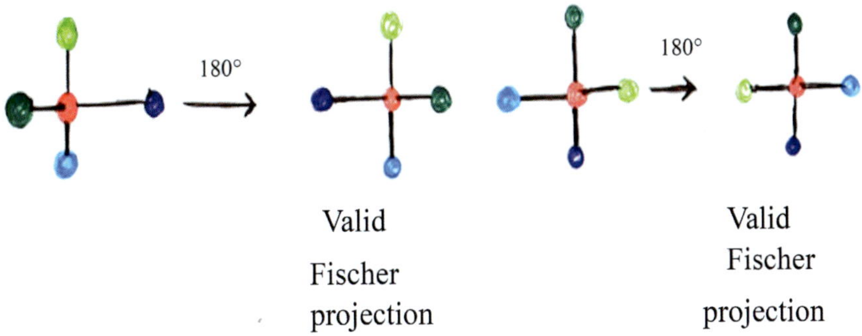

Figure 2.13: Fischer projections.

A 90° rotation inverts the stereochemistry and is not right figure 2.14.

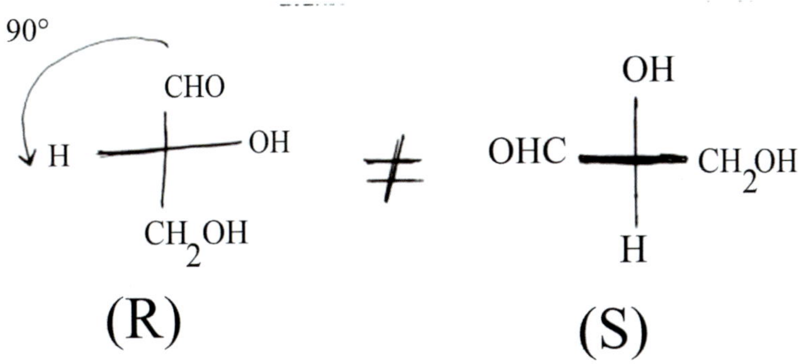

Figure 2.14: Stereochemistry.

If one group of a Fischer projection is held steady, the other three groups can be rotated clockwise or counterclockwise figure 2.15 (data collect from www.scribd.com., www.slideshare.com., www.coursehero.com., www.biocyclopedia.com., www. molecular-plant-biotechnology.info., www.slideplayer.com., www.documents.mx., www.docplayer.net., www.bmb.psu.edu.,www.archive.org.,www.mafiadoc.com., www.powershow.com., www.bioscience.jbpub.com., www.outhorstream.com., www.docslide.com., www.msluay.com., www.scienceprofonlue.com., www.csus.edu., www.krishikosh.egrouth.ac.in., www.tankonyvtor.hu., www.silo.pub., www.studyblue.com., www.Preservearticles.com., www.biotechuniverse.blogspot.com).

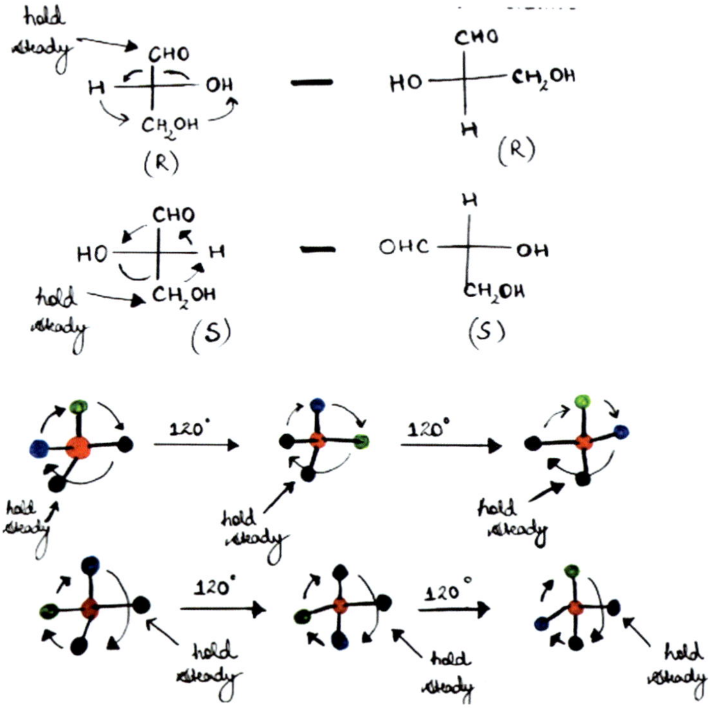

Figure 2.15: Fischer projection is held steady; the other three groups can be rotated clockwise or counterclockwise.

Amino Acid

Amino acid is a compound containing carbon, hydrogen, oxygen and nitrogen and serves as monomers of proteins. It contains both α-Amino acid and carboxylic acids. **Note: except for glycine (R = H), all a-amino acids are chiral molecules.** There are 20 naturally occurring R-groups **figure 2.16.**

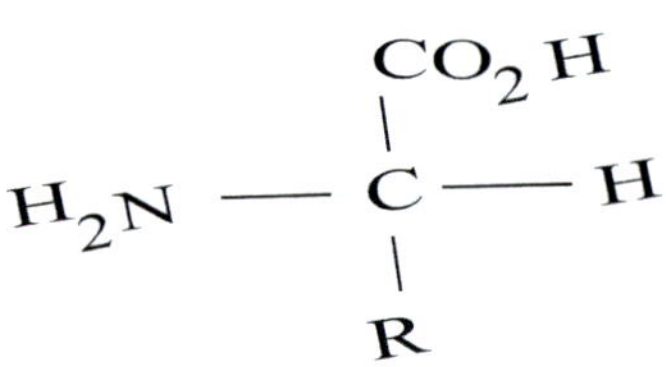

Figure 2.16: Structure of α-amino acid and carboxylic acids.

As with carbohydrates, it is traditional to use the D and L nomenclature with amino acids based on the configuration of glyceraldehyde. Naturally occurring amino acids generally have the same configuration as L-glyceraldehyde **figure 2.17**.

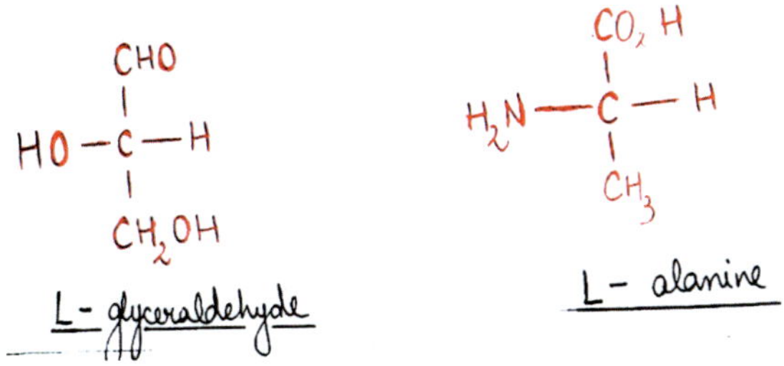

Figure 2.17: *S*-configuration at the α-carbon.

Acid–base Properties

Since amino acids have both an acidic functionality and a basic functionality, we should expect the following equilibrium. In fact, the equilibrium lies to the right all amino acids are charged at any pH. Such species that are overall neutral molecules but contain charged ends are called **zwitterions figure 2.18.**

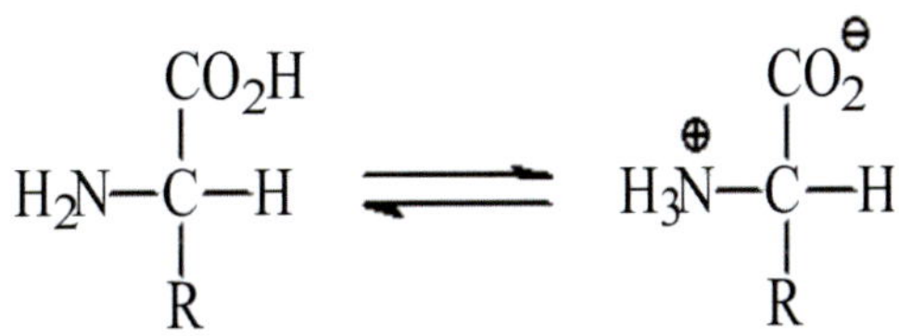

Figure 2.18: Structure of zwitterions.

Amino acids can react as either acids or bases figure 2.19.

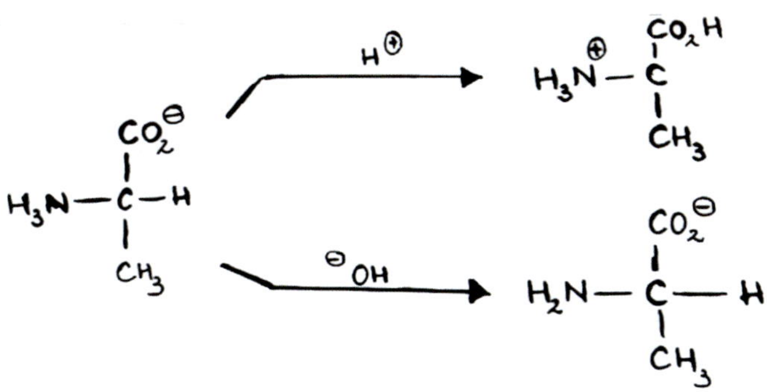

Figure 2.19: Structure of amino acids (acids or bases).

Amino acids can also have side chains containing acidic or basic groups figure 2.20.

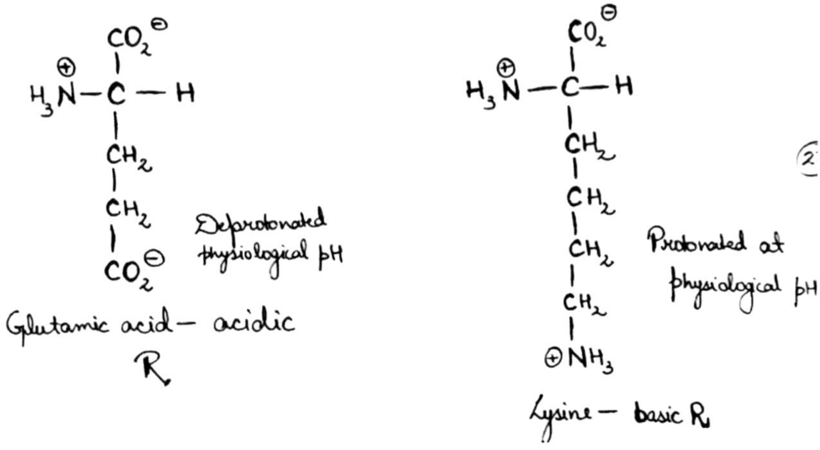

Figure 2.20: Structure of amino acids side chains containing groups (acidic or basic).

Classification

Amino acids are classified on the basis of the structure of R.

- **Aliphatic side chains:** Hydrophobic such as (**Alanine, isoleucine, leucine, proline, and valine) figure 2.21.**
- **Polar side chains:** Text classifies as HO-, S-, and amide containing hydrophilic groups such as **Asparagine (N), Glutamine (Q), Serine (S), Threonine (T) and Tyrosine (Y) figure 2.23.**
- **Acidic** – Hydrophilic such as **aspartic acid or aspartate (Asp) and glutamic acid or glutamate (Glu) figure 2.24**
- **Basic** – Hydrophilic such as **histidine, isoleucine, leucine, lysine, methionine, phenylalanine, threonine, tryptophan, and valine figure 2.22 and 2.25.**
- **Heterocyclic/Aromatic**: Hydrophilic or hydrophobic such as **phenylalanine and tyrosine and tryptophan figure 2.26.**
- **Aliphatic side chains:** Hydrophobic such as **Alanine, isoleucine, leucine, proline, and valine**

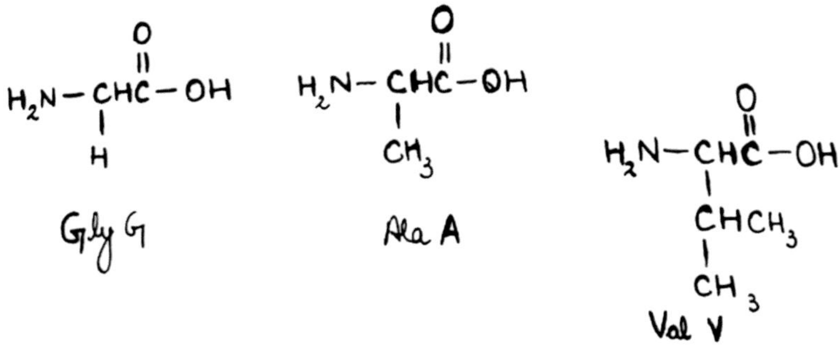

Figure 2.21: Aliphatic side chains.

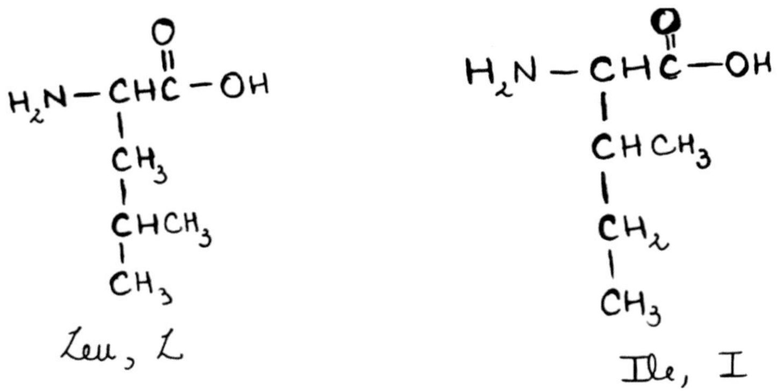

Figure 2.22: Basic side chain.

Polar side chains: Text classifies as HO-, S-, and amide containing hydrophilic groups.

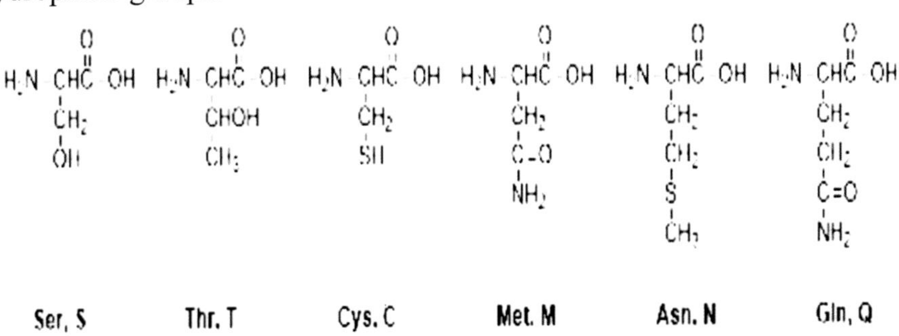

Figure 2.23: Polar side chain.

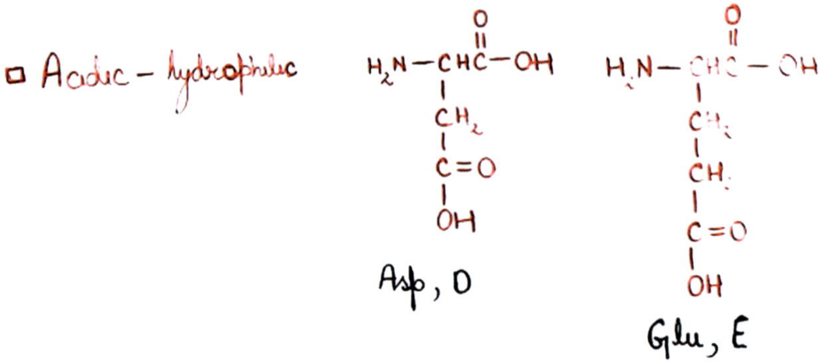

Figure 2.24: Acidic hydrophilic side chain.

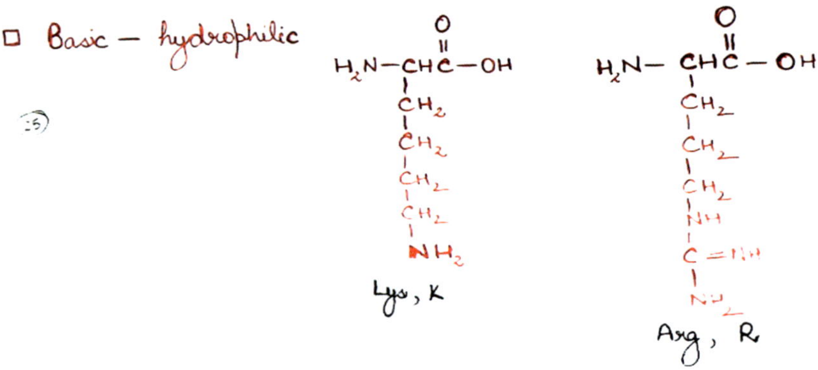

Figure 2.25: Basic Hydrophilic side chain.

Heterocyclic/Aromatic: Hydrophilic or Hydrophobic

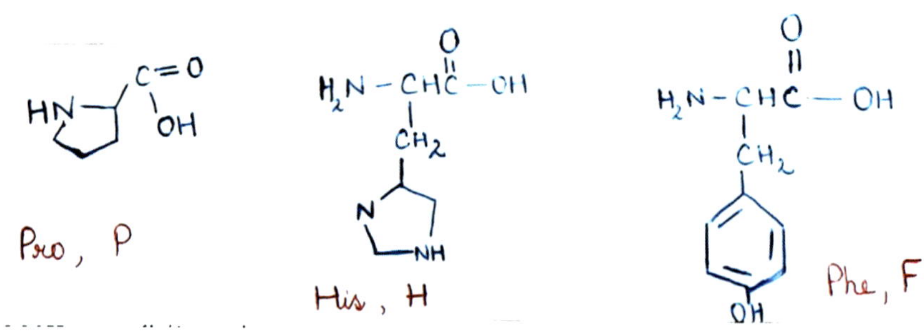

Figure 2.26 Heterocyclic/Aromatic

Acid-Base Properties

Nonpolar & polar side chains	pKa of α–COOH	pKa of $\alpha-NH_3^+$
alanine	2.35	9.87
asparagine	2.02	8.80
glutamine	2.17	9.13
glycine	2.35	9.78
isoleucine	2.32	9.76
leucine	2.33	9.74
methionine	2.28	9.21
phenylalinine	2.58	9.24
proline	2.00	10.60
serine	2.21	9.15
threonine	2.09	9.10
tryptophan	2.38	9.39
valine	2.29	9.72

Acidic Side Chains	pKa of α-COOH	pKa of $\alpha-NH_3$	pKa of side chain	Side Chain Group
aspartic acid	2.10	9.82	3.86	Carboxyl
glutamic acid	2.10	9.47	4.07	carboxyl
cysteine	2.05	10.25	8.00	sulfhydryl
tyrosine	2.20	9.11	10.07	phenolic

Basic Side Chains	pKa of α-COOH	pKa of $\alpha-NH_3^+$	pKa of Side Chain	Side Chain group
arginine	2.01	9.04	12.48	guanidino
histidine	1.77	9.18	6.10	imidazole
lysine	2.18	8.95	10.53	1° amine

Acidity of a-COOH Groups

The average pK_a of a-carboxyl group is 2.19, which makes them considerably stronger acids than acetic acid (pK_a 4.76). The greater acidity is accounted for by the electron-withdrawing inductive effect of the adjacent $-NH_3^+$ group figure 2.27.

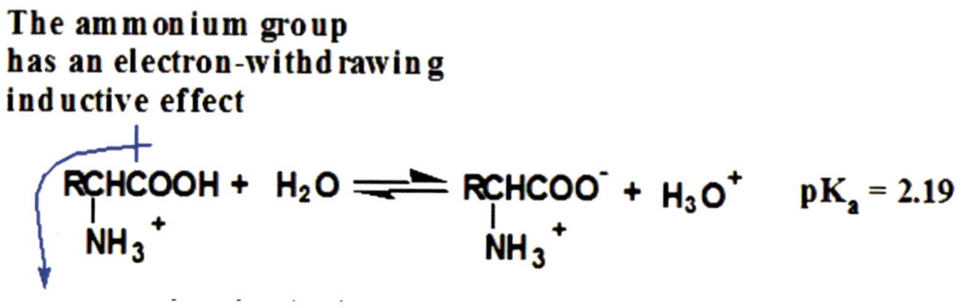

Figure 2.27: Structure of pK_a of a-carboxyl group.

Side-chain COOH Groups

Due to the electron-withdrawing inductive effect of the α-NH_3^+ group and side-chain -COOH groups are also stronger than acetic acid. The effect decreases with distance from the a-NH_3^+ group compared as:

α-COOH group of alanine (pK_a 2.35)

β-COOH group of aspartic acid (pK_a 3.86)

γ-COOH group of glutamic acid (pK_a 4.07)

Acidity of a-NH_3^+ Groups

The average value of pK_a for α-NH_3^+ group is 9.47 as compared with a value of 10.76 for 1° alkylammonium ions figure 2.28.

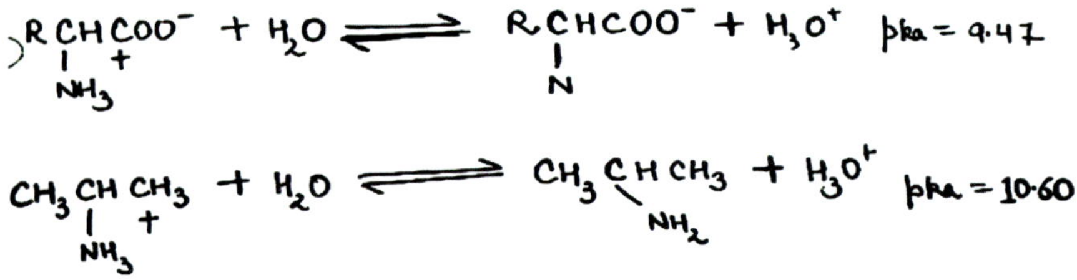

Figure 2.28: Structure of pK_a for α-NH_3^+ group.

Basicity of Guanidine Group

The side chain of arginine contains a guanidine group **figure 2.29**.

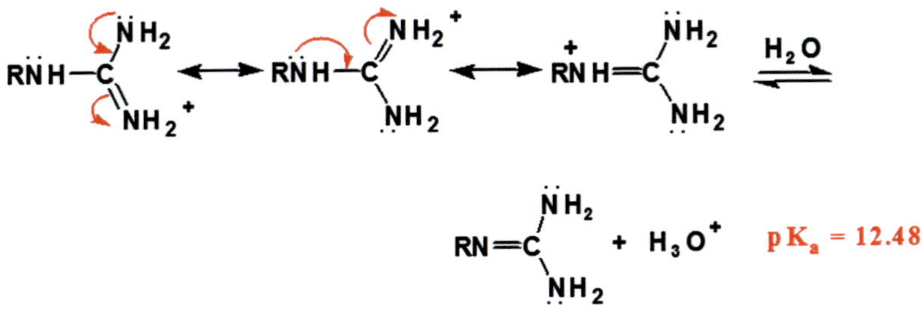

Figure 2.29: Structure of guanidine group.

The Imidazole Groups of His

The imidazole group is a heterocyclic aromatic amine figure 2.30.

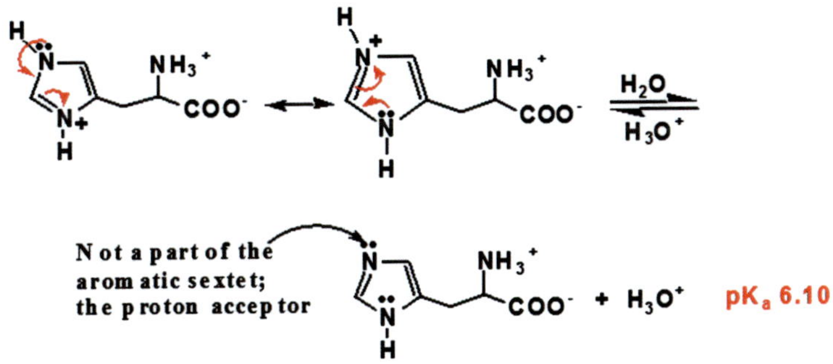

Figure 2.30: Structure of heterocyclic aromatic amine.

Titration of Amino Acids

Titration of glycine with NaOH figure 2.31.

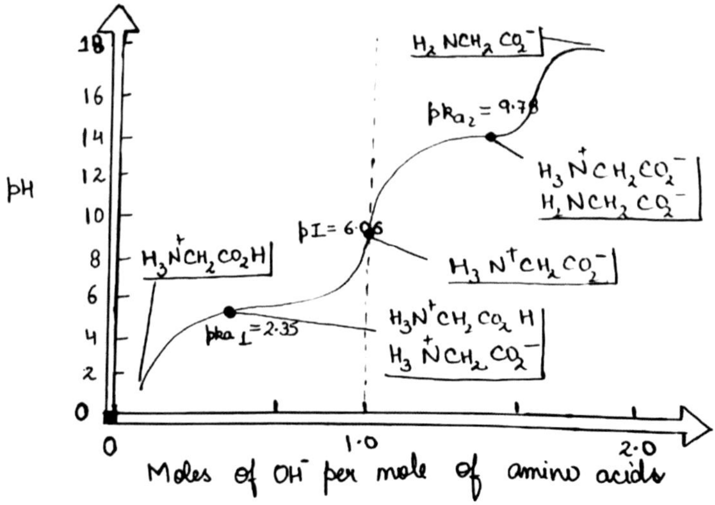

Figure 2.31: Titration of amino acids (glycine with NaOH).

Isoelectric Point (pI)

The isoelectric point, pI: the pH at which the majority of amino acid molecules in solution have no net charge. The pI for glycine, for example, falls between the pK_a values for the carboxyl and amino groups. The following tables give isoelectric points for the 20 protein-derived amino acids **figure 2.32**.

$$pI = \frac{1}{2}\left(pKa\ \alpha\text{-}COOH + pK_a\ \alpha\text{-}NH_3^{+}\right)$$

$$= \frac{1}{2}(2.35 + 9.78) = 6.06$$

Figure 2.32: Isoelectric point (pI).

Nonpolar & polar side chains	pKa of α-COOH	pKa of $\alpha-NH_3^+$	pKa of side Chain	pI
alanine	2.35	9.87	---	6.11
asparagine	2.02	8.80	---	5.41
glutamine	2.17	9.13	---	5.65
glycine	2.35	9.78	---	6.06
isoleucine	2.32	9.76	---	6.04
leucine	2.33	9.74	---	6.04
methionine	2.28	9.21	---	5.74
phenylalanine	2.58	9.24	---	5.91
proline	2.00	10.60	---	6.30
serine	2.21	9.15	---	5.68
threonine	2.09	9.10	---	5.60
valine	2.29	9.39	---	6.00
tryptophan	2.38	9.72	---	5.88

Acidic side chain	pKa of α-COOH	pKa of $\alpha-NH_3^+$	pKa of side chain	pI
aspartic acid	2.10	9.82	3.86	2.98
glutamic acid	2.10	9.47	4.07	3.08
cysteine	2.05	10.25	8.00	5.02
tyrosine	2.20	9.11	10.07	5.63

Basic side chain	pKa of α-COOH	pKa of $\alpha-NH_3^+$	pKa of side chain	pI
arginine	2.01	9.04	12.48	10.76
histidine	1.77	9.18	6.10	7.64
lysine	2.18	8.95	10.53	9.74

A reagent is commonly used to detect amino acids is ninhydrin figure 2.33.

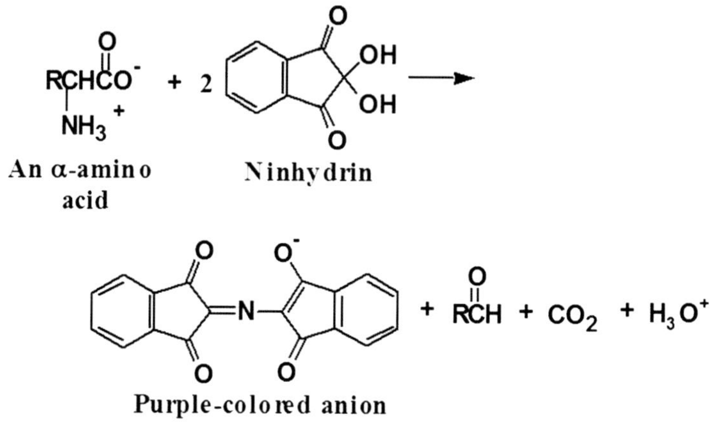

Figure 2.33: Ninhydrin.

Proteins

In 1902, Emil Fischer proposed that proteins are long chains of amino acids joined by peptide bonds. Peptide bond: amide bond between the α-carboxyl group of one amino acid and the a-amino group of another.

Peptides

- **Peptide:** the short polymer of amino acids are joined by peptide bonds; they are classified by the number of amino acids in the chain.
- **Dipeptide:** a molecule containing two amino acids joined by a peptide bond.
- **Tripeptide:** a molecule containing three amino acids joined by peptide bonds.
- **Polypeptide:** a macromolecule containing many amino acids joined by peptide bonds.
- **Protein:** a biological macromolecule of molecular weight 5000 g/mol or greater, consisting of one or more polypeptide chains.

Peptides are amino acid polymers containing 2–50 individual units. Peptides with >50 units are called proteins. By convention, peptide structures are written with the *N*-terminal amino acid on the left and the *C*-terminal amino acid on the right figure 2.34 and 2.35.

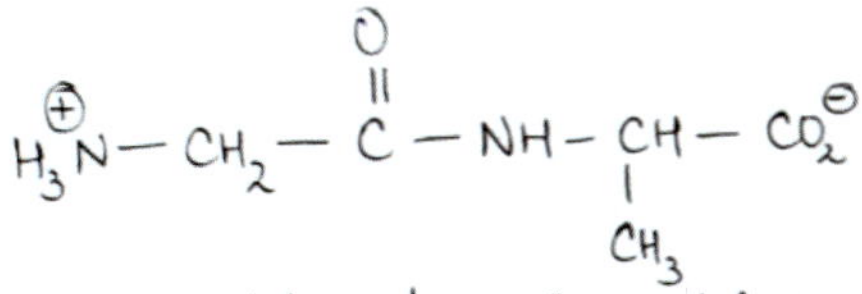

A dipeptide Glycylalanine = gly-ala

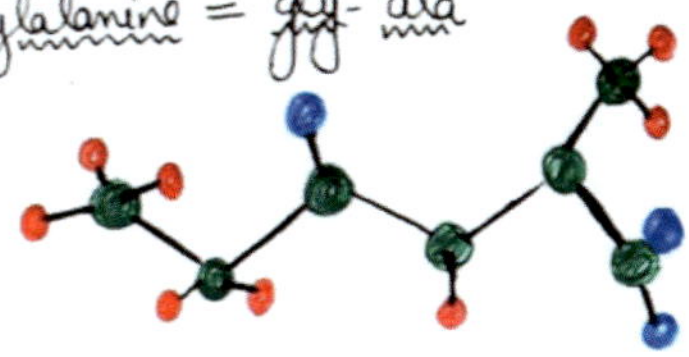

Glycylalanine = gly-ala

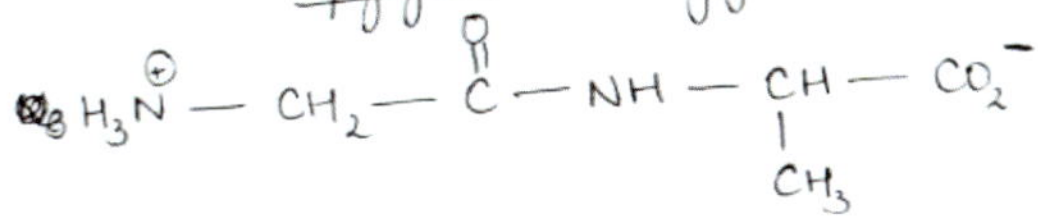

Glycine
amino-terminal amino acid

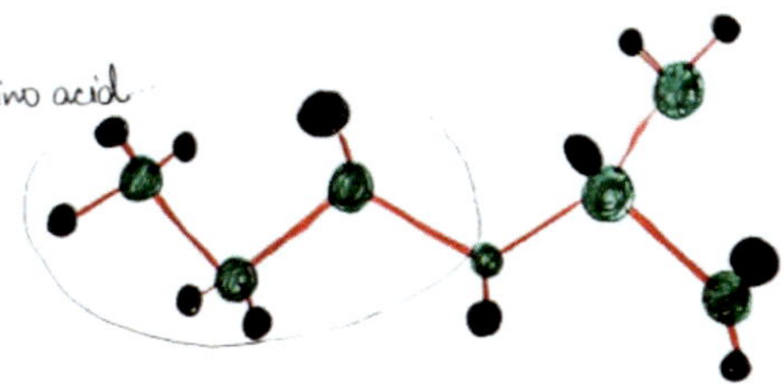

Glycylalanine = gly-ala

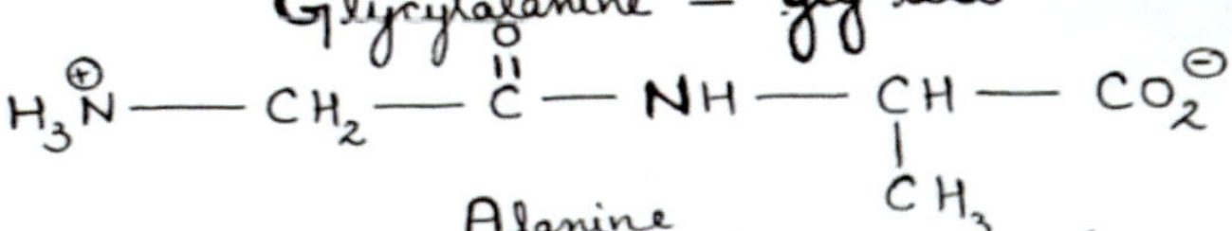

Alanine
carboxy-terminal amino acid

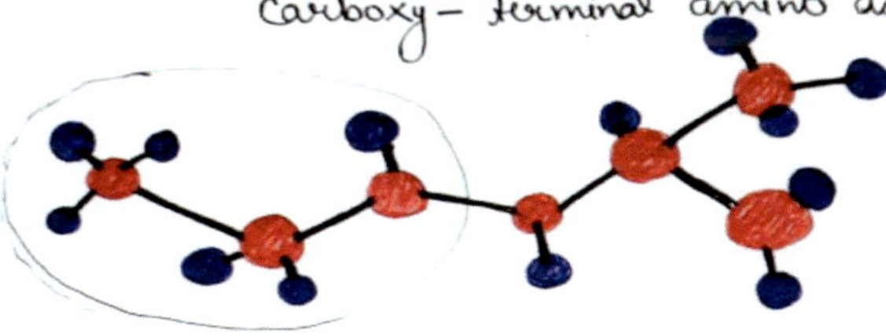

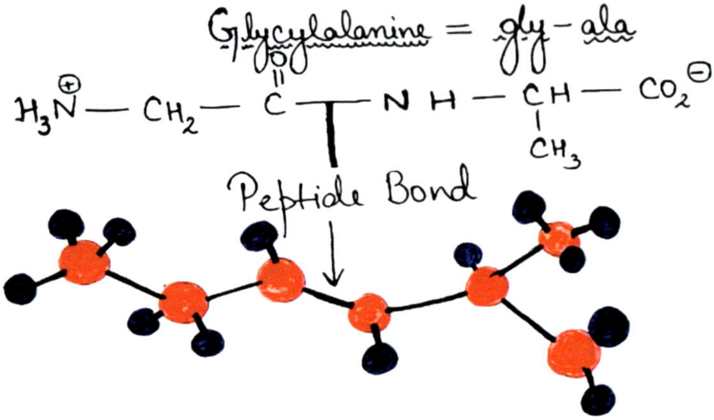

Figure 2.34: Structure of glycylalanine.

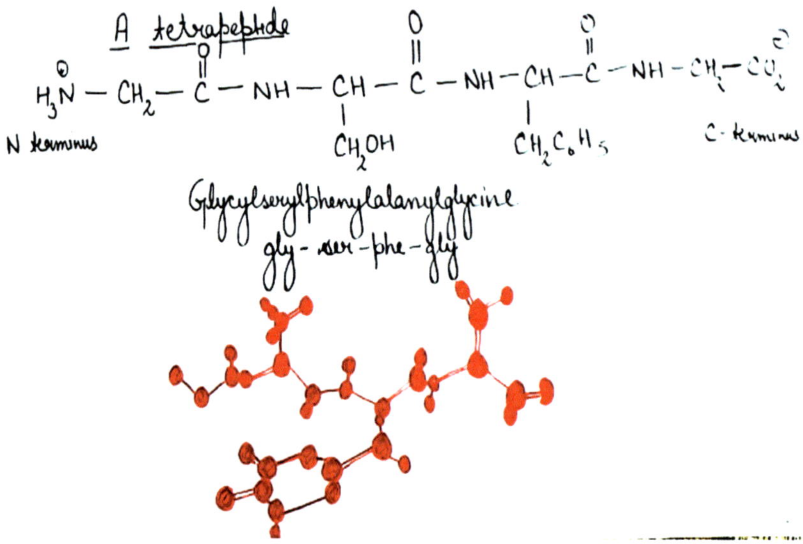

Figure 2.35: Glycylserylphenylalanyl glycine.

Peptide (Amide) Bond

The amide nitrogen is sp^2 hybridized and the lone pair is conjugated with the carbonyl group. There is considerable C–N double-bond character. Rotation about the C–N bond is difficult figure 2.36.

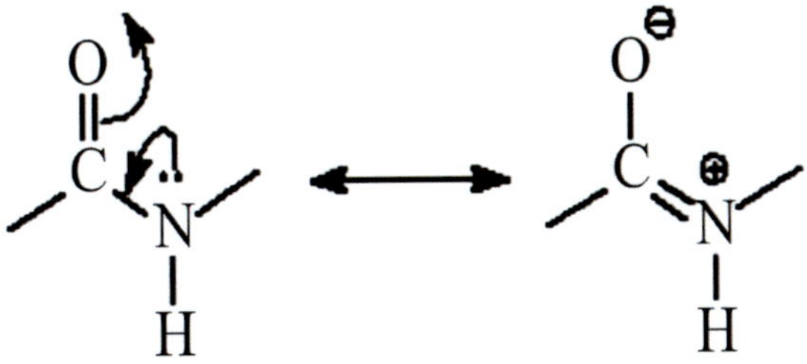

Figure 2.36: Peptide bond.

The amide groups are planar. The R-groups are on opposite sides of the plane. The only other type of covalent bond between amino acids in proteins and peptides is the disulfide linkage between two cysteine units.

Note:

- Thiols are readily oxidized to disulfides.
- Disulfides are readily reduced to thiols.

$$\mathrm{R{-}SH} + \mathrm{HS{-}R} \underset{[\mathrm{H}]}{\overset{[\mathrm{O}]}{\rightleftharpoons}} \mathrm{R{-}S{-}S{-}R}$$

Disulfide links serve to connect polypeptide chains figure 2.37.

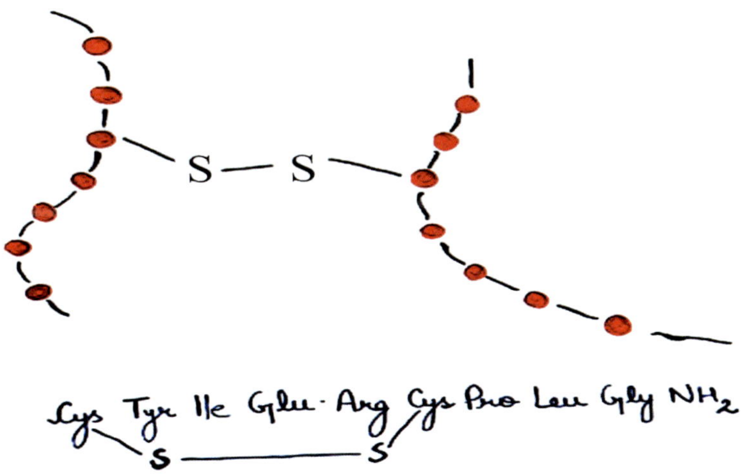

Figure 2.37: Disulfide links serve to connect polypeptide chains.

It can form a macrocycle.

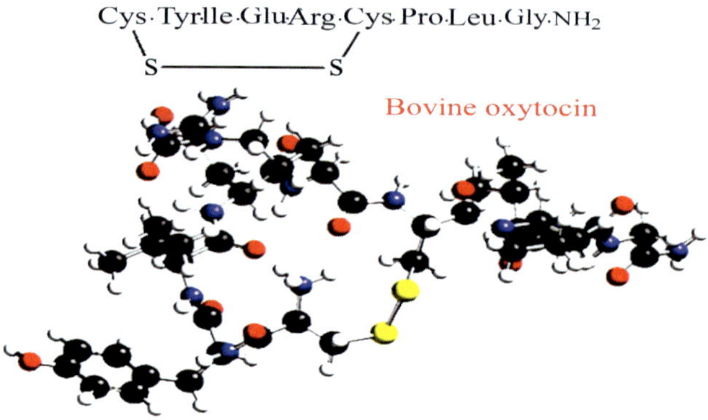

Figure 2.38: Macrocycle.

Peptides

A peptide is a short chain of amino acids. The amino acids in a peptide are connected to one another in a sequence by bonds called peptide bonds. It contains are amino acid analysis, terminal residue analysis, N-Terminal analysis (Edman degradation), C-Terminal analysis (Carboxypeptidase), partial hydrolysis (enzymatic), hydrolyze the peptide into smaller fragments which are, trypsin -cleaves at lys and arg, chymotrypsin - Cleaves at phe, tyr, and trp, pepsin - Cleaves at phe, tyr, trp, leu, asp, glu, cyanogen bromide (not enzymatic) - Cleaves at met, determine the sequence of the fragments and successive Edman degradations (www.scribd.com., www.slideshare.com., www.coursehero.com., www. biocyclopedia.com., www. molecular-plant-biotechnology.info., www.slideplayer.com., www.documents.mx., www. docplayer.net., www.bmb.psu.edu.,www.archive.org.,www.mafiadoc.com., www.powershow.com., www.bioscience.jbpub.com., www.outhorstream.com., www.docslide.com., www.msluay.com., www.scienceprofonlue.com., www.csus.edu., www.krishikosh.egrouth.ac.in., www.tankonyvtor.hu., www.silo.pub., www.studyblue.com., www.Preservearticles.com., www.biotechuniverse.blogspot.com).

Lipids

A lipid is an organic substance found in living systems that is insoluble in water but is soluble in organic solvents. Lipids are organic molecules essential for life that are composed mostly of C, H, O. They serve as a storage form of metabolic fuel. They serve as a transport form of metabolic fuel. They provide the structural components of membranes. They have protective functions in bacteria, plants, insects, and vertebrates serving as a part of the outer coating between the body of the organism and the environment figure 2.39 and 2.40.

There are four types of lipids

- Fats (triglycerides)
- Phospholipids
- Steroids
- Waxes

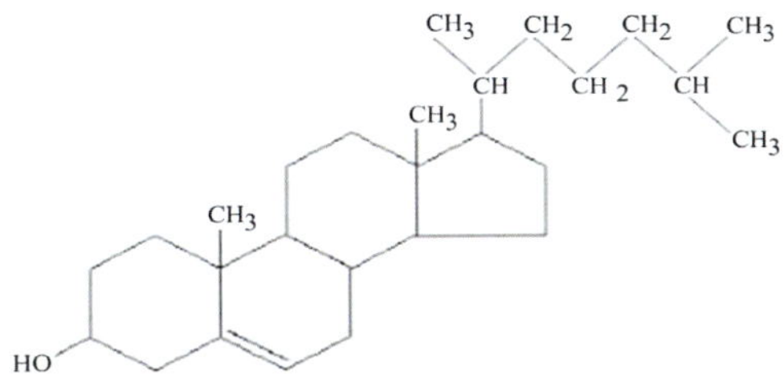

Structure of Lipid

Figure 2.39: Structure of Lipids.

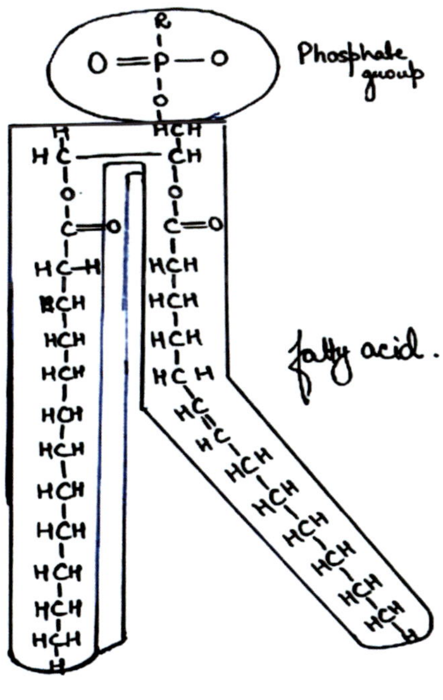

Figure 2.40: Structure of Fatty acid.

Fats & oils make up 95% of the nutritional lipids; the other 5% are steroids. Waxes are functional only. Fats are solid triglycerides and oils are liquid triglycerides figure 2.41.

Representative Fatty Acids

C12-C18 (also C20) Dietary Fats and Oils

$CH_3-(CH_2)_{12}-COOH$ (1)

$CH_3-(CH_2)_{14}-COOH$ (2)

$CH_3-(CH_2)_{16}-CO_2H$ (3)

$CH_3-(CH_2)_7-CH=CH-(CH_2)_7-CO_2H$ (4)

$CH_3-(CH_2)_4-CH=CH-CH_2-CH=CH-(CH_2)_7-CO_2H$ (5)

$CH_3-CH_2-CH=CH-CH_2-CH=CH-CH_2-CH=CH-(CH_2)_7-CO_2H$ (6)

Name	Mp (°C)	Class
(1) Myristic acid	58	Saturated C-14
(2) Palmitic acid	63	Saturated C-16
(3) Stearic acid	71	Saturated C-18
(4) Oleic acid	4	Monounsaturated C-18
(5) Linoleic acid	-5	Polyunsaturated C-18
(6) Linolenic acid	-11	Polyunsaturated C-18

Figure 2.41: List of Fatty acids.

Triglycerides

We consider the length and the number of carbons in the side chains. The number of carbon-carbon double bonds in the side chains (the degree of unsaturation).

Fat Molecules

- Structure = 3 fatty acid chains linked to glycerol figure 2.42.
- Functions = Long-term energy storage and insulation for animals.
- It can be saturated or unsaturated.

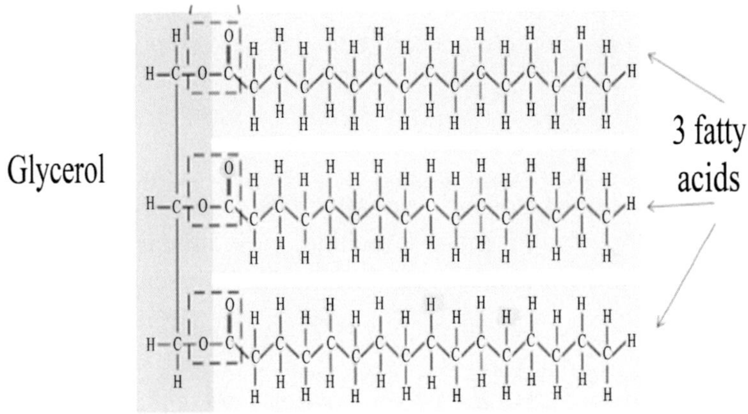

Figure 2.42: Structure of Glycerol.

Body Fat is Stored Energy

The body converts the unused carbohydrates, proteins and triglycerides that make up our macronutrients into small globules of fat that end up in the specialized cells of adipose tissue and the fatty tissue of the body. One pound of adipose tissue stores, and provides when needed, ~3500 Cal of energy. The high energy density of fat and its ability to store energy (9 Cal/gram) compactly in relatively little space and with relatively little weight, compared with carbohydrates and proteins (4 Cal/gram) which allows us to carry stores of energy with us.

Fat

Most of our long-term energy supplies operate via the formation, storage and metabolism of the body fat (triglycerides). Short-term energy storage, from one meal to another, occurs through a starch-like substance called glycogen (a carbohydrate).

Adipose Tissue and Fat

Adipose tissue forms cushioning shields around our major organs, protecting them against damage from physical shock and providing insulation to our bodies, guarding against a rapid loss of body heat to the external environment. Fats carry the flavors and vitamins of many of our foods although fats have no flavors of their own such as Carrying Vitamins A, D, E and K from our foods to our tissues. Fatty acids form not only the triglycerides but other compounds as well, including such vital classes as prostaglandins and phospholipids.

Saturated Fats

All carbon bonded to hydrogen in saturated fats. There are no C=C double bonds. It has a long, straight chain. It includes most animal fats and butter. It is solid at room temperature. It contributes to cardiovascular disease figure 2.43.

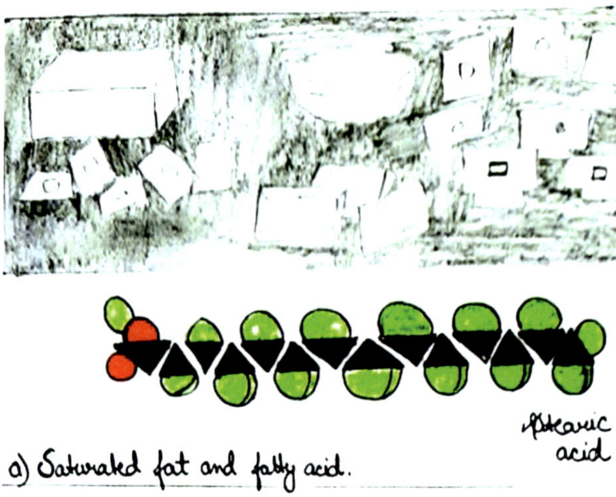

Figure 2.43: Structure of saturated fat and fatty acid.

Unsaturated Fats

At least one C=C double bond in the fatty acids such as plant & fish fats, vegetable oils and it is liquid at room temperature figure 2.44, 2.45 and 2.46.

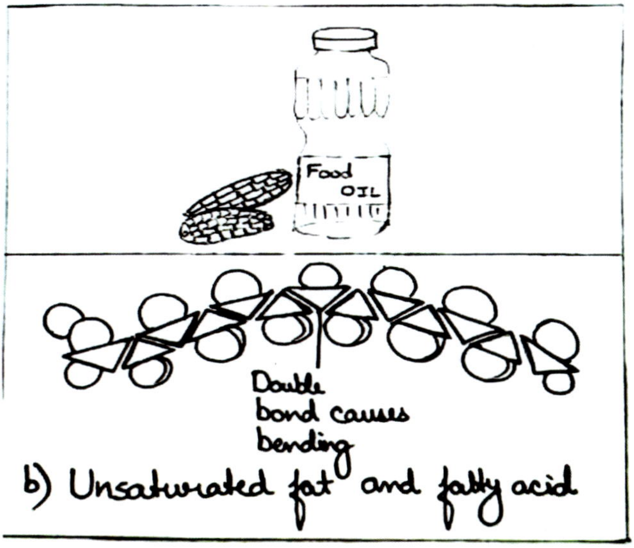

Figure 2.44: Structure of unsaturated fat and fatty acid.

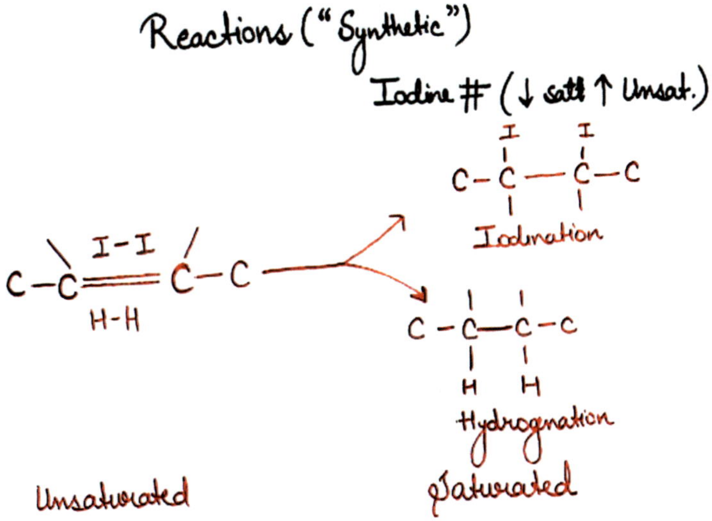

Figure 2.45: Structure of unsaturated and saturated fatty acids.

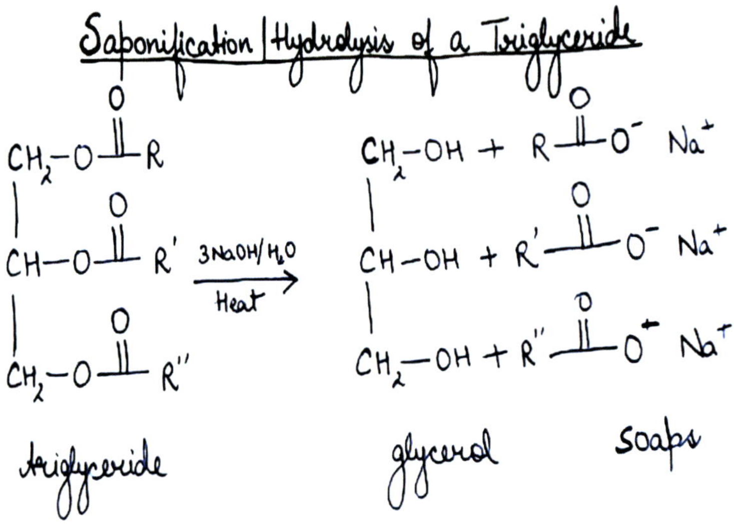

Figure 2.46: Saponification and hydrolysis of a triglyceride.

Omega (3 Fatty Acids, Especially in Fish Oils)

It contains mainly 20 & 22 C's, polyunsaturated and 1st 'ene' is 3 from Ω-end. It is eicosapentaenoic acid (EPA): 20 (5, 8, 11, 14, and 17) mp -50°C. It is Docosahexaenoic acid (DHA): 22 (5,8,11,14,17, 20). It includes α-linolenic acid: 18 (9, 12, 15) figure 2.47.

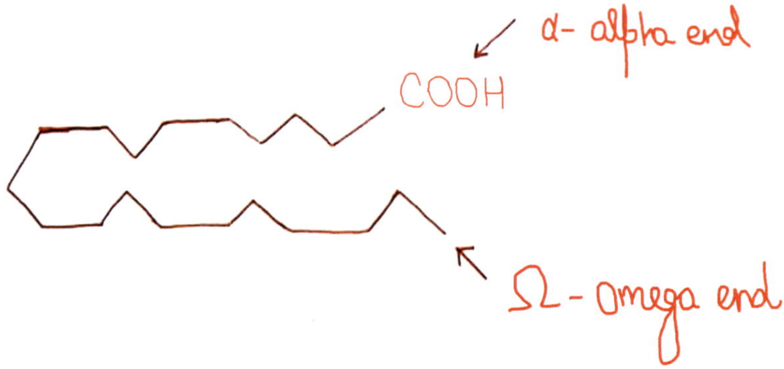

Figure 2.47: Structure of omega fatty acids.

Phospholipids

It has two fatty acids chains with a positive phosphate head. The fatty acid chains are non-polar hydrophobic (water hated) and they have also phosphate head with a polar hydrophilic (water loving) nature figure 2.48.

- **Function:** It makes up the cell membranes (phospholipid bilayer).

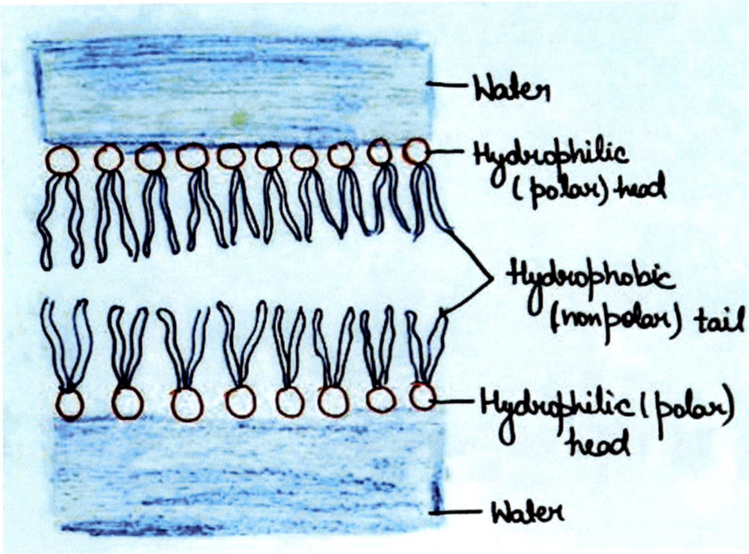

Figure 2.48: Structure of phospholipids.

Steroids

It contains four fused carbon rings figure 2.49.

- Examples:

Cholesterol: It controls the fluidity of the cell membrane.

Hormones: It regulates processes in the body (such as pregnancy).

Vitamins (A, B, D): It supports metabolism and cell processes.

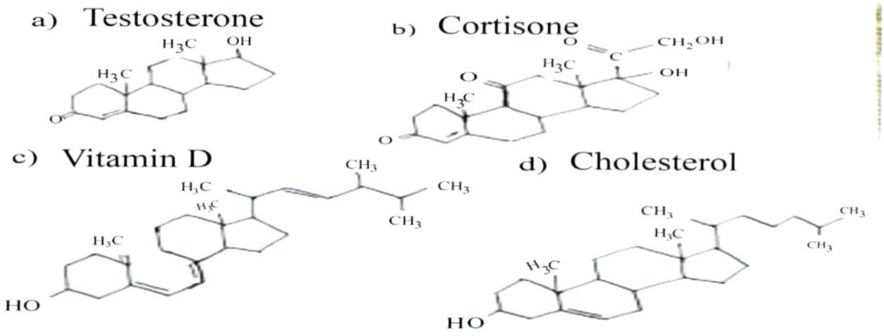

Figure 2.49: Structure of steroids.

Waxes

It is solid at room temp and insoluble in water figure 2.50.

Function

- It is used protective covering for animals and plants.
- **Examples:** Beeswax, earwax, cuticle of leaf (keeps water in).

Figure 2.50: Structure of waxes.

Lipids may be broadly defined as hydrophobic or amphiphilic small molecules that originate entirely or in part from two distinct types of biochemical subunits: *ketoacyl* and *isoprene* groups. Using this approach, lipids may be divided into eight categories: fatty acyls, glycerolipids, glycerophospholipids, sphingolipids, saccharolipids and polyketides (derived from the condensation of ketoacyl subunits); and sterol lipids and prenol lipids (derived from the condensation of isoprene subunits) figure 2.51. (Fahy, E., et al, Journal of Lipid Research, Vol. 46, 839-862, May 2005 and www.slideshare.net), (www.scribd.com., www.slideshare.com., www.coursehero.com., www. biocyclopedia.com., www. molecular-plant-biotechnology.info., www.slideplayer.com., www.documents.mx., www.docplayer.net., www.bmb.psu.edu.,www.archive.org.,www.mafiadoc.com., www.powershow.com., www.bioscience.jbpub.com., www.outhorstream.com., www.docslide.com., www.msluay.com., www.scienceprofonlue.com., www.csus.edu., www.krishikosh.egrouth.ac.in., www.tankonyvtor.hu., www.silo.pub., www.studyblue.com., www.Preservearticles.com., www.biotechuniverse.blogspot.com).

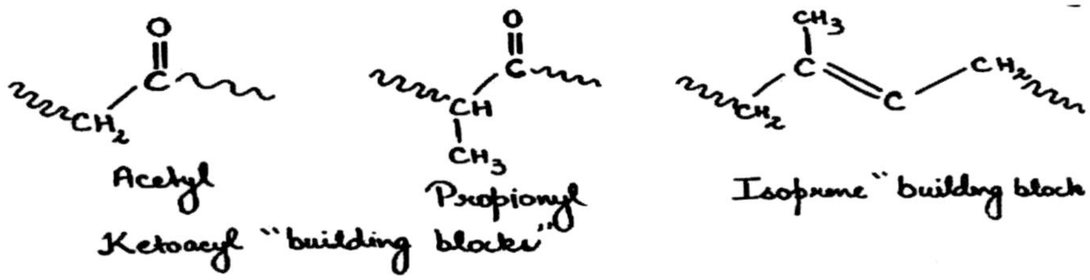

Figure 2.51: Fundamental biosynthetic units of lipids.

Table 2.1: List of Lipids

Category	Abbreviation	Example
Fatty acyls	FA	Dodecanoic acid
Glycerolipids	GL	1-hexadecanoyl-2-(9Z-octadecenoyl)-sn-glycerol
Glycerophospholipids	GP	1-hexadecanoyl-2-(9Z-octadecenoyl)-sn-glycero-3-phosphocholine
Sphingolipids	SP	N-tetradecanoyl-sphing-4-enine
Sterol lipids	ST	Cholest-5-en-3ß-ol
Prenol lipids	PR	2E, 6E-farnesol
Saccharolipids	SL	UDP-3-O-3R-hydroxy-tetradecanyoyl-Ad-n-acetylglucosamine
Polyketides	PK	Aflatoxin B_1

Consistent Structure Representation across Classes of Lipids

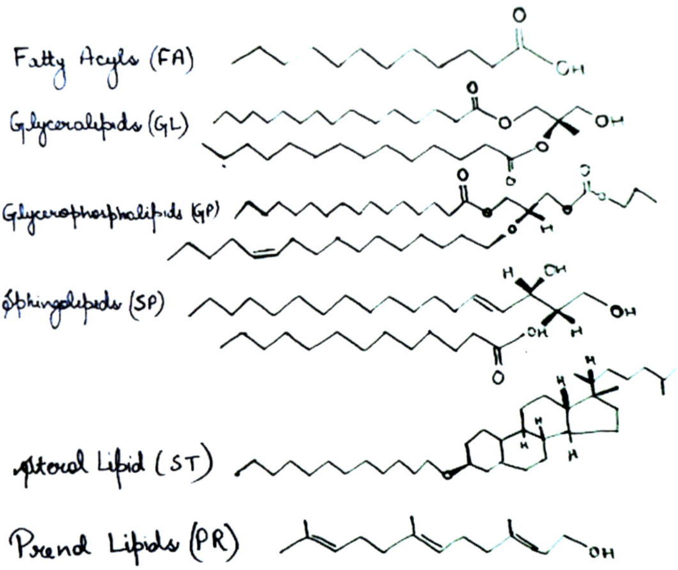

Structural Comparison of Sphingomyelin and Phosphatidylcholine

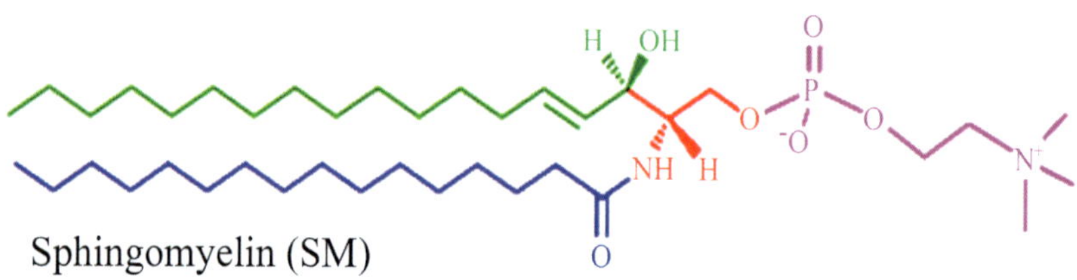

Sphingomyelin (SM)
(Example: N-palmitoyl-sphingomyelin)

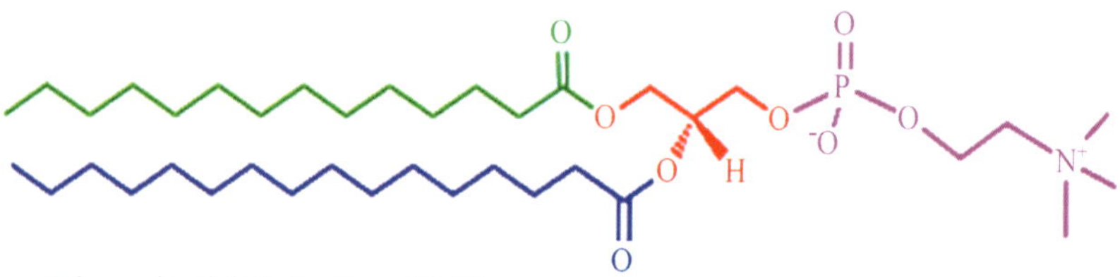

Phosphatidylcholine (PC)
(Example: 1-myristoyl-2-palmitoyl-sn-glycerophosphocholine)

Proteins

Proteins are including a diversity of structures, resulting in a wide range of functions. Proteins are account for more than 50% of the dry mass of most cells. "Protein functions include structural support, storage, transport, cellular communications, movement, and defense against foreign substances. At first, all we have is a string of amino acids bound with peptide bonds. Once the string of amino acid interacts with itself and its environment (often aqueous), then we have a functional protein that consists of one or more polypeptides precisely twisted, folded, and coiled into a unique shape. The sequence of amino acids determines a protein's three-dimensional structure. A protein's structure determines its function.

- A protein is a biologically functional molecule that consists of one or more polypeptides
- Proteins are unbranched polymers constructed from 22 standard α-amino acids.
- They have four levels of structural organization.
- Primary structure: the amino acid sequence is specified by genetic information.
- As the polypeptide chain folds and it forms certain localized arrangements of adjacent amino acids that constitute the secondary structure.
- The overall three-dimensional shape that a polypeptide assumes is called the tertiary structure.
- Proteins that consist of two or more polypeptide chains are said to have a quaternary structure.

Amino Acids

Amino acids are chemically diverse but they all contain a carboxyl acid group and an amino group both linked to a single carbon atom. Proteins are long linear polymers of amino acid joined head to tail by a peptide bond between the carboxyl acid group of one amino acid and the amino group of the next amino acid. There are twenty amino acids in proteins each with a different side chain attached to the α-carbon atom. All amino acids contain nitrogen atoms; therefore, the protein represents a class of nitrogenous compounds. The properties of amino acid side chains in aggregates determine the properties of proteins they constitute. Such amino acids are called standard amino acids. Amino acids that occur naturally in cells but are not incorporated ribosomally into proteins are called nonstandard amino acids. Some nonstandard amino acids are constituents of ribosomally synthesized proteins, but they are generated by post-translational modification of standard amino acids which are incorporated ribosomally into proteins.

- Selenocysteine is the 21st standard amino acid.
- Pyrrolysine is the 22nd standard amino acid.

Primary Structure

The sequence of amino acids linked by peptide bonds along the polypeptide chain is the primary structure. Each polypeptide has its own unique sequence of amino acids. The primary structure of polypeptide determines to a large extent the native secondary and tertiary structure.

Secondary Structure

The structure of a protein in which the polypeptide chain is folded in a regular manner with the helical or pleated structure is called secondary structure. Much of the folding is the result of linking carbonyl and amide groups of the peptide chain backbone by hydrogen bonds. In many proteins, the hydrogen bonding produces a regular coiled arrangement called the α-helix and the ß-sheet figure 2.52 and 2.53.

Tertiary Structure

A final structure of a protein in three-dimensional from a single polypeptide chain is formed. The final confirmation of a protein from a single polypeptide that is bent or tightly folded and compact is determined by a variety of interactions including hydrogen bonds, salt bonds, hydrophobic or nonpolar bonds and van der Waals forces. It is by virtue of their tertiary structure that proteins adopt a globular shape figure 2.54 and 2.55.

Quaternary Structure

It contains more than a single polypeptide chain to indicate the spatial relationships among the separate chains or subunits. It means how two or more chains are arranged in relation to each other. Most large proteins contain two or more polypeptide chains, and in between, there are no covalent linkages. Proteins with two or more polypeptide chains are known as oligomeric proteins. A well-known example of an oligomeric protein is hemoglobin.

Enzymatic Proteins

They are used for the acceleration of chemical reactions.

- **Example:** Digestive enzymes catalyze the hydrolysis of bonds in food molecules.

Storage Proteins

They function in storage

- **Examples:** Casein, the protein of milk, is the major source of amino acids for baby mammals. Plants have storage proteins in their seeds. Ovalbumin is the protein of egg white, used as an amino acid source for the developing embryo.

Hormonal Proteins

They allow coordination of an organism's activities.

- **Example:** Insulin, a hormone secreted by the pancreas, causes other tissues to take up glucose, thus regulating blood sugar concentration

Defensive Proteins

They are used for protection against disease.

- **Example:** Antibodies inactivate and help destroy viruses and bacteria.

Transport Proteins

They are used for the transport of substances.

- **Examples:** Hemoglobin, the iron-containing protein of vertebrate blood, transports oxygen from the lungs to other parts of the body. Other proteins transport molecules across cell membranes.

Receptor Proteins

They are responses of the cell to chemical stimuli.

- **Example:** Receptors built into the membrane of a nerve cell detect signaling molecules released by other nerve cells.

Structural Proteins

They are used for support.

- **Examples:** Keratin is the protein of hair, horns, feathers, and other skin appendages. Insects and spiders use silk fibers to make their cocoons and webs, respectively. Collagen and elastin proteins provide a fibrous framework in animal connective tissues.

The primary structure consists of its unique sequence of amino acids. Secondary structure, found in most proteins, consists of coils and folds in the polypeptide chain. Tertiary structure is determined by interactions among various side chains (R groups). Quaternary structure results when a protein consists of multiple polypeptide chains. Primary structure, the sequence of amino acids in a protein, is like the order of letters in a long word. The primary structure is determined by inherited genetic information.

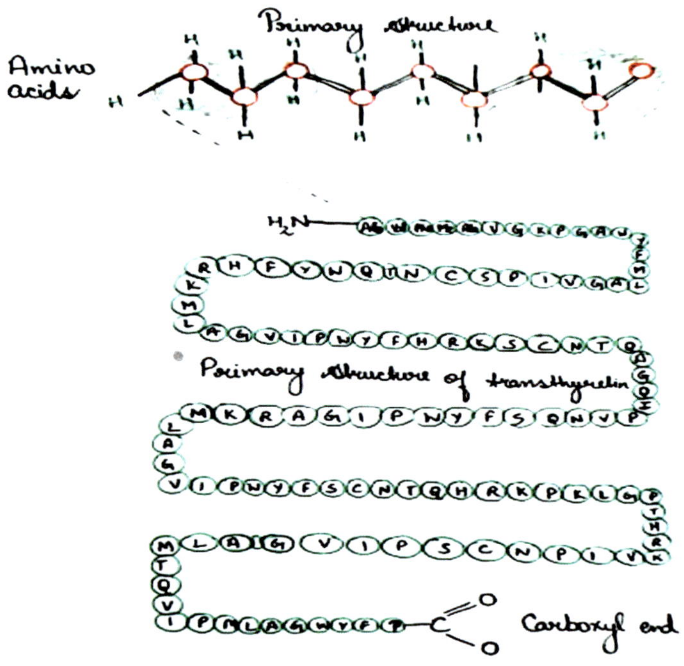

Figure 2.52: Hydrogen bonding arrangement in the α-helix.

The coils and folds of the secondary structure result from hydrogen bonds between repeating constituents of the polypeptide backbone. Typical secondary structures are a coil called an α helix and a folded structure called a β pleated sheet.

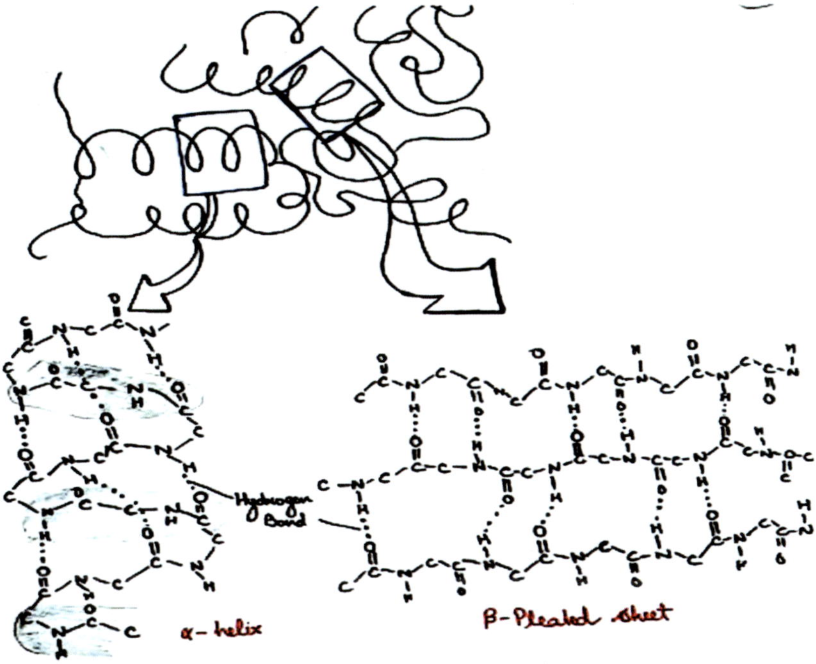

Figure 2.53: Geometry of α-helix and ß-sheet.

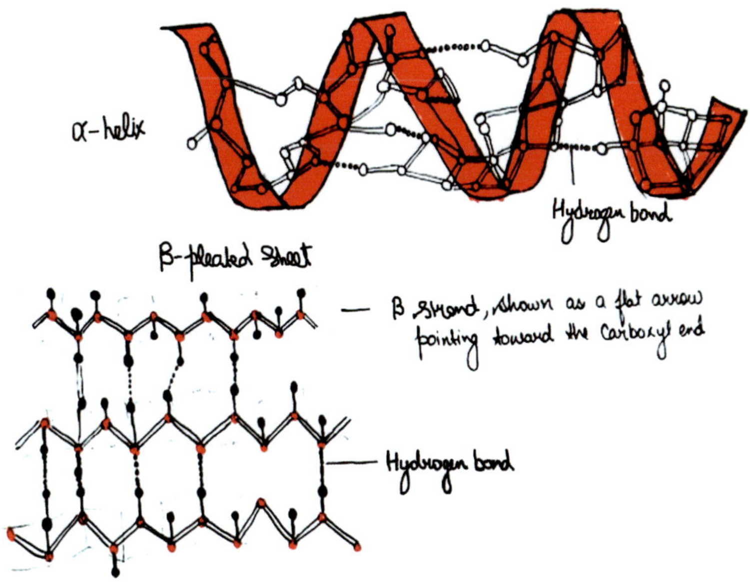

Figure 2.54: Hydrogen bonding arrangement in the α-helix and ß-sheet. The helix is defined by the pitch and rise per residue.

Tertiary structure is determined by interactions between R groups, rather than interactions between backbone constituents. These interactions between R groups include actual ionic bonds and strong covalent bonds called disulfide bridges figure 2.55 that may reinforce the protein's structure. IMFs such as London dispersion forces (LDFs a.k.a. and van der Waals interactions), hydrogen bonds (IMFs), and hydrophobic interactions (IMFs) may affect the protein's structure.

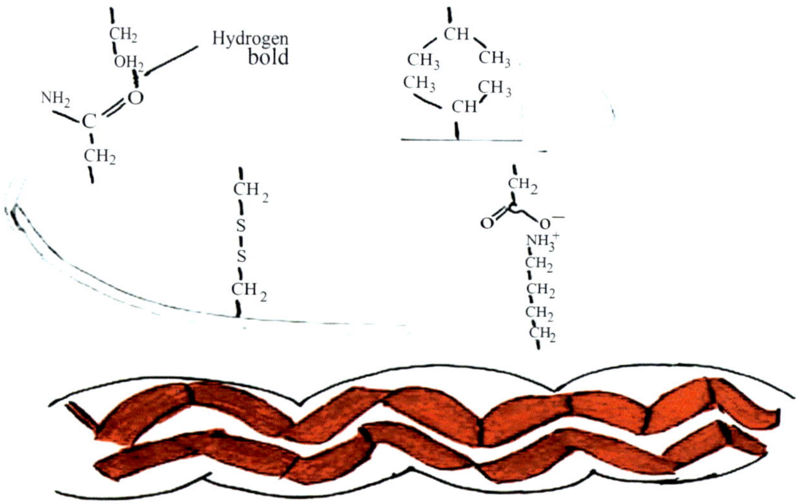

Figure 2.55: Disulfide bridges.

Quaternary structure results when two or more polypeptide chains form one macromolecule. Collagen is a fibrous protein consisting of three polypeptides coiled like a rope figure 2.56.

Figure 2.56: Collagen fiber.

Hemoglobin is a globular protein consisting of four polypeptides: two alpha and two beta chains figure 2.57.

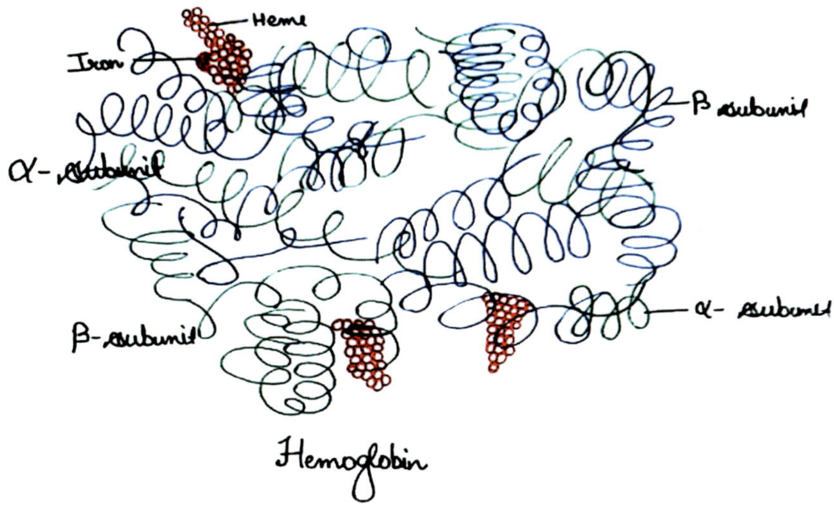

Figure 2.57: Structure of hemoglobin.

Sickle-cell Disease

A slight change in the primary structure can affect a protein's structure and ability to function. Sickle-cell disease, an inherited blood disorder, results from a single amino acid substitution in the protein hemoglobin. A slight change in primary structure can affect a protein's structure and ability to function. Sickle-cell disease, an inherited blood disorder, results from a single amino acid substitution in the protein hemoglobin. In addition to the primary structure, physical and chemical conditions can affect the structure. Alterations in pH, salt concentration, temperature, or other environmental factors can cause a protein to unfold figure 2.58. This loss of a protein's native structure is called denaturation. A denatured protein is biologically inactive.

Protein Subunits & Domains

- A single polypeptide may have regions along its length that fold up separately - domains
- More than one polypeptide coming together to form a protein - subunit

Protein Shape

- Globular proteins – The polypeptide folds tightly together on itself – most of the enzymes
- Fibrous proteins – They have an elongated structure that span great distances - hair - keratin
  - Actin & collagen
  - Strong & rigid (Collagen - ligaments)
  - Strong & flexible (elastin - skin)

Proteins linked together

- Domain to domain links (same P')
- Protein to protein links (between P's)
- Covalent bonds used to stabilize
- The disulfide bond is the most common
  - Cysteine AA
  - No impact of the final shape, but stabilize.

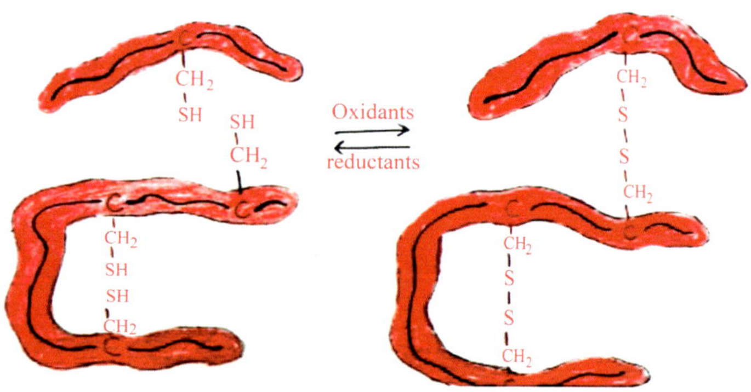

Figure 2.58: Disulfide bond.

Proteins Bind

- The biological properties of proteins result from interactions with other molecules.
  - Antibodies, enzymes, structure, etc.
- The binding is always very specific.
- Ligand - Anything that is bound by a protein.

- Ligand binding is by ionic bonds only.
- Many ionic bonds are required to stabilize the link - matching configurations needed specificity **figure 2.59**.

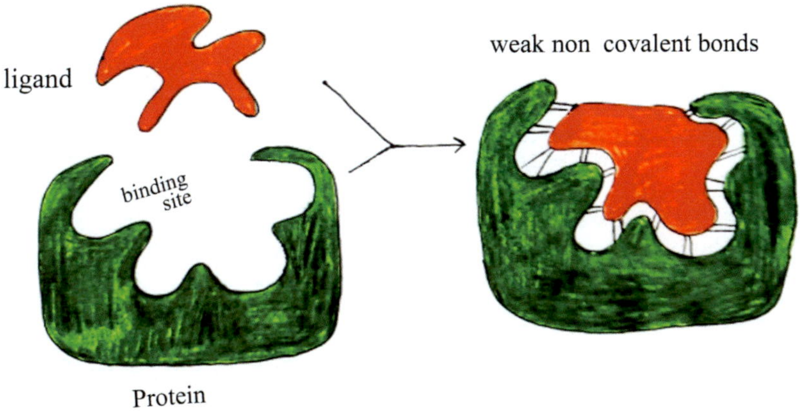

Figure 2.59: Ligand binding with protein.

- A ligand binds to a protein at a binding site.
- A protein may have more than one binding site.
- It may bind the same ligand many times or it may bind different ligands.
- Ligands can bind to regulate the activity of the protein.
- Interior amino acids have a say in the conformation of the molecule too.

Proteins Pair with Other Molecules

Some biologically active molecules are a result of a very convenient relationship.

- Rhodopsin & Opsin.
- Retinal (a light sensing pigment) + opsin protein.
- Heme (Fe containing molecule) + globin protein.

Protein Folding and Protein Modification

Protein folding is the physical process by which a "polypeptide folds into its characteristics and functional three-dimensional conformations. The correct three-dimensional structures are essential to protein function. Failure to fold into the native structure produces inactive proteins. A protein molecule folds spontaneously during or after biosynthesis. However, the process also depends on the nature of the solvent and the concentration of salts the temperature and the presence of molecular chaperones. To understand the complete three-dimensional structure of a protein, it is important to analyze its folding pattern. Two terms are used to describe a folding pattern-motif and domain. A motif is a recognizable folding pattern involving two or more elements of secondary structure and the connections between them. Motifs can be simple, such as the ß-α-ß loop motif; motifs also can be elaborate, such the ß barrel motif. In some cases, a single large motif may comprise the entire protein. Motif and fold can be used interchangeably; however, the term fold is more commonly applied to describe more complex folding patterns". For example, the distinctive arrangement of eight α helices in myoglobin is found in all globins and is called the globin fold figure 2.60 and 2.61.

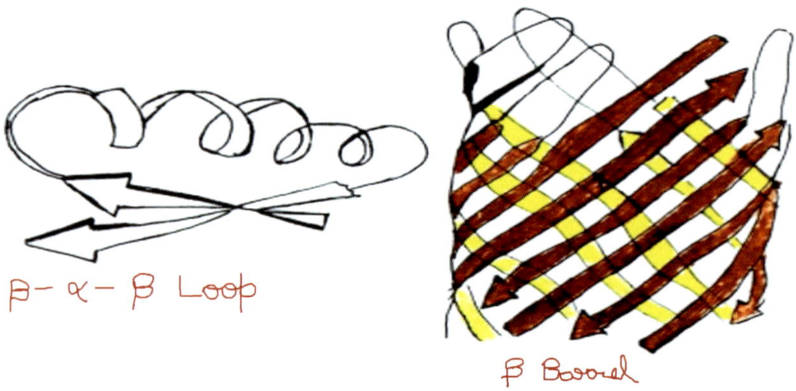

Figure 2.60: Protein folding.

A domain is a part of a "polypeptide chain that is independently stable or could undergo movements as a single entity with respect to the entire protein. Each domain can appear as a distinct globular lobe region as is true for the calcium binding protein of muscle, troponin C. However, extensive contacts can occur between domains and make individual domains hard to discern. The two or more domains of a protein often have different functions, which are coordinated in the overall function of the protein. Small proteins such as myoglobin usually only have one domain".

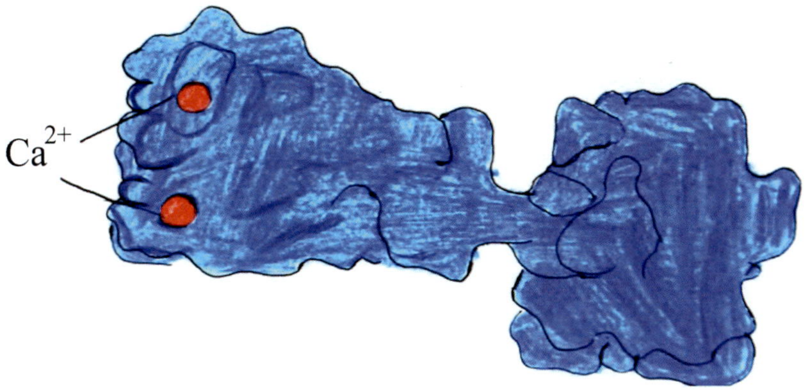

Figure 2.60: Calcium binding protein of muscle.

Several Rules have emerged from Studies of Common Protein Folding Patterns.

The "hydrophobic interactions are important for protein stability. At least two layers of secondary structure are required for shielding hydrophobic residues from water. The α helices and ß sheets usually are found in different structural layers of a protein. Segments adjacent to each other in the primary structure are usually stacked next to each other in the folded structure". The ß conformation is most stable when the individual strands are twisted slightly in a right-handed sense figure 2.61.

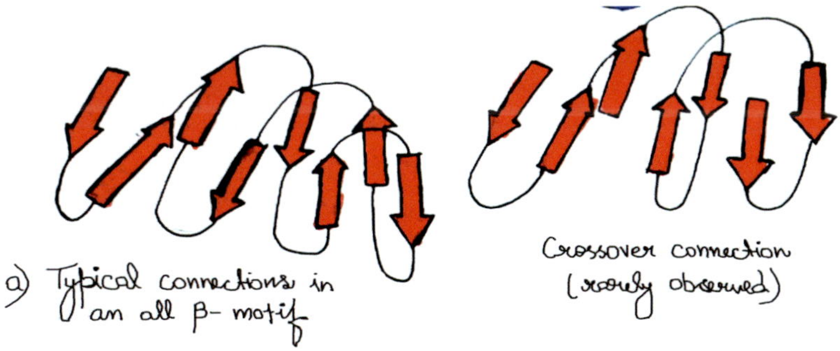

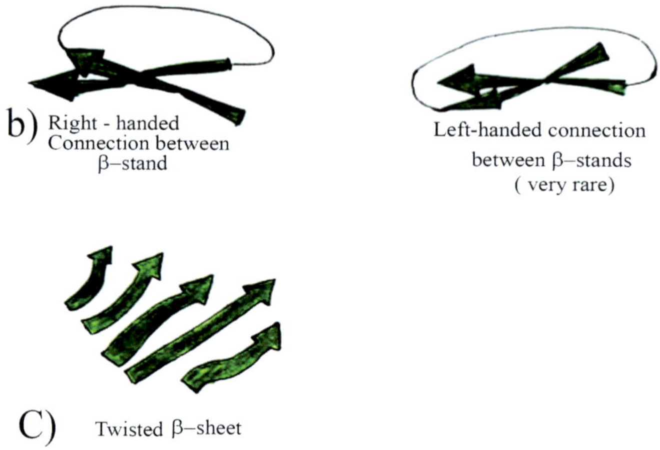

Figure 2.61: Secondary structures are required for shielding of hydrophobic residues.

Protein Denaturation

The conformation of a "native protein is only marginally stable. The conditions are different from those in a cell, therefore can cause protein structural changes resulting in the loss of three-dimensional structure needed for function (i.e., denaturation). In the denatured state, the conformation of the protein need not be completely randomized. A number of physical and chemical agents can cause protein denaturation. Strong acids and bases, organic solvents (ethanol), detergents, reducing agents (ß-mercaptoethanol) and heavy metals (mercury and lead) denature the proteins. A classic agent is heat, which has complex effects on many weak interactions in a protein (primarily on the hydrogen bonds). On heating, a protein's conformation generally remains intact until an abrupt loss of structure occurs over a relatively narrow temperature range (the T_m). The unexpectedness of the loss of structure suggests a cooperative process in which loss of structure in one or more parts of the protein rapidly destabilizes the structure" of other parts.

Renaturation of Ribonuclease

The early studies are examining the "renaturation (refolding) of the enzyme ribonuclease which provided experimental proof that the tertiary structure of a protein is determined by its primary structure. In this experiment, purified ribonuclease was denatured to its unfolded inactive state by exposure to a concentrated solution of urea and the reducing agent, mercaptoethanol. This unfolded the protein and reduced all disulfide cross-links to cysteine residues. When urea and mercaptoethanol were removed, the randomly-coiled, denatured ribonuclease spontaneously refolded to its catalytically active state. Even the protein's four unique disulfide bonds reformed correctly. This experiment provided the first evidence that the amino acid sequence of a polypeptide chain contains all the information required to fold the chain into its native" three-dimensional structure figure 2.62.

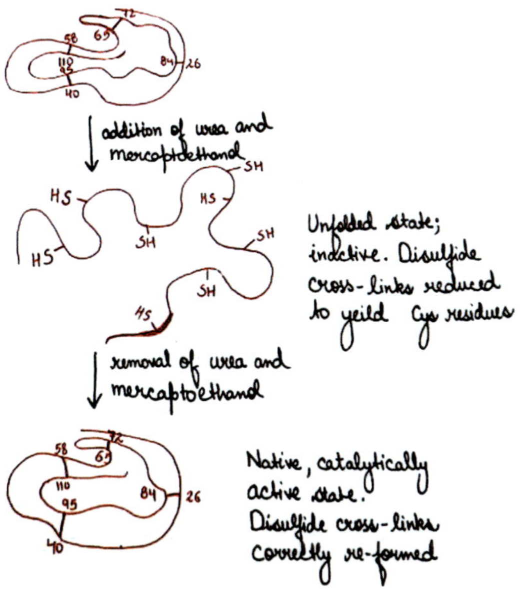

Figure 2.62: Renaturation of ribonuclease.

Protein Folding

Protein folding is "supposed to occur via a hierarchical pathway. The certain amino acid sequences are folded first into local secondary structures. Assemblies of local structures are followed by the formation of longer-range interactions between these secondary structure elements to obtain folding motifs. Throughout the process, hydrophobic interactions play a significant role in aggregating non-polar amino acid side chains in the core of folding intermediates and the final tertiary structure. The computer programs have now been developed that can often predict the structures of small proteins on the basis of their amino acid sequences".

Chaperones

Many proteins do not fold spontaneously as they are synthesized by a cell. The folding of these "proteins are requiring the general class of proteins called chaperones. Chaperones interact with partially folded or improperly folded polypeptides, facilitating correct folding pathways or providing microenvironments in which folding can occur. Several types of chaperones are found in organisms ranging from bacteria to humans. The two major classes of chaperones are the molecular chaperones (e.g., the Hsp70 family of proteins) and chaperonins (e.g., the GroEL/Hsp60 family of proteins) and (Hsp stands for heat shock protein). These proteins are induced in bacterial cells by exposure to high temperatures". Molecular chaperones act as monomers, whereas chaperonins are large multi-subunit complexes.

Chaperone-assisted Protein Folding

Molecular chaperones such as "Hsp70, bind to regions of unfolded polypeptides that are rich in hydrophobic residues. They protect both proteins denatured by heat and newly synthesized and not yet folded proteins from aggregation. They also block the folding of certain proteins that must remain unfolded until they have been translocated across a membrane. Some chaperones also facilitate the quaternary assembly of oligomeric proteins. The Hsp70 proteins bind to and release polypeptides in a cycle that uses energy from ATP hydrolysis and involves several other proteins, such as Hsp40 and NEF (nucleotide-exchange factor). When the bound polypeptide is released after a cycle of ATP hydrolysis, it has a chance of folding properly. If not, the protein may be bound again, and the process repeated. Hsp70 can also deliver unfolded polypeptides to chaperonins. Chaperonins are elaborate protein complexes required for the folding of many cellular proteins that do not fold spontaneously. In *E. coli*, 10-to-15% of newly translated proteins require the GroEL/GroES chaperonin system for folding under normal conditions. This increases to 30% when cells are subjected to heat stress. The mechanism of protein folding by the GroEL/GroES chaperonin is shown in the next slide figure 2.63. Each GroEL complex consists of two large chambers formed by two heptameric rings. GroES (also a heptamer) blocks one of the GroEL chambers after an unfolded protein is bound inside. The chamber with the unfolded protein is referred to as cis; the opposite one is trans. Folding occurs within the cis chamber during the time it takes to hydrolyze the 7 ATPs that are bound to subunits in the heptameric GroEL ring. The GroES and the ADP molecules then dissociate and the protein is released. The two chambers of the GroEL system are alternate in the binding and facilitated folding of client proteins. Inside the chamber, a protein has about 10 seconds to fold. The constraining the protein within the chamber prevents inappropriate protein aggregation and also restricts the conformational space that a polypeptide chain can explore as it folds".

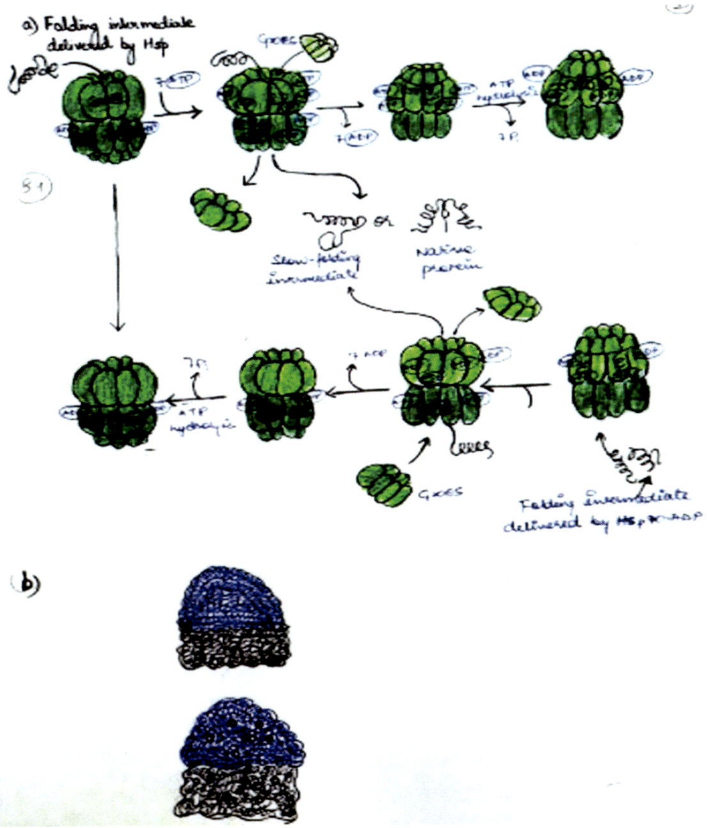

Figure 2.63: Structure and function of the chaperonin like GroEL.

Protein Folding Disorders: Amyloidoses

In a number of human disorders such "amyloidoses; including Alzheimer's disease a soluble protein that is normally produced by a cell adopts a misfolded state and converts into an insoluble extracellular amyloid fiber. Commonly, fiber formation is promoted by the aggregation of regions of proteins that have a ß conformation. In Alzheimer's disease, proteolytic cleavage of a neuronal membrane protein (amyloid-ß precursor protein, APP) produces an α-helical membrane-spanning peptide (amyloid-ß peptide) that converts to the ß conformation and aggregates into amyloid fibrils. The extracellular deposition of amyloid fibrils is associated with plaque formation and ultimately death of the nearby neurons".

Protein Folding Disorders: Prion Diseases

The "neurodegenerative diseases are known as spongiform encephalopathies and a misfolded form of a normal neuronal protein PrP is responsible for the disease. Spongiform encephalopathies occur in many species of animals. In humans, the disorders are known as kuru and Creutzfeld-Jacob disease. In cows, the disorder is known as mad cow disease. In sheep, it is called scrapie, and in deer and elk, it is called chronic wasting disease. The diseased brain becomes riddled with holes. Progressive deterioration of the brain leads to a spectrum of neurological symptoms and is always fatal. Prion protein (PrP) is a normal constituent of brain tissue in all mammals. The protein is thought to possibly function in cell signaling. Illness occurs when the normal cellular PrP (PrP^C) folds into an altered conformation called PrP^{Sc} (Sc denotes scrapie). In PrP^{Sc}, part of the α-helical region of PrP^C is converted to the ß conformation. PrP^{Sc} monomers then associate forming amyloid-like ß sheets. Importantly, the interaction of PrP^{Sc} with PrP^C converts the latter to PrP^{Sc}, initiating a domino effect in which more and more of the brain protein converts to the disease-causing form. For this reason, PrP^{Sc} is infectious. The mechanism by which PrP^{Sc} leads to spongiform encephalopathy is not understood". In inherited forms of prion diseases, a mutation in the gene encoding PrP produces a change in one amino acid residue that is believed to make the conversion of PrP^C to PrP^{Sc} more likely.

Protein Modification

Proteins undergo a variety of modifications that are critical for function. There are numerous amino acid modifications such as collagen. Thirty percent of the residues are glycine. The ~ 30 of residues are Proline or hydroxyproline (HyPro). The 5-hydroxylysine (HyLys) also occurs; a site for glycosylation. The Hydroxylation of Pro, Lys is a post-translational modification, requires vitamin C as a reactant. The sequence of collagen bears long stretches of Gly--Pro/HyPro-X repeats.

Structure of Collagen

Collagen makes up 25 to 35% of the total protein of mammals. It is found in all forms of connective tissue. "Collagen itself is an insoluble protein because of extensive cross- linking. The structural unit of collagen is a tropocollagen, a supercoil made up of 3 helices, with a molecular mass of ~285 kDa. Each collagen helix consists of ~ 1000 amino acid residues. The helix is left-handed. It is not an α-helix. The helix contains 3 amino acids per turn, with a pitch of 0.94 nm. Each unit of tropocollagen is about 1.5 nm wide and 300 nm long. The bundles of these 3-stranded supercoils can be seen as collagen fibers in the electron microscope. Mature collagen has extensive covalent cross-linkages between individual collagen molecules. There is no intra-helical H-bonding in collagen helices. Rather, H-bonding occurs between the amide N of glycine residues in the central axis and the carbonyls of other residues in the adjacent chains. Often proline and hydroxyproline" are involved.

Genetic Disorders of Collagen are Numerous

One such group of 10 different collagen deficiency diseases is the Ehlers-Danlos (E-D) syndrome. The "India-rubber man" of circus fame had an E-D syndrome. Osteogenesis imperfecta: abnormal (fragile) bone formation is seen in human babies" (www.scribd.com., www.slideshare.com., www. coursehero.com., www. biocyclopedia.com., www. molecular-plant-biotechnology.info., www.slideplayer.com., www.documents.mx., www. docplayer.net., www.bmb.psu.edu.,www.archive.org.,www.mafiadoc.com., www.powershow.com., www.bioscience.jbpub.com., www.outhorstream. com., www.docslide.com., www.msluay.com., www.scienceprofonlue. com., www.csus.edu., www.krishikosh.egrouth.ac.in., www.tankonyvtor. hu., www.silo.pub., www.studyblue.com., www.Preservearticles.com., www.biotechuniverse.blogspot.com).

Vitamins

It is an organic compound required by an organism in trace amounts to perform specific cellular functions. These compounds serve the "same roles in nearly all forms of life, but higher animals have lost the capacity to synthesize them in the course of evolution. Vitamins are not synthesized by humans. Therefore, they must be supplied by diet. They can be classified according to the solubility and their functions in metabolism. The requirement for any given vitamin depends on the organism. Not all vitamins are required by all organisms. It may be water-soluble or fat-soluble. There are nine vitamins such as thiamine, riboflavin, niacin, biotin, pantothenic acid, folic acid, cobalamin, pyridoxine, and ascorbic acid that are classified as water-soluble, whereas four vitamins (Vitamins A, D, E and K) are termed as fat-soluble. The water-soluble vitamins are all precursors" coenzymes, except vitamin C. Certain vitamins and minerals are needed for the body to function.

- 13 vitamins
- 22 minerals

Two tyPes of Vitamins

- Water-soluble
- Fat-soluble

Fat-Soluble Vitamins

Fat-soluble vitamins are A, D, E and K and It has excess stored in the liver and in body fat. It is possible to build up to a toxic level.

Vitamin A (Retinol)

Beta-carotene is converted into vitamin A. It promotes good vision. It helps healthy skin. It helps with the growth and maintenance of bones, teeth, and cell structure

- RDA: 900 micrograms for males; 700 micrograms for females.

Too much Vitamin A

It may turn your skin orange. It may cause fatigue, weakness, severe headache, blurred vision, hair loss and joint pain.

Toxicity:

- May cause severe liver or brain damage
- Birth defects

Too Little Vitamin A

It may cause night blindness and there is a lowered immune system

Foods Rich in Vitamin A

- Foods
- Only animal products
  - Liver
  - Eggs
  - Milk, butter and cheese
- Carotenoids
  - Orange/Yellow fruits and vegetables
    - Cantaloupes, carrots, sweet potatoes, winter squash
  - Leafy green vegetables
    - Spinach, broccoli

Vitamin D

It is essential for building and maintaining bones and teeth. It is responsible for the absorption and utilization of calcium. It may boost the immune system. It may also help decrease certain cancers.

- RDA: 5 micrograms until age 50
  - 10 micrograms / day until 70; 15 mcg for 70+ age

Vitamin D deficiency has been in the news a lot lately. Deficiency may occur from the:

- Inadequate diet
- Vegetarianism, lactose intolerance and milk allergy.
- The body is unable to absorb needed vitamin D. Limited exposure to sunlight.
- It may lead to osteomalacia and/or osteoporosis.
- Getting vitamin D from sun exposure for 10 minutes a day

Foods

- Fortified milk
- Tuna
- Salmon
- It may need a supplement

Vitamin E

It is important in red blood cells, muscles and other tissues. Deficiency is rare.

- Toxicity is rare.
  - But, Vitamin E acts as a blood thinner.
- Foods
  - Vegetable oils, salad dressings, whole grain cereals, green leafy vegetables, nuts, seeds, peanut butter and wheat germ.

Vitamin K

It is important for blood clotting and also has a role in bone health. It is mostly made in the intestines.

- Foods
  - Turnip greens, cauliflower, spinach, liver, broccoli, kale and cabbage.

Water-soluble Vitamins

Water-soluble vitamins are B and C

There are eight vitamins B such as

- Thiamin (B-1)
- Riboflavin (B-2)
- Niacin (B-3)
- Pyridoxine (B-4)
- Cobalamin (B-12)
- Folic acid
- Pantothenic acid
- Biotin

Thiamin or B-1

It helps to convert carbohydrates into energy.

- Deficiency
  - Fatigue, nausea, depression and nerve damage.
  - Foods: Pork, beef, liver, peas, seeds, legumes, whole-grain products, and oatmeal

Riboflavin or B-2

It is important to metabolism and red blood cells.

- Deficiency
  - Dry, scaly skin
- Foods: Milk, yogurt, cheese, whole-grain bread, green leafy vegetables, meat, and eggs

Niacin or B-3

It is involved in energy production. It also helps the skin, nerves and digestive system.

- Deficiency:
  - Rare but causes: diarrhea, dermatitis, dementia and death
- Foods: Meat, poultry, liver, eggs, brown rice, baked potatoes, fish, milk, and whole-grain foods.

Pyridoxine or B-6

It is involved in the chemical reactions of proteins and amino acids.

- **Deficiency:** Skin changes, dementia, nervous system disorders and anemia.
- **Foods:** Lean meats, fish, legumes, green leafy vegetables, raisins, corn, bananas and mangos.

Cobalamin or B-12

It helps the nervous system, red blood cells and DNA synthesis.

- Deficiency
  - ➢ Nervous system disorders and pernicious anemia
  - ➢ Foods: Only found in animal products
    - Meat, fish, poultry, eggs, milk products and clams

Folic acid (Folacin, Folate)

It has an important role in red blood cell formation and cell division.

- Deficiency:
  - Anemia, digestive disorders
  - Foods: Leafy, dark green vegetables and also found in liver, beans, peas, asparagus, oranges, avocados

Pantothenic Acid and Biotin

It helps metabolism and formation of some hormones.

- Deficiencies are rare
- Foods: Almost any food, plant-based or animal-based

Vitamin C

It is important in bone health, blood vessel health, cell structure and absorption of iron. Too much vitamin C is found in foods: Melons, berries, tomatoes, potatoes, broccoli, fortified juices, kiwi, mangos, yellow peppers and citrus fruits.

Table 2.1: Roles of vitamins

Vitamin	**Role of action**
Water-soluble	
Thiamine (Vit B1)	Cofactor in decarboxylations Deficiency disease: Beri-beri
Riboflavin (Vit B2)	Constituent of flavoproteins
Niacin (Vit B3)	Constituent of NAD^+ and $NADP^+$ Deficiency disease: Pellagra
Pantothenic acid (Vit B5)	Constituent of CoA and acyl-group transfer
Pyridoxine (Vit B6)	Group transfer to or from amino acids
Biotin (Vit H)	ATP-dependent carboxylation and carboxyl group transfer
Folate (Folic acid)	Coenzymes for 1-carbon transfer Deficiency disease: Megaloblastic anemia
Cyanocobalamin (Vit B12)	Coenzyme in amino acid metabolism Deficiency disease: Pernicious anemia
Ascorbic acid (Vit C)	Antioxidant Deficiency disease: Scurvy
Fat Soluble	
Vitamin A	Role in vision, growth, reproduction. Deficiency disease: Night blindness
Vitamin D	Regulation of calcium and phosphate metabolism Deficiency disease: Rickets (children), osteomalacia (adults).
Vitamin E	Antioxidants; cofactors in electron transport in cyto-chrome chain
Vitamin K	Blood coagulation, Post-translational modification of blood clotting factors.

Minerals

- Calcium
- Chromium
- Folate
- Iron
- Magnesium
- Selenium
- Zinc
- Selenium
- Phosphorus
- Sodium
- Potassium
- Chloride
- Sulfur
- Copper
- Iodine
- Cobalt
- Fluoride
  - Our body is needed 22 minerals.
  - There are two categories such as
    - ✓ Include calcium, chloride, magnesium, phosphorus, potassium, sodium, and sulfur
    - ✓ Include iron, zinc, iodine, selenium, copper, manganese, fluoride, chromium, molybdenum, arsenic, nickel, silicon, boron and cobalt.

Sodium

- It helps maintain fluid balance.
- It helps transmit nerve impulses.
- It influences the contraction and relaxation of muscles.

Sodium & Health

High Amount of Sodium

It causes high blood pressure and it may lead to fluid retention.

Sodium Savvy

The human body requires about 500 mg of sodium per day, while the average American usually ingests between 2,300-6,900 mg each day. It is recommended to stay in the range of 1,500 to 2,400 mg/day.

Sodium & Food

- Monosodium glutamate (MSG)
- Baking soda
- Baking powder
- Disodium phosphate
- Sodium alginate
- Sodium nitrate or nitrite

Reducing Sodium in Your Diet

- Eat more fresh foods.
- Eat less processed foods.
- Look for low-sodium products.
- Limit the salt you add to foods.
- Experiment with other seasonings.
- Use salt substitutes with caution.

Calcium

The most abundant mineral in your body and 99% is stored in the bones. It is known for bone health.

- Males 19-50 years old: 1,000 mg / day
- Females 19-50 years old: 1,000 mg / day

We can take calcium from dairy products, fortified juices and sardines.

Food	Calcium
Yogurt, plain (low-fat) Yogurt, flavored (low-fat)	1 cup - 415 mg 1 cup – 345 mg
Milk, skim Milk, 1-2%	1 cup – 302 mg 1 cup – 300 mg
Ice cream	½ cup – 88 mg
Broccoli, cooked	½ cup – 68 mg
Salmon, canned	3 oz – 165 mg
Fortified orange juice	8 oz – 300 mg

Iron

Iron deficiency is the most widespread vitamin or mineral deficiency in the world.

- 70% of your body's iron is in your hemoglobin
- Too little iron = too little oxygen

Iron & Foods

It found in animal products

- Red meats, liver, poultry and eggs

Non-heme iron: It found in plant products

- Beans, nuts, seeds, dried fruits, fortified bread and cereals

Iron Supplements

- Strict vegetarians
- Those who do not eat a balanced diet
- Those who are over 60
- Smokers and those who regularly drink alcohol
- Chronic dieters
- Those who suffer from food allergies, intolerances

Potassium

It keeps muscles and the nervous system working properly and is an electrolyte that conducts electricity in the body.

- Potassium-rich foods: bananas, broccoli, tomatoes, potatoes with skins, leafy green vegetables, citrus fruits, dried fruits, and legumes.

Zinc

It supports normal growth and development during pregnancy, childhood, and adolescence. It is required for the proper sense of taste and smell. It is involved in numerous aspects of cellular metabolism. Foods rich in zinc: beef, pork, lamb, & legumes.

Magnesium

Approximately 50% of the total body magnesium is found in bone. The other half is found predominantly in cells of body tissue and organs. It helps to maintain muscle and nerve function, keeps heart rhythm steady, supports a healthy immune system, keeps bones strong. It helps to regulate blood sugar levels, promotes normal blood pressure. Eating a wide variety of legumes, nuts, whole grains, and vegetables will help meet your daily need for magnesium.

Trace Minerals

There are several trace minerals such as Iron, Manganese, Copper, Iodine, Zinc, Fluoride, and Selenium. Manganese is important for the production of enzymes and antioxidants. Copper is necessary for your body to make connective tissue and the metabolism of iron. Iodine is used to synthesize thyroid hormones. Fluoride is used for strong bones and teeth. It reduces the number of cavities in teeth. Selenium combines with proteins to make antioxidants and is also essential for normal thyroid function.

Simple Salmon Supper

- 1 each zucchini, yellow squash, and red pepper, chopped
- 4 skinless salmon fillets
- 1 can diced tomatoes, drained
- 1/2c Kraft Sun Dried Tomato Vinaigrette Dressing

- 2-2/3 cups hot cooked long-grain brown rice
  - **Heat** oven to 375° F.
  - **Combine** zucchini, squash and peppers in a 13x9-inch baking dish; top with fish. Mix tomatoes and dressing; spoon over fish.
  - **Bake** 20 to 25 min. or until the fish flakes easily with a fork.
  - **Serve** with rice.

Water

It is essential for life. It makes up about 45-75% of your body weight. It is important for transport. It is used for mechanical functions. It helps to break substances down. It helps to maintain body temperature/pH

- For men: 3.7 L/ day
- For women: 2.7 L / day
- Perfectly 80% of water should come from drinking fluids.
- The balance 20% of water should come from food (www.scribd.com., www.slideshare.com., www.coursehero.com., www.biocyclopedia.com., www. molecular-plant-biotechnology.info., www.slideplayer.com., www.documents.mx., www.docplayer.net., www.bmb.psu.edu.,www.archive.org.,www.mafiadoc.com., www.powershow.com., www.bioscience.jbpub.com., www.outhorstream.com., www.docslide.com., www.msluay.com., www.scienceprofonlue.com., www.csus.edu., www.krishikosh.egrouth.ac.in., www.tankonyvtor.hu., www.silo.pub., www.studyblue.com., www.Preservearticles.com., www.biotechuniverse.blogspot.com).

Enzyme

An enzyme is a biocatalyst that increases the rate of a chemical reaction without itself being changed in the overall process. Virtually all cellular reactions or processes are mediated by enzymes. Enzymes have several properties that make them unique. An enzyme is a globular protein that functions as a biological catalyst, speeding up reaction rate by lowering activation energy without being affected by the reaction it catalyzes.

Enzymes are Proteins in Nature

It is a globular protein. Ribozymes are RNA molecules with enzymatic activity. The catalytic behavior of any enzyme depends upon its primary, secondary, tertiary or quaternary structure. Enzymes of the digestive tract and those found in blood are present in an inactive form called zymogen or proenzymes.

Active Site

Enzymes are composed of long chains of amino acids that have folded into a very specific three-dimensional shape that contains an active site. An active site is a region on the surface of an enzyme to which substrates bind to catalyze a chemical reaction figure 2.64.

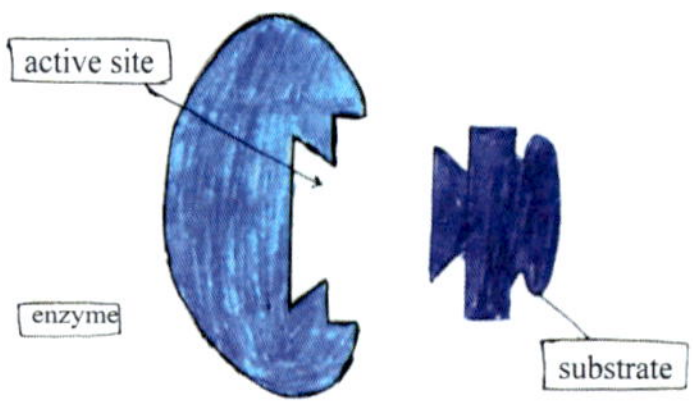

Figure 2.64: Structure of enzyme-substrate.

Enzymes are highly specific for the type of reaction they catalyze and for their substrate. Enzymes with highest specificity and accuracy are involved in copying and expression of genome. These enzymes have proof reading function e.g. DNA polymerase figure 2.65. Enzymes make reactions easier to occur at reasonable temperature by lowering the Activation energy E_A of the reaction.

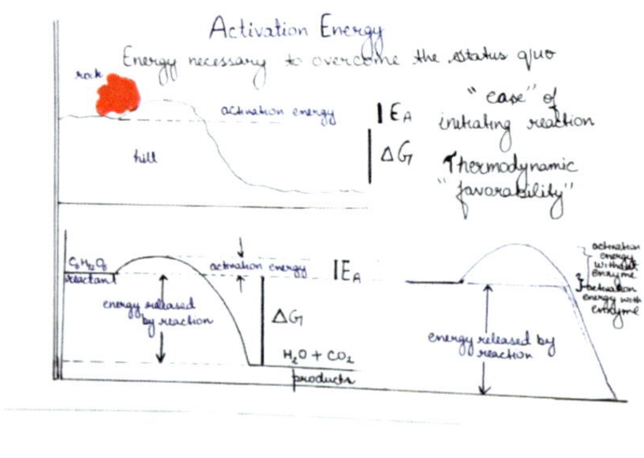

Figure 2.65: Activation energy E_A of the reaction.

Some enzyme requires a non-protein molecule for their activity. So enzymes are conjugated proteins known as Holoenzymes.

- Holoenzyme = Apoenzyme +cofactor/coenzyme.

Enzyme without non-protein part is inactive and called apoenzyme. Cofactors can be either inorganic or organic compounds.

Mechanism of Enzyme Action

The enzymatic reactions take place by binding of the substrate with the active site of the enzyme molecule by several weak bonds figure 2.66.

- E + S ‹--------› ES --------› E + P

Formation of the ES complex is the first step in the enzyme-catalyzed reaction then the ES complex is subsequently converted to the product and free enzyme figure 2.67 and 2.68.

Figure 2.66: Lock and key model.

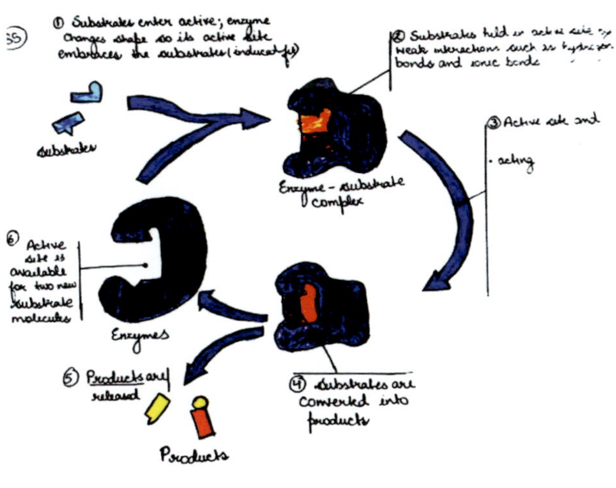

Figure 2.67: Induced-fit model.

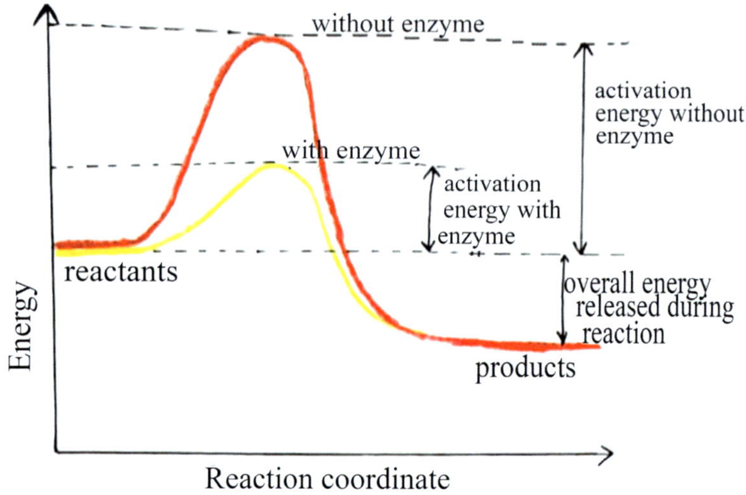

Figure 2.68: Activation energy with and without enzyme.

Nomenclature Enzyme Classification

The **IUBMB** has recommended a system of nomenclature for enzymes & according to them, each enzyme is allotted with two names.

- **Trivial name** (common name, recommended name).
- **Systemic name** (official name).

Systemic Name

Each enzyme is characterized by a code number called enzyme Code number or **EC number** and it contains four digits separated by a dot.

- E.g. **EC M.N.O.P**

The first digit represents the class. The second digit stands for sub-class. The third digit stands for the sub-sub class or subgroup. The fourth digit gives the serial number of the particular enzyme in the list.

- E.g. **EC 2.7.1.1 for hexokinase**.

According to the IUBMB system of enzyme nomenclature, enzymes are grouped into 6 major classes.

- EC 1 Oxidoreductases
- EC 2 Transferases

- EC 3 Hydrolases
- EC 4 Lyases
- EC 5 Isomerases
- EC 6 Ligases

Factors affecting Reaction Velocity

- Temperature
- Hydrogen ion concentration (pH)
- Substrate concentration
- Enzyme concentration
- Products of the reaction
- Presence of activator/inhibitor
- Allosteric effects
- Time

Effect of Temperature

The rate of an enzymes catlysed reaction increases as a temperature increases, enzyme rate of reaction is decraeses at high temperature as due to denature enzyme figure 2.69.

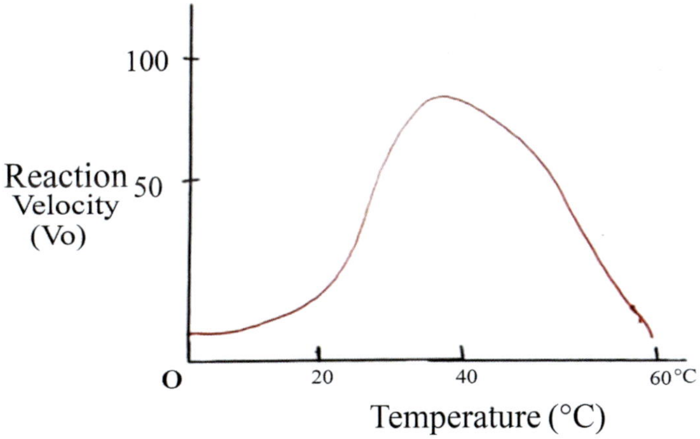

Figure 2.69: Temperature effect on the reaction.

Effect of pH

Every enzyme has an optimum range of pH, slow activity of rate of reaction due to changing the outside range of pH, the enzyme is denaturing at extremely high value of pH figure 2.70.

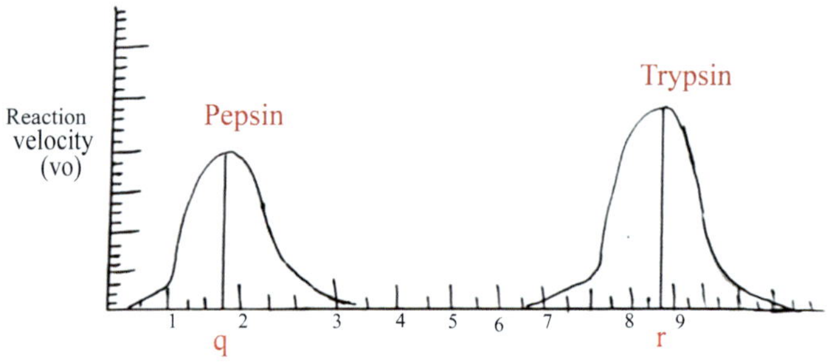

Figure 2.70: Effect of pH on the rate of reaction.

The rate of the reaction or velocity is directly proportional to the enzyme concentration when sufficient substrate is present figure 2.71. Accumulation of product in a reaction causes inhibition of the enzyme activity.

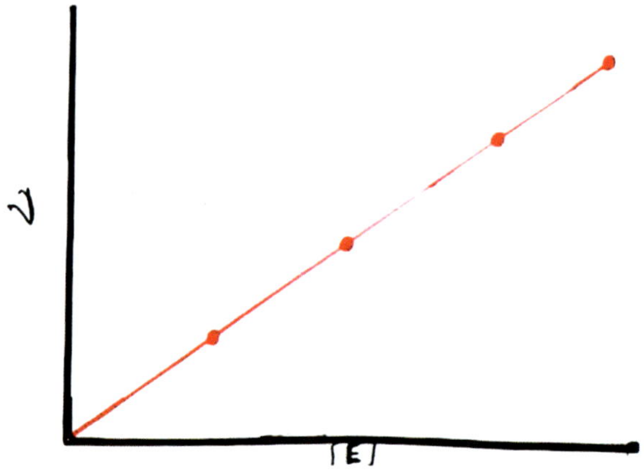

Figure 2.71: Effect of velocity on the rate of reaction.

Effect of Substrate Concentration

The rate of reaction increases with substrate concentration til a specific point. As the amount of substrate increases, it has no effect on the rate of reaction once all the enzymes have bound. The available enzymes will be present in asaturated state and it will work at maximum rate figure 2.72.

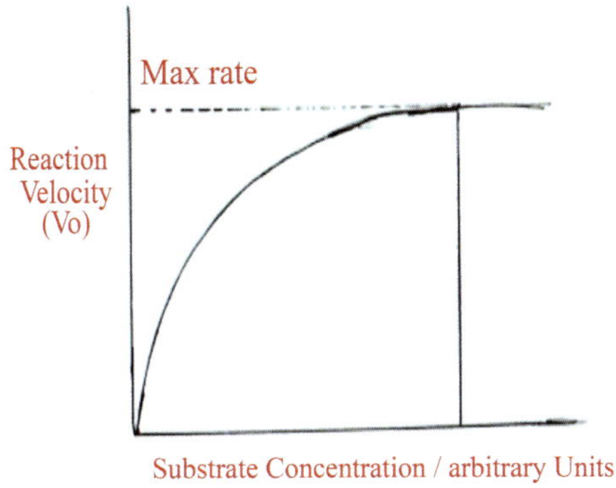

Figure 2. 72: Effect of substrate on the rate of reaction.

Enzyme Kinetics

The study of the rate of reaction and how it changes in response to change in experimental parameters is known as kinetics. The amount of substrate present is one of the key factors affecting the rate of the reaction catalyzed by an enzyme *in vitro*.

Effect of Substrate Concentration on Reaction Velocity

The initial velocity V_0 progressively increases with the increasing concentration of the substrate during enzyme-substrate reaction. The increase in V_0 will not depend on the susbstrate. The substrate concentration on the X-axis and reaction velocity on Y-axis are plotted on graph figure 2.73.

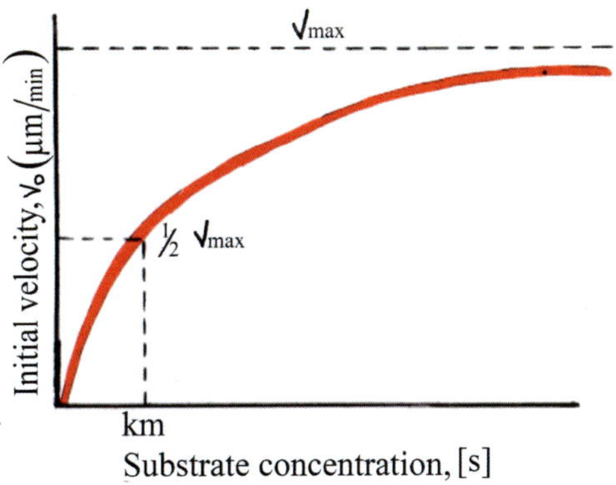

Figure 2.73: Effect of substrate concentration on the initial velocity.

Michaelis- Menten Kinetics

The model involves one substrate molecule.

$$E + S \underset{k_{-1}}{\overset{k_1}{\longleftrightarrow}} ES \xrightarrow{k_2} E + P$$

Where

- S is the substrate
- E is the enzyme
- K_1, k_{-1} and k_2 are the rate constants
- The mathematical equation that defines the quantitative relationship between the rate of an enzyme reaction and the substrate concentration is the *Michaelis-Menten equation*:

$$V_0 = \frac{V_{max}[S]}{K_m + [S]}$$

- V_0 is the observed velocity at the given [S]
- K_m is the Michaelis-Menten constant
- $K_m = (K_{-1} + K_2) / K_1$
- V_{max} is the maximum velocity at saturating [S] conc.

Lineweaver-Burk (Double Reciprocal) Plot

A linear representation is more accurate and convenient for determining V_{max} and K_m. This equation is obtained by taking the reciprocal of both sides of the Michelis-Menton equation.

1/[S] vs. $1/V_o$

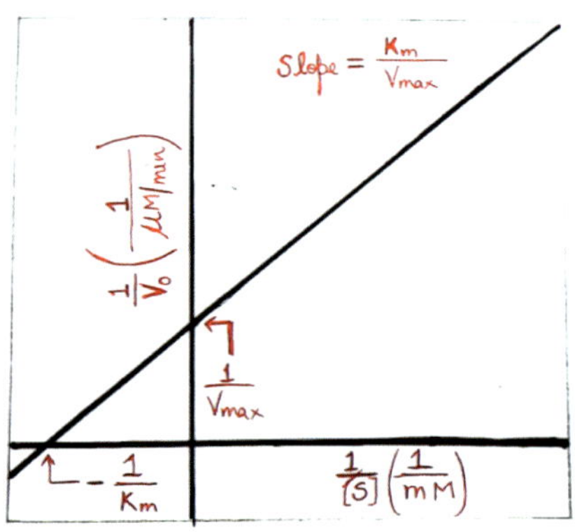

Figure 2.74: Lineweaver-Burk (Double Reciprocal) plot.

Enzyme Inhibition

Any substance that can reduce the velocity of an enzyme catalyzed-reaction is an inhibitor. These include drugs, antibiotics, poisons, and anti-metabolites. The sequence of enzyme-catalyzed reactions has been useful in understanding, metabolic regulation, studying the mechanism of cell toxicity produced by toxicants.

Types of Enzyme Inhibition

- Reversible inhibitors
- Irreversible inhibitors

Reversible Inhibitors can be classified as

- Competitive
- Non-competitive
- Un-competitive

Competitive Inhibition

The structure of a competitive inhibitor closely resembles that of the enzyme's normal substrate. The competitive inhibitor binds reversibly to the enzyme's active site. The inhibitors form an enzyme-inhibitor complex (EI) that is equivalent to the ES complex figure 2.75.

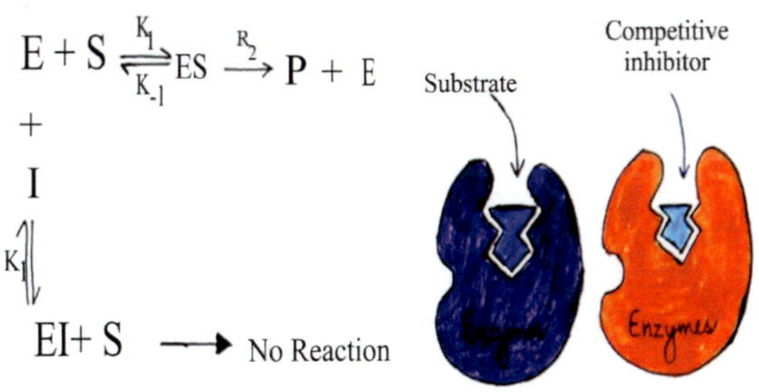

Figure 2.75: Competitive inhibition.

Non-competitive Inhibition

The inhibitor binds to the enzyme at a site other than the active site. Inhibitor binding alters the enzymes three-dimensional configuration and blocks the reaction figure 2.76.

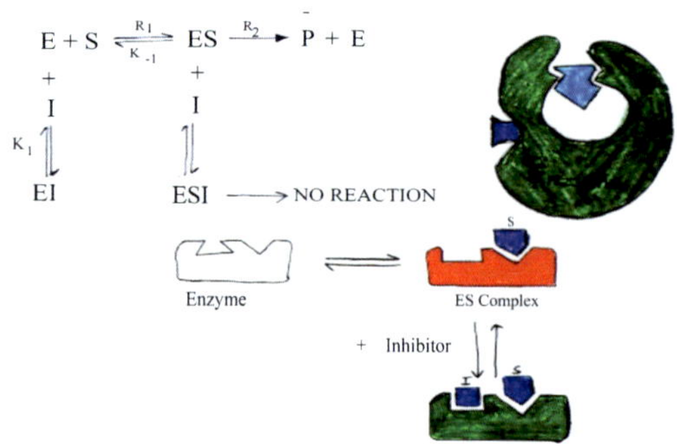

Figure 2.76: Non-competitive inhibition.

Un-competitive Inhibition

It binds only to the enzyme-substrate complex. It does not have the capacity to bind to the free enzyme. Not overcome by increasing substrate concentration. Both the K_m and V_{max} are reduced figure 2.77 and 2.78.

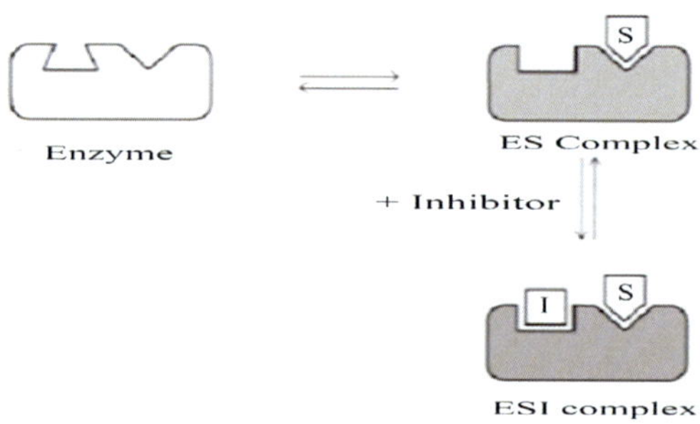

Figure 2.76: Un-competitive inhibition.

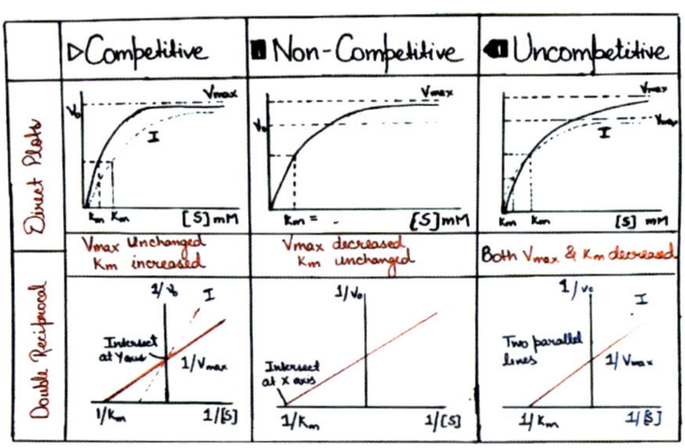

Figure 2.78: Details of competitive, non-competitive and uncompetitive graphs.

Isoenzymes

Isoenzymes (Isozymes): Some metabolic processes are regulated by enzymes that exist in different molecular and multiple forms. All forms demonstrate enzymatic activity and catalyze the same biochemical reaction, but they may differ in kinetics (different K, M and Vmax) and regulatory properties (different effectors).

Ribozymes

Catalytic RNA: Some forms of RNA can serve as a biological catalyst was acknowledged by the awarding of 1989 Nobel Prize in Chemistry Sidney Altman Thomas Cech and Ribozyme: RNA enzymes Ribonuclease P: it is used as self-splicing RNA introns.

Significance of Ribozymes

Ribozymes are used as tools in specific cleavage of RNA and have been designed to catalyze the *in vitro* cleavage of the RNA of HIV.

Abzymes: Catalytic Antibodies

It was reported in 1986 by Peter Schultz and Richard Lerner. It has catalyzed the hydrolysis of esters and carbonates (www.scribd.com., www.slideshare.com., www.coursehero.com., www. biocyclopedia.com., www. molecular-plant-biotechnology.info., www.slideplayer.com., www. documents.mx., www.docplayer.net., www.bmb.psu.edu.,www.archive. org.,www.mafiadoc.com., www.powershow.com., www.bioscience.jbpub. com., www.outhorstream.com., www.docslide.com., www.msluay.com., www.scienceprofonlue.com., www.csus.edu., www.krishikosh.egrouth. ac.in., www.tankonyvtor.hu., www.silo.pub., www.studyblue.com., www. Preservearticles.com., www.biotechuniverse.blogspot.com).

3. NUCLEIC ACIDS AND CHROMOSOMES: NUCLEIC ACID STRUCTURE

Introduction

It was first discovered by Friedrich Miescher from the nuclei of the pus cells (Leukocytes) from discarded surgical bandages and was called nuclein. "Nuclein was chemically a nucleoprotein. Later, Richard Altman in 1889 obtained the first protein-free material, to which he gave the name nucleic acid. It is the polymer of nucleotides. There are two types of nucleic acids: ribonucleic acid (RNA) and deoxyribonucleic acid (DNA).

Genetic Material

The cell is the basic unit of life. All living organisms including bacteria, plants and animals are built from cells. Generally, cells are very small, with diameters much less than 1 mm, so they are invisible to the naked eye. In the simplest cells, bacteria, a cell wall surrounds a very thin fatty acid containing outer plasma membrane, which in turn surrounds a superficially unstructured inner region located in the bacterial DNA that carries the genomic information. These nuclei-free bacteria and their close relatives are known as prokaryotes. In prokaryotes, the nuclear material is not separated from the cytoplasm by a discrete membrane. These also lack an extensive cellular architecture: membranous organelles are absent and the genetic material is not enclosed in a distinct structure. In all the cells other than bacteria, the inner cellular mass is partitioned into a membrane-bound spherical body called the nucleus and an outer surrounding cytoplasm.

In the nucleus is located the cellular DNA in the form of coiled rods known as chromosomes. Cells that contain a nucleus and with other distinctive membranous organelles such as mitochondria, vesicles and Golgi bodies are referred to as eukaryotes. In the cells of eukaryotes, a nuclear membrane separates the genetic material from the cytoplasm, which is then further subdivided by other distinct membranous structures. Eukaryotes can be unicellular and multicellular and comprise all macroscopic forms of life such as plants, animals and fungi. Plant cells differ from animal cells in the presence of outer cell walls surrounding the cell membrane and chloroplasts, whereas animal cells contain centrosomes, which are involved in cell division and lysosomes. The essence of cells is the ability to grow and divide to produce progeny, which are capable of generating new cellular molecules and undergoing replicating themselves. Thus, cells are in effect tiny factories that grow by taking up simples molecular building blocks. Considering the complexity of an organism, one is struck by the relatively small number of organic components used in the formation of substances Figure 3.1. Cells contain molecules of great variety, but living cells are composed of restricted sets of elements six of which (C, H, N, O, P, and S) make up more than 99% of their weight. Certain simple combinations of atoms such as methyl (-CH3), hydroxyl (OH), carboxyl (-COOH), and amino groups (-NH2) groups occur repeatedly in biological molecules. Each such group has distinct chemical and physical properties. Molecules can be broadly grouped into two categories.

Small Organic Molecules

These are compounds with molecular weight in the 100-1000 range containing up to 30 or so carbon atoms. Molecules of this kind are usually found in free solution in the cytoplasm where they form a pool of intermediates from which large molecules are made. Some of the small organic molecules are sugars, amino acids and fatty acids. At least 750 different types of small molecules are found.

Macromolecules

These are invariably polymeric molecules that are formed by the joining together of specific types of small molecules. A large polymeric molecule, or more than one polymer associated into large molecules can be referred to as a macromolecule. Some of the macromolecules such as carbohydrates, proteins and nucleic acids" are the most important. Three simple molecules such as sugar, amino acids and nucleotides that form macromolecules have been briefly described.

Sugars

There are three types of sugar molecules that are involved in different roles in organisms.

- Monosaccharide $(CH_2O)n$
- Disaccharide $(C_{12}H_{22}O_{11})$
- Polysaccharide (C6H10O5)n

Nucleotides

Nucleotides have 3 basic parts:

1. A 5-carbon sugar.
2. A phosphate group.
3. A nitrogenous (nitrogen-containing) base Figure 3.2.

A. Nucleic acids

Nucleic acids are containing instructions that cells need to carry out all the functions of life. Nucleic acid: It is a very large organic molecule made up of carbon, oxygen, hydrogen, nitrogen, and phosphorus.

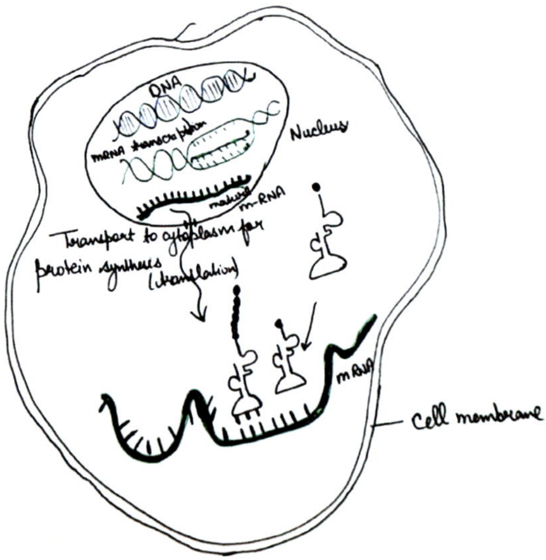

Figure 3.1: Cell structure.

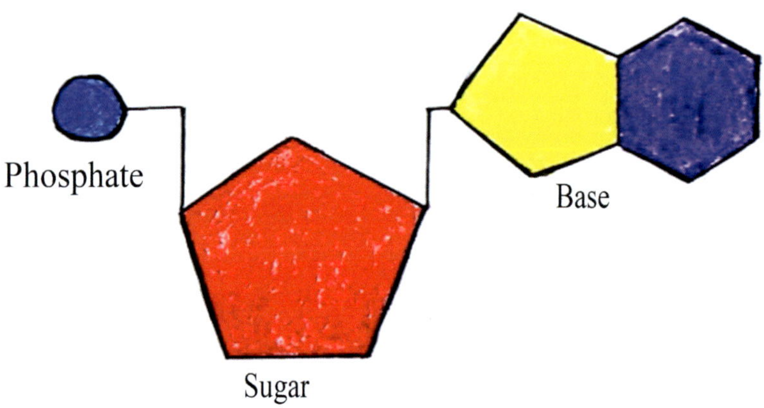

Figure 3.2 Nucleotides.

Types of Nucleic Acid

- DNA (deoxyribonucleic acid) - the genetic material that carries information about an organism. It is passed from the parent to the offspring.
- DNA directs all cellular activities and is only found in the nucleus of a eukaryotic cell.
- RNA (ribonucleic acid) – a nucleic acid found in the nucleus and cytoplasm of cells that plays an important role in the production of proteins Figure 3.3.

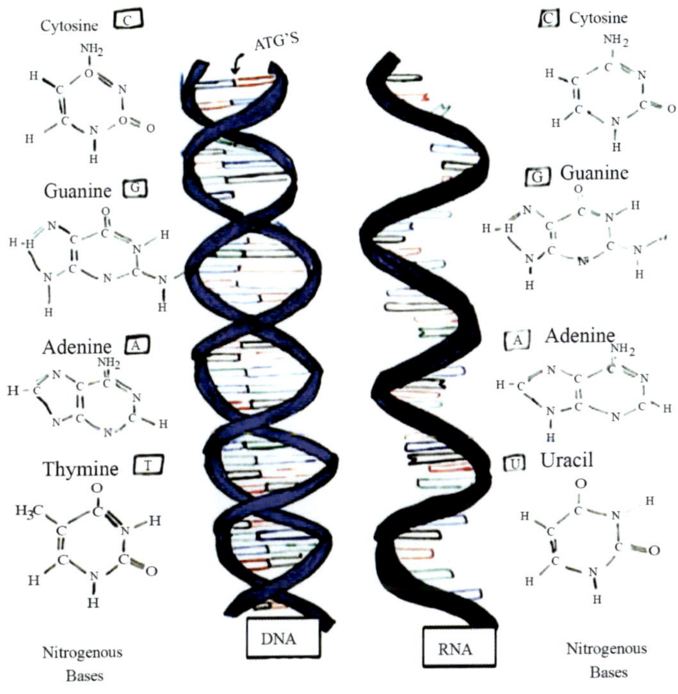

Figure 3.3: DNA and RNA sugars.

Nucleotides & Genetic Organization

Nucleic acids are made up of nucleotides.

Nucleotides have 3 basic parts

1. A 5-carbon sugar.
2. A phosphate group.

3. A nitrogenous (nitrogen-containing) base.

Levels of Genetic Organization

Nucleotide	Gene	DNA	Chromosomes	Genome
Smallest building block of gene	Code for traits	Made of many genes	Tightly coiled strands of DNA	The entire genetic make-up of an organism

Composition of DNA

- ➢ Major parts of a DNA nucleotide:
- ✓ A 5-carbon sugar called deoxyribose.
- ✓ Phosphates
- ✓ Four different nitrogenous bases (rungs of the ladder) Figure 3.4.

a. Adenine (A)
b. Thymine (T)
c. Guanine (G)
d. Cytosine (C)

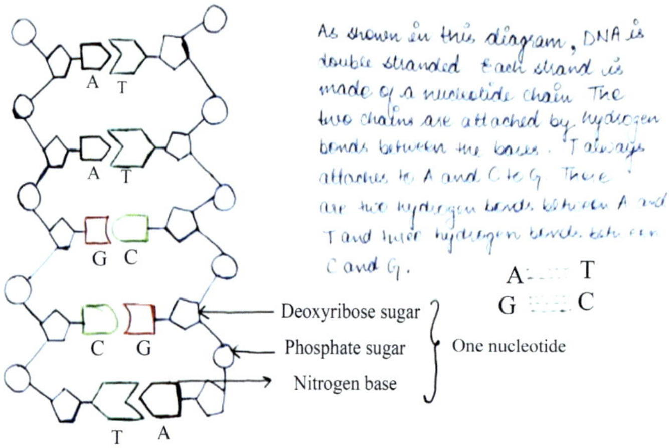

Figure 3.4: Nucleotide chain.

Pairing the Nitrogenous bases in DNA

Adenine is always pairs (matches up) with thymine and guanine always pairs (is complementary to) with cytosine. These nitrogenous bases make up four different possible nucleotides of DNA. Many nucleotides (basic unit of nucleic acids) placed together to make one DNA molecule Figure 3.5.

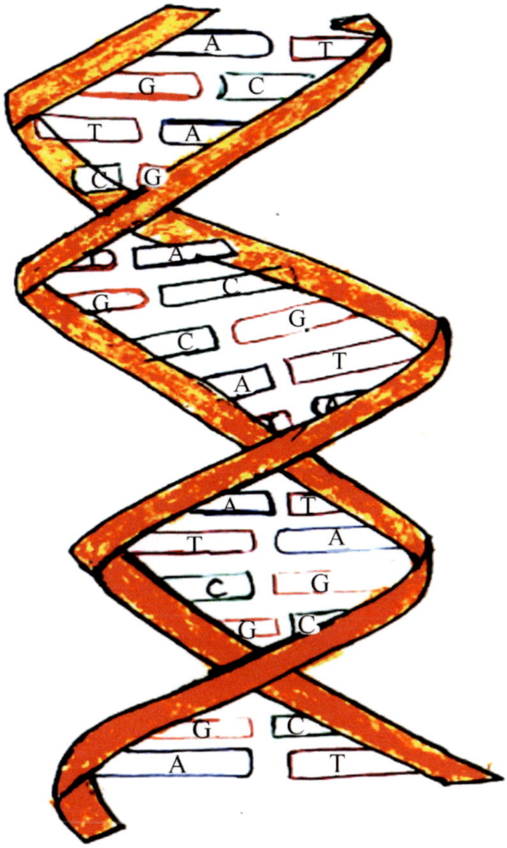

Figure 3.5: DNA structure.

Structure of DNA

- **James Watson & Francis Crick** – an American biologist & a British physicist who built the first accurate structural model of DNA (Won the Nobel Prize in 1962).
- **Watson & Crick's** model of DNA is a double helix (twisted ladder), in which 2 strands are wound around each other.
- The outside/sides of the ladder are made of deoxyribose sugar and phosphates.
- The stages of the ladder are the nitrogenous bases (connect to deoxyribose sugar in the outside/side of the ladder).

Base Pairs

DNA strands are complementary to one another (match up with each other). The bases are A & T and G & C. The following sequences of DNA are given, and also the complementary strands.

AGT TCA	CAG GTC
CTT GAA	AGC TCG

DNA Replication

They make copies of their DNA before cell divide. DNA Replication ensures that each daughter cell will have all of the genetic information and it needs to carry out its activities. DNA replication is semiconservative, meaning that when DNA copies, each new strand has ½ of the original strand.

- DNA unwinds (untwists) exposing the nitrogenous bases.
- DNA unzips – weak hydrogen bonds between nitrogenous bases break.
- Hydrogen bonds are the weakest type of bond that occurs in molecules.
- New complementary strands form nitrogenous bases floating in the nucleus of the cell pair up with the bases on each half of the DNA molecule.
- Two new DNA molecules that are exactly alike are formed.
- Both new strands are identical to the original strand of DNA.

Genes in DNA

Proteins determine the size, shape, and many other traits of an organism, e.g. Eye color, hair color, type of hairline, etc. Each gene in a DNA molecule codes for the production of a particular protein. Gene is a small portion of the DNA molecule that codes for a particular trait or protein.

Nitrogenous Bases

The order of the nitrogenous bases along a gene forms a genetic code that specifies what type of protein will be produced. The order of the nitrogenous bases tells the cell which amino acids are to be put together to make a protein. If one amino acid is changed or out of place, a different protein is made.

Since DNA cannot leave the nucleus to make proteins at the ribosomes, RNA directs protein synthesis; see Figures 3.6 and 3.7.

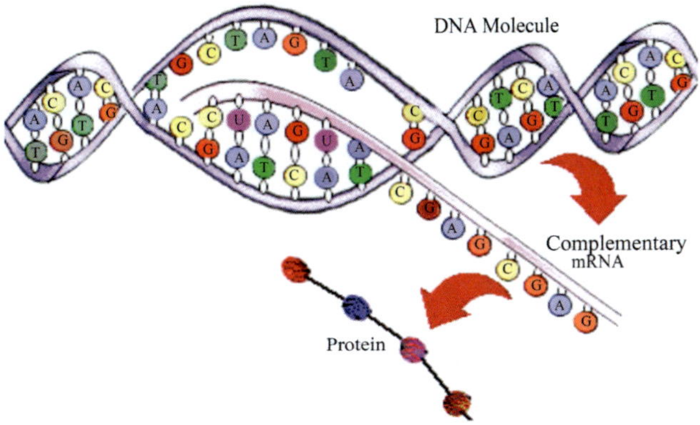

Figure 3.6: Nitrogenous bases involved to make protein.

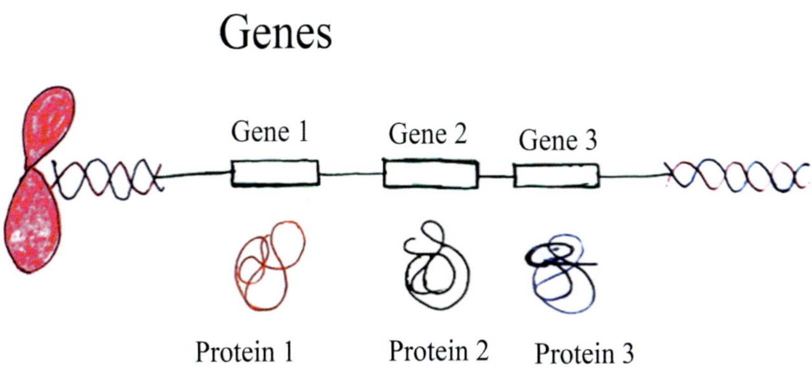

Figure 3.7: Sequence of genes.

It all started with Mendel

The beginning of our thinking of the possibility of genetic material begins with Mendel. He described the genetics of pea crosses, although he did not know what was the genetic material, he referred to "factors" as the things that gave his pea plants their physical characteristics. While Mendel's work gave a logical description of heredity it was largely ignored for many decades. His work was finally reproduced in the early 1900s and was thereafter regarded as facts read from genetics (www.scribd.com., www.slideshare.com., www.coursehero.com., www. biocyclopedia.com., www. molecular-plant-biotechnology.info., www.slideplayer.com., www. documents.mx., www.docplayer.net., www.bmb.psu.edu.,www.archive. org.,www.mafiadoc.com., www.powershow.com., www.bioscience.jbpub. com., www.outhorstream.com., www.docslide.com., www.msluay.com., www.scienceprofonlue.com., www.csus.edu., www.krishikosh.egrouth. ac.in., www.tankonyvtor.hu., www.silo.pub., www.studyblue.com., www. Preservearticles.com., www.biotechuniverse.blogspot.com).

4. BIOENERGETICS: PRINCIPLES OF THERMODYNAMICS AND THEIR APPLICATIONS IN BIOCHEMISTRY

Introduction

Living cells and organisms must perform work to stay alive, to grow, and to reproduce themselves. The ability to "harness energy from various sources and to channel it into biological work is a fundamental property of all living organisms; it must have been acquired very early in the process of cellular evolution. Modern organisms carry out a remarkable variety of energy transductions, conversions of one form of energy to another. They use chemical energy in fuels to bring about the synthesis of complex molecules from simple precursors, producing macromolecules with highly ordered structures. They also convert the chemical energy of various fuels into concentration gradients and electrical gradients, motion, heat, and even, in a few organisms such as fireflies, light. Photosynthetic organisms transduce light energy into all of these other forms of energy. In this century, biochemical studies have revealed much of the chemistry of energy transductions in living organisms. Biological energy transductions obey the same physical laws that govern all other natural processes. It is therefore essential for a student of biochemistry to understand these laws and the ways in which they apply to the flow of energy in the biosphere. In this chapter, we first review the laws of thermodynamics and the quantitative relationships among free energy, enthalpy, and entropy. We then describe the special role of ATP in biological energy exchanges". Finally, we consider the importance of oxidation-reduction reactions in living cells, the energetics of such electron transfer reactions, and the electron carriers commonly employed as cofactors of the enzymes that catalyze these reactions.

Bioenergetics is the quantitative study of the energy transductions that occur in living cells and of the nature and function of the chemical processes underlying these transductions. Although many of the principles of thermodynamics have been introduced in earlier chapters and may be familiar to you, it is worth reviewing the quantitative aspects of these principles.

Laws of Thermodynamics

Many quantitative observations made by physicists and chemists on the interconversion of different forms of energy led to the formulation, in the nineteenth century, of two fundamental laws of thermodynamics. The first law is the principle of the "conservation of energy: in any physical or chemical change, the total amount of energy in the universe remains constant, although the form of the energy may change. The second law of thermodynamics, which can be stated in several forms, says that the universe always tends toward more and more disorder: in cell natural processes, the entropy of the universe increases.Living organisms consist of collections of molecules much more highly organized than the surrounding materials from which they are constructed, and they maintain and produce order, seemingly oblivious to the second law of thermodynamics. Living organisms do not violate the second law; they operate strictly within it. To discuss the application of the second law to biological systems, we must first define those systems and the universe in which they occur. The reacting system is the collection of matter that is undergoing a particular chemical or physical process; it may be an organism, a cell, or two reacting compounds. The reacting system and its surroundings together constitute the universe. Some chemical or physical processes can be made to take place in isolated or closed systems, in which no material or energy is exchanged with the surroundings. Living cells and organisms are open systems, which exchange both material and energy with their surroundings; living systems are never at equilibrium with their surroundings.

We have defined earlier in this text three thermodynamic quantities that describe the energy changes occurring in a chemical reaction. Gibbs free energy (G) expresses the amount of energy capable of doing work during a reaction at constant temperature and pressure (p. 8). When a reaction proceeds with the release of free energy (i.e., when the system changes so as to possess less free energy), the free-energy change, ΔG, has a negative sign and the reaction is said to be exergonic. In endergonic reactions, the system gains free energy and ΔG is positive. Enthalpy, *H*, is the heat content of the reacting system. It reflects the number and kinds of chemical bonds in the reactants and products. When a chemical reaction releases heat, it is said to be exothermic; the heat content of the products is less than that of the reactants and ΔH has a negative value. Reacting systems that take up heat from their surroundings are endothermic and have positive values of ΔH (p. 66). Entropy, S, is a quantitative expression for the randomness or disorder in a system (Box 13-1). When the products of a reaction are less complex and more disordered than the reactants, the reaction is said to proceed with a gain in entropy (p. 72). The units of ΔG and ΔH are joules/mole or calories/mole (recall that 1 cal equals 4.18 J); units of entropy are joules/mole•degree Kelvin" (J/mol•K).

Under the conditions existing in biological systems (at constant temperature and pressure), changes in free energy, enthalpy, and entropy are related to each other quantitatively by the equation

$\Delta G = \Delta H - T\Delta S$ ....

in which ΔG is the change in Gibbs free energy of the reacting system, ΔH is the change in enthalpy of the system, "T is the absolute temperature, and ΔS is the change in entropy of the reacting system. By convention, ΔS has a positive sign when entropy increases and ΔH has a negative sign when heat is released by the system to its surroundings. Either of these conditions, which are typical of favorable processes, will tend to make ΔG negative. In fact, ΔG of a spontaneously reacting system is always negative. The second law of thermodynamics states that the entropy of the universe increases during all chemical and physical processes, but it does not require that the entropy increase take place in the reacting system itself. The order produced within cells as they grow and divide is more than compensated for by the disorder they create in their surroundings in the course of growth and division. In short, living organisms preserve their internal order by taking from the surroundings free energy in the form of nutrients or sunlight, 2 and returning to their surroundings an equal amount of energy as heat and entropy.

Free Energy

Cells are isothermal systems-they function at essentially constant temperature (and at constant pressure). Heat flow is not a source of energy for cells because heat can do work only as it passes from a zone or object at one temperature to a zone or object at a lower temperature. The energy that cells can and must use is free energy, described by the Gibbs free-energy function G, which allows prediction of the direction of chemical reactions, their exact equilibrium position, and the amount of work they can in theory perform at constant temperature and pressure. Heterotrophic cells acquire free energy from nutrient molecules, and photosynthetic cells acquire it from absorbed solar radiation. Both kinds of cells transform this free energy into ATP and other energy-rich compounds, capable of providing energy for biological work at a constant temperature.

Equilibrium Constant

The composition of a reacting system (a mixture of chemical reactants and products) will tend to continue changing until equilibrium is reached. At the equilibrium concentration of reactants and products, the rates of the forward and reverse reactions are exactly equal and no further net change occurs in the system. The concentrations of reactants and products at equilibrium define the equilibrium constant (p. 90). In the general reaction aA + bB $\longleftrightarrow$ cC + dD, where a, b, c, and d are the number of molecules of A, B, C, and D participating, the equilibrium constant is given by:

$$K_{eq} = \frac{[C]^c[D]^d}{[A]^a[B]^b}$$

where [A], [B], [C], and [D] are the molar concentrations of the reaction components at the point of equilibrium. When a reacting system is not at equilibrium, the tendency to move toward equilibrium represents a driving force, the magnitude of which can be expressed as the free-energy change for the reaction, ΔG. Under standard conditions (298 K (25 °C)), when reactants and products are initially present at 1 M concentrations or, for gases, at partial pressures of 101.3 kPa (1 atm), the force driving the system toward equilibrium is defined as the standard free-energy change, ΔG° By this definition, the standard state for reactions that involve hydrogen ions is $[H^+]$ = 1 M, or pH is 0. Most biochemical reactions occur in well-buffered aqueous solutions near pH 7; both the pH and the concentration of water (55.5 M) are essentially constant. For the convenience of calculations, biochemists, therefore, define a slightly different standard state, in which the concentration of H^+ is 10^{-7}M (pH is 7) and that of water is 55.5 M. Physical constants based on this biochemical standard state are written with a prime (e.g., ΔG°' and K'_{eq}) to distinguish them from the constants used by chemists and physicists. Under this convention, when H_2O or H^+ are reactants or products, their concentrations are not included in equations such as Equation 13-2, but are instead incorporated into the constants" ΔG°› and K'_{eq}.

Just as K'_{eq} is a physical constant characteristic for each reaction, so too is ΔG°› a constant. There is a simple relationship between K'_{eq} and ΔG°›:

ΔG°› = -RT ln K'_{eq}

The standard free-energy change of a chemical reaction is simply an alternative mathematical way of expressing its equilibrium constant. The relationship between ΔG°› and K'_{eq}. If the equilibrium constant for a given chemical reaction is 1.0, the standard free energy change of that reaction is 0.0 (the natural logarithm of 1.0 is zero). If K'_{eq} of a reaction is greater than 1.0, its ΔG°› is negative. If K'_{eq} is less than 1.0, ΔG°› is positive. Because the relationship between ΔG°› and K'_{eq} is exponential, relatively small changes in ΔG°› correspond to large changes in K'_{eq}. "It may be helpful to think of the standard free-energy change in another way. $\Delta G^{\circ\prime}$ is the difference between the free-energy content of the products and the free-energy content of the reactants under standard conditions. When $\Delta G^{\circ\prime}$ is negative, the products contain less free energy than the reactants. The reaction will therefore proceed spontaneously to form the products under standard conditions because all chemical reactions tend to go in the direction that results in a decrease in the free energy of the system. A positive value of $\Delta G^{\circ\prime}$ means that the products of the reaction contain more free energy than the reactants. The reaction will therefore tend to go in the reverse direction if we start with 1.0 M concentrations of all components. As an example, let us make a simple calculation of the standard free-energy change of the reaction catalyzed by the enzyme phosphoglucomutase:

$$\text{Glucose-1-phosphate} \longleftrightarrow \text{glucose-6-phosphate}$$

Chemical analysis shows that whether we start with, say, 20 mM glucose-1-phosphate (but no glucose-6-phosphate) in the presence of phosphoglucomutase, or with 20 mM glucose-6-phosphate, the final equilibrium mixture in either case will contain 1 mM glucose-1-phosphate and 19 mM glucose-6-phosphate at 25 °C and pH 7.0. (Remember that enzymes do not affect the point of equilibrium of a reaction; they merely hasten its attainment.) From these data we can calculate the equilibrium constant:

$$K'_{eq} = \frac{[\text{glucose-6-phosphate}]}{[\text{glucose-1-phosphate}]} = 19\ \text{mM} / 1\ \text{mM}$$

From this value of K'_{eq} we can alculate the standard free-energy change:

$\Delta G°› = -RT \ln K'_{eq} = -7{,}296$ J/mol = -7.3 kJ/mol

Because the standard free-energy change is negative when the reaction starts with 1.0 M glucose-1-phosphate and 1.0 M glucose-6-phosphate, the conversion of glucose-1-phosphate into glucose-6-phosphate proceeds with a loss (release) of free energy. For the reverse reaction (the conversion of glucose-6-phosphate to glucose-1-phosphate), ΔG°' has the same magnitude but the opposite sign. The standard free-energy changes for several representative chemical reactions. Note that hydrolysis of simple esters, amides, peptides, and glycosides, as well as rearrangements and eliminations, proceed with relatively small standard free-energy changes, whereas hydrolysis of acid anhydrides occurs with relatively large decreases in standard free energy. The oxidation of organic compounds to CO_2 and H_2O proceeds with especially large decreases in standard free energy. However, standard free-energy changes tell how much free energy is available from a reaction under *standard conditions*. To describe the energy released under the conditions that exist within cells, an expression for the actual free-energy change is essential".

Free-energy Change Depends on the Reactant and Product Concentrations

We must be careful to distinguish between two different quantities, the free-energy change, "ΔG, and the standard free-energy change, ΔG°'. Each chemical reaction has a characteristic standard free-energy change, which may be positive, negative, or zero, depending on the equilibrium constant of the reaction. The standard free-energy change tells us in which direction and how far a given reaction will go to reach equilibrium when the initial concentration of each component is 1.0 M, the pH is 7.0, and the temperature is 25 °C. Thus ΔG°' is a constant: it has a characteristic, unchanging value for a given reaction. But the actual free-energy change, ΔG, of a given chemical reaction is a function of the concentrations and of the temperature actually prevailing during the reaction, which is not necessarily the standard conditions as defined above. Moreover, the ΔG of any reaction proceeding spontaneously toward its equilibrium is always negative, becomes less negative as the reaction proceeds, and is zero at the point of equilibrium, indicating that no more work can be done by the reaction.

ΔG and ΔG°› for any reaction $A + B \longleftrightarrow C + D$ are related by the equation

$$\Delta G = \Delta G^{\circ\prime} + RT \ln[C][D]/[A][B]$$

in which the terms in red are those actually prevailing in the system under observation. The concentration terms in this equation express the effects commonly called mass action. As an example, let us suppose that the reaction $A + B \longleftrightarrow C + D$ is taking place at the standard conditions of temperature (25 °C) and pressure (101.3 kPa) but that the concentrations of A, B, C, and D are not equal and that none of them is present at the standard concentration of 1.0 M. To determine the actual free-energy change, ΔG, that will occur under these nonstandard conditions of concentration as the reaction proceeds from left to right, we simply put in the actual concentrations of A, B, C, and D; the values of R, T, and ΔG°' are the standard values". ΔG will be negative and will approach zero as the reaction proceeds because the actual concentrations of A and B will be getting smaller and the concentrations of C and D will be getting larger. Notice that when a reaction is at equilibrium, where there is no force driving the reaction in either direction and ΔG is equal to zero,

$$0 = \Delta G^{\circ\prime} + RT \ln \frac{[C]_{eq}[D]_{eq}}{[A]_{eq}[B]_{eq}}$$

or

$$\Delta G^{\circ\prime} = -RT \ln K'_{eq}$$

the equation that, as we noted above (p. 368), relates the standard free-energy change and the equilibrium constant. Even a reaction for which ΔG°' is positive can go in the forward direction if ΔG is negative. This is possible if the term RT ln ([products]/[reactants]) is negative and has a larger absolute value than ΔG. For example, the immediate removal of the products of a reaction can keep the ratio [products]/[reactants] well below 1, giving the term RT ln ([products]/[reactants]) a large, negative value. ΔG and ΔG°' are expressions of the maximum amount of free energy that a given reaction can theoretically deliver. This amount of energy could be realized only if there were a perfectly efficient device available to trap or harness it. Given that no such device is available, the amount of work done by the reaction at constant temperature and pressure is always less than the theoretical amount. It is also essential to understand that some reactions that are thermodynamically favorable (i.e., for which ΔG is large and negative) nevertheless do not occur at measurable rates. For example, firewood can be converted into CO_2 and

H_2O by combustion in a reaction that is very favorable thermodynamically. Nevertheless, firewood is stable for years, because the activation energy (see Fig. 8-4) for its combustion is higher than that provided by room temperature. If the necessary activation energy is provided (with a lighted match, for example), combustion will begin, converting the wood to the more stable products CO_2 and H_2O and releasing energy as heat and light. In living cells, reactions that would be extremely slow if uncatalyzed are caused to occur, not by supplying additional heat but by lowering the activation energy with an enzyme. The free energy change ΔG for a reaction is independent of the pathway by which the reaction occurs; it depends only on the nature and concentration of the initial reactants and the final products. An enzyme provides an alternative reaction pathway with lower activation energy, so that at room temperature a large fraction of the substrate molecules have enough thermal energy to overcome the activation barrier, and the reaction rate increases dramatically". Enzymes cannot change equilibrium constants, but they can and do increase the rate at which a reaction proceeds in the direction dictated by thermodynamics.

Standard Free-energy

In the case of two sequential "chemical reactions, A $\longleftrightarrow$ B and B $\longleftrightarrow$ C, each reaction has its own equilibrium constant and each has its characteristic standard free-energy change, $\Delta G°›_1$ and $\Delta G°›_2$. As the two reactions are sequential, B cancels out and the overall reaction is A $\longleftrightarrow$ C. Reaction A $\longleftrightarrow$ C will also have its own equilibrium constant and thus will also have its own standard free-energy change, $\Delta G°›_{total}$. The ΔG° values of sequential chemical reactions are additive. For the overall reaction A $\longleftrightarrow$ C, $\Delta G°›_{total}$ is the algebraic sum of the individual standard free-energy changes, $\Delta G°›_1$ and $\Delta G°'_2$, of the two separate reactions: $\Delta G°›_{total} = \Delta G°›_1 + \Delta G°›_2$. This principle of bioenergetics explains how a thermodynamically unfavorable (endergonic) reaction can be driven in the forward direction by coupling it to a highly exergonic reaction through a common intermediate». For example, the synthesis of glucose-6-phosphate is the first step in the utilization of glucose by many organisms:

Glucose + Pi $\longrightarrow$ glucose-6-phosphate + H_2O . . . ΔG°' = 13.8 kJ/mol

The positive value of ΔG°' predicts that under standard conditions the reaction will tend not to proceed spontaneously in the direction written.

Another cellular reaction, the hydrolysis of ATP to ADP and P_i, is very exergonic:

$ATP + H_2O \longrightarrow ADP + P_i$ . . . ΔG°› = -30.5 kJ/mol

These two reactions share the common intermediates P_i and H_2O and may be expressed as sequential reactions:

(1) Glucose + P_i ⟶ glucose-6-phosphate + H_2O

(2) ATP + H_2O ⟶ ADP + P_i

Sum: ATP + glucose ⟶ ADP + glucose-6-phosphate

The overall standard free-energy change is obtained by adding the ΔG°' values for individual reactions:

ΔG°› = +13.8 kJ/mol + (-30.5 kJ/mol) = -16.7 kJ/mol

Subjective Questions

1. Describe common structural and functional features of the eukaryotic cell?
2. Write a note on the chemical constituents of a cell?
3. Write about the structure of the cell membrane?
4. Describe transport of molecules across cell membrane with examples?
5. Define ionophores and ion channels. Give examples?
6. Write differences between facilitated transport and active transport?
7. What are the functions of the cell membrane?
8. Write a note on mitochondria structure and functions?
9. Write briefly on nucleus/ nucleolus?
10. Write about cytomembranes of a eukaryotic cell?
11. Facilitated transport?
12. Write a note on monosaccharides?
13. Define disaccharides. Give examples?
14. How do sucrose and maltose differ with respect to a. Structure b. Source c. Function?
15. Write briefly about the structure and functions of starch and glycogen?
16. What are polysaccharides? Classify. Give examples?
17. Write briefly about mucopolysaccharides?

18. Write four functions of carbohydrates.?
19. Write composition and functions of a. Hyaluronicacid b. Heparin?
20. Write composition and function of lactose?
21. Write a note on immunoglobulins.?
22. Write briefly about conjugated proteins?
23. What are derived proteins? Give examples?
24. What is meant by the secondary structure of proteins? Give examples?
25. Write about the quaternary structure of proteins giving examples?
26. Write about the structural feature of the α –helix?
27. Write a note on ß -pleated sheet?
28. Draw the general structure of immunoglobulins. Label its parts?
29. Write briefly about Ig E and Ig M?
30. Write briefly about the tertiary structure of a protein?
31. Define the primary structure of the protein. Write a method for its determination?
32. Write a note on albumin?
33. Write very briefly about aromatic aminoacids?
34. Write note on α 1- globulins?
35. Write clinical importance of alpha-fetoprotein and macroglobulin?
36. Name sulfur-containing amino acids and basic amino acids?
37. Classify immunoglobulins. Give structural and functional aspects of each class?
38. Name three charged forms of amino acids. When does the amino acid assume these states?
39. Write a note on α 2- globulins?
40. Write a note on ß-globulins?
41. Write a note on the general structure of immunoglobulins?
42. Define nucleoside. Give examples?
43. Define nucleotide. Give examples?
44. Write functions of nucleosides and nucleotides?
45. What are nucleic acids? Classify. Write their composition?
46. Describe the structure and function of DNA?
47. Define nucleoside. List purine and pyrimidine nucleosides?
48. Give examples for cyclic nucleotides. Write their importance?
49. Write a note on tRNA.
50. List purine nucleotides and their functions?
51. Give examples for pyrimidine nucleotides. Mention their functions?
52. Write a note on mRNA?

53. Define chromatin and nucleosome. Explain how they are related?
54. Write the functions of nucleic acids?
55. Purine and pyrimidine bases?
56. Define oxidative phosphorylation and substrate-level phosphorylation. Write their differences?
57. Write the principle of the chemiosmotic hypothesis?
58. What are uncouplers? Give examples?
59. Write about inhibitors of oxidative phosphorylation?
60. Write about a) ATP synthase b) Free radicals?

c) Substrate level phosphorylation?

61. Write about the P: O ratio?
62. Describe de novo synthesis of purine nucleotides.?
63. Write about pyrimidine nucleotide synthesis de novo?
64. Write about salvage pathways of purine nucleotides?
65. Define vitamins. Classify. Give examples for each class?
66. Write differences between water-soluble and fat-soluble vitamins?

Objective Questions

1. **Which of the following is required for crystallization and storage of the hormone insulin?**

(A) Mn++

(B) Mg++

(C) Ca++

(D) Zn++

Answer: D

2. **Oxidation of which substance in the body yields the most calories**

(A) Glucose

(B) Glycogen

(C) Protein

(D) Lipids

Answer: D

3. **Milk is deficient in which vitamins**

(A) Vitamin C

(B) Vitamin A

(C) Vitamin B2

(D) Vitamin K

Answer: A

4. **Milk is deficient in which mineral?**
   (A) Phosphorus
   (B) Sodium
   (C) Iron
   (D) Potassium

Answer: C

5. **Synthesis of prostaglandins is inhibited by**
   (A) Aspirin
   (B) Arsenic
   (C) Fluoride
   (D) Cyanide

Answer: A

6. **HDL is synthesized and secreted from**
   (A) Pancreas
   (B) Liver
   (C) Kidney
   (D) Muscle

Answer: B

7. **Which are the cholesterol esters that enter cells through the receptor-mediated endocytosis of lipoproteins hydrolyzed?**
   (A) Endoplasmic reticulum
   (B) Lysosomes
   (C) Plasma membrane receptor
   (D) Mitochondria

Answer: B

8. **Which of the following phospholipids is localized to a greater extent in the outer leaflet of the membrane lipid bilayer?**
   (A) Choline phosphoglycerides
   (B) Ethanolamine phosphoglycerides
   (C) Inositol phosphoglycerides
   (D) Serine phosphoglycerides

Answer: A

9. All the following processes occur rapidly in the membrane lipid bilayer except?

(A) Flexing of fatty acyl chains

(B) Lateral diffusion of phospholipids

(C) Transbilayer diffusion of phospholipids

(D) Rotation of phospholipids around their long axes

Answer: C

10. Which of the following statement is correct about membrane cholesterol?

(A) The hydroxyl group is located near the center of the lipid layer

(B) Most of the cholesterol is in the form of a cholesterol ester

(C) The steroid nucleus form forms a rigid, planar structure

(D) The hydrocarbon chain of cholesterol projects into the extracellular fluid

Answer: C

11. Which one is the heaviest particulate component of the cell?

(A) Nucleus

(B) Mitochondria

(C) Cytoplasm

(D) Golgi apparatus

Answer: A

12. Which one is the largest particulate of the cytoplasm?

(A) Lysosomes

(B) Mitochondria

(C) Golgi apparatus

(D) Endoplasmic reticulum

Answer: B

13. The degradative Processess are categorized under the heading of

(A) Anabolism

(B) Catabolism

(C) Metabolism

(D) None of the above

Answer: B

14. The exchange of material takes place

(A) Only by diffusion

(B) Only by active transport

(C) Only by pinocytosis

(D) All of these

Answer: D

15. The average pH of Urine is

(A) 7.0

(B) 6.0

(C) 8.0

(D) 0.0

Answer: B

16. The pH of blood is 7.4 when the ratio between H2CO3 and NaHCO3 is

(A) 1 : 10

(B) 1 : 20

(C) 1 : 25

(C) 1 : 30

Answer: B

17. The phenomenon of osmosis is opposite to that of

(A) Diffusion

(B) Effusion

(C) Affusion

(D) Coagulation

Answer: A

18. The surface tension in the intestinal lumen between fat droplets and the aqueous medium is decreased by

(A) Bile Salts

(B) Bile acids

(C) Conc. H2SO4

(D) Acetic acid

Answer: A

19. Which of the following is located in the mitochondria?

(A) Cytochrome oxidase

(B) Succinate dehydrogenase.

(C) Dihydrolipoyl dehydrogenase

(D) All of these

Answer: D

20. The most active site of protein synthesis is the

(A) Nucleus

(B) Ribosome

(C) Mitochondrion

(D) Cell sap

Answer: B

21. The fatty acids can be transported into and out of mitochondria through

(A) Active transport

(B) Facilitated transfer

(C) Non-facilitated transfer

(D) None of these

Answer: B

22. Mitochondrial DNA is

(A) Circular double-stranded

(B) Circular single-stranded

(C) Linear double helix

(D) None of these

Answer: A

23. The absorption of intact protein from the gut in the fetal and newborn animals takes place by

(A) Pinocytosis

(B) Passive diffusion

(C) Simple diffusion

(D) Active transport

Answer: A

24. The cellular organelles called "suicide bags" are

(A) Lysosomes
(B) Ribosomes
(C) Nucleolus
(D) Golgi's bodies

Answer: A

25. The power house of the cell is

(A) Nucleus
(B) Cell membrane
(C) Mitochondria
(D) Lysosomes

Answer: C

26. The digestive enzymes of cellular compounds are confined to

(A) Lysosomes
(B) Ribosomes
(C) Peroxisomes
(D) Polysomes.

Answer: A

27. The general formula of monosaccharides is

(A) CnH2nOn
(B) C2nH2On
(C) CnH2O2n
(D) CnH2nO2n

Answer: A

28. The general formula of polysaccharides is

(A) (C6H10O5)n
(B) (C6H12O5)n
(C) (C6H10O6)n
(D) (C6H10O6)n

Answer: A

29. The aldose sugar is

(A) Glycerose

(B) Ribulose

(C) Erythrulose

(D) Dihydoxyacetone

Answer: A

30. A triose sugar is

(A) Glycerose

(B) Ribose

(C) Erythrose

(D) Fructose

Answer: A

31. A pentose sugar is

(A) Dihydroxyacetone

(B) Ribulose

(C) Erythrose

(D) Glucose

Answer: B

32. The pentose sugar present mainly in the heart muscle is

(A) Lyxose

(B) Ribose

(C) Arabinose

(D) Xylose

Answer: A

33. Polysaccharides are

(A) Polymers

(B) Acids

(C) Proteins

(D) Oils

Answer: A

34. The number of isomers of glucose is

(A) 2
(B) 4
(C) 8
(D) 16

Answer: D

35. Two sugars that differ from one another only in configuration around a single carbon atom are termed

(A) Epimers
(B) Anomers
(C) Optical isomers
(D) Stereoisomers

Answer: B

36. Isomers differing as a result of variations in the configuration of the —OH and —H on carbon atoms 2, 3 and 4 of glucose are known as

(A) Epimers
(B) Anomers
(C) Optical isomers
(D) Stereoisomers

Answer: A

37. Polysaccharides

(A) Contain many monosaccharide units which may or may not be of the same kind
(B) Function mainly as storage or structural compounds
(C) Are present in large amounts in connective tissue
(D) All of these

Answer: D

38. The absorption of glucose in the digestive tract

(A) Occurs in the small intestine
(B) Is stimulated by the hormone Glucagon
(C) Occurs more rapidly than the absorption of any other sugar
(D) Is impaired in cases of diabetes mellitus

Answer: A

39. UDP-Glucose is converted to UDP-Glucuronic acid by

(A) ATP

(B) GTP

(C) NADP+

(D) NAD+

Answer: B

40. The enzymes involved in Phosphorylation of glucose to glucose 6-phosphate are

(A) Hexokinase

(B) Glucokinase

(C) Phosphofructokinase

(D) Both (A) and (B)

Answer: D

45. Cori cycle is

(A) Synthesis of glucose

(B) reuse of glucose

(C) uptake of glycose

(D) Both (A) & (B)

Answer: D

46. Cane sugar is known as

(A) Galactose

(B) Sucrose

(C) Fructose

(D) Maltose

Answer: B

47. Which of the following is not a reducing sugar?

(A) Lactose

(B) Maltose

(C) Sucrose

(D) Fructose

Answer: C

48. α–D-Glucose and β–D-glucose are related by

(A) Epimers

(B) Anomers

(C) Multirotation

(D) Ketoenol pair

Answer: B

49. The stable ring formation in D-Glucose involves

(A) C-1 and C-4

(B) C-1 and C-2

(C) C-1 and C-5

(D) C-2 and C-5

Answer: C

50. Reduction of Glucose with Ca++ in water produces

(A) Sorbitol

(B) Dulcitol

(C) Mannitol

(D) Glucuronic acid

Answer: A

51. Starch and glycogen are polymers of

(A) Fructose

(B) Mannose

(C) α–D-Glucose

(D) Galactose

Answer: C

52. Reducing ability of carbohydrates is due to

(A) Carboxyl group

(B) Hydroxyl group

(C) Enediol formation

(D) Ring structure

Answer: A

53. Which of the following is not a polymer of glucose?

(A) Amylose

(B) Inulin

(C) Cellulose

(D) Dextrin

Answer: B

54. Invert sugar is

(A) Lactose

(B) Mannose

(C) Fructose

(D) Hydrolytic product of sucrose

Answer: D

CHAPTER 3

CELL ENZYMOLOGY

1. HISTORY, ENZYME CLASSIFICATION AND NOMENCLATURE

Introduction

It is the most remarkable and highly specialized protein. Catalyze hundreds of stepwise reactions in biological systems. Regulate many different metabolic activities necessary to sustain life. Living systems use catalysts called enzymes to increase the rate of chemical reactions. The study of enzymes has immense practical importance. In medical science: to know the epidemiology, to diagnose, and to treat diseases (inheritable genetic disorders).

- In chemical industries
- In food processing
- In agriculture
- In everyday activities in the home (food preparation, cleaning, beauty care, etc.)

Nomenclature of Enzymology

The International Enzyme Commission (EC) has recommended a systematic nomenclature for enzymes. This commission assigns names and numbers to enzymes according to the reaction they catalyze. An example of a systematic enzyme name is EC 3.5.1.5 urea aminohydrolases for the enzyme that catalyzes the hydrolysis of urea. The name of an enzyme frequently provides a clue to its function. In some cases, an enzyme is named by incorporating the suffix *-ase* into the name of its substrate, e.g., *pyruvate decarboxylases* catalyze the removal of a CO_2 group from pyruvate. Certain protein cutting digestive enzymes are an exception to this general rule of enzyme nomenclature, e.g., *pepsin, trypsin, chymotrypsin and thrombin.* Isozyme: Within an organism, more than one enzyme may catalyze a given reaction. Multiple enzymes catalyzing the same reaction are called isozymes. Isozymes differ in their catalytic properties. Consequently, the various isozymes that are present in different tissues or at different developmental stages can carry out slightly different metabolic functions.

Pyruvate Decarboxylase

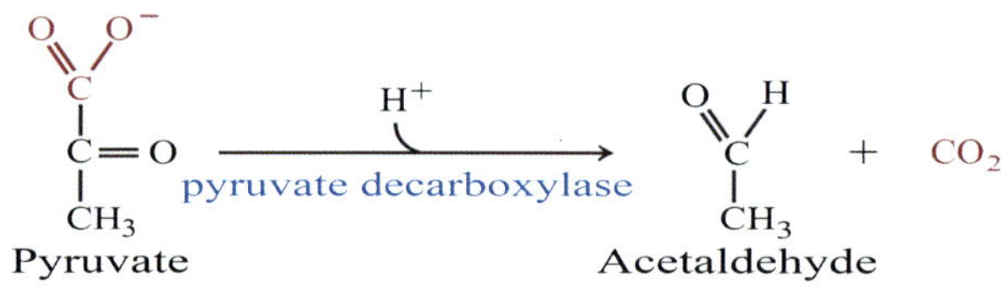

Alanine Aminotransferase

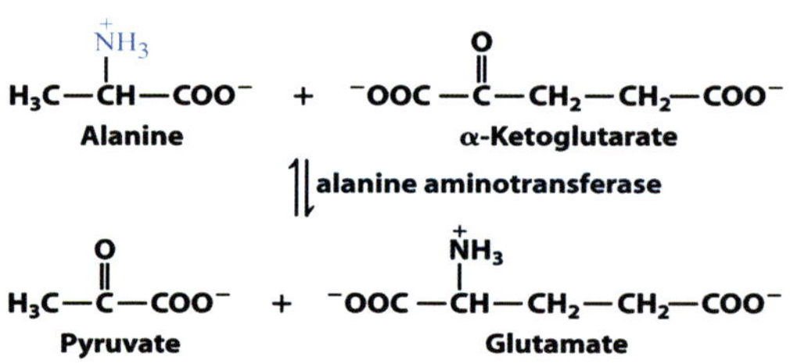

Enzyme Classification

Class of Enzyme

1. **Oxidoreductases: Oxidation-reduction reactions**
2. **Transferases: Transfer of functional groups**
3. **Hydrolases: Hydrolysis reactions**
4. **Lyases: Group elimination to form a double bond**
5. **Isomerases: Isomerization reactions**
6. **Ligases: Bond formation coupled with ATP hydrolysis**

Enzymes vs. Non-biological Catalysts

Enzymes have their reaction specificity. The activities of many enzymes are regulated so that the organism can respond to the changing condition or follow genetically determined developmental programs. Enzyme activity is affected by the temperature and pH of the medium.

Properties of Enzymes

All enzymes are proteins or protein in nature (except a small group of catalytic molecules). Enzymes have a molecular weight ranging from 20,000 - 1 million. Some enzymes require no chemical group other than their amino acid residue for activity. Some enzymes may require an additional chemical compound called cofactors (Zn^{2+}, Mg^{2+}, Fe^{2+}, Mn^{2+}etc.). Some enzymes may require a complex organic or metallorganic molecule called coenzymes. Sometimes they need both cofactors and coenzymes. A coenzyme or metal ion that is covalently bound to the enzyme protein is called a prosthetic group. A complete catalytically active enzyme together with its coenzymes and or metal ions is called a holoenzyme. The protein part of such an enzyme is called the apoenzyme or apoprotein. Enzymes are classified by the reactions they catalyze. Many enzymes have been named by adding the suffix "-ase" to the name of their substrate or to a word or phrase describing their activities. Thus urease catalyzes the hydrolysis of urea. An enzyme circumvents many problems providing a specific environment within which a given reaction is energetically more favorable. An enzyme-catalyzed reaction occurs within the confinement of a pocket on the enzymes called the **active site**. The molecule that is bound by the active site and acted upon by the enzyme is called the **substrate.** The enzyme-substrate complex is central to the action of enzymes.

Enzyme Specificity

Enzymes are specific and efficient catalysts. In nonenzymatic reactions, substrates are transformed into a product by bond-making and bond-breaking processes. In an enzymatic reaction, these processes occur through interactions of the substrates with appropriate side chains of the amino acid residues of the enzyme or with the coenzyme of the enzyme. One property of many enzymes is specificity: catalyzing one chemical reaction with only one substrate. An example is carbonic anhydrase. This enzyme catalyzes only one reversible reaction, the breakdown of carbonic acid (H_2CO_3) to water and carbon dioxide. At ideal conditions, a single molecule of Carbonic anhydrase is capable of catalyzing the breakdown of about 36 million molecules of carbonic acid in one minute.

4In a biological reaction, the reacting species must come together and undergo electronic rearrangement that results in the formation of a product. Let us consider an idealized transfer reaction in which compound A—B reacts with compound C to form two new products, A and B—C. In order for the first two compounds to react, they must approach closely enough for their constituent atoms to interact. Normally, atoms that approach too closely repel each other; but if the groups have sufficient free energy, they can pass this point and react with each other to form products. The energy-requiring step of the reaction is known as an energy barrier, called free energy of activation or activation energy ($G^{\ddagger}$).

Idealized Transfer Reaction

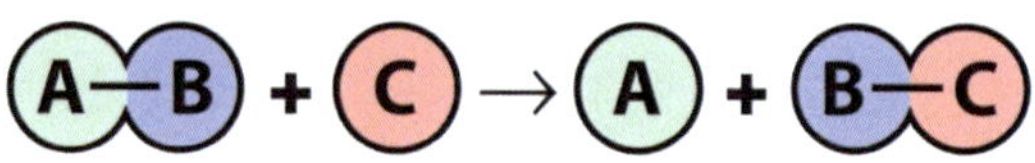

Reaction coordinate diagram for the reaction A—B + C A + B→C

The progress of the reaction (reaction coordinate) is shown on the horizontal axis, and the free energy(*G*) is shown on the vertical axis. The transition state of the reaction, represented as a..b..c, is the point of highest free energy. The free energy difference between the reactant and the transition state is the free energy of activation (*$G^{\ddagger}$*).

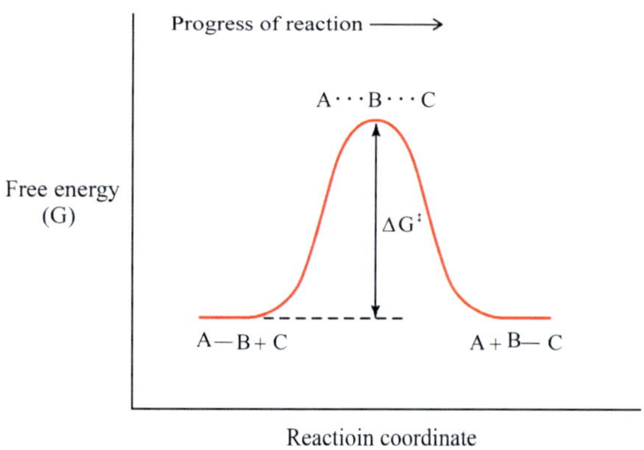

Reaction Coordinate of a Reaction: Reactants and Products have different Free Energies

When the free energy of the reactant is greater than that of the product, the free energy changes for the reaction is negative, so the reaction proceeds spontaneously (a). When the free energy of the product is greater than that of the reactant, the free energy change of the reaction is positive, so the reaction does not proceed spontaneously (however the reverse reaction does proceed).

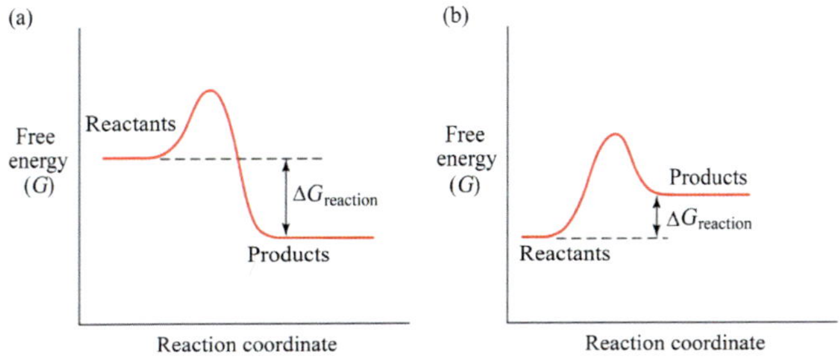

Effect of a Catalyst on a Chemical Reaction

Here the reactants proceed through a transition state ($X^{‡}$) during their conversion to products. A catalyst lowers the free energy of activation () $G^{‡}$) for the reaction so that)$G^{‡}_{cat}$ < $G^{‡}_{uncat}$. Lowering the free energy of the transition state ($X^{‡}$) accelerates the reaction because more reactants are able to achieve the transition state per unit time.

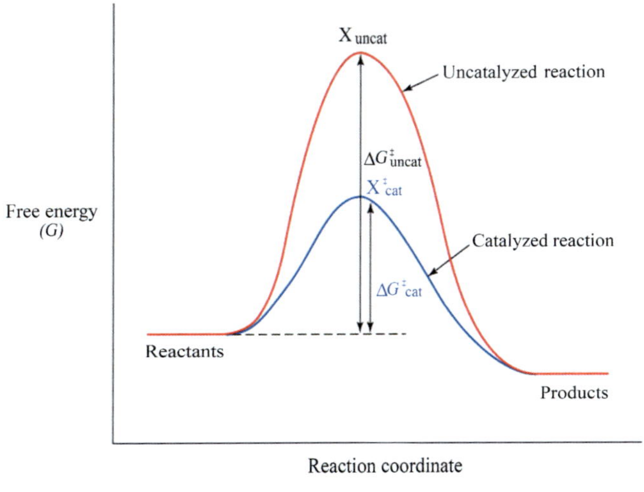

Enzymes use a Chemical Catalytic Mechanism

There are three basic kinds of chemical catalytic mechanisms used by enzymes:

- Acid-base catalysis
- Covalent catalysis
- Metal ion catalysis

Acid-base Catalysis

In acid-base catalysis, a proton is transferred between the enzyme and the substrate. It may be either acid catalysis or base catalysis or both. Many enzymes use both acid and base catalysis.

Tautomerization of a Ketone

Tautomers are interconvertible isomers that differ in the placement of hydrogen and a double bond. The dotted lines represent bonds in the process of breaking or forming.

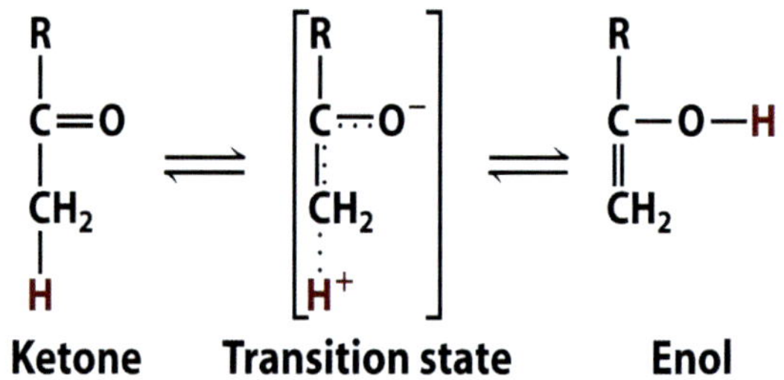

Acid Catalysis

If a catalyst (H—A) donates a proton to the ketone's oxygen atom, it reduces the unfavorable character of the transition state, thereby lowers its energy. Hence, it also lowers the activation energy barriers for the reaction. This is an example of an acid catalyst since the catalyst acts as an acid by donating a proton.

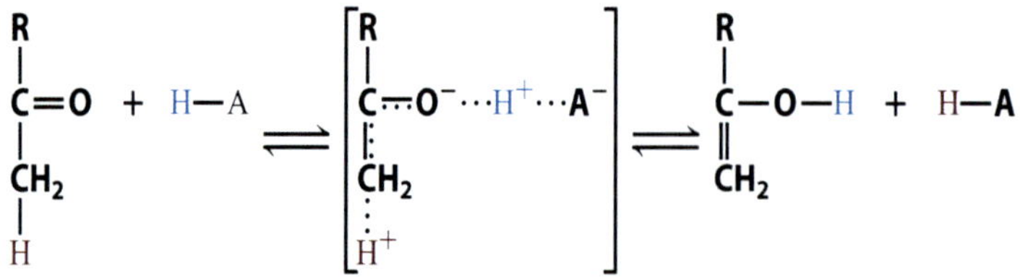

Ketoenol Tautomerization (Base Catalysis)

The same keto-enol tautomerization reaction shown in the previous slide can be accelerated by a catalyst that can accept a proton. In this case, it is a base catalyst (B), where the dots represent unpaired electrons.

Base catalysis lowers the energy of the transition state and thereby accelerates the reaction.

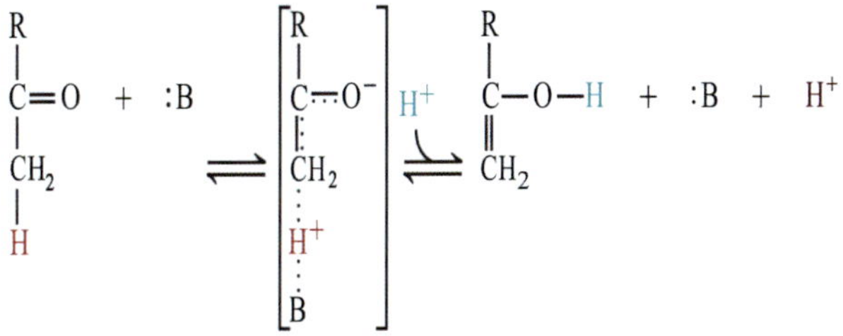

Amino Acid Side Chains that can act as Acid-base Catalysts

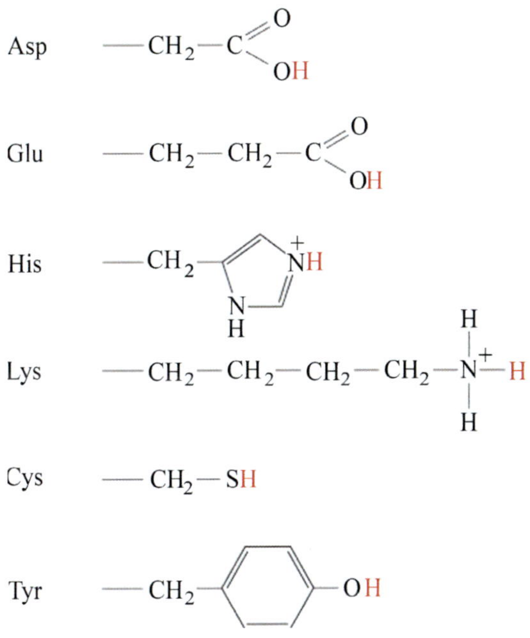

Covalent Catalysis

In covalent catalysis, a covalent bond is formed between the catalyst and the substrate during the formation of the transition state. In covalent catalysis, an electron-rich group in the enzyme active site forms a covalent adduct with a substrate. This covalent complex is much more stable than a transition state and can be isolated. Covalent catalysis is also called nucleophilic catalysis because the catalyst is a nucleophile (an electron-rich group in search of an electron -poor center (a compound with an electron deficiency is known as an electrophile). Enzymes that use covalent catalysis undergo a two-part reaction process. So the reaction coordinate diagram contains two energy barriers with the reaction intermediate between them.

Reaction Coordinate Diagram for a Reaction Accelerated by Covalent Catalysis

Two transition states flank the covalent intermediate. The relative heights of the activation energy barriers (the energies of the two transition states, $X^{\ddagger}_1$ and $X^{\ddagger}_2$) vary depending on the reactions.

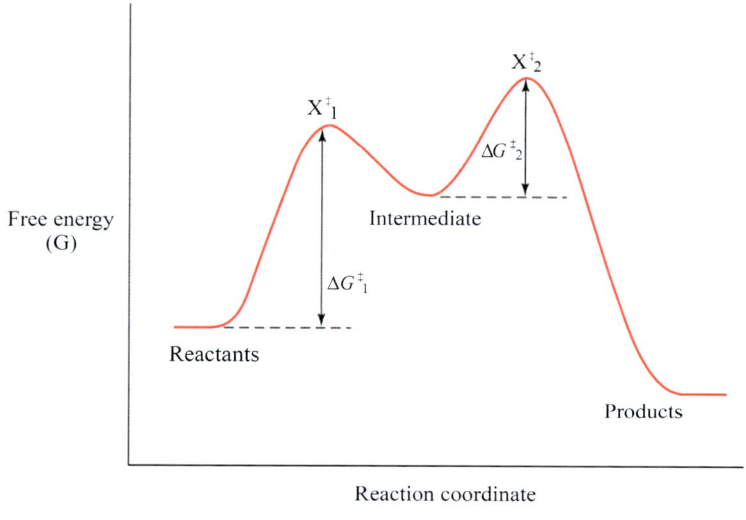

Metal Ion Catalysis

Metal ions participate in enzymatic reactions by mediating oxidation-reduction reactions, or by promoting the reactivity of other groups in the enzyme's active site through electrostatic effects. A protein-bound metal ion can interact directly with the reacting substrate. The conversion of acetaldehyde to ethanol is catalyzed by the liver enzyme alcohol dehydrogenase. Here a zinc ion stabilizes the developing negative charge on the oxygen atom during the formation of the transition state.

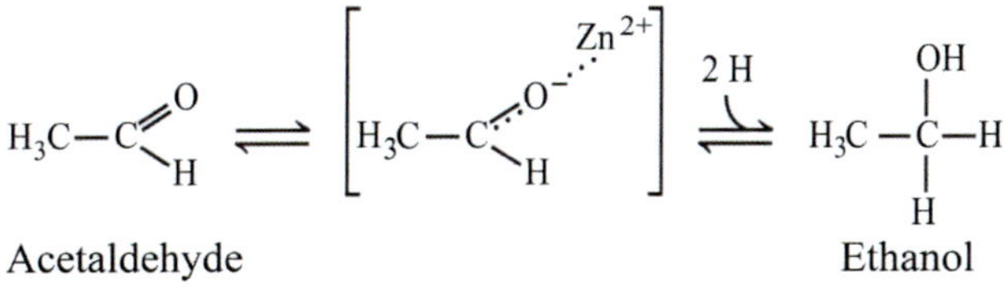

Enzyme–substrate Complexes

Enzymes are specific and efficient catalysts. Lock and key fit of enzymes and substrates help produce enzyme specificity.

Lock-and-key Model of Enzyme Action

In this early model, the substrate (key) was believed to bind tightly to a site on the enzyme (lock) that was exactly complementary to it in its size, shape and charge.

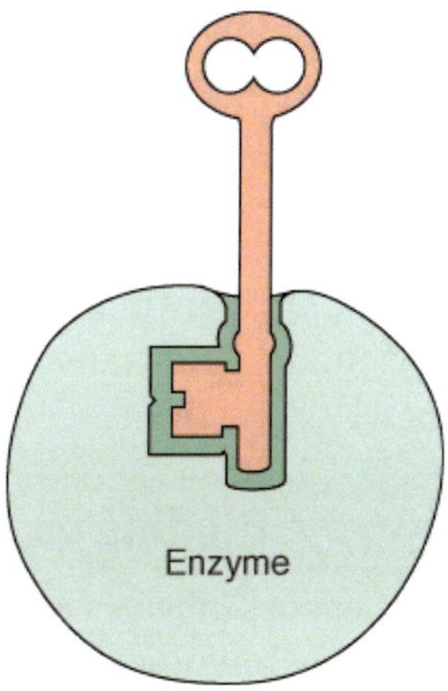

Enzyme-substrate Complexes

Lock and key fit of enzymes and substrates helps produce enzyme specificity: A substrate molecule must contact an enzyme molecule before it can be transformed into products. Once the substrate has made contact, it must bind to the enzyme at the active site: a place where the conversion of substrate to a product can take effect.

The active site is usually a dimple, pocket or crevice formed by a fold in the chain conformation of the protein. When the substrate key fits the enzyme lock, an enzyme-substrate complex is formed. The fit between enzyme and substrate in binding helps to produce enzyme specificity. Complementarity between the enzyme and the substrate is governed by factors besides shape– e.g., the electrical charge. The substrate may be positively charged and the enzyme's active site negatively charged and vice versa. Hydrogen bonds and other weak forces may also hold the enzyme-substrate complex together. Hydrophobic regions of the substrate associate with similar regions on the active site of the enzyme. For most enzymes contributions from shape, charge, and a large number of weak forces are responsible for the formation and strength of the complex.

Transition State Stabilization

Linus Pauling (1946) formulated that an enzyme increases the reaction rate not by binding tightly to the substrates, but by binding tightly to the reaction's transition state i.e., a substrate that has been strained toward the structure of the product. In other words, the tightly bound key of the lock and key model is the 'transition state', not the substrate. Pauling's theory of enzyme's actions is consistent with the chemical catalytic mechanisms, in which the catalyst stabilizes the transition state of the reaction. In an enzyme, tight binding(stabilization) of the transition state occurs in addition to acid-base, covalent, or metal ion catalysis.

Effect of very tight Substrate Binding on Enzyme Catalysis

The red line indicates the free energy changes as the reactants pass through a transition state to products. The blue line indicates free energy changes for the same reaction when the reactants (substrates) are very tightly bound to the enzyme (they are at a very low free energy level) before they react. The activation energy barrier ($G^{\ddagger}$) is therefore increased because the tight-binding must be overcome before the substrates can reach the transition state.

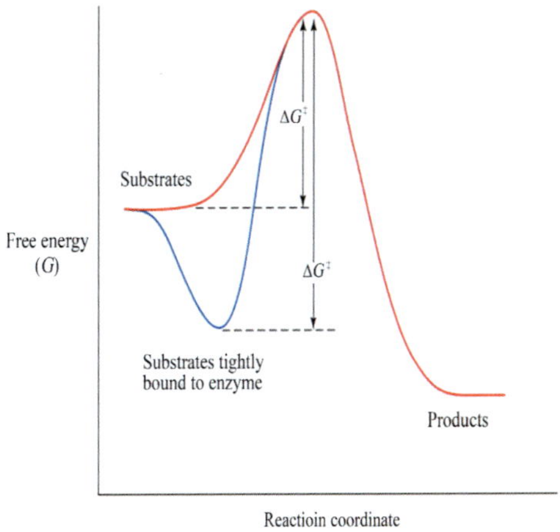

Enzyme Assay

Enzyme activity is expressed as the rate at which the enzyme catalyzes the conversion of a substrate to products. Activity depends on the concentration of the enzyme and substrate, and reaction conditions (pH, temperature). Experiments in which the enzyme activity is measured are called **enzyme assays.** Many diagnostic procedures consist of assays for blood serum enzymes. The concept of this diagnostic procedure is simple: slight leakage of enzymes from tissues and organs to the circulatory system is normal; additional leakage is abnormal and indicates damage or disease. Damage to the heart muscle in heart attacks results in the release of the enzyme creatine phosphokinase (CPK) into the bloodstream. Over 40 other blood serum enzymes can be subjected to clinical testing. \

Introduction

Steps involved in Isolation of a Protein or Enzyme

- Cell disruption (this can be done via a number of different processes of choice e.g.: Detergents lysis, Osmolysis, freeze-thaw cycles, enzymatic lysis, ultrasonication, Homogenisation).
- Centrifugation (at a specific speed depending on the organ, tissue, organelle or fluid).
- Removal of supernatant (Decantation to obtain supernatant).

Huge number of enzymes more than the number of genes encoded in a genome). Most cell types in multicellular organisms express tens of thousands of different proteins. Relative abundances of various proteins vary widely.

Goals of Protein Purification

It obtains a particular protein free of others and other cell components. It obtains a good yield (absolute amount and proportion of starting amount). They maintain the activity of the protein.

Problems during Purification

- Denaturation
- Proteolysis
- *In vitro* mixing

A measurement of how well the process worked. Some characterization of the purified protein.

Make an Extract from the Source

Almost always want to keep it very cold (just above freezing). Gentle breaking. In some cases can obtain big purification in one step by separating cellular compartments (e.g., purify nuclei from cytoplasm before extracting nuclear proteins).

Begin Separating Components

Remove nucleic acids, polysaccharides, cell membrane debris. Ammonium sulfate precipitations, other crude fractionations (pH or other salt precipitations, antibody clearing, "autolysis"). These crude steps are often needed to avoid ruining or overloading chromatography agents.

Fractionate by Chromatography

Several steps are almost always needed. Need to assay for amount and purity at each step. Need a way to decide when you are finished.

Separation Principles

Sizing

The principle is based on the exclusion of larger molecules from pores in resin; smaller molecules require longer times of transit. Sizing resins = "gel filtration". Different matrices have different size ranges.

Examples: Sephadex, Sephacryl, Ultragel, some HPLC resins.

Ion Exchange

Separate on the basis of the net charge. Wash and elute with higher salt concentrations. Determine the highest salt that your protein binds to the resin, and use this to load or to pre-wash before eluting your protein.

Examples: DEAE cellulose, phosphocellulose, some HPLC resins.

Affinity Chromatography

Separate based on binding to residues specific or semi-specific to the protein you are purifying. Easy to overload with non-specific proteins **ex.** sequence-specific DNA sites and antibody resin.

Enzyme Characterization

Enzyme characterization simply refers to the determination of the various chemical and physical properties (characteristics) of an enzyme. It involves the use of a series of laboratory procedures. Examples of assays for commonly characterized enzyme properties in biochemical researches include:

- Effect of changes in temperature on the enzyme's activity and optimum temperature.
- Enzyme's thermal stability.
- Effect of changes in pH on the enzyme's activity and optimum pH.
- pH stability.
- Effect of changes in substrate concentration on enzyme's activity and kinetic constants e.g Vmax, Km, Kcat, Km/Kcat, etc.
- Substrate specificity
- Molecular weight (Mw) of enzyme
- Effect of metal ions, chelating agent or denaturing agents.
- Enzyme's isoelectric point (pI)

- Effect of duration of incubation.
- Active site fractional saturation (V/Vmax) at a particular substrate concentration.
- Enzyme's turnover number (Vmax/ET)
- Activation energy (Ea)
- Salt tolerance

Within the cell, enzymes are generally found along with other proteins, nucleic acids, polysaccharides and lipids. The activity of the enzyme in relation to the total protein present (i.e. the specific activity) can be determined and used as a measure of enzyme purity. A variety of methods can be used to remove contaminating material in order to purify the enzyme and increase its specific activity. Enzymes that are used as diagnostic reagents and in clinical therapeutics are normally prepared to a high degree of purity because the great emphasis is placed on the specificity of the reaction that is being catalyzed. Clearly the higher the level of purification, the greater the cost of enzyme production. In the case of many bulk industrial enzymes, the degree of purification is less important, and such enzymes may often be sold as very crude preparations of culture broth containing the growth medium, organisms (whole or fragmented) and enzymes of interest. However, even when the cheapest bulk enzymes are utilized (e.g. proteases for use in washing powders), the enzyme cost can contribute around 5–10% of the final product value.

Pre-treatment

At the end of a fermentation in which a microorganism rich in the required enzyme has been cultured, the broth may be cooled rapidly to 5°C to prevent further microbial growth and stabilize the enzyme product. The pH may also be adjusted to optimize enzyme stability. If the enzyme-producing organism is a fungus, this may be removed by centrifugation at low speed. If the enzyme source is bacterial, the bacteria are often flocculated with aluminum sulfate or calcium chloride, which negate the charge on the bacterial membranes, causing them to clump and thus come out of suspension.

Treatment

Extracellular enzymes are found in the liquid component of the pretreatment process. However, intracellular enzymes require more extensive treatment. The biomass may be concentrated by centrifugation and washed to remove medium components. The cellular component must then be ruptured to release the enzyme content. This can be done using one or more of the following processes:

- Ball milling (using glass beads)
- Enzymic removal of the cell wall
- Freeze–thaw cycles
- Liquid shearing through a small orifice at high pressure (e.g. within a French press)
- Osmotic shock
- Sonication.

The separation of enzymes from the resulting solution may then involve a variety of separation processes, which are often employed in a sequential fashion. The first step in an enzyme purification procedure commonly involves the separation of the proteins from the non-protein components by a process of salting out. Proteins remain in an aqueous solution because of interactions between the hydrophilic (water-loving) amino acids and the surrounding water molecules (the solvent). If the ionic strength of the solvent is increased by adding an agent such as ammonium sulfate, some of the water molecules will interact with the salt ions, thereby decreasing the number of water molecules available to interact with the protein. Under such conditions, when protein molecules cannot interact with the solvent, they interact with each other, coagulating and coming out of the solution in the form of a precipitate. This precipitate (containing the enzyme of interest and other proteins) can then be filtered or centrifuged and separated from the supernatant. Since different proteins vary in the extent to which they interact with water, it is possible to perform this process using a series of additions of ammonium sulfate, increasing the ionic strength in a stepwise fashion and removing the precipitate at each stage. Thus such fractional precipitation is not only capable of separating protein from non-protein components, but can also enable separation of the enzyme of interest from some of the other protein components. Subsequently, a wide variety of techniques may be used for further purification, and steps involving chromatography are standard practice.

Ion-exchange Chromatography

It is often effective during the early stages of the purification process. The protein solution is added to a column containing an insoluble polymer (e.g. cellulose) that has been modified so that its ionic characteristics will determine the type of mobile ion (i.e. cation or anion) it attracts. Proteins whose net charge is opposite to that of the ion-exchange material will bind to it, whereas all other proteins will pass through the column. A subsequent change in pH or the introduction of a salt solution will alter the electrostatic forces, allowing the retained protein to be released into the solution again.

Gel Filtration

It can be utilized in the later stages of a purification protocol to separate molecules on the basis of molecular size. Columns containing a bed of cross-linked gel particles such as Sephadex are used. These gel particles exclude large protein molecules while allowing the entry of smaller molecules. Separation occurs because the larger protein molecules follow a path down the column between the Sephadex particles (occupying a smaller fraction of the column volume). Larger molecules , therefore, have a shorter elution time and are recovered first from the gel filtration column.

Affinity Chromatography

The procedures can often enable purification protocols to be substantially simplified. Typically, with respect to enzyme purification, a column would be packed with a particulate stationary phase to which a ligand molecule such as a substrate analog, inhibitor or cofactor of the enzyme of interest would be firmly bound. As the sample mixture is passed through the column, the enzyme interacts with, and binds, to the immobilized ligand, being retained within the column as all of the other components of the mixture pass through the column unrewarded. Subsequently, a solution of the ligand is introduced to the column to release (elute) and thereby recover the bound enzyme from the column in a highly purified form. Nowadays numerous alternative affinity chromatography procedures exist that are able to separate enzymes by binding to areas of the molecule away from their active site. Advances in molecular biology enable us to purify recombinant proteins, including enzymes, through affinity tagging. In a typical approach, the gene for the enzyme of interest would be modified to code for a further short amino acid sequence at either the N- or C- terminal.

For example, a range of polyhistidine tagging procedures is available to yield protein products with six or more consecutive histidine residues at their N- or C- terminal end. When a mixture containing the tagged protein of interest is subsequently passed through a column containing a nickelnitrilotriacetic acid (Ni-NTA) agarose resin, the histidine residues on the recombinant protein bind to the nickel ions attached to the support resin, retaining the protein, whilst other proteins and non-protein components pass through the column. Elution of the bound protein can then be accomplished by adding imidazole to the column, or by reducing the pH to 5-6 to displace the His-tagged protein from the nickel ions. Such techniques are therefore capable of rapidly and highly effectively isolating an enzyme from a complex mixture in only one step, and typically provide protein purities of up to 95%. If more highly purified enzyme products are required, other supplemental options are also available, including various forms of preparative electrophoresis e.g. disc-gel electrophoresis and isoelectric focusing.

Finishing of Enzymes

Enzymes are antigenic, and since problems occurred in the late 1960s when manufacturing workers exhibited severe allergic responses after breathing enzyme dust, procedures have now been implemented to reduce dust formation. These involve supplying enzymes as liquids wherever possible or increasing the particle size of dry powders from 10 μm to 200–500 μm by either prilling (mixing the enzyme with polyethylene glycol and preparing small spheres by atomization) or marumerizing (mixing the enzyme with a binder and water, extruding long filaments, converting them into spheres in a marumerizer, drying them and covering them with a waxy coat).

2. ENZYME ASSAYS, FACTORS AFFECTING THE RATE OF ENZYMATIC REACTIONS AND MICHAELIS-MENTEN EQUATION

Introduction

Kinetics is the study of the rates of reactions. Enzymes provide cells with the remarkable capacity to exert kinetic control over thermodynamic potentiality. Enzymes are the agents of metabolic function that increase the rate of reaction.

It is able to Determine:

- Maximum velocity
- Substrate affinity

- Inhibitor affinity

It can tell us:

- Flow-through metabolic pathways
- Utilization of substrates

It can we do with the Information:

- Control and manipulate metabolic events

The Michaelis-Menten Equation

The ES complex is the key to understanding this kinetic behavior. In 1903, this kinetic pattern led Victor Henry to propose that an enzyme combines with its substrate molecule to form the ES complex as a necessary step in enzyme catalysis. In 1913, Leonor Michaelis and Maud Mentenexpanded the idea of Victor Henry into a general theory of enzyme actions.

Michaelis and Menten postulated that:

The enzyme first combines reversibly with its substrate to form an ES complex in a relatively fast reversible step. The ES complex then breaks down in a slower second step (rate-limiting step) to yield the free enzyme and the product. Michaelis-Menten equation is the rate equation for an enzyme-catalyzed reaction and is the mathematical description of the hyperbolic curve we have discussed earlier. The formula is:

$$v_o = \frac{V_{max}[S]}{K_m + [S]}$$

Where,

V_0 is the initial velocity

V_{max} is the maximum velocity

[S] is the substrate concentration

Km (Michaelis-Menten constant) is the substrate concentration at which the reaction velocity is half of the maximum velocity.

$$E + S \underset{k_{-1}}{\overset{k_1}{\rightleftharpoons}} ES \xrightarrow{k_2} P + E$$

Conclusions of Michaels - Menten Kinetics

- When [S]= K_M, the equation reduces to

$$V = \frac{V_{max}[S]}{K_M + [S]} = \frac{V_{max}[S]}{[S] + [S]} = \frac{V_{max}}{2}$$

- when [S] >> K_M, the equation reduces to

$$V = \frac{V_{max}[S]}{K_M + [S]} = \frac{V_{max}[S]}{[S]} = V_{max}$$

- when [S] << K_M, the equation reduces to

$$V = \frac{V_{max}[S]}{K_M + [S]} = \frac{V_{max}[S]}{K_M} = \frac{V_{max}}{K_M}[S]$$

Lineweaver – Burk Double Reciprocal Plots

- It is difficult to determine Vmax experimentally
- The equation for a hyperbola can be transformed into the equation for a straight line by taking the reciprocal of each side
- The formula for a straight line is y = mx + b
- A plot of 1/V versus 1/[S] will give a straight line with the slope of KM/Vmax and y-intercept of 1/Vmax
- Such a plot is known as a Lineweaver-Burk double reciprocal plot

$$\frac{1}{V} = \frac{K_M}{V_{max}} \cdot \frac{1}{[S]} + \frac{1}{V_{max}}$$

$$y = m \cdot x + b$$

Significance of K_m

- Km is a constant
- Small Km means tight binding; high Km means weak binding
- Useful to compare Km for different substrates for one enzyme
  Hexokinase: D-fructose – 1.5 mM
  D-glucose – 0.15 mM
- Useful to compare Km for a common substrate used by several enzymes
  Hexokinase: D-glucose – 0.15 mM
  Glucokinase: D-glucose – 20 mM

Kinetic vs Chemical Mechanism:

- **An enzyme kinetic mechanism is the order of substrate addition and product release in an enzyme-catalyzed reaction**
- **A chemical mechanism is the chemical pathway of conversion of S → P, including the structures of any intermediates**

Bi-substrate Reactions:

- **The Michaelis –Menten model of enzyme kinetics was derived for single substrate reactions**
- **The majority of enzymatic reactions have multiple substrates and products**
- **Bi-substrate reactions account for ~ 60% of the known enzymatic reactions.**

$$A + B \overset{E}{\rightleftharpoons} P + Q$$

Substrate Addition / Product Release

- **The order of substrate addition and product release in most enzymatic reactions follow two reaction mechanism**

– Sequential reaction - all substrates must bind to the enzyme before the reaction occurs and products are released

- **Ordered sequential**
- **Random sequential**

– Ping-pong reaction - one or more products are released before all substrates have been added and an alternate stable enzyme form, F, is produced in the half reaction

Sequential Reaction

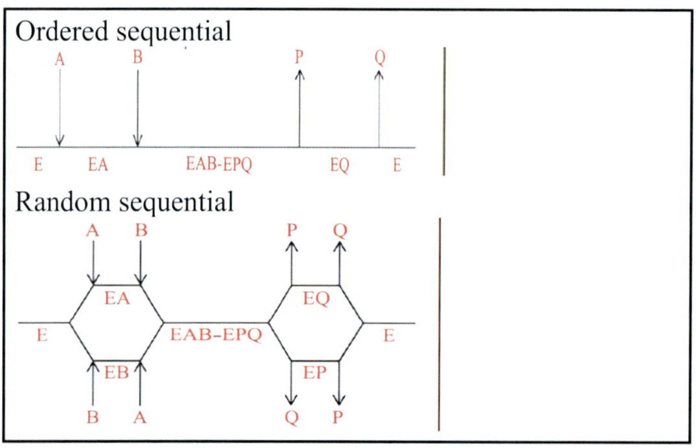

Ping-pong Reaction

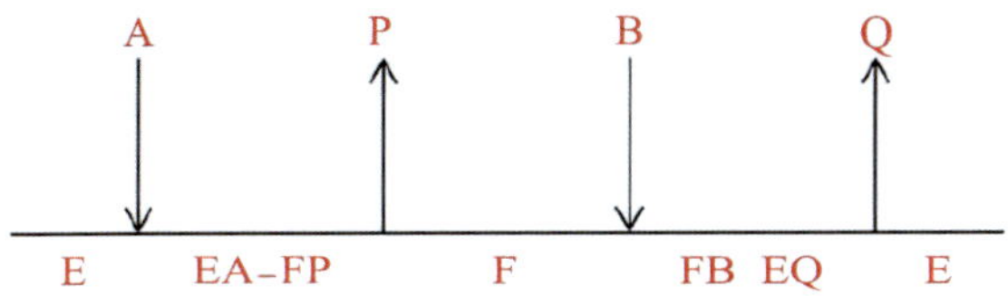

Initial Velocity Plots

Sequential reaction exhibits an intersecting pattern of lines. Order and random substrate additions cannot be distinguished in this type of plot.

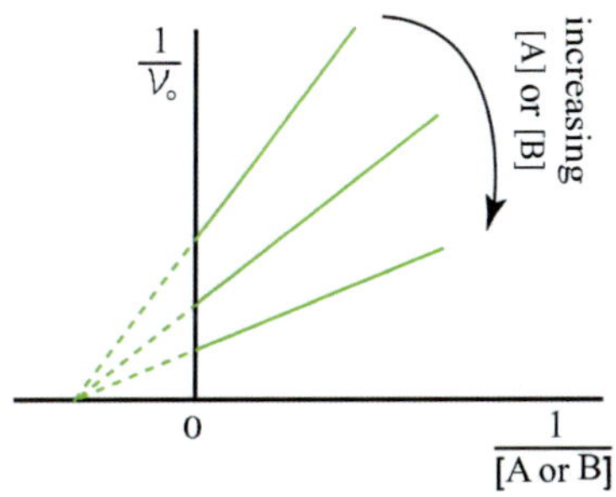

Ping-pong Reaction shows Parallel or Non-intersecting Lines

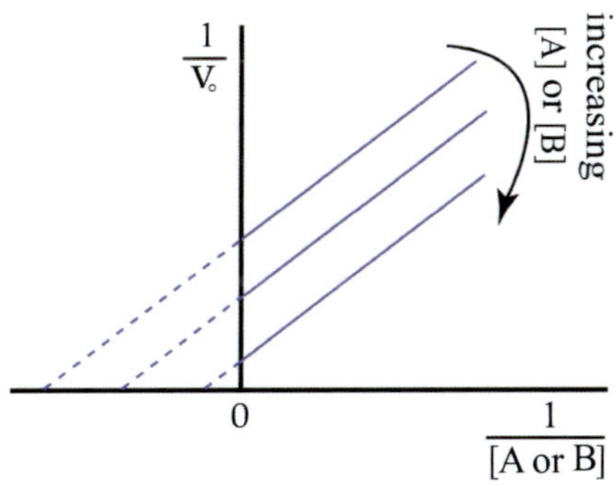

Influence of Enzyme Concentration

$$V = \frac{K_3[E][S]}{K_m + [S]}$$

$v = k_3$ [E], as [S]>>[E]

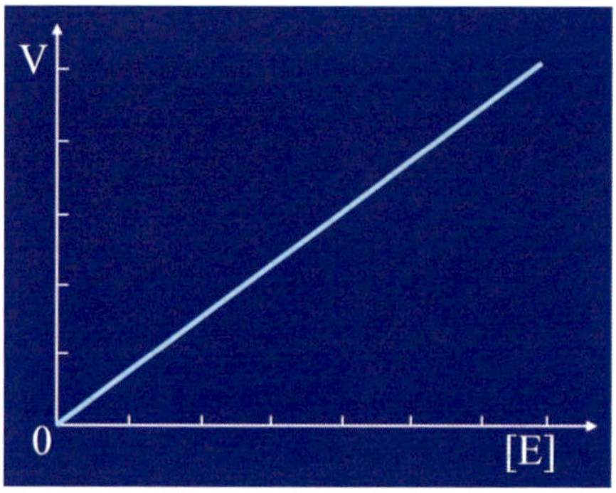

Influence of Temperature

Optimum temperature, most of them are in the range from 35 to 45°C.

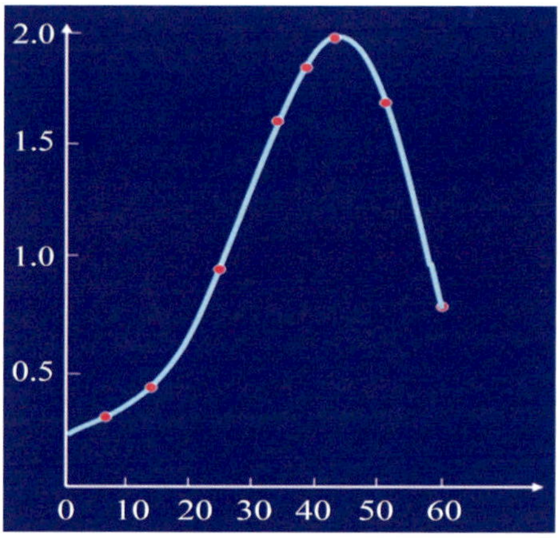

Influence of pH

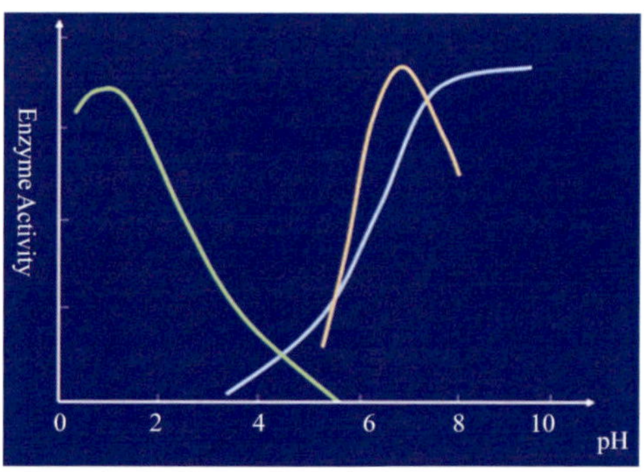

4. ENZYME INHIBITION

Introduction

Enzyme inhibitors are important

- They can be used to gain information about the shape of the enzyme active site and the amino acid residues in the active site.
- They can be used to gain information about the chemical mechanism.
- They can be used to gain information about the regulation or control of a metabolic pathway.
- They can be very important in drug design.

Reversible Inhibitor

A substance that binds to an enzyme to inhibit it, but can be released. It usually involves the formation of non-covalent bonds.

Irreversible Inhibitor

A substance that causes inhibition that cannot be reversed. It is usually involving the formation or breaking of covalent bonds to or on the enzyme.

Inhibitors

- Reversible
- Irreversible inhibition

Irreversible Inhibition

The inhibitor combines with an essential group of enzyme active centers by a covalent bond, resulting in enzymatic activity loss.

Inhibition Patterns

An inhibitor may bind at the same site as one of the substrates. These inhibitors structurally look like the substrate. An inhibitor may bind at an alternate site affecting catalytic activity without affecting substrate binding. Many inhibitors do both

- Competitive
- Mixed or Non-competitive
- Uncompetitive

Competitive Inhibition

A competitive inhibitor is competing with a substrate for the enzyme-substrate binding site. Malonate is a competitive inhibitor of succinate for succinate dehydrogenase.

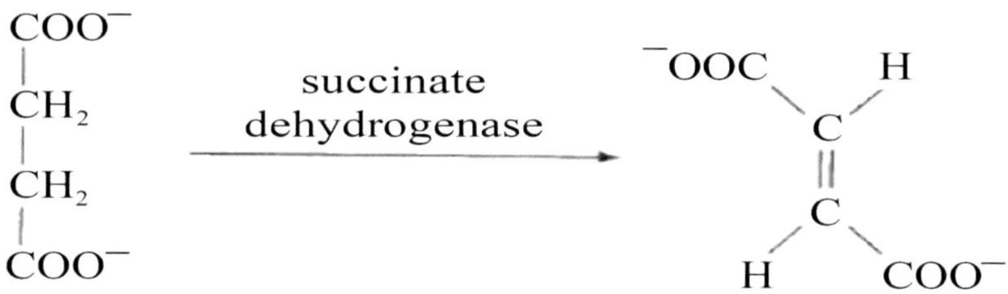

Succinate **Fumarate**

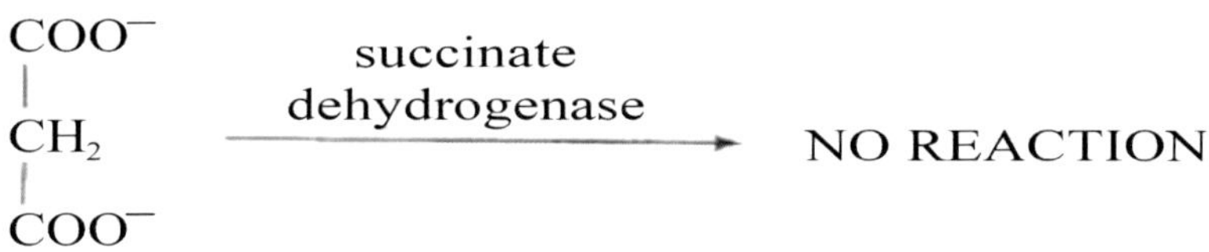

Malonate

Competitive Inhibition

A competitive inhibitor reduces the amount of free enzyme available for substrate binding thus increasing the Km for the substrate. The effect of a competitive inhibitor can be overcome with high concentrations of the substrate.

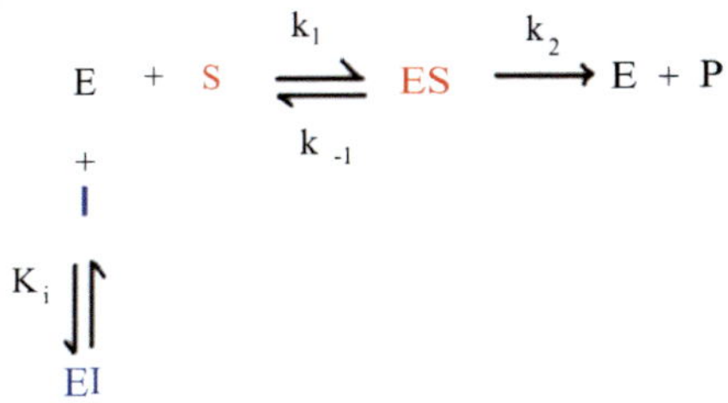

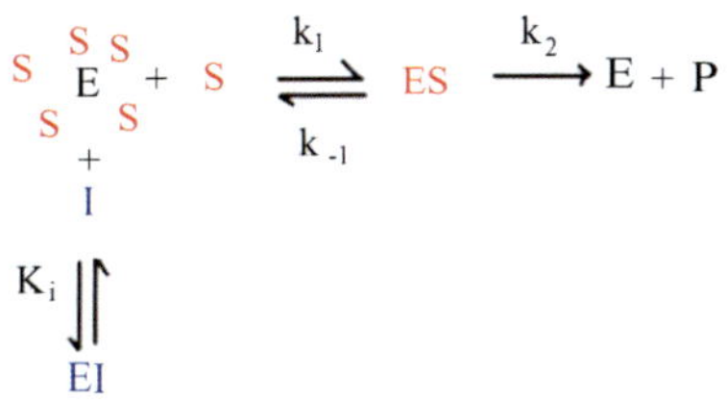

Uncompetitive Inhibition

An uncompetitive inhibitor binds to the enzyme-substrate complex but not to the free enzyme. The result is a decrease in V_{max} and Km. The effect of an uncompetitive inhibitor can not be overcome by high concentrations of the substrate.

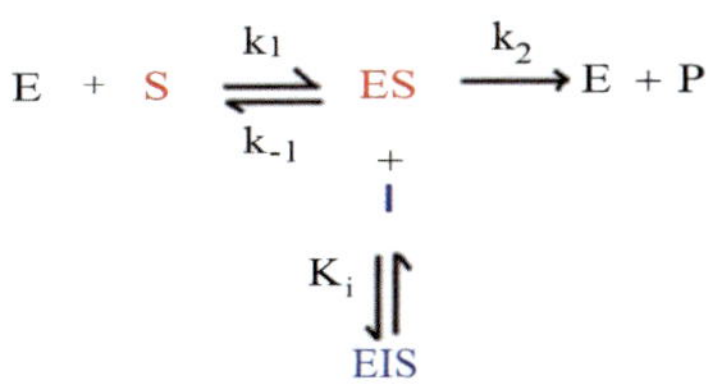

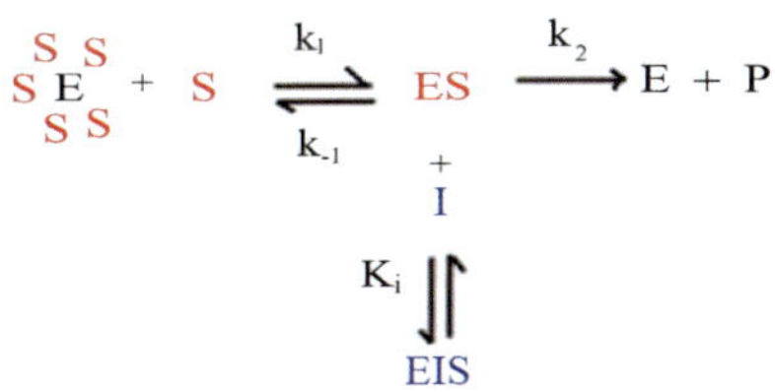

Mixed or non-nompetitive inhibition

- **The inhibitor can bind to both free enzyme and the ES complex**
- **The affinity of the inhibitor to the two complexes might be different**

– If the binding of inhibitor changes the affinity for the substrate, Km will be changed and called mixed inhibition

– If only Vmax affected called Non-competitive inhibitor

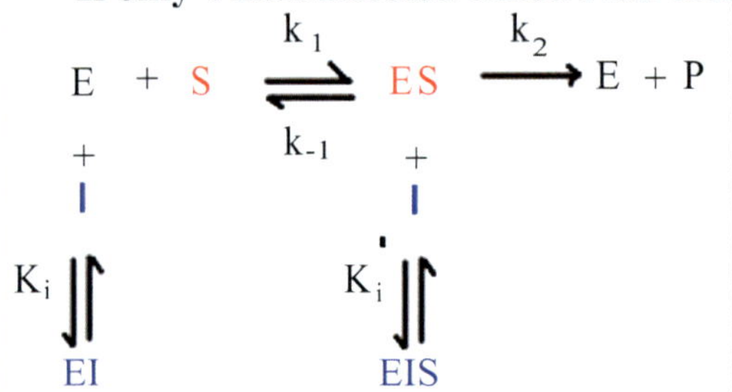

Allosteric Inhibition

Those with multiple active sites in one multi-subunit protein are commonly subjected to allosteric inhibition. Catalytic proteins containing both an active site and a modulator site are called allosteric enzymes. In allosteric inhibition, the inhibitor binding to one subunit of an enzyme may decrease the catalytic activity of all the subunits.

Feedback Inhibition

Cells have other ways of controlling the activity of enzymes. A fine slowdown control of enzyme activity built into some enzymes relies on **feedback inhibition** i.e., the concentration of a product at the end of a long series of steps builds up and then "feedback" to inhibit the enzyme in a preceding step.

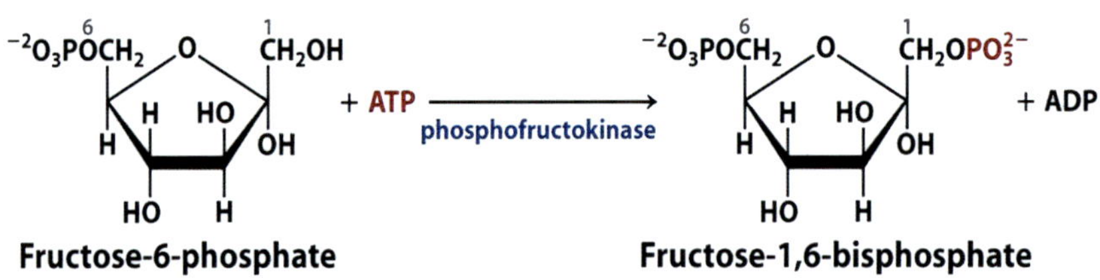

Denaturation of Enzymes

Enzymes are sensitive to pH and temperature. Enzyme activity varies with changes in pH. Most enzymes exhibit maximum activity at a pH near 7, the pH of most body fluids. The pH where maximum activity occurs is called the optimum pH of the enzyme. Enzyme activity usually varies with temperature. At a temperature near 20 ^{0}C, most enzymes are active and relatively stable. An enzyme's activity decreases beyond the optimum temperature. The decrease in activity is usually due to heat denaturation of the enzyme. In the case of a non-biological catalyst, an increase in temperature increases the reaction rate.

Introduction

Enzymes provide functional groups that act as proton donors or acceptors.

General Acid-base Catalysis

Enzymes provide functional groups that act as proton donors or acceptors.

- Amino groups
- Carboxyl groups
- Sulfhydryl groups

Example – Lysozyme (donates proton)

Metal Ion Catalysis

Metal ion acts as an electrophilic catalyst. Stabilizes are negative charges on intermediate. The metal ions may generate a nucleophile by increasing the acidity of nearby molecules.

Example – Carbonic anhydrase

Metal ions may bind to a substrate and increase binding energy.

Example – NMP kinases

Binding to enzyme brings two substrates together.

Example – NMP kinases

Binding to enzyme orients substrate.

The susceptible bond is close to catalytic groups of the active site

Orbital steering (Koshland and Storm)

Proteases

Protein degrading enzymes

- Recycling of amino acids
- Digestion of proteins in the diet
- Enzyme regulation

Chymotrypsin

Covalent Catalysis

A nucleophilic group in enzyme attacks the unreactive carbonyl group of the substrate forming a covalent bond.

- Serine 195

The three-dimensional structure of enzymes has helped to reveal their catalytic mechanism. Three amino acid residues are involved in catalysis: serine 195, histidine 57, aspartate 102. The unstable tetrahedral intermediate is stabilized by interactions with NH groups from protein at oxyanion hole.

Serine Proteases

Trypsin

- Cleaves at peptide bond after residue with long, positively charged side chain

Elastase

- Cleaves at the peptide bond after residue with small side chains

Specificity is depending upon residues in a pocket. Subtilisin from *Bacillus amyloliquefaciens* also uses catalytic triad and oxyanion hole. Carboxypeptidase II which has a very different structure also uses the catalytic triad and oxyanion hole.

Cysteine proteases – cysteine residue acts as a nucleophile.

- Papain

Aspartyl proteases – pair of aspartate residues act to enable water molecules to attack peptide bonds.

- Renin

Metalloproteases – active site contains a metal ion that activates water to act as a nucleophile.

- Metal is usually zinc
- Carboxypeptidase A

Protease Inhibitors

Several useful drugs

- Catopril – ACE inhibitor
  - Regulator of blood pressure
- Crixivan – HIV protease
  - AIDS treatment

Introduction

Cells of our body are accomplished by making different enzymes that can be regulated by other molecules that increase and decreases molecule's activity. The molecule increases the activity of reaction is called activators although molecules that decrease the activity of an enzyme are called inhibitors.

Regulation by Modification

- Proteolytic cleavage
- Covalent modification
- Protein-protein interaction

Allosteric Regulation

- Properties of allosteric enzymes (important).
- Sigmoid kinetics (what does Km mean in this case) (important).
- Positive and negative modulators (where do they act and how do they modify activity at constant substrate concentration) (most important).

Models of Allosteric Transitions (Important)

It has more than one site, where effector binding at one site induces a conformational change in the enzyme, altering its affinity for a substrate. An allosteric activator increases the enzyme rate of activity; an allosteric inhibitor decreases its activity.

Regulation Mechanism

Reversible, non-covalent binding of allosteric effectors. Covalent modification (phosphorylation, adenylation, etc.). Binding by separate regulatory proteins. Proteolytic activation (irreversible). In most cases, the first enzyme of the multi-reaction pathway (catabolism and anabolism) is a regulatory enzyme to avoid an unneeded accumulation of the intermediates.

Regulatory Enzymes are Generally More Complex than other Enzymes

- ✓ **Aspartate transcarbamoylase – the first step in CTP synthesis, converts Asp to *N*-carbamoyl Asp**

CO_2 + Gln + ATP → H_2N-(C=O)-OPO_3^{2-}
(carbamoyl phosphate)

- ✓ **Asp transcarbamoylase catalyzes the following reaction:**
- ✓ **Carbamoyl phosphate + Asp →*N*-carbamoylAspartate →→→ CTP (building block of DNA) CTP, the end product of the reaction, decreases the rate of enzyme activity – allosteric inhibitor.**
- ✓ **ATP increases the rate of enzyme activity – an allosteric activator.**

Normal metabolic control may be exerted in a variety of ways. Examples are:

1. **Proteolytic Cleavage of inactive Proenzymes to active enzymes**
   Pepsinogen pepsin + small peptide

in the gastrointestinal tract for protein digestion

1. **Coagulation cascade—a series of proenzymes are converted to active enzymes. The last step is**
   Fibrinogen Fibrin

Enzymes are biological catalysts (also known as biocatalysts) that speed up biochemical reactions in living organisms, and which can be extracted from cells and then used to catalyze a wide range of commercially important processes. For example, they have important roles in the production of sweetening agents and the modification of antibiotics, they are used in washing powders and various cleaning products, and they play a key role in analytical devices and assays that have clinical, forensic and environmental applications. The word 'enzyme' was first used by the German physiologist Wilhelm Kuhne in 1878 when he was describing the ability of yeast to produce alcohol from sugars, and it is derived from the Greek words *en* (meaning 'within') and *zume* (meaning 'yeast').

Immobilization of Enzymes

Enzyme immobilization can be defined as restricting the enzyme molecule in a distinct phase from the one in which substrates and products are present. The materials used for the immobilization of enzymes are called carrier matrices, these are usually inert polymers or inorganic materials.

Properties

- Low cost
- Inert
- Stable
- Physical strength
- Enhancement of enzyme specificity
- Regenerability
- Reduction in product inhibition
- Reduction in microbial contamination
- Non-specific adsorption

Methods for Immobilization

- Adsorption
- Covalcnt binding
- Entrapment
- Membrane confinement

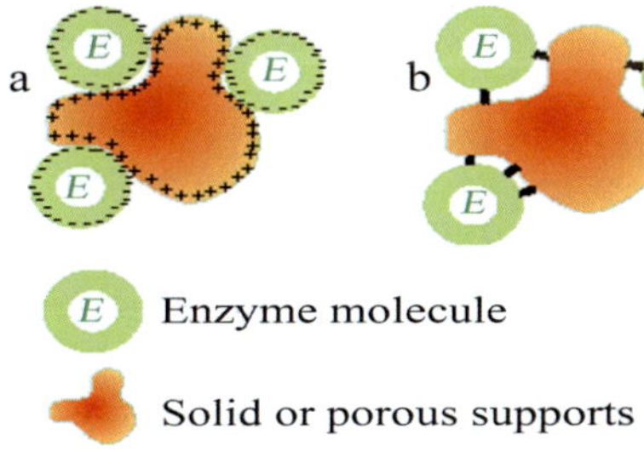

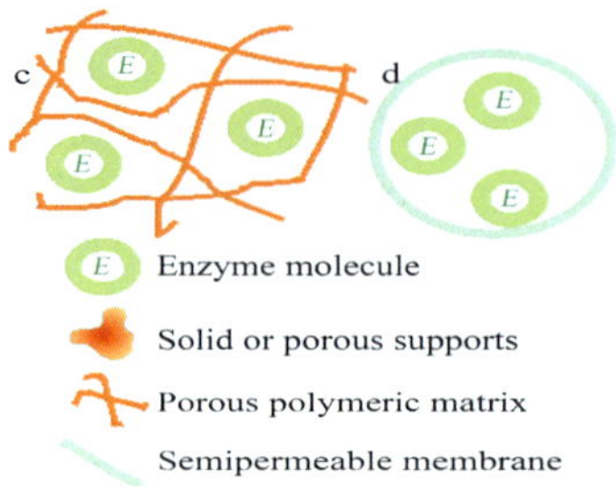

Adsorption

The enzyme is adhering to the carrier matrix due to hydrophobic effects or by forming salt links. Binding is strong but maybe weakened by substrate and , pH etc. The commonly used matrices are ion-exchange matrices, porous carbon, clays, hydrous metal oxides, polymeric aromatic resins, etc.

Covalent Binding

The enzyme is attached to the certain matrix by forming covalent bonds. The covalent binding is very strong. It generally occurs with the side chain of the enzyme. Lysine is very useful, as they are very reactive, usually exposed to the surface and very rarely occur at the active site of an enzyme and commonly used matrices: -Agarose, cellulose, polyacrylamides, etc.

Entrapment

The enzyme is entrapped within suitable gels or fibers, with or without covalent bonds and commonly used matrices: Cellulose acetate fibers, calcium alginate, etc.

Membrane Confinement

The enzyme molecules are confined to a semipermeable membrane. Thus in an aqueous phase of the membrane is free to move without allowing an enzyme to come out.

Advantages

Enzymes are costly and immobilization permits their repeated use. The product can be easily separated without any additional cost. Immobilized enzymes can be used in a non-aqueous system. It continuous production systems can be used. Thermo stability of some enzymes can be increased by immobilization. An enzyme can be used at a much higher concentration than a free enzyme.

Disadvantages

Some enzymes are very expensive. It sometimes affects the stability and activity of an enzyme. This approach can not be used when one of the substrates is insoluble. Some immobilization methods are restricting the diffusion of substrate.

Applications of Immobilized Enzymes

Enzyme	Substrate	Product
Glucose isomerase	D-glucose	HFCS
Invertase	Sucrose	Invert sugar
Lipase	Vegetable oil	Cocoa butter substitute
Lactase	Milk and whey	Lactose-free milk &whey
Nitrile hydratase	Acrylonitrile	Acrylamide
Raffinase	Raffinose	Raffinose free solution
Glucoamylase	Dextrins	D-glucose

Subjective Questions

1. Classify enzymes. Give examples for each class along with reaction and cofactors involved?
2. Define enzymes. Write an enzymatic reaction and properties of enzymes?
3. Explain how enzymes accelerate reactions in living organisms**?**
4. Write about nomenclature and EC number of enzymes?
5. Define the active site of the enzyme. Write its characteristics?
6. Define cofactors. Classify. Give examples. Write reaction for each example?
7. Explain competitive inhibition with examples?
8. Describe enzyme inhibition?
9. How competitive inhibitors are useful in clinical medicine?
10. Write a note on the influence of substrate concentration on enzymatic reaction?
11. Define Km. Write its importance?
12. Explain effect of the following on enzymatic reaction?

13. Hydrogen ion concentration. B. Temperature?
14. Write about competitive inhibitors used as chemotherapeutic agents?
15. Write the effect of substrate concentration on enzyme reaction?
16. What is meant by feedback inhibition? Give examples?
17. Write about cardiac enzymes and isoenzymes?
18. Define functional enzymes and non-functional enzymes. Give examples?

Objective Questions

1. A Holoenzyme is

(A) Functional unit
(B) Apo enzyme
(C) Coenzyme
(D) All of these

Answer: D

2. Gaucher's disease is due to the deficiency of the enzyme

(A) α-Fucosidase
(B) β-Galactosidase
(C) β-Glucosidase
(D) Sphingomyelinase

Answer: C

3. Neimann-Pick disease is due to the deficiency of the enzyme

(A) Hexosaminidase A and B
(B) Ceramidase
(C) Ceramide lactosidase
(D) Sphingomyelinase

Answer: D

4. Krabbe's disease is due to the deficiency of the enzyme

(A) Ceramide lactosidase
(B) Ceramidase
(C) β-Galactosidase
(D) GM1 β-Galactosidase

Answer: C

5. Fabry's disease is due to the deficiency of the enzyme

(A) Ceramide trihexosidase
(B) Galactocerebrosidase
(C) Phytanic acid oxidase
(D) Sphingomyelinase

Answer: A

6. Farber's disease is due to the deficiency of the enzyme

(A) α-Galactosidase
(B) Ceramidase
(C) β-Glucocerebrosidase
(D) Arylsulphatase A.

Answer: B

7. A synthetic nucleotide analog, used in organ transplantation as a suppressor of immunologic rejection of grafts is

(A) Theophylline
(B) Cytarabine
(C) 4-Hydroxypyrazolopyrimidine
(D) 6-Mercaptopurine

Answer: D

8. Example of an extracellular enzyme is

(A) Lactate dehydrogenase
(B) Cytochrome oxidase
(C) Pancreatic lipase
(D) Hexokinase

Answer: C

9. Enzymes, which are produced in an inactive form in the living cells, are called

(A) Papain
(B) Lysozymes
(C) Apoenzymes
(D) Proenzymes

Answer: D

10. An example of the ligase is

(A) Succinate thiokinase

(B) Alanine racemase

(C) Fumarase

(D) Aldolase

Answer: A

11. An example of lyases is

(A) Glutamine synthetase

(B) Fumarase

(C) Cholinesterase

(D) Amylase

Answer: B

12. Activation or inactivation of certain key regulatory enzymes is accomplished by covalent modification of the amino acid

(A) Tyrosine

(B) Phenylalanine

(C) Lysine

(D) Serine

Answer: D

13. The enzyme which can add water to a carbon-carbon double bond or remove water to create a double bond without breaking the bond is

(A) Hydratase

(B) Hydroxylase

(C) Hydrolase

(D) Esterase

Answer: A

14. Fischer's 'lock and key model of the enzyme action implies that

(A) The active site is complementary in shape to that of substance only after the interaction.

(B) The active site is complementary in shape to that of the substance

(C) Substrates change conformation prior to active site interaction

(D) The active site is flexible and adjusts to the substrate

Answer: B

15. From the Lineweaver-Burk plot of Michaelis-Menten equation, Km and Vmax can be determined when V is the reaction velocity at substrate concentration S, the X-axis experimental data are expressed as

(A) 1/V

(B) V

(C) 1/S

(D) S

Answer: C

16. A sigmoidal plot of substrate concentration ([S]) verses reaction velocity (V) may indicate (A) Michaelis-Menten kinetics

(B) Co-operative binding

(C) Competitive inhibition

(D) Non-competitive inhibition

Answer: B

17. The Km of the enzyme giving the kinetic data as below is

(A) –0.50

(B) –0.25

(C) +0.25

(D) +0.33

Answer: D

18. The kinetic effect of a purely competitive inhibitor of an enzyme

(A) Increases Km without affecting Vmax

(B) Decreases Km without affecting Vmax

(C) Increases Vmax without affecting Km

(D) Decreases Vmax without affecting Km

Answer: A

19. If curve X in the graph (below) represents no inhibition for the reaction of the enzyme with its substrates, the curve representing the competitive inhibition, of the same reaction is

(A) A

(B) B

(C) C

(D) D

Answer: A

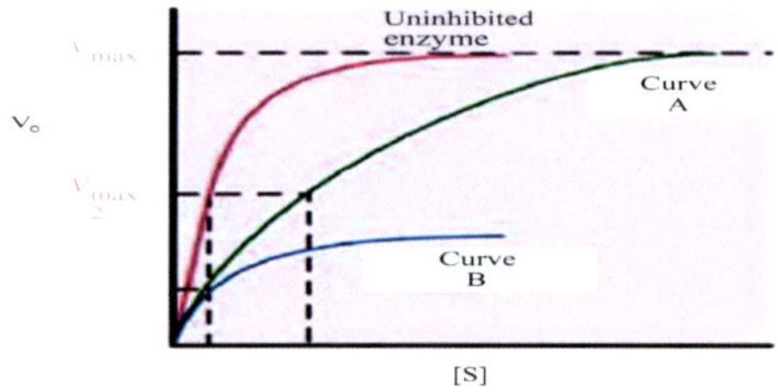

20. An inducer is absent in the type of enzyme

(A) Allosteric enzyme

(B) Constitutive enzyme

(C) Co-operative enzyme

(D) Isoenzymic enzyme

Answer: B

21. A demonstrable inducer is absent in

(A) Allosteric enzyme

(B) Constitutive enzyme

(C) Inhibited enzyme

(D) Co-operative enzyme

Answer: B

22. A coenzyme containing a nonaromatic hetero ring is

(A) ATP

(B) NAD

(C) FMN

(D) Biotin

Answer: D

23. A coenzyme containing an aromatic hetero ring is

(A) TPP

(B) Lipoic acid

(C) Coenzyme Q

(D) Biotin

Answer: A

24. Isoenzymes are

(A) Chemically, immunologically and electrophoretically different forms of an enzyme

(B) Different forms of an enzyme similar in all properties

(C) Catalyzing different reactions

(D) Having the same quaternary structures like the enzymes

Answer: A

25. Isoenzymes can be characterized by

(A) Proteins lacking enzymatic activity that are necessary for the activation of enzymes

(B) Proteolytic enzymes activated by hydrolysis

(C) Enzymes with identical primary structure

(D) Similar enzymes that catalyze different reaction

Answer: B

26. The isoenzymes of LDH

(A) Differ only in a single amino acid

(B) Differ in catalytic activity

(C) Exist in 5 forms depending on M and H monomer contents

(D) Occur as monomers

Answer: C

27. The normal value of CPK in serum varies between

(A) 4–60 IU/L

(B) 60–250 IU/L

(C) 4–17 IU/L

(D) > 350 IU/L

Answer: A

28. Factors affecting enzyme activity

(A) Concentration

(B) pH

(C) Temperature

(D) All of these

Answer: D

29. The normal serum GOT activity ranges from

(A) 3.0–15.0 IU/L

(B) 4.0–17.0 IU/L

(C) 4.0–60.0 IU/L

(D) 0.9–4.0 IU/L

Answer: B

30. The normal GPT activity ranges from

(A) 60.0–250.0 IU/L

(B) 4.0–17.0 IU/L

(C) 3.0–15.0 IU/L

(D) 0.1–14.0 IU/L

Answer: C

31. The normal serum acid phosphatase activity ranges from

(A) 5.0–13.0 KA units/100 ml

(B) 1.0–5.0 KA units/100 ml

(C) 13.0–18.0 KA units/100 ml

(D) 0.2–0.8 KA units/100 ml

Answer: B

32. The normal serum alkaline phosphatase activity ranges from

(A) 1.0–5.0 KA units/100 ml

(B) 5.0–13.0 KA units/100 ml

(C) 0.8–2.3 KA units/100 ml

(D) 13.0–21.0 KA units/100 ml

Answer: B

33. In early stages of myocardial ischemia the most sensitive indicator is the measurement of the activity of

(A) CPK

(B) SGPT

(C) SGOT

(D) LDH

Answer: A

34. Serum acid phosphatase level increases in

(A) Metastatic carcinoma of prostate

(B) Myocardial infarction

(C) Wilson's disease

(D) Liver diseases

Answer: A

35. Serum alkaline phosphatase level increases in

(A) Hypothyroidism

(B) Carcinoma of prostate

(C) Hyperparathyroidism

(D) Myocardial ischemia

Answer: C

36. Serum lipase level increases in

(A) Paget's disease

(B) Gaucher's disease

(C) Acute pancreatitis

(D) Diabetes mellitus

Answer: C

37. The pH optima for salivary analyse is

(A) 6.6–6.8

(B) 2.0–7.5

(C) 7.9

(D) 8.6

Answer: A

38. The pH optima for pancreatic analyse is

(A) 4.0

(B) 7.1

(C) 7.9

(D) 8.6

Answer: D

39. The pH optima for sucrase is

(A) 5.0–7.0
(B) 5.8–6.2
(C) 5.4–6.0
(D) 8.6

Answer: A

40. The pH optima for maltase is

(A) 1.0–2.0
(B) 5.2–6.0
(C) 5.8–6.2
(D) 5.4–6.0

Answer: C

41. The pH optima for lactase is

(A) 1.0-2.0
(B) 5.4–6.0
(C) 5.0–7.0
(D) 5.8–6.2

Answer: B

42. The substrate for amylase is

(A) Cane sugar
(B) Starch
(C) Lactose
(D) Ribose

Answer: B

43. The ion which activates salivary amylase activity is

(A) Chloride
(B) Bicarbonate
(C) Sodium
(D) Potassium

Answer: A

44. The pancreatic amylase activity is increased in the presence of

(A) Hydrochloric acid
(B) Bile salts

(C) Thiocyanate ions

(D) Calcium ions

Answer: B

45. A carbohydrate that cannot be digested in the human gut is

(A) Cellulose

(B) Starch

(C) Glycogen

(D) Maltose

Answer: A

46. The sugar absorbed by facilitated diffusion and requiring Na independent transporter is

(A) Glucose

(B) Fructose

(C) Galactose

(D) Ribose

Answer: B

47. In the intestine the rate of absorption is highest for

(A) Glucose and galactose

(B) Fructose and mannose

(C) Fructose and pentose

(D) Mannose and pentose

Answer: A

48. The hydrogen acceptor used in the pentose phosphate pathway is

(A) NAD

(B) NADP

(C) FAD

(D) FMN

Answer: B

49. The enzymes of the pentose phosphate pathway are found in the

(A) Cytosol

(B) Mitochondria

(C) Nucleus

(D) Endoplasmic reticulum

Answer: A

50. In the pentose phosphate pathway, D-ribulose5-phosphate is converted to D-ribose-5- phosphate by the enzyme

(A) Fumarase

(B) Ketoisomerase

(C) G-6-PD

(D) Epimerase

Answer: B

51. The transketolase enzyme in the pentose phosphate pathway requires the B vitamin.

(A) Pantothenic acid

(B) Thiamin

(C) Riboflavin

(D) Nicotinic acid

Answer: B

CHAPTER 4

CELL IMMUNOLOGY

1. INTRODUCTION TO IMMUNOLOGY, IMMUNE SYSTEMS-OVERVIEW, IMMUNITY WITH HISTORICAL PERSPECTIVE, MOLECULAR AND CELLULAR BASIS OF THE IMMUNE SYSTEM

Introduction

It is the science that is concerned with the immune response to foreign challenges or simply study of the body defense against infection. It addresses the questions such as how does the body defend itself against infection, when an infection does occur, how does the body eliminate the pathogens and how does long lasting immunity to many infectious diseases develop? The ability of an organism to resist infections by pathogens or a state of protection against foreign organisms or substances is called immunity. The array of cells, tissues and organs which carry out this activity constitute the immune system. The immune system is a complex process and it is derived into two categories such as innate (native) and adaptive (acquired) immunity.

Innate Immunity: It is a general, non-specific immune response that presents in all individuals at all times.

Adaptive immunity: It is highly specific to the particular pathogen that induces it.

Historical Events in Immunology

Emil von Behring and Kitasato (1890): Serum from vaccinated animals was protective (diptheria).

Metchnikoff (1880): Cell based Immunity and Merrill Chase (1940)-Transfer of WBC (immunity to tuberculosis).

1796	Edward Jenner (smallpox)
1881	Loius Pasteur (vaccines)
1884	Elie Metchnikoff (phagocytes)
1890	Emil von Behring* (antibodies)
1895	Jules Bordet* (complement)
1906	August Wasserman (syphilis)
1959	Rodney Porter Gerald Edelman* (antibodies)
1960	F McFarlane Burnet* (tolerance)
1975	Cesar Milstein*(monoclonal Ab)
1987	Susumu Tonegawa* (genetics)
1996	Peter Doherty

Immunity

Immunity (immunis- Latin-exempt, state of protection from infectious diseases). Immunity is the body's ability to resist or eliminate potentially harmful foreign materials or abnormal cells consists of the following activities such as defense against invading pathogens (viruses & bacteria), removal of 'worn-out' cells (e.g., old RBCs) & tissue debris (e.g., from injury or disease), Identification & destruction of abnormal or mutant cells (primary defense against cancer), Rejection of 'foreign' cells (e.g., organ transplant), Inappropriate responses: Allergies - response to normally harmless substances Autoimmune diseases figure 1.1 and 1.2.

Immune System

- Innate (Nonspecific): 1° line of defense (Pre-existing and Non-specific).
- Adaptive (Specific) 2° line of defense (Responsive and Specific).

Innate Immunity vs Adaptive Immunity

Cells of the immune system

- Phagocytic cells
- Nonspecific
- Specific

The first line of defense ("not connected with activity of immune system directly) Epithelial surfaces External physical barriers which restrict the entry of pathogen. Examples: skin, mucous membranes and hair, blood clotting. The intact skin and mucous membrane covering the body gives considerable protection against bacteria on it. They provide mechanical barriers. They also provide bactericidal secretions (for the second line). Movement out of the body. Examples: respiratory tract ciliated epithelium (moves particles trapped in mucus toward the pharynx), bleeding, urinating, vomiting, diarrhea, tears, coryza. Chemical defense. Examples: the wax in the ears, the sticky mucus lines of nose and other openings, the acid in the stomach. Normal microflora Intact normal flora allows the protection of the colonized surfaces in a healthy host body. It has been discussed in several contexts. There are three main ways that the normal flora protects the surfaces: 1. Competition with non-indigenous species for binding (colonization) sites. The normal flora is highly-adapted to the tissues of their host; 2. Nonspecific antagonism to non-indigenous species. The normal flora produces a variety of metabolites and ends products that inhibit other microorganisms. These include fatty acids (lactate, propionate, etc.) and peroxides; 3. Specific antagonism to non-indigenous species. Members of the normal flora may produce highly specific proteins called bacteriocins, which kill or inhibit other (usually closely-related) species of bacteria. A reactivity of cells and tissues This concept is about specific innate immunity. A reactivity of cells and tissues is connected with the absence of cells receptors for pathogen absorption or toxins fixation. A pathogen should attach to some cells surfaces by specific receptors of the host. Such interaction is necessary for the next stage - multiplying of microbe. It means the disease is not possible without attachment. In another situation, it can be because the cellular machinery has no need of the microorganism. In both examples the question is that the pathogen of one species can be not pathogenic for another one.

The second line of defense (connected with the activity of immune system directly) Innate non-specific immune defenses provide a rapid local response to pathogen after it has entered host (tissue factors). If the barrier of the body is overcome by the microorganisms, a number of factors play its role in normal tissue and body fluid. Tissue factors may be divided into: cellular and humoral factors. Examples: Cellular factors - leukocytes (phagocytes:

macrophages and neutrophils), basophils, eosinophils; lymphatic system. Humoral factors - organic substances present in the extracellular fluids of the body - complement cascade responses, interferon responses, lysozyme in tears, swear and saliva; lysines, interleukins, antiviral inhibitors, C-reactive protein, cationic and hydrolytic proteins, lactoferrins, etc. in blood. Fever and inflammatory responses are provided by the interaction of the cellular and humoral factors. The third line of defence Antigen-specific immune responses specifically target and attack invaders that go through first two lines of defence. Examples: antibodies and lymphocytes".

Lymphocytes

- B lymphocytes
- T lymphocytes
- T helper
- T cytotoxic

Lymphocytes Lymphocytes (lymphoid cells) are one of the principal leukocytes. "There are three major types of lymphocytes B, NK and T. The T- and B-cells in the lymph nodes are confined to discrete zones. They have distinctive surface markers and functions. B cells are called so because they are connected with a special organ "bursa" in birds. Mature B cells (Plasma cells) are the producers of antibodies or immunoglobulins - Ig (classes A, M, G, E and D). B-memory cells are capable of reacting in the future. It is the essence of humoral adaptive immunity. Additionally, B cells present antigens (they are also classified as professional antigen-presenting cells (APCs). NK (normal killer cells) carry out the extracellular killing of the target cell – tumor cells or virus-infected cells with antibody-independent cytolytic granule mediated or cytokine-induced NK activated cell apoptosis. It is possible for antibody-dependent cytotoxicity resulting in NK activation, release of cytolytic granules and consequent cell apoptosis. Antibody plays here recognizing role while the NK-cell is affecting participants. NK cells are representatives of innate cellular immunity that distinguishes them from other lymphocytes. T cells are called so because they are predominantly produced in the thymus. T cells are factors of cellular adaptive immunity. The main of them are T cytotoxic (CD8 T cells) and CD4 T cells. T cytotoxic cells are the effector cells that recognize specific antigen on the surface of the target cell and after attachment kill it extracellularly. The special proteins with enzymatic activity (perforins) are released by the cell for that. Perforins destroy the cellular membrane. Myeloid cells First subgroup of this is granulocytic cells (granulocytes). There are neutrophils, basophils, mast cells and eosinophils. All these types of cells work on the innate cellular area of defense. Their activity can be enhanced by feedback with the adaptive level of immunity (by opsonins and IgE). Neutrophils (microphages) are cells with phagocytic activity. They are present in blood and tissues. They migrate in injured zones very fast and take part in the inflammation reactions. Basophils and mast cells release histamine, provide allergic response, are antiparasitic. Eosinophils are involved in parasite defense and allergic response". The second subgroup is monocytic cells (macrophages). This group of cells defends at the innate cellular level. In contradistinction to neutrophils, macrophages are directly connected with adaptive level of defense (with CD4 T cells). The macrophages are the main antigen-presenting cells (APCs). It has a place in process of cooperation between immune cells in the process of forming an adaptive immune response.

Two Arms of the Adaptive Immune System

- Humoral immunity
- Antibodies
- Cell-mediated immunity
- T-cells

Immunological Memory

It is more rapid, intense and prolonged.

Chemical Mediators

Extracellular

- Cytokines
- Complement

Intracellular

- Antigens and Antibodies

Antigens

- Chemical makeup
- Types of antigens

Antibodies

- Chemical makeup
- Roles

Innate Immune System: Components of Blood

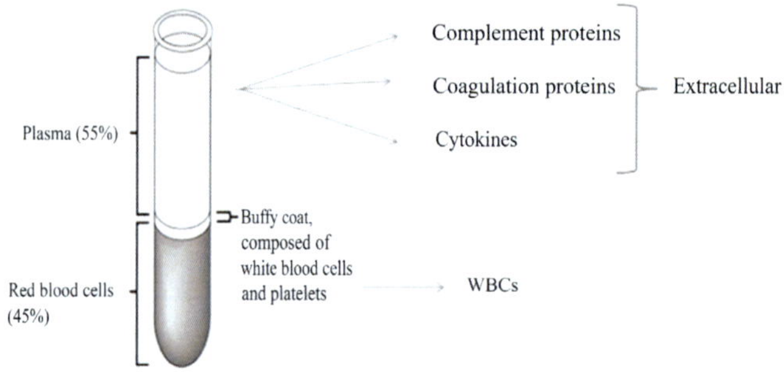

Figure 1.1: Innate immune system: components of the blood.

White Blood Cells (WBCs)

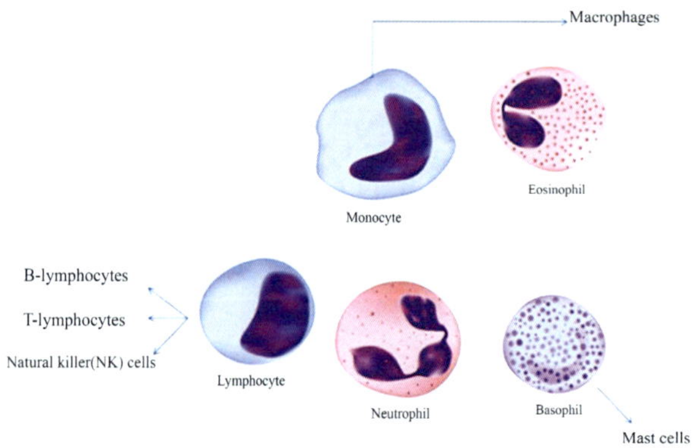

Figure 1.2: Structure of white blood cells.

Neutrophils in Innate Immune Response

- Most abundant WBCs (~50-60%)
- Efficient phagocytes
- Most important cells of the innate immune system

Phagocytosis

- Phago = to eat
- Cyte = cell
- WBCs (eg. Neutrophils) – find, eat and digest microbes.

Monocytes

- Monocytes (~5% of WBCs).
- Migrate into the tissues and become macrophages.

Macrophages ("Big eaters")

- Phagocytosis of microbes in tissue (neutrophils are present only in the blood).
- Antigen presentation.

Natural Killer Cells

- It is not B-lymphocytes / T-lymphocytes.
- Important part of the innate immune system.
- Kill virus /bacteria-infected cells (Intracellular pathogens).
- Kills cancer cells.

Toll-like Receptors (TLRs)

- Transmembrane proteins.
- Present on macrophages / few other cells.
- Conserved across vertebrates.
- Important part of the innate immune system.
- They look out for microbes (or their components).
- They bind to the microbes (or their components).
- They trigger a cascade of events to kill or protect against pathogens.
- They are innate immune sensors.

Innate Response – Internal Defenses – Cellular (WBCs)

- Come into play when the external defenses are breached.
- Neutrophils.

- Monocytes /macrophages.
- NK cells.
- TLRs.

Cytokines

- Small proteins secreted by cells of the immune system.
- Affect the behavior of other cells signalling molecules.
- Key players in innate and acquired immunity.

Cytokines

- Cells of the immune system.
- Neutrophils – when they encounter a pathogen.
- Macrophages – when they encounter a pathogen.
- TLRs – bind to microbe/components of a microbe.
- NK cells – on encountering a microbe infected cell /tumor cell.
- Lymphocytes – when they are activated.

Cytokines

- Interferons
- Interleukins
- Tumor necrosis factor (TNF)

Interferons (IFN)

- The signaling proteins produced by virus-infected monocytes and lymphocytes.
- It is secreted proteins – Key anti-viral proteins.
- "Interfere" with virus replication.
- It is warning the neighboring cells that a virus is around.
- If we did not have IFNs – most of us may die of influenza virus infection.

Interleukins

- Interleukins – 1-37.
- It is not stored inside cells.

- It is quickly synthesized and secreted in response to infection.
- The key modulators of behavior of immune cells.
- It is mostly secreted by T-lymphocytes & macrophages.

Complement

- A large number of distinct plasma proteins that react with one another (C1 thro' C9).
- The complement can bind to microbes and coat the microbes.
- It is essential part of innate immune response.
- It enhances adaptive immune resposne (taught later).

Coagulation Proteins

- The mechanism to stop bleeding after injury to blood vessels.
- The complex pathway involves.
- Platelets.
- The coagulation factors.
- Vitamin K

Internal Defenses

Cellular

- Neutrophils
- Monocytes /macrophages
- NK cells
- TLRs

Extracellular

- Cytokines
- Complement
- Coagulation

Inflammation

Complex biological process by which body responds to pathogens and irritants. Associated with swelling of tissue. Key player in innate immune response figure 1.3.

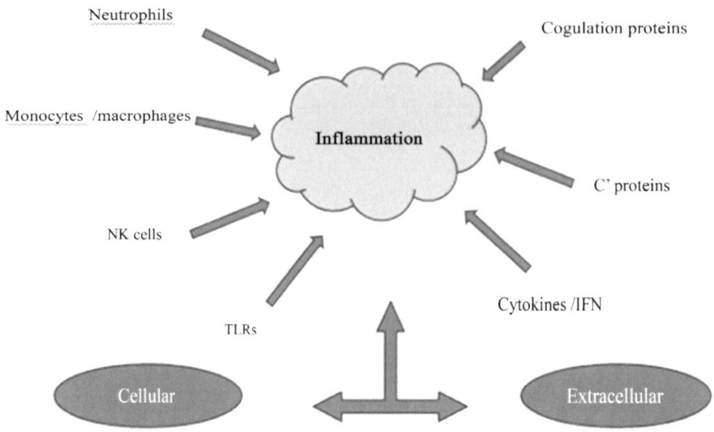

Figure 1.3: Structure of inflammation.

Role of Inflammation in Innate Immunity

- It is the initiation of phagocytosis – killing of pathogen.
- It is limiting the spread of infection.
- It stimulates the adaptive immune response.
- It initiates tissue repair.

Chronic Inflammation = Tissue Damage

- Chronic inflammation - macrophages in the injured tissue.
- Macrophages release toxins (including reactive oxygen species or ROS) that injure tissues.
- Chronic inflammation is almost always accompanied by tissue destruction.
- Immunogens and antigens

Immunogen/antigen: a substance that elicits an immune response [i.e. a humoral (antibody response) or cell-mediated immune response]. **Epitope:** the portion of an antigen that is recognized and bound by an antibody (Ab) or a T-cell receptor (TCR) epitope = antigenic determinant figure 1.4.

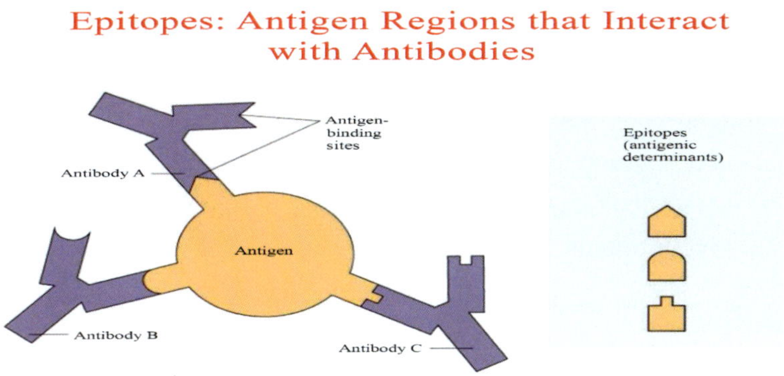

Figure 1.4: Antibody-antigen reaction.

Epitope: the portion of an antigen that is recognized and bound by an Ab or a T Cell receptor. One protein may have multiple antigenic determinants.

B-cell epitopes – recognized by B-cells

T-cell epitopes – recognized by T cells

Immunogenicity

Immunogenicity: is the ability to induce a humoral antibody and/or cell-mediated immune response.

- Weak immunogens
- Strong immunogens

Determines Immunogenicity

Foreignness: It is essential for immunogenicity self-responsive immune cells are eliminated during lymphocyte development.

Chemical composition: Proteins > nucleic acids / polysaccharides / lipids

Structure: Primary /secondary /tertiary structures play a role

Physical form: Particulate> Soluble

Host factors affecting immunogenicity

- Differences across the species (interspecies).
- Differences within a species (intraspecies): Responders/non-responders to vaccine and differences in disease severity in epidemics.

Isoantigens

- **Isoantigens:** Antigens present in some but not all members of a species.
- Blood group antigens – basis of blood grouping.
- MHC (major histocompatibility complex)- cell surface glycoproteins

Autoantigens

Autoantigens are substances capable of immunizing the host from which they are obtained. Self-antigens are ordinarily non-antigenic. Modifications of self-antigens are capable of eliciting an immune response.

Haptens are small molecules that are non-immunogenic, thus could never induce an immune response by themselves.

Antibody

It is produced by Plasma cells (B-lymphocytes producing Ab). It is an essential part of adaptive immunity. It specifically binds a unique antigenic epitope (also called an antigenic determinant). It possesses antigen-binding sites. The members of the class of proteins are called immunoglobulins **figure 1.5**.

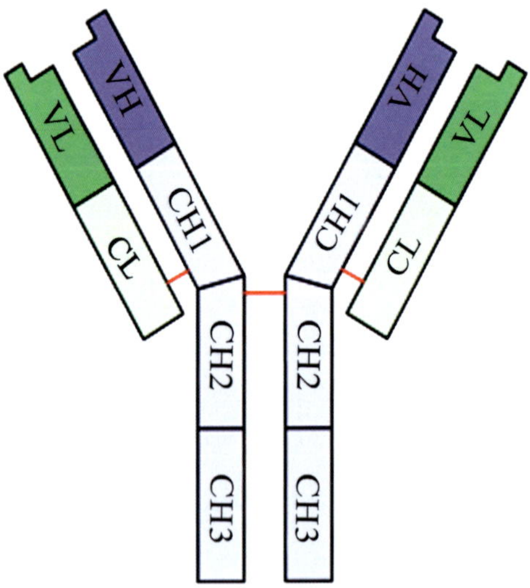

Figure 1.5: Structure of immunoglobulin.

It has 2 identical heavy chains and 2 identical light chains. Each heavy chain has a constant and a variable region. Each light chain has a constant and a variable region figure 1.6.

Antibody: Structure and Function

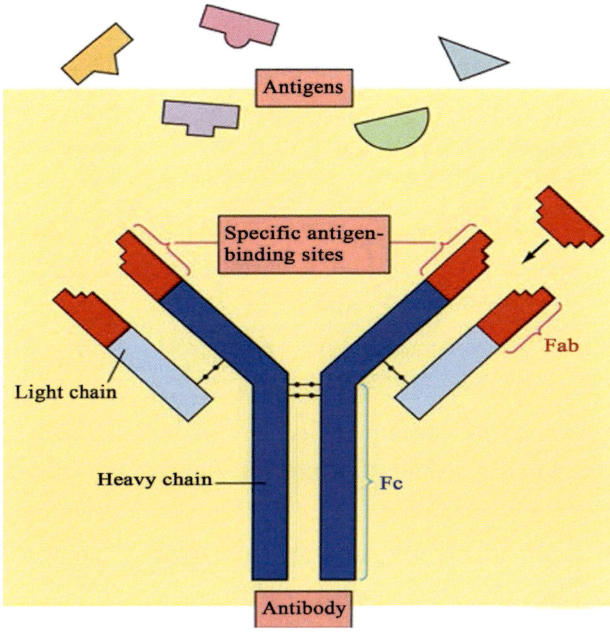

Figure 1.6: Antibody-specific antigen-binding sites.

Antibodies can also be divided into two regions based on their functions. The tip of the "Y", consisting of one constant and one variable domain from each heavy and light chain of the antibody. It contains the region that binds the antigen. It is for this reason that we call it the Fab region (fragment and antigen-binding). The base of the "Y", composed of two heavy chains, plays a role in modulating immune cell activity. This region is called the Fc (Fragment, crystallizable) region. The Fc region can bind to cell receptors, complement proteins, as well as to other immune molecules, thereby initiating different physiological effects including opsonization, cell lysis, and degranulation.

Fab Region

It has the variable region of the antibody. Tip of the antibody and it binds the antigen. The specificity of antigenbinding are determined by V_H and V_L.

Antibody: Fc

It has Fc region, constant region and base of the antibody and it can bind cell receptors and complement proteins.

Antibodies exist in two forms

Antibodies occur in 2 forms soluble Ag: secreted in blood and tissue and membrane-bound Ag: found on the surface of B-cell, also known as a B-cell receptor (BCR).

Antigen-presenting Cells

The process of antigen and "present" it to T cells, dendritic cells, phagocytic and motile. The resident in epithelial surfaces and lymphoid tissues and mature in response to molecular patterns.

Organs of the Immune System

Primary immune organs: immune cells develop and mature. Bone marrow: Differentiation, B cell maturation. Thymus: T cell maturation and Bursa of Fabricius: B cell maturation.

Secondary immune organs: immune cells become activated such as the spleen, Lymph nodes and Mucosa-associated lymphoid tissue.

Communicate Immune Cells

- Soluble mediators
- Extracellular
- Receptor/ligand-specific
- Paracrine vs endocrine
- Signal transduction
- Direct contact:
- Antigen presentation
- Cytotoxicity

Soluble Mediators of the immune response

Signaling Molecules

Innate: Cytokines, complement

Adaptive: Cytokines, antibodies, chemokines, interleukins

Cytotoxic Molecules

Innate: Enzymes, reactive oxygen mediators

Adaptive: Enzymes, reactive oxygen mediators

Cytokines

It has low molecular weight proteins that stimulate or inhibit the differentiation, proliferation, or function of cells. It is multifunctional, overlapping functions. It is produced by immune and non-immune cells. One cell may produce many cytokines. One cytokine may be produced by many cells. The interleukins: a subset of cytokines that communicate between leukocytes and named consecutively by il- designation.

Chemokines

A subset of cytokines (chemotactic cytokines). A family of structurally related cytokines that selectively induce chemotaxis and activation of leukocytes". The old nomenclature: by function mostly. It has new nomenclature: according to N-terminal amino acid sequence such as

CXC= one amino acid separating two N-terminal cysteine residues

CX3C = three amino acids

CC= cysteines adjacent

C= only one cysteine present

Antibody

Antibodies are made up of: 2 light chains (identical) ~25 kDa and 2 heavy chains (identical) ~50 kDa. each light chain is bound to a heavy chain by disulfide (Heavy and light). Heavy chain bound to heavy chain (Heavy-Heavy). The first 100 a/an of the amino-terminal vary of both heavy and light chain are variable. It is referred to as v_l, v_h, c_h and c_l. DRS (complementarity determining regions) are what bind Ags and the remaining regions are very similar within the same class figure 1.7.

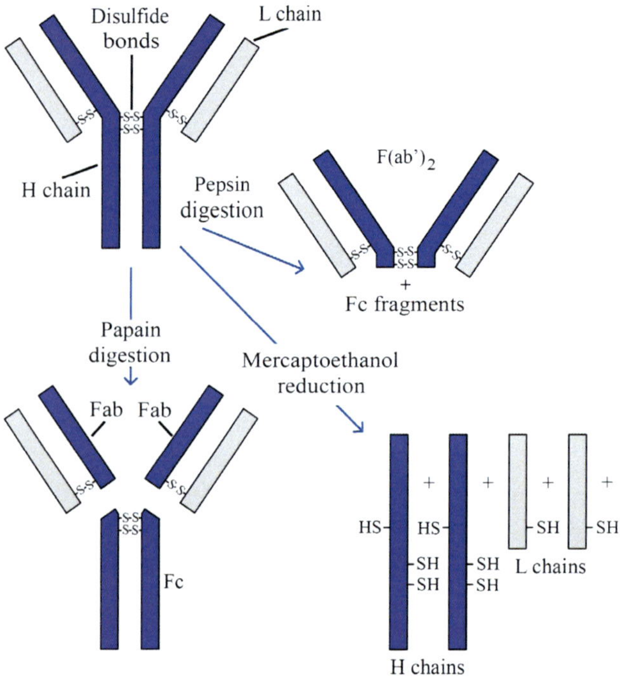

Figure 1.7: Two light chains and two heavy chains (Light chain bound to a heavy chain by a disulfide bond).

Repeating domains of ~110 a/a, intrachain disulfide bonds within each domain. Heavy chains. 1 V_H and either 3 or 4 C_H (C_H1, C_H2, C_H3, C_H4). Light chains: 1 V_L and 1 C_L.

Hinge Region

It is rich in proline residues (flexible). It has a Hinge found in IgG, IgA and IgD. It has proline residues that are target for proteolytic digestion (papain and pepsin). It is rich in cysteine residues (disulfide bonds) and IgM and IgE lack hinge regions. They instead have an extra C_H4 Domain.

Enzymatic Digestion of Antibodies

The digestion with papain yields 3 fragments and 2 identical Fab and 1 Fc. Fab is antigen binding. Fc is found to crystallize in cold storage. It has pepsin digestion F(ab`)2. No Fc recovery, digested entirely and mercaptoethanol reduction (eliminates disulfide bonds) and alkylation showed.

Sequencing of Heavy Chains

sequencing of several immunoglobulins revealed 100-110 amino terminus, highly variable (V). Five basic sequence patterns α,γ, δ, ε, μ (IgA, IgG, IgD, IgE and IgM). The above classes are called isotypes. Each class can have either κ or λ light chains. Minor differences led to sub-classes for IgA and IgG, IgA1, IgGA2 and IgG1, IgG2, IgG3, IgG4.

Monoclonal Antibodies

Immunizing an animal with antigen and has multiple clones are generated, good for *in vivo* for clinical diagnosis, research, one clone that reacts to single epitope is preferred. The solution by Kohler and Milstein fuse a myeloma cell (cancerous) with a normal plasma cell. The resulting clones can be cultured indefinitely and produce an antibody recognizing one epitope figure 1.8.

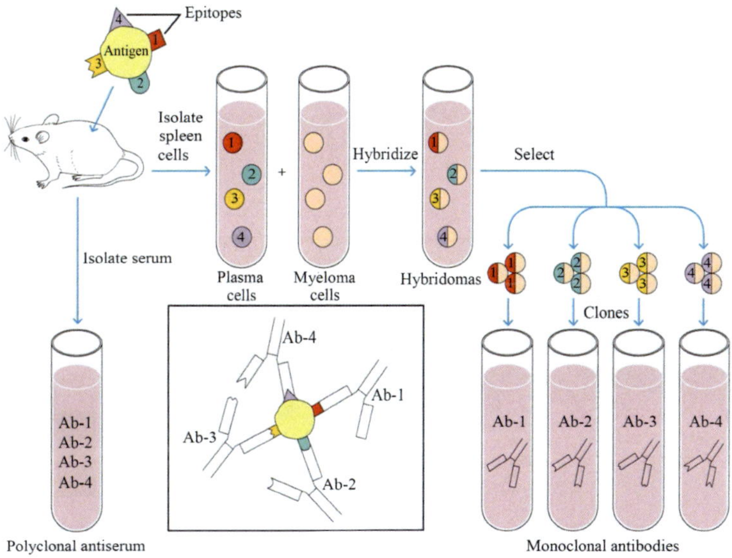

Figure 1.8: Monoclonal antibody.

Monoclonal Antibody Applications

Diagnostic Tests

Antibodies are capable to detect tiny amounts (pg/mL) of molecules eg. pregnancy hormones and diagnostic imaging: mAbs that recognize tumor antigens are radiolabeled with iodine I-131. Immunotoxins: mAbs conjugated with toxins and mAbs to clear pathogens.

B-cell Receptor

BCR is an antibody on the surface of cell mIg. It has a very short cytoplasmic tail and cannot transduce signal heterodimeric molecule Ig-α/Ig-β transduces (long cytoplasmic tail) figure 1.9.

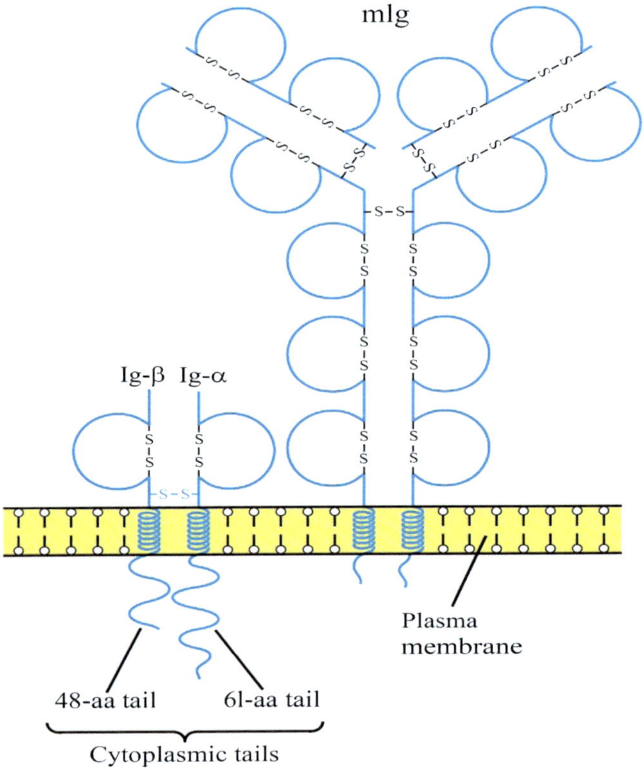

Figure 1.9: Heterodimeric molecule Ig-α/Ig-β transduces.

Antibody Classes and Biological Activities

IgG

- Most abundant immunoglobin- 80% of serum Ig~10mg/mL.
- IgG1,2,3,4 (decreasing serum concentration).
- IgG1, IgG3 and IgG4 cross the placenta.
- IgG3 Most effective complement activator.
- IgG1 and IgG3 High affinity for FcR on phagocytic cells and good for opsonization.

IgM

- Five and ten percent of serum immunoglobulin 1.5mg/Ml.
- mIgM (also IgD) expressed on B-cells as BCR.
- Its pentameric version is secreted.
- First Ig of the primary immune response.
- It has high valence Ig (10 theoretical) and 5 empirical.
- It is more efficient than IgG in complement activation.

IgA

- Ten to 15 percent of serum IgG.
- It is the predominant Ig in secretions.
- It is in milk, saliva, tears and mucus.
- It has 5-15 g of IgA released in secretions.
- The serum mainly contains monomeric and polymers.
- The secretions, as a dimer or tetramer+J-chain polyptetide+secretory component (Poly IgR).

IgE

- It has a very low serum concentration, 0.3μg/mL.
- It participates in immediate hypersensitivities reactions eg. asthma, anaphylaxis, hives

- It s binds mast cells and blood basophils.
- Its binding causes degranulation (Histamine Release).

IgD

- It is expressed on the B-cell surface.
- IgM and IgD are expressed on the B-cell surface.

Antibodies act as Immunogens

Antigenic determinants on Abs fall into three categories.

- Isotypic
- Allotypic
- Idiotypic

Isotypic

- It has constant region of Ab.
- If you inject Ab in a different species anti-isotype is generated.
- If within the same species and no anti-isotype.

Allotype

Even though same isotypes within one species small differences (1-4 a/a) arise in different individuals (form of polymorphism). If injected with such Ab you generate anti-allotype Ab eg. during pregnancy and blood transfusion.

Idiotype

Unique V_H and V_L binds antigen but can also behave as antigenic determinants. If you inject a monoclonal antibody into a genetically identical recipient, then anti-idiotypic antibodies are generated. No anti-isotypic and no anti-allotypic Abs will be generated figure 1.10.

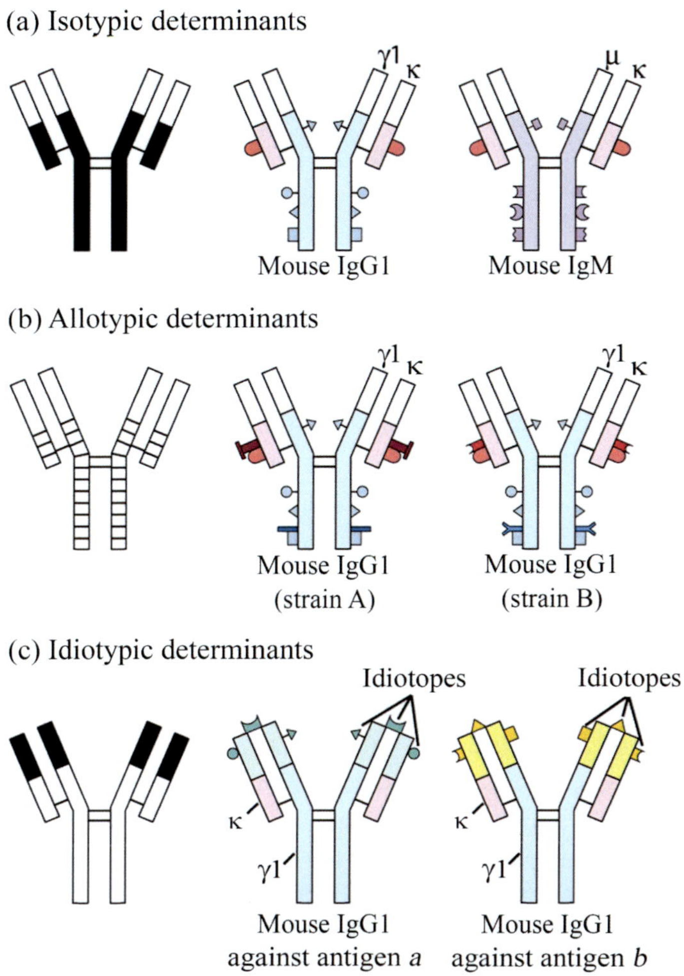

Figure 1.10: Antibodies act as immunogens.

Antigen Recognition

B cells can recognise antigens via their surface Ig molecules. T cells can only recognise antigen in association with a Major Histocompatibility Complex (MHC) molecule figure 1.11.

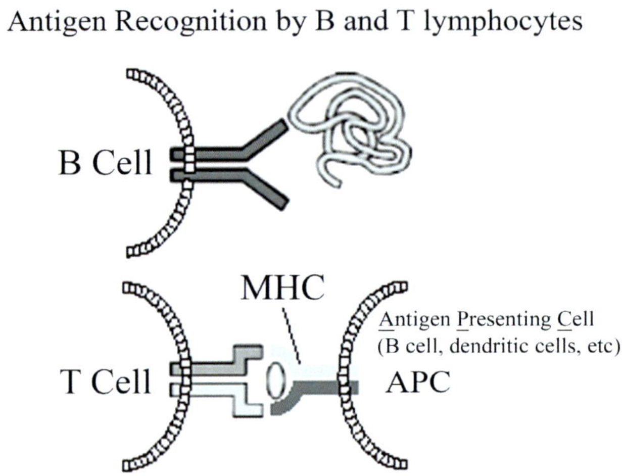

Figure 1.11: Antigen recognition by B and T lymphocytes.

Immunoglobin Fold

V and C domains share the basic Ig fold. It has differences between the two domains. C domain is built of seven b-strands arranged so that four strands form one sheet and three strands form a second sheet. The strands are closely packed together and joined by a single disulfide bond. It has three most of the invariant residues of the constant domain are in the sheets. It has the overall structure of the V domain very similar but there are nine strands instead of seven. The two additional strands harbor CDR2.

Complementarity determining Regions

Six loops of the VH (H1, H2 and H3) and VL (L1, L2 and L3) domains create a great variety of surfaces. It has deep binding cavities: such as those seen in some antibody-hapten complexes. It has wide pockets: seen in certain antibody-peptide complexes. It has flat surfaces: seen in antibody-protein interactions and H3 is the most variable of the loops and in all crystallographically solved antibody-antigen complexes makes several contacts with antigen.

Antigen: Antibody Complex

Antibodies bind to antigens by recognizing a large surface, and through surface complementarity. Thus, these complexes have a very high affinity for each other.

Weak Forces vs High Affinity

The interaction between an antigen and antibody can be very strong, and yet all of the forces involved are considered to be relatively weak. How can weak hydrogen bonds, electrostatic attractions, hydrophobic forces, and van der Waals contacts lead to a high affinity. Its contact between antigen and antibody occurs over a wide surface area, allowing multiple weak interactions that give a strong affinity. The hydrogen bonds join the antibody and antigen over a wide surface area. Other weak forces, including van der Waals forces, electrostatic attractions and hydrophobic forces, add to the strength and specificity of antibody/antigen binding.

Antibody-hapten Complex

Haptens, having a limited total surface area, deeply embed themselves into the VL/VH dimer interface. Hapten binding antibodies frequently show a deep central cavity, long CDR L1 loops and a CDR H3 loop with an "open" conformation, allowing the hapten to bind as much as 80% of its total surface in the interaction.

Protein Antibody Complex

In contrast, proteins preferentially to a relatively flat binding surface. In a "closed" CDR H3conformation, the CDR H3 loop packs down onto the central cavity, and the protein antigen binds on top of it.

Effector response is mediated via Ig-FcR Complex Formation

Antibodies not only must recognize antigens but also must invoke responses – effector functions – that will remove the antigen and kill the pathogen. The variable regions of the antibody are the sole agents of binding to antigen. The heavy chain constant region (C_H) is responsible for interactions with other proteins (e.g. complement), cells (elements of the innate immune system), and tissues that result in the effector functions of the humoral response. FcR recognize the Fc portion of antibodies not antigens.

The Fc-Fc Receptor Complex

FcR plays important role in antibody-mediated immune responses. Ig and FcR binding activates effector functions. Fc receptor interacts with the CH2 and CH3 domains of immunoglobulins.

Immune Recognition: MHC and TCR Interactions

T cells can only recognize antigen in association with a major histocompatibility complex (MHC) molecule.

T Cells

T cells display TCR as their antigen recognition protein. When stimulated they become Cytotoxic.

Helper T Cells

Secrete cytokines that recruit other cells of the IS. TCR's only recognize short peptides.

MHC & T Cells

T cells have a requirement to recognize both the antigen and the MHC molecule. This is because the molecular structure of the MHC-antigen complex is arranged so that some of the polymorphic amino acids of the MHC molecule are in direct contact with the TCR Therefore T cell recognition of antigen is said to be MHC 'restricted'.

Antigen Processing and Presentation

There is the fragmentation of protein into peptides and association of peptide with an MHC molecule and transport to the cell surface for expression. It has different cellular pathways for the association of peptides with MHC class I and class II molecules.

MHC & Antigens

MHC Class I

It has present endogenously derived peptides. These can be either self or derived from viruses because MHC Class I is present on all cells any cell can interact with T cells if infected by a virus. It is detected on all nucleated cells. It has very highly polymorphic. It has a tight fit for peptides of only about 9 aa. It has consisted of an a-chain of 3 domains associated with b-2 microglobulin.

MHC Class II

It has present exogenous antigen that has been phagocytosed and processed eg. bacteria. This is performed by professional antigen-presenting cells eg. macrophages. It is seen only on the 'professional antigen processing cells' e.g macrophage slightly less polymorphic accepts peptides of up to fifteen aa acids.

MHC Class I and Class II Proteins

Class I

Alpha Chain

It has three external domains, one transmembrane, one cytoplasmic tail and encoded in MHC.

Beta-2 Microglobulin

It has one external domain, no transmembrane, no cytoplasmic tail and is not encoded in MHC.

Class II

Alpha Chain

It has two external domains, one transmembrane, one cytoplasmic tail and encoded in MHC.

Beta Chain

It has two external domains, one transmembrane, one cytoplasmic tail and encoded in MHC.

Peptide Binding by Major Histocompatibility Complex (MHC) Antigen-presenting Proteins

MHC I

It has peptides of intracellular origin. It has peptides nine-ten residues long. It has deep pockets that bind peptide sidechains. It has deep pockets that bind peptide N- and C-termini.

MHC II

It has peptides of extracellular origin. It has peptides fifteen residues or longer. It has shallow pockets that bind peptide sidechains and peptide termini free. It has H-bonds to peptide backbone.

MHC Polymorphism

It has both class I and class II genes that are highly polymorphic. It has most polymorphic residues of class I are in the alpha 1 and alpha 2 domains. It has most polymorphic residues of Class II are in the alpha 1 and beta 1 domains.

2. CELLS INVOLVED IN IMMUNITY AND HUMORAL IMMUNE RESPONSE

Introduction

Antibodies are protein molecules secreted by immune system cells that help the body fight invaders. Each antibody-producing cell makes a single, unique antibody which will be in one of the 5 major classes – IgA, IgD, IgE, IgG, and IgM. These classes are determined by the constant region of heavy chains of the antibody, and have different immunologic targets and mechanisms of action. They can also bind their targets with different degrees of strength, depending on the fit between the antibody and its target figure 2.1.

Nature of Ag/Ab Reactions: It has **the** lock and key concept

- Non-covalent bonds
- Hydrogen bonds
- Electrostatic bonds
- Van der Waal forces
- Hydrophobic bonds
- Multiple bonds
- Reversible

Strength of the reaction between a single antigenic determinant and a single Ab combining site.

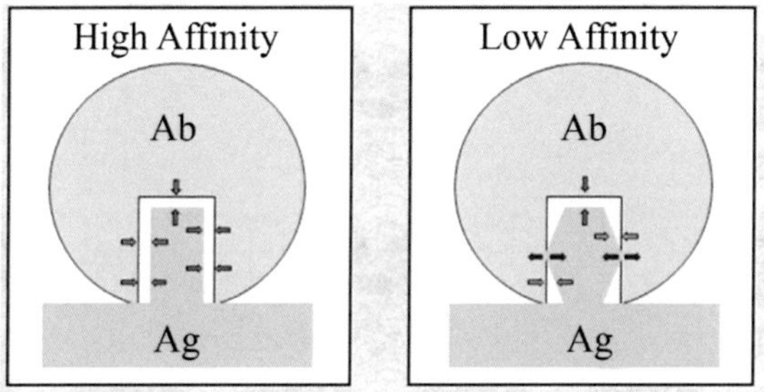

Figure 2.1: High and low affinity of antibody-antigen reactions.

Affinity = ∑ attractive and repulsive forces

Calculation of Affinity

Ag + Ab   Ag-Ab

Applying the Law of Mass Action

$$K_{eq} = \frac{[Ag\text{-}Ab]}{[Ag] \times [Ab]}$$

Avidity

The overall strength of binding between an Ag with many determinants and multivalent Abs specificity. The ability of an individual antibody combining site to react with only one antigenic determinant. The ability of a population of antibody molecules to react with only one antigen figure 2.2.

Cross Reactivity

The ability of an individual Ab combining site to react with more than one antigenic determinant. The ability of a population of Ab molecules to react with more than one Ag.

Factors affecting the Measurement of Ag/Ab Reactions

- Affinity.
- Avidity.
- Ag: Ab ratio.
- The physical form of Ag.

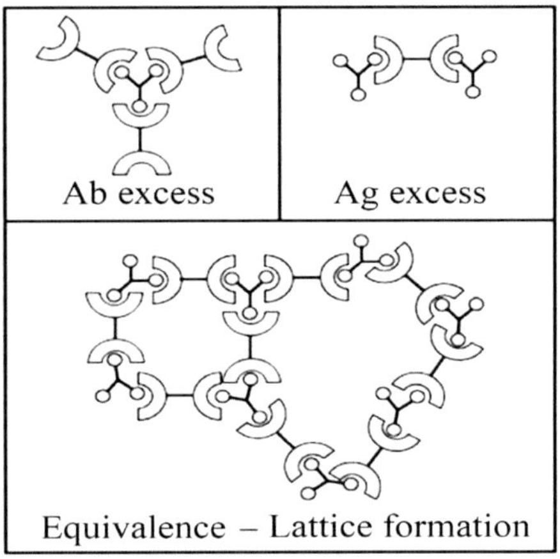

Figure 2.2: Antibody-anitgen reactioin.

Tests Based on Ag/Ab Reactions

All tests based on Ag/Ab reactions will have to depend on lattice formation or they will have to utilize ways to detect small immune complexes. All tests based on Ag/Ab reactions can be used to detect either Ag or Ab

Agglutination/Hemagglutination

Definition - tests that have as their endpoint the agglutination of a particulate antigen: Agglutinin/hemagglutinin

Qualitative agglutination test: Ag or Ab

Quantitative agglutination test: Titer and Prozone

Precipitation Tests

- Radial immunodiffusion (Mancini)
- Immunoelectrophoresis
- Radioimmuoassays (RIA)
- Enzyme-Linked Immunosorbent Assays (ELISA)

Cells and Organs of the Immune System

Cells – Leukocytes

- Lymphocytes – specialized for adaptive immunity.
- Granulocytes/Agranulocytes – function in accessory roles in adaptive immunity.

Immune Organs – 2 Major Groups

- 1° lymphoid organs – where lymphocytes develop and mature.
- 2° lymphoid organs – where lymphocytes interact with an Ag.

Hematopoiesis

It begins with hematopoietic stem cells (IISC): It is less in the bone marrow; difficult to culture and pluripotent able to produce RBC's, WBC's, megakaryocytes, HSC differentiates to become either a) myeloid progenitor cell b) Lymphoid progenitor cell figure 2.3.

Myeloid →RBC's and WBC's and dendritic cells

Lymphoid →B and T cells and dendritic cells

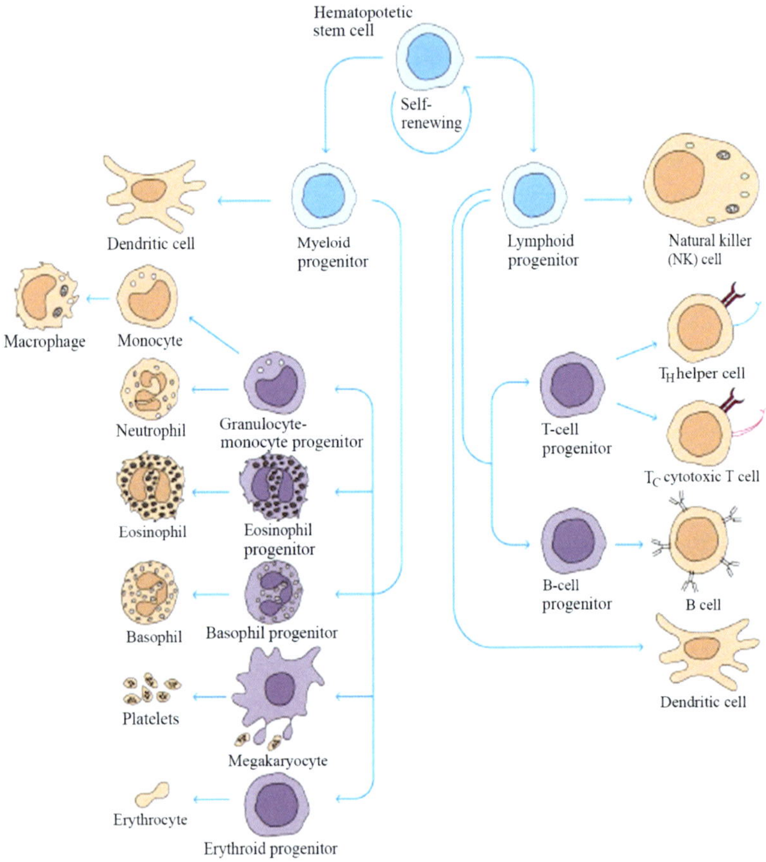

Figure 2.3: Hematopoiesis process.

Hematopoietic Influences

Stromal cells influence diff of HSC by providing hematopoietic-inducing microenvironment (HIM) with colony stimulating factor (CSF) and -erythropoietin (EPO).

Cytokine influences > stimulated by activated MØ and T cells

Genetic Regulation

Hematopoietic homeostasis is required to maintain certain cells balance between cells removed by cell death and those produced for WBC's: a human must produce ~3.7 x 10^{11}/day. Each WBC has a specific life span + dies by apoptosis; neutrophils live only a few days.

Apoptosis – Programmed Cell Death

Cells die in this process + later phago'd decrease in cytoplasmic volume; apoptotic bodies, clumping/break up of DNA, phago'd by MØ → blocks release of cytoplasmic contents no local inflammatory response. It is influenced by gene expression and chief among the genes is bcl2 - inhibits cell death figure 2.4.

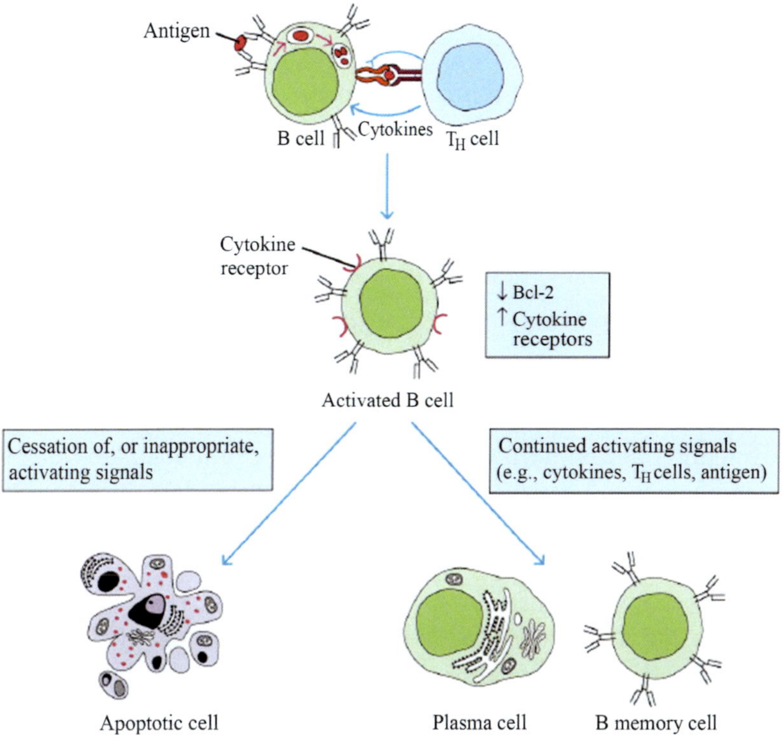

Figure 2.4: Apoptosis.

Pathogens, agents that cause disease, infect a wide range of animals, including humans. The immune system recognizes "foreign bodies and responds with the production of immune cells and proteins. All animals have innate immunity, a defense active immediately upon infection. Vertebrates also have adaptive immunity. Innate immunity is present before any exposure to pathogens and is effective from the time of birth. It involves nonspecific responses to pathogens. Innate immunity consists of external barriers plus internal cellular and chemical defenses. Adaptive immunity, or acquired immunity, develops after exposure to agents such as microbes, toxins, or other foreign substances.

It involves a very specific response to pathogens. The immune system recognizes bacteria and fungi by structures on their cell walls. An immune response varies with the class of pathogen encountered. The immune system of mammals is the best understood of the vertebrates. Innate defenses include barrier defenses, phagocytosis, antimicrobial peptides. Additional defenses are unique to vertebrates: natural killer cells, interferons, and the inflammatory response. The barrier defenses include the skin and mucous membranes of the respiratory, urinary, and reproductive tracts. Mucus traps and allows for the removal of microbes. Many body fluids including saliva, mucus, and tears are hostile to many microbes. The low pH of skin and the digestive system prevents the growth of many bacteria. Pathogens entering the mammalian body are subject to phagocytosis. Phagocytic cells recognize groups of pathogens by TLRs, Toll-like receptors. A white blood cell engulfs a microbe, then fuses with a lysosome to destroy the microbe. There are different types of phagocytic cells

- **Neutrophils** engulf and destroy pathogens
- **Macrophages** are found throughout the body
- **Dendritic cells** stimulate development of adaptive immunity

Eosinophils discharge destructive enzymes cellular innate defenses in vertebrates also involve natural killer cells. These circulate through the body and detect abnormal cells. They release chemicals leading to cell death, inhibiting the spread of virally infected or cancerous cells. Many cellular innate defenses involve the lymphatic system. The adaptive response relies on two types of lymphocytes, or white blood cells. Lymphocytes that mature in the thymus above the heart are called T cells, and those that mature in bone marrow are called B cells. Antigens are substances that can elicit a response from a B or T cell. Exposure to the pathogen activates B and T cells with antigen receptors specific for parts of that pathogen. The small accessible part of an antigen that binds to an antigen receptor is called an epitope. B cells and T cells have receptor proteins that can bind to foreign molecules. Each individual lymphocyte is specialized to recognize a specific type of molecule

Antigen Recognition by B Cells and Antibodies

Each B cell antigen receptor is a Y-shaped molecule with two identical heavy chains and two identical light chains. The constant regions of the chains vary little among B cells, whereas the variable regions differ greatly. The variable regions provide antigen specificity. The binding of a B cell antigen receptor to an antigen is an early step in B cell activation. This gives rise to cells that secrete a soluble form of the protein called an antibody or immunoglobulin (Ig). Secreted antibodies are similar to B cell receptors but lack transmembrane regions that anchor receptors in the plasma membrane. Immunological memory is responsible for long-term protections against diseases, due to either a prior infection or vaccination. The first exposure to a specific antigen represents the primary immune response. During this time, selected B and T cells give rise to their effector forms. In the secondary immune response, memory cells facilitate a faster, more efficient response. Acquired immunity has two branches: the humoral immune response and the cell-mediated immune response. In the humoral immune response, antibodies help neutralize or eliminate toxins and pathogens in the blood and lymph. In the cell-mediated immune response, specialized T cells destroy affected host cells. Antibodies do not kill pathogens; instead they mark pathogens for destruction. In neutralization, antibodies bind to viral surface proteins preventing infection of a host cell. Antibodies may also bind to toxins in body fluids and prevent them from entering body cells. In opsonization, antibodies bind to antigens on bacteria creating a target for macrophages or neutrophils, triggering phagocytosis". Antigen-antibody complexes may bind to a complement protein—which triggers a cascade of complement protein activation. Ultimately a membrane attack complex forms a pore in the membrane of the foreign cell, leading to its lysis. B cells can express five different forms (or classes) of immunoglobulin (Ig) with similar antigen-binding specificity but different heavy chain C regions.

- IgD: Membrane bound
- IgM: First soluble class produced
- IgG: Second soluble class; most abundant
- IgA and IgE: Remaining soluble classes

Effector Functions of Antibodies

Opsonization: is the promotion of phagocytosis of antigens by macrophages and neutrophils. Protein molecules are called Fc receptors (FcR), which bind the constant region of an antibody, are present on the surfaces of phagocytes.

Activation of Complement: IgM (most effective) and most subclasses of IgG can activate the classical complement pathway. Antibody-dependent cell-mediated cytotoxicity (ADCC) – The linking of antibody bound to target cells (virus-infected cells, or some tumor cells) with FcR of natural killer cells (NK cells), neutrophils, macrophages, or eosinophils can result in the killing of the target cell. Neutralization of viruses and bacteria can prevent attachment to cell receptors and neutralization of toxins.

3. CELL-MEDIATED IMMUNE RESPONSE, IMMUNITY TO INFECTIOUS DISEASES, VACCINES AND HYBRIDOMA TECHNOLOGY

Introduction

In the cell-mediated immune response, specialized T cells destroy affected host cells. A type of T cell called a "helper T cell triggers both the humoral and cell-mediated immune responses. Signals from helper T cells initiate the production of antibodies that neutralize pathogens and activate T cells that kill infected cells. Antigen-presenting cells have class I and class II MHC molecules on their surfaces. Class II MHC molecules are the basis upon which antigen-presenting cells are recognized. Antigen receptors on the surface of helper T cells bind to the antigen and the class II MHC molecule; then signals are exchanged between the two cells. The helper T cell is activated, proliferates, and forms a clone of helper T cells, which then activate the appropriate B cells.

Cytotoxic T Cells: a Response to Infected Cells

Cytotoxic T cells are the effector cells in the cell-mediated immune response. Cytotoxic T cells recognize fragments of foreign proteins produced by infected cells and possess an accessory protein that binds to class I MHC molecules. The activated cytotoxic T cell secretes proteins that disrupt the membranes of target cells and trigger apoptosis figure 3.1.

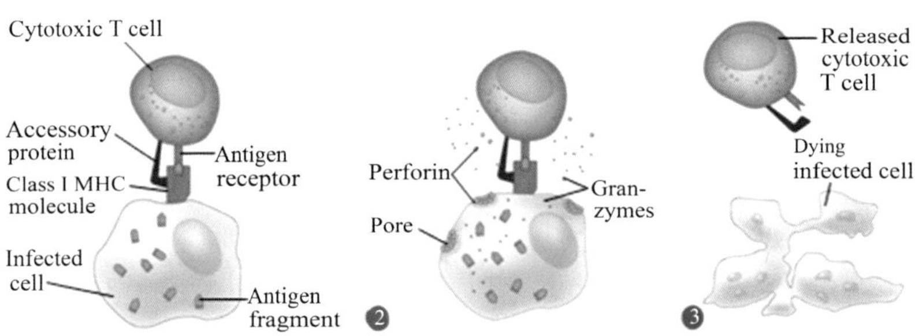

Figure 3.1: Cytotoxic T cells.

Effector T Cells

The membrane bound molecules eg. FASL (CD8+). The soluble molecules CTLs secrete cytotoxins perforins and granzymes also secret IFN-γ and TNF-β.

Cytotoxic T Cells

CTLs recognize cells that have been infected viruses or transformed to a tumor. CTL activation is divided into 2 phases. Activation and differentiation of naïve CTL. Effector recognizes Class I MHC/peptide and destroys target naïve CTLs cannot kill. It is referred to as CTL-Ps (precursors). It has 3 signals needed for activation. Ag specific signal thru TCR/MHC I+Ag. The co-stimulatory signal CD28(CTL)/B7 (APC). IL-2 signaling inducing proliferation (CTL-P does not express IL-2 R). IL-2 is provided by T_{H1} or CTL-P itself. IL-2R is expressed only after activation. T_{H1} and CTL collaborate to induce effector CTL. IL-2 seems to be crucial (knock out data). Lack of IL-2R in CTL-P ensures Ag specificity. Upon clearance of antigen CTLs undergo apoptosis. T_{H1} induces up-regulation of Co-stimulatory molecules on APCs enhancing APC/CTL co-stimulation.

Conjugate Formation

LFA-1 (CTL) binds ICAMs (Target), LFA-1 changes to high avidity if Ag Is Recognized, Activated LFA-1 persists for 5-10 mins, Membrane attack, Requires Ca^{2+} and energy. Granules release Perforins (65 kDa) and Granzymes (serine proteases) at the junctional space. Perforins polymerize forming cylindrical pores (5-20 nm), Ca^{2+} is needed". Granzymes enter target cell. Granzyme B can enter through the mannose-6-phosphate receptor in a vesicle. Lytic enzymes (MΦ, Neutrophils, Eosinophils, NK).

Cytokines

It has low molecular weight proteins, <30kD, high affinity for receptors and active in picomole amounts. Stimulus>cytokine-producing cell>cytokine>target cell>receptor>biological effect(s).

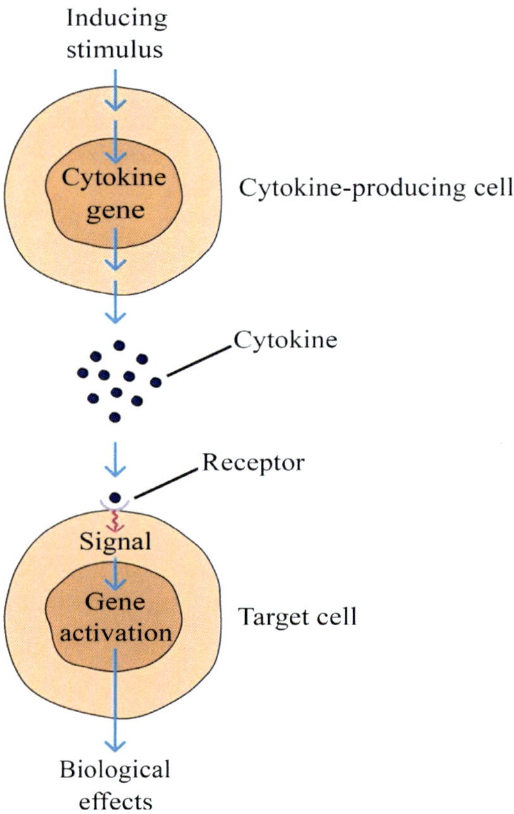

Figure 3.2: Chain of cytokine action.

Types of Cytokines

- Lymphokines
- Function e.g., Chemokines

Intercellular action e.g., Interleukins

Action of Cytokines

Autocrine: Affects the generating cell (self)

Paracrine: Affects cells in the immediate vicinity

Endocrine: affects cells remote from the secreting cell figure 3.3.

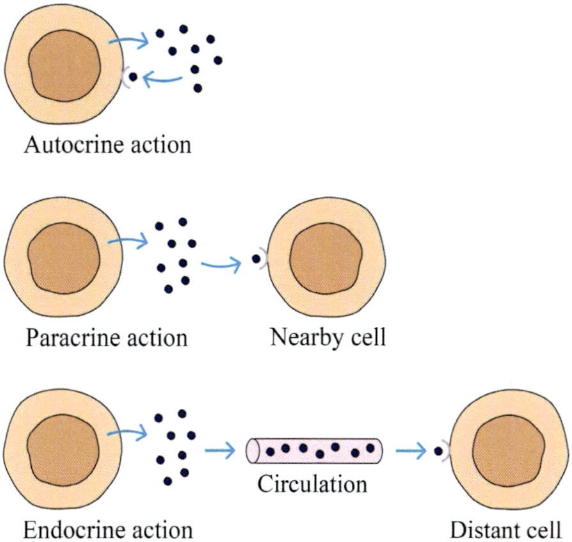

Figure 3.3: Action of cytokines.

Pleiotropy: affects multiple cell types.

Redundancy: multiple cytokines affect cells of the same type.

Synergy: cytokines acting in concert on the same cell.

Antagonism: competing actions.

Cascading: cytokines acting sequentially figure 3.4.

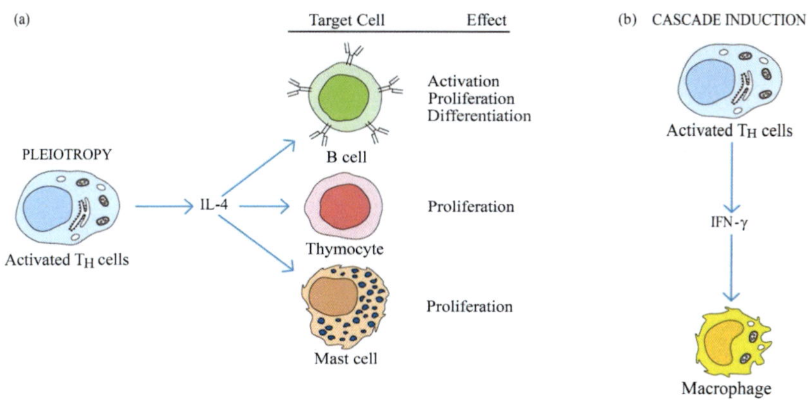

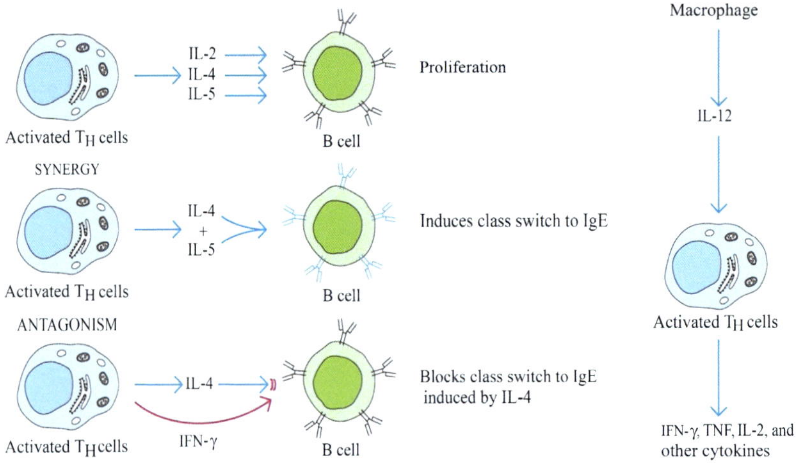

Figure 3.4: Cascade induction.

Cytokine-generating Cells

- Innate immunity
- Macrophages
- Endothelial cells
- Fibroblasts
- Adaptive immunity
- T lymphocytes
- Macrophages
- NK cells

Cytokine Receptors

- Ig super family
- Class I- hematopoietic
- Class II-interferon
- TNF
- Chemokine
- It has multimeric receptors
- It is common signal-transducing subunits
- It has unique high-affinity subunits

- It has high-affinity subunits associated with activation of the target cell

Signal Transduction

- It is initiated by cytokine binding.
- The activates of JAK (Janus kinase) signaling.
- Phosphorylation of tyrosine process.
- The binding of STAT (signal transducers and activators of transcription) pathway.
- The translocation of STAT to cell's DNA.
- The transcription of specific target genes.

Role of Cytokines in Hematopoiesis

- SCF (Stem cell factor)
- GM-CSF (Granulocyte-macrophage colony-stimulating factor)
- IL-3 (Interleukin 3)
- IL-5 (Interleukin 5)
- IL-7 (Interleukin 7)
- IL-11(Interleukin 11)

Cytokines in the Immune Response

Innate immune response

- IL 1-(Macrophage)-fever, capillary effects
- IL 6-(Macrophage)-adaptive immunity via B cells
- IL 12(Macrophage)-adaptive immunity via T helper cells
- TNF (Macrophage)-capillary effects, activates neutrophils
- IFN alpha (Macrophage)-multiple effects
- IFN beta (Fibroblasts)-multiple effects

Adaptive Immune Response

- IL 2- (T cells)-multiple effects).
- IL 4- (T cells & mast cells)-T cell differentiation, IgE production.
- TGF beta – (T cells, macrophages)-inhibits the adaptive immune response.

- IFN gamma- (T cells, NK cells)-Macrophage activation.

Subsets of T Helper Cells

- T helper 1
- IgG production/Complement fixation
- Macrophage activation
- DTH
- CTL production
- T helper 2
- IgE production
- Eosinophils

Major Histocompatibility Complex, aka MHC

A complex of genes encoding "cell-surface molecules that are required for antigen presentation to T-cell.

The basis of self / not self distinction, presentation of processed antigen. MHC-I (on nearly all nucleated cells)-Class I and MHC-II (on B-cells, macrophages, dendritic cells) Class II.

Major Histocompatibility Complex

The cluster of genes found in all mammals, its products play role in discriminating self/non-self and participant in both humoral and cell-mediated immunity. MHCs act as antigen presenting structures, in human MHC is found on chromosome 6 referred to as HLA complex, in mice mhc is found on chromosome 17 referred to as H-2 complex.

MHC-I

Heavy chain (alpha) and "microglobulin" (beta two). Heavy is 45 kilodaltons, has three domains + a transmembrane component (40 aa) + a cyto- plasmic tail (30 aa). The three alpha domains are called: $\alpha1$, $\alpha2$, & $\alpha3$, $\alpha1$ and $\alpha2$ interact to present processed Ag. The process of Ag is optimally a nonomer, Microglobulin (12 kDa) associates non-covalently with $\alpha3$, Microglobulin and $\alpha3$ are part of the immunoglobulin superfamily, microglobulin is the only member of the superfamily that does not have a component linking it to a membrane.

MHC-II

An alpha and beta chain, 33 kDA and 28 kDa, respecitvely. The chains are non-covalently associated. Each chain has two domains like α1-β1 interact to present processed Ag. It processed Ag is optimally 13-18 aaα2 & β2 are part of immunoglobulin super family figure 3.5.

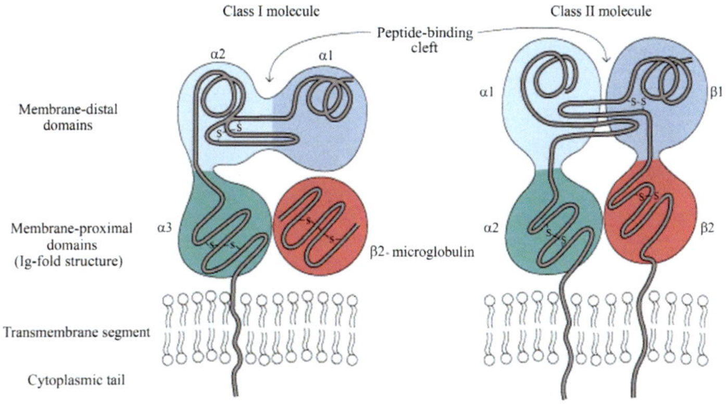

Figure 3.5: Major histocompatibility complex.

Genes of MHC Organized in 3 Classes

Class I MHC Genes

- Glycoproteins expressed on all nucleated cells.
- Major function to present processed Ags to T_C.

Class II MHC Genes

Glycoproteins expressed on MΦ, B-cells, DCs.

Major function to present processed Ags to $T_{H.}$

Class III MHC Genes

The products include secreted proteins that have immune functions eg. complement system, inflammatory molecules. Class I MHC genes found in regions A, B and C in humans (K and D in mice). Class II MHC genes found in regions DR, DP and DQ (IA and IE in mice). Class I and Class II MHC share structural features. Both involved in APC". Class III MHCs have no structural similarity to Class I and II eg. TNF, heat shock proteins and complement components figure 3.6.

Mouse H-2 complex

Complex	H-2						
MHC class	I	II		III		I	
Region	K	IA	IE	S		D	
Gene products	H-2K	IA $\alpha\beta$	IE $\alpha\beta$	C′ proteins	TNF-α TNF-β	H-2D	H-2L

Human HLA complex

Complex	HLA							
MHC class	II			III		I		
Region	DP	DQ	DR	C4, C2, BF		B	C	A
Gene products	DP $\alpha\beta$	DQ $\alpha\beta$	DR $\alpha\beta$	C′ proteins	TNF-α TNF-β	HLA-B	HLA-C	HLA-A

Figure 3.6: Classes of MHC genes.

MHC Expression

Expression is regulated by many cytokines: IFNα, IFNβ, IFNγ and TNF increase MHC expression. Transcription factors that increase MHC gene expression: CIITA (transactivator), RFX (transactivator). Some viruses decrease MHC Expression: CMV, HBV, Ad12 reduction of MHC may allow for immune system evasion.

Lymphocytes are derived in bone marrow from stem cells include both T cells and B cells. "T cells: stored & mature in thymus-migrate throughout the body killer cells. It performs the lysis (infected cells), cell-mediated immune response, Helper cells are enhancing T killer or B cell activity. Suppressor cells reduce/suppress immune activity and it may help prevent auto immune disease.

B-cells are stored and mature in the spleen, secrete highly specific Ab to bind foreign substance (antigen: Ag), form Ab-Ag complex, responsible for humoral response, perform antigen processing and presentation, differentiate into plasma cells (large Ab secretion). Neutrophils are found throughout body, in blood phagocytosis of Ab-Ag CX. Macrophages are produced throughout the body, blood, lymphatics phagocytose non-specifically (non Ab coated Ag). Phagocytose entails Ab-Ag CX. These have a large number of lysosomes (degradative enzymes). It performs Ag processing and presentation. It present Ag to T helper cell. It secretes lymphokines/ cytokines to stimulate T helper.

Natural killer cells in blood throughout the body, destroy cancer cells and are stimulated by interferons.

Severe Combined Immunodeficiency Disease (SCID)

It is generally caused by a defect of a single gene needed for T-cell and B-cell function and the subject exhibits no cell-mediated response**, the** subject cannot make antibodies**.** It is about 25% of cases defective gene for the enzyme adenosine deaminase (required for purine breakdown) are involved"**.**

Adaptive Immune Responses

Despite innate and adaptive immune responses to pathogens, infectious diseases have plagued human populations throughout history still cause millions of deaths per year. There are four main types of pathogens that cause infectious diseases such as

- Viruses
- Bacteria
- Protozoa
- Helminths

One of the four main pathogens are responsible for infectious diseases which responsible for smallpox, the common cold, chickenpox, influenza, shingles, herpes, polio, rabies, Ebola, hanta fever, and AIDS. Several specific immune effector & nonspecific defense mechanisms. Viruses act to subvert one or more of these mechanisms to prolong their survival.

Viruses: Structure & Function

Viruses depend on host cells for reproduction. The viruses remain metabolically inert outside of host cells. They exist as a protein coat or capsid, sometimes enclosed within a membrane. The capsid encloses either DNA or RNA which codes for the virus elements.

Innate Immunity Cells are Express Receptors that Recognize Bacterial Molecules

- TLR-4 (Recognizes LPS, gram negative bacteria)
- TLR-2 (Recognizes PGN, gram positive bacteria)
- TLR-3,7,9 (viral nucleic acids)

Viruses induce the production of interferons (IFN-α, IFN-β and IFNγ): Interferons produce an anti-viral state. The pathogens use several tricks to avoid the immune system.

Induce a specific type of immunity that allows it to survive. tThere is a continuous variation of antigens eg. HIV

It expresses antigens that resemble our own genes. Infections are prevalent in developing countries 4,000,000 of deaths are due to respiratory infections.

Viral Infections

The immune response against a virus is primarily mediated through interferons. The double-stranded RNA induces the production of IFN, It is the main producer of IFNα and IFNβ are pDCs. TLR-3 (dsRNA); TLR-7 (ssRNA). Interferons produce an anti-viral state. A state that inhibits viral replication. A state that inhibits viral infection.

IFN Signaling

Upon ligand binding, the following events occur "expression of oligo-adenylate synthetase [2-5 (A) synthetase, activation of RNAase, degradation of viral RNA. In addition, activation of dsRNA-dependent Protein Kinase (PKR), Phosphorylates eIF-2, protein synthesis is inhibited and viral replication is inhibited. IFN signaling activates NK cells and start eliminating virally infected cells.

Antibody Protection against Viruses

Antibodies bind to viral surface antigens protect against re-infection and huge amounts of secretory IgA in lumen block viral attachment. Viral entry into cells is mediated through receptors.

Vaccines

The term was coined by Pasteur to honor Jenner's work. Vaccines are cost-effective uses of our immune system, dramatic reduction of Diptheria, Measles, Mumps, Pertussis, Polio and Tetanus. No more naturally acquired cases of smallpox. **For other diseases:** TB, malaria and HIV increase the safety of the present vaccine, lower cost, and dissemination. Vaccine development is long and laden with: side effects, exacerbation of disease state and acquisition of disease state.

Immunization

Passive immunity & short-term protection, **Transient** protection (remedy for current problem), involves the transfer of preformed Ig: between ♀ and fetus (trans-placental) & colostrum or by injection. It is given to those exposed to botulism, tetanus, diptheria, hepatitis, rabies, measles, snake/insect bites. It provides immediate protection to healthcare/travelers. It has passive immunity does not activate and produces no memory.

Immunization Risks of passive immunity

If an antibody is produced in another spp, the human recipient can produce a response. In some → IgE production vs isotypic Ab systemic mast cell cause anaphylaxis. In others IgM or IgG vs isotype complement activation Type III reaction. If human gammaglobulin results can be less severe.

Active Immunization and Long-term Protection

It promotes protective immunity and memory is achieved by: Natural infection and artificial introduction of whole cells/antigens". The immune system plays an active role in stimulating Ag-reactive T/B cells, Immunizations have played a sig role in the decrease of infectious disease in children.

Childhood Vaccines

There are 7 major vaccines such as:

1. HepB
2. DTaP
3. IPV
4. MMR
5. Hib

6. Var
7. PCV
8. Types

It is usually requiring repeated boosters. It has predominantly humoral immune response risks of containing the active pathogen.

Recombinant Vector Vaccines

Genes are encoding significant Ags from pathogens may be transferred to attenuated viruses/bacteria, include: vaccinia, polio, adenoviruses, salmonella, BCG strain of *M. bovis*. Other vectors may prove to be safer.

Monoclonal Antibodies

Monoclonal antibodies are monospecific antibodies that are "identical because they are produced by one type of immune cell (a single parent cell). Antibodies are produced from a single clone of B cells. It has produced by fusing a B cell secreting the desired antibody with a myeloma cell capable of growing indefinitely in tissue culture. Monoclonal antibodies all have identical antigen-binding sites. They all bind to the same epitope with the same affinity. They are all of the same antibody class (isotype). Polyclonal antibodies are a mixture of antibodies with different antigen-binding sites that may bind to different epitopes or antigens with varying affinities. They may be of different antibody classes. The serum obtained from an immunized animal is referred to as a polyclonal antiserum.

Hybridoma

Hybridoms are cells that have been engineered to produce a desired antibody in large amounts. To produce monoclonal antibodies. B-cells are removed from the spleen of an animal that has been challenged with the relevant antigen. These B-cells are then fused with myeloma tumor cells that can grow indefinitely in culture. This fusion is performed by making the cell membranes more permeable. The fused hybrid cells (hybridomas), being cancer cells, will multiply rapidly and indefinitely and will produce large amounts of the desired antibodies.

Monoclonal Antibody Production

Immunize animal, Isolate spleen cells (containing antibody-producing B cells), Fuse spleen cells with myeloma cells" (eg using PEG - polyethylene glycol). It allows unfused B cells to die. Add **aminopterin** to culture to kill unfused myeloma cells. Clone remaining cells (place 1 cell/well and allow each cell to grow into a clone of cells). The supernatant of each clone for the presence of the desired antibody. The chosen clone of cells in tissue culture grows and indefinitely. antibody is harvested from the culture supernatant.

Hybridoma Selection

Myeloma cells lack certain enzymes so that they can not use hypoxanthine, aminopterin, and thymidine (**HAT** medium) as a source for nucleic acid biosynthesis and will die in culture. Only B cells that have fused with the engineered myeloma cells will survive in culture when grown in HAT medium.

Methods of Purification

- Filteration for larger particles
- Ultrafiltration esp. for low concentration samples
- Chromatography:

Ion exchange chromatography (either cation or anion) can be used (Most impurities are usually anions). Size exclusion chromatography: recombinant monoclonal antibodies. The term recombinant and especially recombinant DNA refers to an artificial method of producing DNA (synthetic DNA and proteins). Recombinant DNA is produced through the addition of relevant DNA into an existing organismal genome, such as the plasmid of bacteria. Recombinant antibody engineering involves the use of viruses or yeast to create antibodies, rather than mice. A segment of DNA is recombined (recombinant) with the original DNA of the bacteria, these genetically modified bacteria rapidly reproduce and by that reproduce the DNA segment that was recombined with the original genome.

Uses of Monoclonal Abs

The measuring protein and drug levels in serum. Identifying infectious agents and clusters of differentiation (CDs) for classification and follow-up therapy of leukemias and lymphomas identifying tumor metastasis. Identifying and quantifying hormones and immunoaffinity purification.

Subjective Questions

1. Explain the overview of the immune system?
2. What is innate immunity?
3. What is adaptive immunity?
4. Differentiate between the Immunogens and antigens?
5. Differentiate between the Antibody structure and function?
6. Explain the Antigen-antibody interactions, immune assays, and experimental systems?
7. Describe the genetic basis of antibody structure?
8. Explain the biology of the B lymphocyte?
9. How T cells recognize antigen: the role of the major histocompatibility complex?
10. Explain the biology of the T lymphocyte
11. Definition of Activation and function of T cells?
12. What is cytokines and their function?
13. Differentiate between Tolerance and autoimmunity?
14. What is a complement system?
15. Explain hypersensitivity: type I?
16. Explain Hypersensitivity: types II and III?
17. Explain Hypersensitivity: type IV?
18. Describe the immunodeficiency disorders and neoplasias of the lymphoid system?
19. What is transplantation?
20. Explain the tumourtumor immunology?
21. What is resistance and immunization to infectious diseases?

Objective Questions

1. Naturally acquired active immunity would be most likely acquired through which of the following processes?

a. vaccination

b. drinking colostrum

c. natural birth

d. infection with disease-causing organism followed by recovery.

Answer: D

2. Which of the following convey the longest-lasting immunity to an infectious agent?

a. Naturally acquired passive immunity

b. Artificially acquired passive immunity

c. Naturally acquired active immunity

d. All of these

e. None of these

Answer: C

3. Which of the following substances will not stimulate an immune response unless they are bound to a larger molecule?

a. Antigen

b. Virus

c. Hapten

d. Miligen

e. Antibody

Answer: C

4. B and T cells are produced by stem cells that are formed in:

a. Bone marrow

b. The liver

c. The circulatory system

d. The spleen

e. The lymph nodes

Answer: A

5. B cells mature in the……….. while T cells mature in the

a. Thymus/bone marrow and gut-associated lymphoid tissue (GALT)

b. Spleen/bone marrow and GALT

c. Bone marrow and GALT/Thymus

d. Liver/Kidneys

Answer: C

6. Which of the following immune cells/molecules are most effective at destroying intracellular pathogens?

a. T helper cells

b. B cells

c. Antibodies

d. Complement

e. T cytolytic cells

Answer: E

7. **A living microbe with reduced virulence that is used for vaccination is considered:**

a. A toxoid

b. Dormant

c. Virulent

d. Attenuated

e. Denatured

Answer: D

8. **B cells that produce and release large amounts of antibody are called:**

a. Memory cells

b. Basophils

c. Plasma cells

d. Killer cells

e. Neutrophils

Answer: C

9. **The specificity of an antibody is due to**

a. Its valence

b. The heavy chains

c. The Fc portion of the molecule

d. The variable portion of the heavy and light chain

Answer: D

10. **In agglutination reactions, the antigen is a.........in precipitation reactions, the antigen is a...............**

a. whole cell/soluble molecule

b. Soluble molecule/whole cell

c. Bacterium/virus

d. Protein/carbohydrates

e. Protein/Antibody

Answer: A

11. B Cells are activated by

a. Complement

b. Antibody

c. Interferon

d. Memory cells

e. Antigen

Answer: E

12. Fusion between a plasma cell and a tumor cell creates a

a. Myeloma

b. Natural killer cell

c. Lymphoblast

d. Lymphoma

e. Hybridoma

Answer: E

13. Monoclonal antibodies recognize a single:

a. Antigen

b. Bacterium

c. Epitope

d. B cell

e. Virus

Answer: C

14. Cell-mediated immunity is carried out by………….. while humoral immunity is mainly carried out by………………..

a. B cells/T cells

b. Epitopes/Antigens

c. T cells/B cells

d. Antibodies/Antigens

e. Antibodies/Phagocytes

Answer: C

15. The ability of the immune system to recognize self-antigens versus nonself antigen is an example of:

a. Specific immunity

b. Tolerance

c. Cell-mediated immunity
d. Antigenic immunity
e. Humoral immunity

Answer: B

16. The branch of biology, which involves the study of immune systems in all organisms is called_________.

(a) Zoology
(b) Microbiology
(c) Immunology
(d) Biotechnology

Answer: C

17. Which of the following immunity is obtained during a lifetime?

(a) Acquired immunity
(b) Active immunity
(c) Passive immunity
(d) None of the above.

Answer: A

18. How many types of antibodies are there?

(a) Five.
(b) Three.
(c) Two.
(d) Four.

Answer: A

19. Which of the following cells is involved in cell-mediated immunity?

(a) Leukaemia
(b) T cells
(c) Mast cells
(d) Thrombocytes

Answer: B

20. Which of the following protects our body against disease-causing pathogens?

(a) Respiratory system

(b) Immune system

(c) Digestive system

(d) Respiratory system

Answer: B

21. Hepatitis is an example of _______.

(a) Subunit Vaccine

(b) Killer Vaccine

(c) Toxoids Vaccine

(d) Recombinant Vaccine

Answer: D

22. Which of the following statements is true about the IgM of humans?

(a) IgM can cross the placenta

(b) IgM can protect the mucosal surface

(c) IgM is produced by high-affinity plasma cells

(d) IgM is primarily restricted in the circulation

Answer: D

23. Interferons are

(a) Cytokine barriers

(b) Physical barriers

(c) Cellular barriers

(d) Physiological barriers

Answer: A

24. Which of the following cells of the immune system do not perform phagocytosis?

(a) Macrophage

(b) Neutrophil

(c) Eosinophil

(d) Basophil

Answer: D

24. Monocytes differentiate into which kind of phagocytic cells?

(a) Neutrophil

(b) B cell

(c) Macrophage

(d) T cell

Answer: C

CHAPTER 5

CELL MOLECULAR

1. CHROMOSOME STRUCTURE, NUCLEOSOMES, EUCHROMATIN AND HETEROCHROMATIN

Chromosomes

It contains genetic material such as DNA and RNA. "Chromosomes are thread-like structures located inside the nucleus of animal and plant cells. Each chromosome is made of protein and a single molecule of deoxyribonucleic acid (DNA). Passed from parents to offspring, DNA contains the specific instructions that make each type of living creature unique".

Chromatin

The chromosomal material in its decondensed and threadlike state. "Chromatin is a complex of DNA and proteins that forms chromosomes within the nucleus of eukaryotic cells. Nuclear DNA does not appear in free linear strands; it is highly condensed and wrapped around nuclear proteins in order to fit inside the nucleus. Chromatin exists in two forms".

Mitosis

It forms asexual reproduction. It occurs when an organism grows or replaces damaged cells. Prior to mitosis, a cell undergoes replication. The process in which chromatin is copied. It produces diploid cells. "Mitosis is a process where a single cell divides into two identical daughter cells (cell division). During mitosis one cell? divides once to form two identical cells. The major purpose of mitosis is for growth and to replace worn out cells".

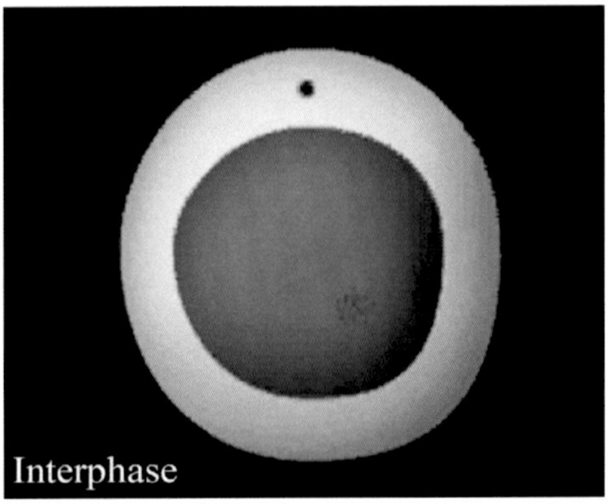

Prophase

Start of mitosis. The chromatin condenses into rod-like chromosomes. Each chromosome is consisting of sister chromatids, connected at the centromere and nuclear membrane disappears."Prophase is **the first phase of mitosis**, the process that separates the duplicated genetic material carried in the nucleus of a parent cell into two identical daughter cells. During prophase, the complex of DNA and proteins contained in the nucleus, known as chromatin, condenses".

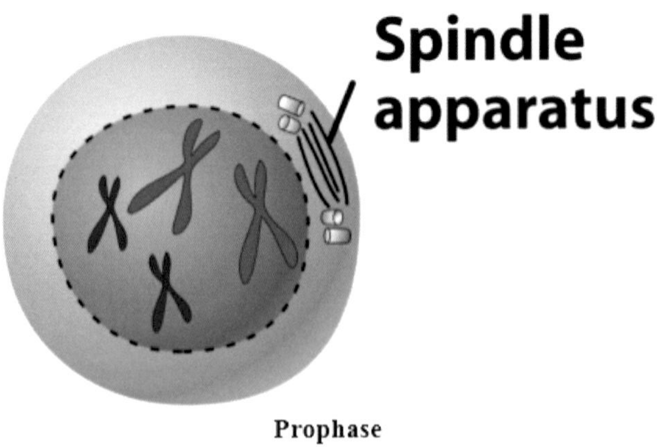

Prophase

Metaphase

Chromosomes align themselves at the flat plane at the cell equator in metaphase. "Metaphase is **the third phase of mitosis**, the process that separates duplicated genetic material carried in the nucleus of a parent cell into two identical daughter cells. Long protein filaments called kinetochore microtubules extended from poles on either end of the cell and attached to the kinetochores".

Metaphase

Anaphase

Centromeres are split. Sister chromatids-now chromosomes- are pulled to the opposite poles of the cell. "Anaphase is the fourth phase of mitosis, the process that separates the duplicated genetic material carried in the nucleus of a parent cell into two identical daughter cells. The separated chromosomes are then pulled by the spindle to opposite poles of the cell".

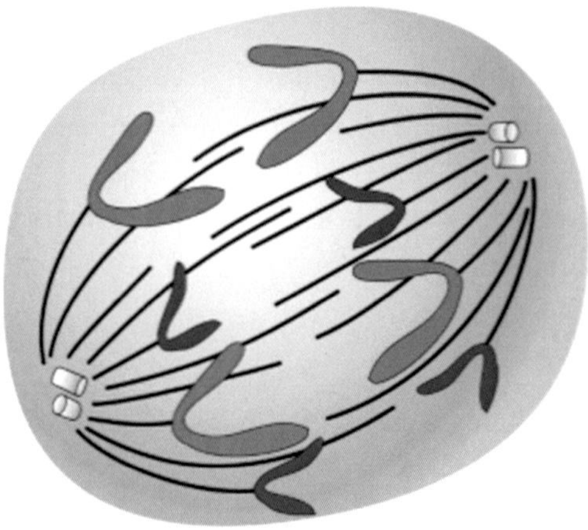

Anaphase

Telophase

Chromosomes unravel, returning the chromatin to its non-dividing thread-like state. The nuclear membrane is assembling."Telophase is the fifth and final phase of mitosis, the process that separates the duplicated genetic material carried in the nucleus of a parent cell into two identical daughter cells. ... During telophase, a nuclear membrane forms around each set of chromosomes to separate the nuclear DNA from the cytoplasm".

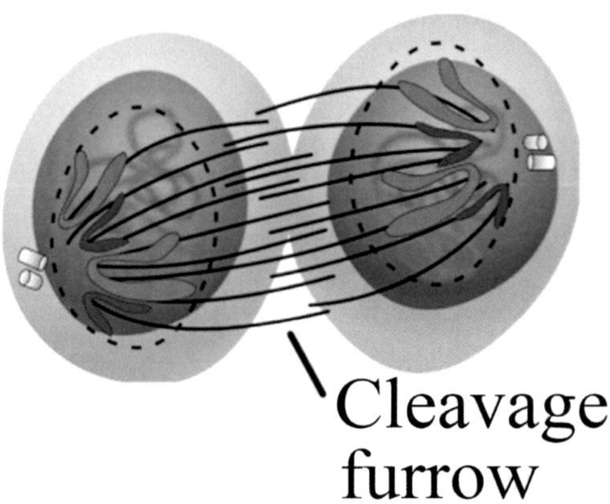

Telophase

Cytokinesis

Division of the cytoplasm in cytokinesis and it initiates during anaphase and telophase.

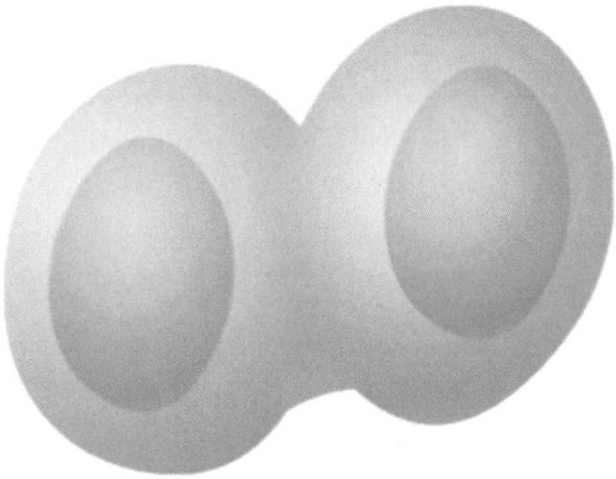

Cytokinesis

Differs in animals and plant cells. Plant cells form a cell plate. Membranous vesicles congregate at the center of the cell. Vesicles contain cell wall material. Animal cells form a cleavage furrow. It forms around the periphery of the dividing cell. Its furrows becomes deeper and deeper until the membrane pinches off forming two cells.

Chromosomes come in Matched Pairs

Homologous pairs: chromosomes that are closely matched in size and shape. It determines the same traits. **Sex chromosomes:** Those that determine the gender of the organism.

Chromosomal Theory of Inheritance

The two members of each pair of a homologous pair of chromosomes carry alleles for the same genes and, therefore, affect the same traits. It was proposed in 1903.

Mapping Genes

The location of a gene on a chromosome. It is currently trying to map all human genes to the appropriate chromosome. **Example:** the gene responsible for sickle cell anemia is located on chromosome 11.

Sexual Reproduction

The type of reproduction in which genetic information from females combines with males. It requires fertilization. The fusion of gametes (egg and sperm). These cells need to have half the amount of genetic information.

Meiosis

It produces haploid cells (gametes-eggs and sperm). It possesses only one member of each pair of homologous chromosomes. The chromosomes replicate before meiosis. It requires two rounds of division.

Prophase I

The chromatin is condensing into compact chromosomes. The nuclear envelope disappears. The synapsis occurs. The homologous pairs of chromosomes closely align allowing the exchange of chromosome segments

Metaphase I

The aligned pairs of replicated chromosomes move to the equator of the dividing cell.

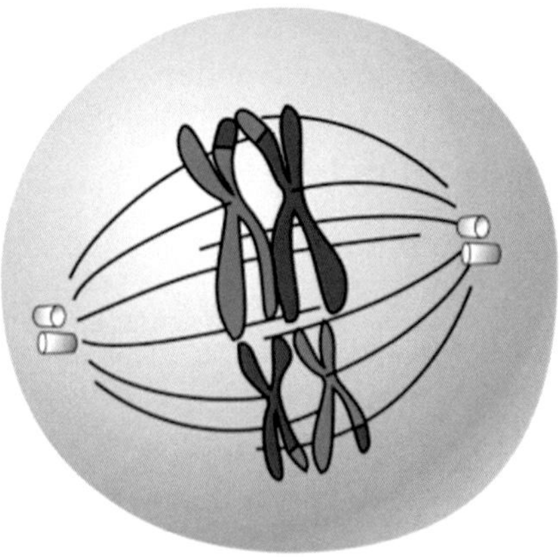

Metaphase I

Anaphase I

The members of homologous pairs of chromosomes are separate from each other. They move to the opposite poles of the cell. The chromosomes experience independent assortment.

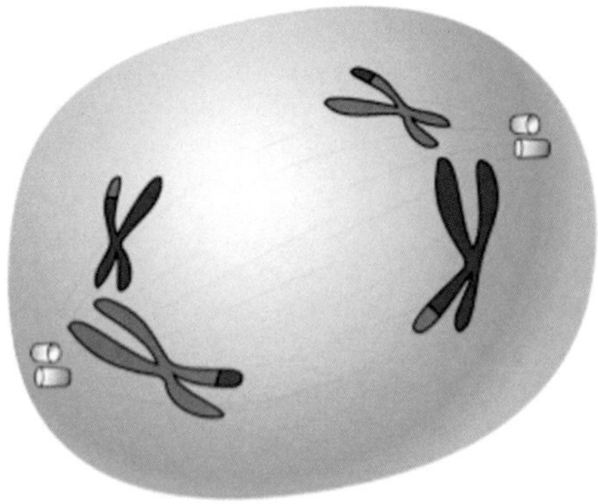

Anaphase I

Telophase I

The chromosomes are cluster at opposite poles of the cell and begin to decondense. The nuclear envelope may reform. The cytokinesis occurs Interkinesis: phase between meiosis I and II. The daughter cells are now haploid.

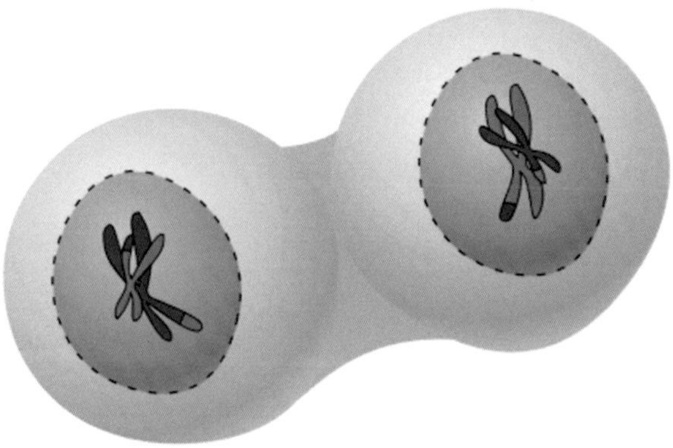

Telophase I

Prophase II

The partially unraveled chromosomes condense again.

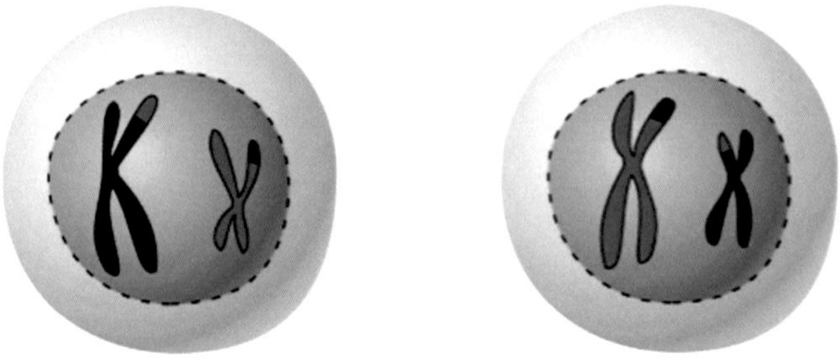

Prophase II

Metaphase II

The chromosomes move to the cell equator. It has no longer homologous pairs, so chromosomes line up singly in the middle of the cell.

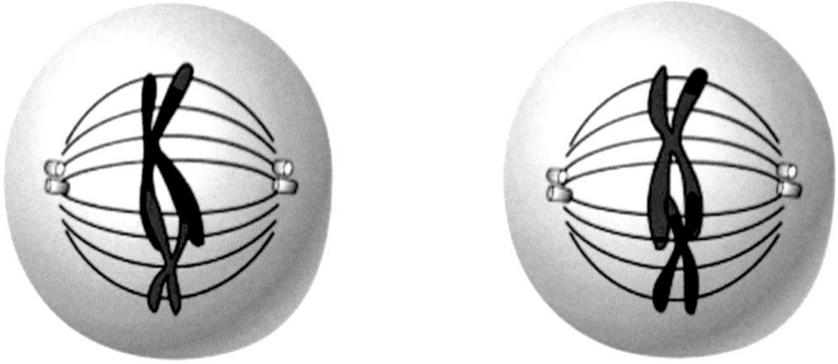

Metaphase II

Anaphase II

The centromeres divide and chromosomes separate. They move to the opposite poles of the cell.

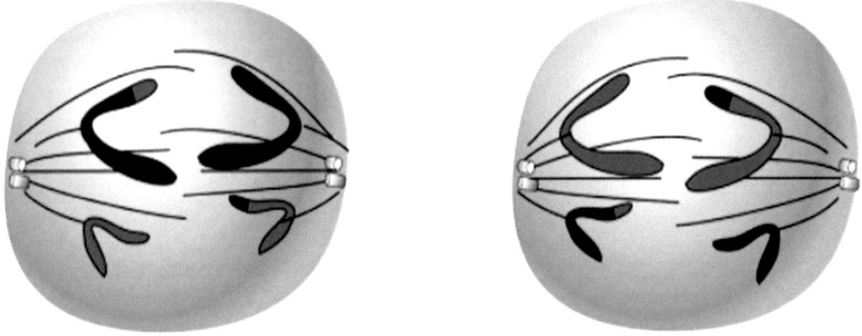

Anaphase II

Telophase II

The clustered chromosomes at the cell pole begin to decondense. The nuclear membrane develops and cytokinesis occurs. It produces four non-identical haploid cells.

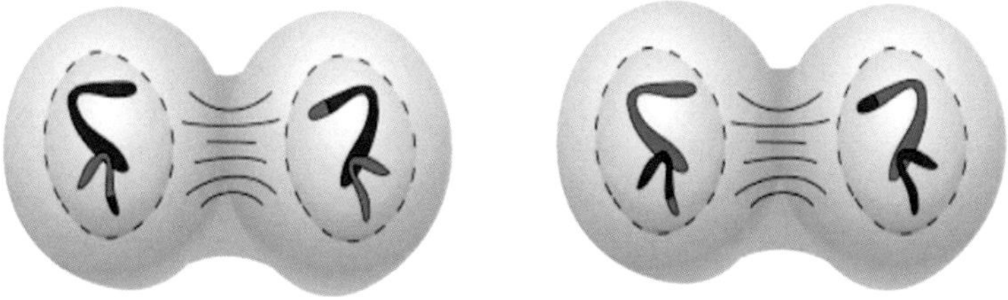

Telophase II

Eukaryotic Chromosome Structure

Chromatin: It is a complex of DNA and chromosomal proteins and it is twice as much protein as DNA.

Two Major Types of Proteins

Histones: They are abundant and basic proteins with a positive charge that bind to DNA 5 main types: H1, H2A, H2B, H3, H4, equal in mass to DNA, evolutionarily conserved.

Non-histones: All the other proteins associated with DNA differ markedly in type and structure amounts vary widely, >> 100% DNA mass and << 50% DNA mass.

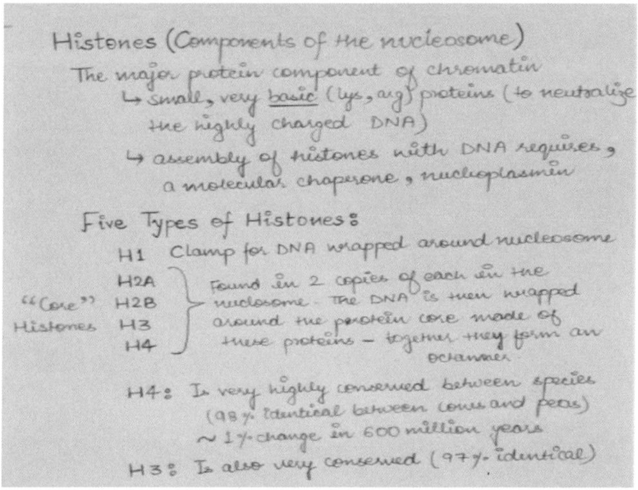

Histones (Components of the nucleosome)

The major protein component of chromatin

↳ small, very basic (lys, arg) proteins (to neutralize the highly charged DNA)

↳ assembly of histones with DNA requires, a molecular chaperone, nucleoplasmin

Five Types of Histones:

H1 Clamp for DNA wrapped around nucleosome

"Core" Histones H2A, H2B, H3, H4 } Found in 2 copies of each in the nucleosome. The DNA is then wrapped around the protein core made of these proteins – together they form an octamer

H4: Is very highly conserved between species (98% identical between cows and peas) ~ 1% change in 600 million years

H3: Is also very conserved (97% identical)

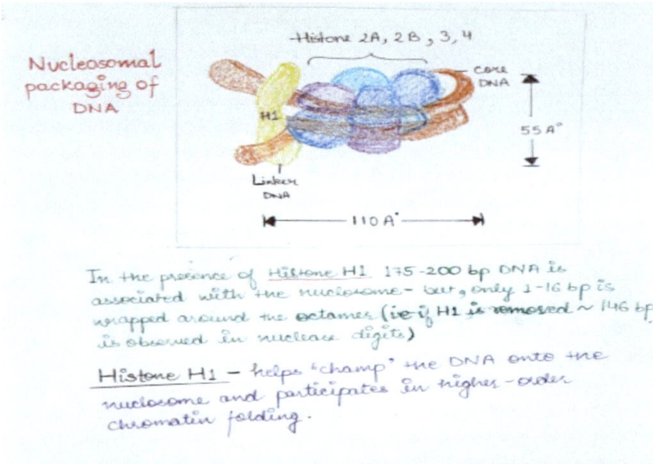
Nucleosomal packaging of DNA
Histone 2A, 2B, 3, 4
core DNA
H1
55 A°
Linker DNA
110 A°
In the presence of Histone H1 175-200 bp DNA is associated with the nucleosome - but, only 1-16 bp is wrapped around the octamer (ie if H1 is removed ~ 146 bp is observed in nuclease digets)
Histone H1 - helps 'champ' the DNA onto the nucleosome and participates in higher-order chromatin folding.

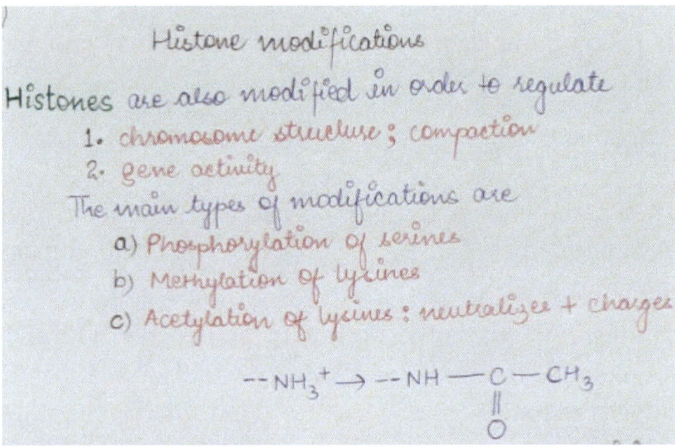
Histone modifications
Histones are also modified in order to regulate
1. chromosome structure; compaction
2. gene activity
The main types of modifications are
a) Phosphorylation of serines
b) Methylation of lysines
c) Acetylation of lysines : neutralizes + charges
$--NH_3^+ \rightarrow --NH-C(=O)-CH_3$

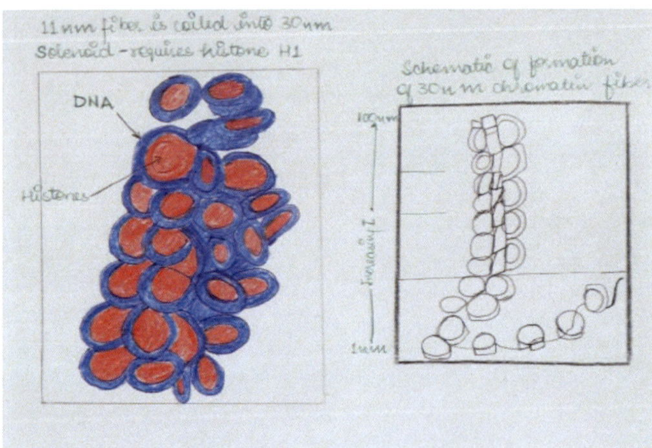
11 nm fiber is coiled into 30nm Solenoid - requires histone H1
DNA
Histones
Schematic of formation of 30nm chromatin fiber
100nm
1nm

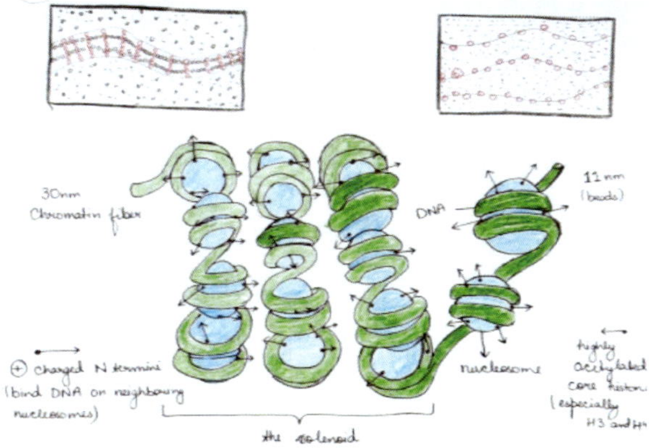

Solenoid:

- HIGH level of histone H1 "pull" nucleosomes into solenoid
- NO gene transcription in solenoid

Short extended regions can exist between solenoids:

- Reduced level of histone H1
- Gene transcription possible

Packing of DNA into chromosomes:

1. Level 1 Winding of DNA around histones to create a nucleosome structure.
2. Level 2 Nucleosomes connected by strands of linker DNA like beads on a string.

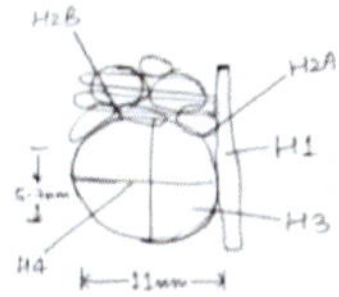

3. Level 3 Packaging of nucleosomes into 30-nm chromatin fibre.
4. Level 4 Formation of looped domains.

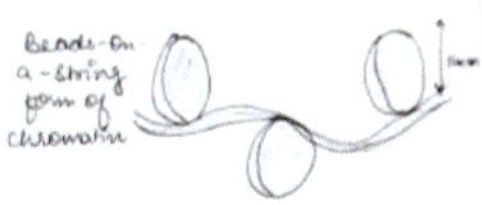

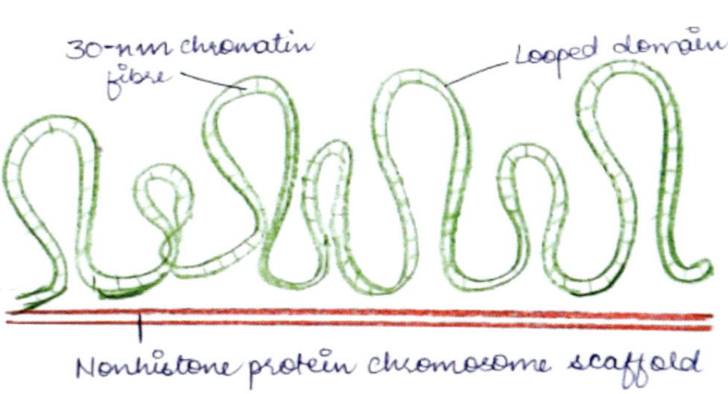

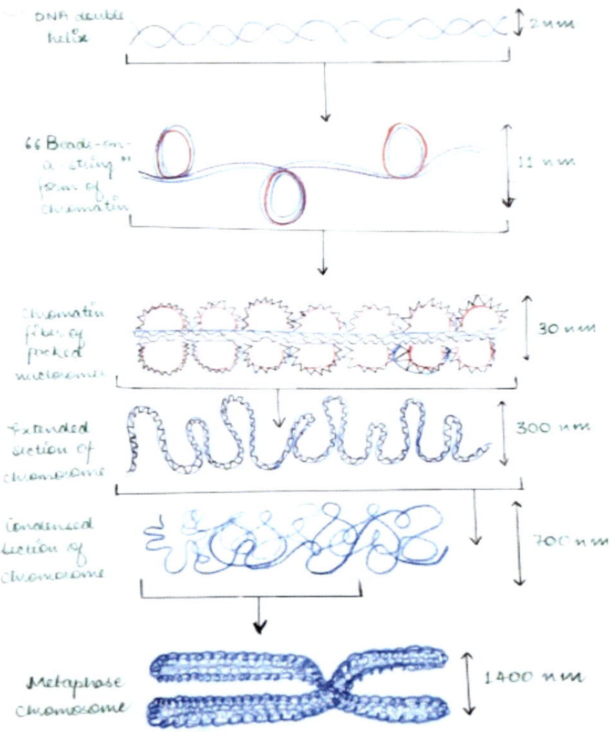

Chromatin is Divided into Euchromatin and Heterochromatin

Individual chromosomes can be seen only during mitosis. During interphase, the general mass of chromatin is in the form of euchromatin. Euchromatin is less tightly packed than mitotic chromosomes. Regions of heterochromatin remain densely packed throughout the interphase.

Chromosomes have Banding Patterns

Certain staining techniques cause the chromosomes to have the appearance of a series of striations called G-bands. The bands are lower in G • C content than the interbands. Genes are concentrated in the G • C-rich interbands.

Eukaryotic DNA has Loops and Domains attached to a Scaffold

DNA of interphase chromatin is negatively supercoiled into independent domains of ~85 kb. Metaphase chromosomes have a protein scaffold to which the loops of supercoiled DNA are attached. DNA is attached to the nuclear matrix at specific sequences called MARs or SARs. The MARs are

A • T-rich but do not have any specific consensus sequence. The centromere is essential for segregation. A eukaryotic chromosome is held on the mitotic spindle by the attachment of microtubules to the kinetochore that forms in its centromeric region. The centromeres often have heterochromatin that is rich in satellite DNA sequences.

Telomeres

It seals the chromosome ends. The protein TRF2 catalyzes a reaction in which the 3 repeating unit of the G+T-rich strand forms a loop by displacing its homolog in an upstream region of the telomere.

Lampbrush Chromosomes

Sites of gene expression on lampbrush chromosomes show loops that are extended from the chromosomal axis.

Polytene Chromosomes

Polytene chromosomes of dipterans have a series of bands that can be used as a cytological map.

Bands that are sites of gene expression on polytene chromosomes expand to give "puffs." The nucleosome is the subunit of all chromatin. Micrococcal nuclease releases individual nucleosomes from chromatin as 11S particles. A nucleosome contains:

- ~200 bp of DNA
- It has two copies of each core histone (H2A, H2B, H3, and H4).
- One copy of H1.
- DNA is wrapped around the outside surface of the protein octamer.

DNA is coiled in arrays of nucleosomes. Greater than 95% of the DNA is recovered in nucleosomes or multimers when micrococcal nuclease cleaves DNA of chromatin. The length of DNA per nucleosome varies for individual tissues in a range from 154-260 bp.

Nucleosomes

Nucleosomal DNA is divided into the core DNA and linker DNA depending on its susceptibility to micrococcal nuclease. The core DNA is the length of 146 bp that is found on the core particles produced by prolonged digestion with micrococcal nuclease. Linker DNA is the region of 8-114 bp that is susceptible to early cleavage by the enzyme. Changes in the length of linker DNA account for the variation in the total length of nucleosomal DNA. H1 is associated with linker DNA and may lie at the point where DNA enters and leaves the nucleosome.

Polytene Chromosome and Lambrush Chromosome

Giant chromosomes were first observed by E. G. Balbiani in 1881 in the salivary gland cells of Dipteran species. Polytene chromosomes are giant chromosomes found in the salivary glands, malpighian tubules, gut lining and adipose tissue cells of some Dipteran larvae. These chromosomes may reach a size up to 200 times (or more) the size of corresponding chromosomes at meiosis or in nuclei of normal mitotic cells. These giant chromosomes are formed by repeated cycles of endo-reduplication of single chromatids. This means that the chromatids replicate without cell division, as a result, the number of chromatids keeps increasing. This process is called Polyteny and therefore, these chromosomes are called polytene chromosomes.

Polytene chromosomes show another characteristic feature i.e., somatic pairing. As a result of this, chromosomes always appear to be half of that in normal somatic cells. Lampbrush chromosomes were first seen in salamander oocytes by Fleming in 1882. The name lampbrush was given by Ruckert in 1892. The nomenclature comes from its similarity in appearance to the brushes used to clean lamp chimneys in old days. These chromosomes are transitory structures that exist during the extended Diplotene in oocytes of most invertebrates and vertebrates except mammals. The loops in these chromosomes are sites of RNA transcriptions. The condensed chromatin in the chromomeres is generally not transcribed. The position of each loop is fixed. A given loop always contains the same DNA sequence. They have been widely used in studies of chromosome organization, gene expression, molecular morphology of RNA transcription, etc.

2. DNA REPLICATION: SEMICONSERVATIVE MECHANISM OF DNA REPLICATION

Introduction

The DNA is found in all plants, animals, prokaryotes and some viruses. DNA is a double-helical structure that consists of genetic material. It is joined by a phosphodiester bond. It is acidic in nature. The entire genetic message that controls the chemistry of every cell of the body acting in a specific way is actually written in the language of four nitrogenous bases of DNA i.e., purines and pyrimidines.

Composition of DNA

It is a polymeric compound containing four monomers known as deoxyribonucleotide monomers or deoxyribotides. Each DNA consists of a pentose sugar, a phosphate group and nitrogenous bases: **Figure 2.1.** Purines bases (adenine and guanine) are heterocyclic and two ringed bases and the pyrimidines (thymine and cytosine) are one ringed bases.

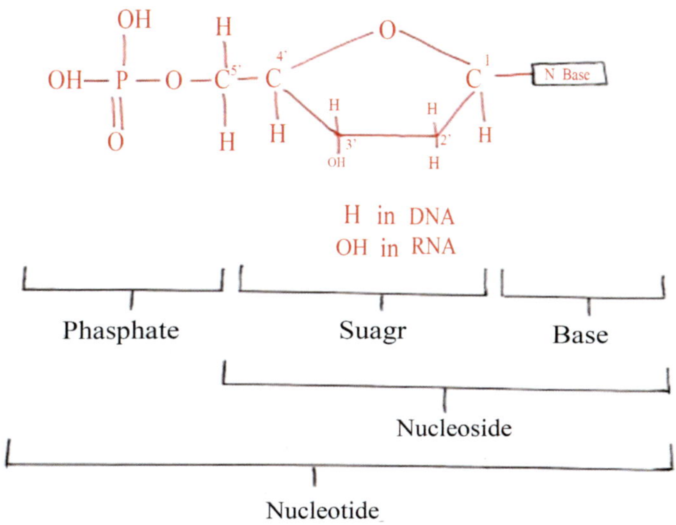

Figure 2. 1: Nucleotides.

Nucleosides

The nitrogenous bases are combined with a pentose sugar are called nucleosides.

Nucleotides

A nucleoside linked with phosphate forms a nucleotide. Nucleoside has pentose sugar + nitrogenous base nucleoside + phosphate Nucleotide. On the basis of different nitrogenous bases the deoxynucleotides are of following types: (i) Adenine (A) deoxyadenosine-3'/5'-monophosphate (3'75'-d AMP) deoxyguanosine -5'-monophosphate (5'-d GMP) deoxythymidine

-5-monophosphate (5'-d TMP) (ii) Guanine (G) (iii) Thymine (T) (iv) Cytosine (C) deoxycytidine -5-monophosphate (5'-d CMP). In addition to the presence of nucleosides in the DNA helix, these are also present in nucleoplasm and cytoplasm in the form of deoxyribonucleotide phosphates eg. deoxyadenosine triphosphate (dATP), deoxyguanosine triphosphate (dGTP), deoxycytidine triphosphate (dCTP), deoxythymidine triphosphate (dTTP) figure 2.2. The advantage of these four deoxyribonucleotides in triphosphate forms is that the DNA polymerase acts only on triphosphates of nucleotides during DNA replication. Similarly, the ribonucleotides contain ribose sugar, nitrogenous bases and phosphate. Except for sugar, the other components are similar. However, uracil (U) is found in RNA instead of thymine. Generally, an RNA molecule is single-stranded besides some exceptions.

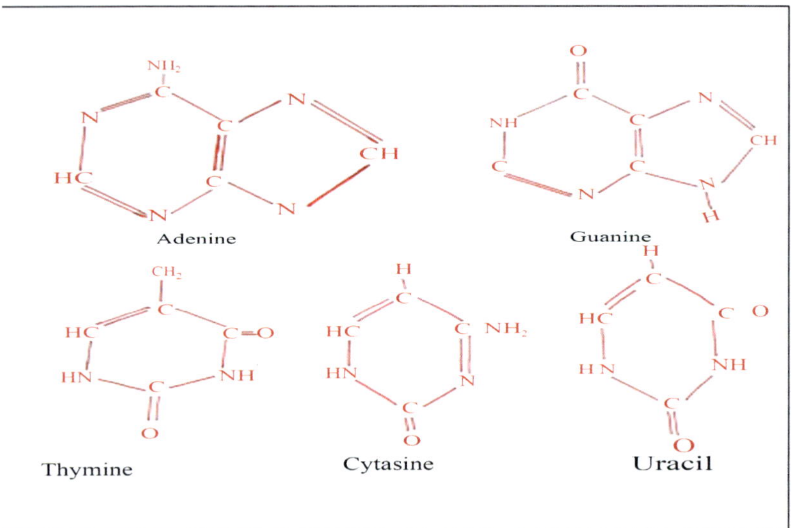

Figure 2.2: Nitrogenous base.

Polynucleotide

The nucleotides undergo the process of polymerization to form a long chain of polynucleotides. The nucleotides are designated by prefixing 'poly' to each repeating unit such as poly A (polyadenylic acid), poly T (polythymidilic acid), poly G (polyguanidylic acid), poly C (polycytidilic acid) and poly U (poly uridylic acid). The polynucleotides that consist of the same units are called homopolynucleotides such as poly A, poly T, poly G, poly C and poly U.

Chargaff-equivalence Rule

By 1948, chemist Erwin Chargaff started using paper chromatography to analyze the base composition of DNA from a number of studies. In 1950, Chargaff discovered that in the DNA of different types of organisms: the total amount of purines is equal to the total amount of pyrimidines i.e. The total number of A is equal to the total number of T (A-T), and the total number of G is equal to the total number of C (G-C). It means that A/T = G/C i.e. A+T/ G+C 1. In the DNA molecules isolated from several organisms, regularity exists in the base composition.

Physical Nature of DNA

Watson and Crick's Model of DNA

J.D.Watson and F.H.C. Crick (1953) combined the physical and chemical data generated by earlier workers and proposed a double-helix model for DNA molecules. This model is widely accepted. According to this model, the DNA molecule consists of two strands that are connected together by hydrogen bonds and helically twisted. Each step on one strand consists of a nucleotide of purine base that alternates with that of a pyrimidine base. Thus, a strand of DNA molecule is a polymer of four nucleotides i.e. A. G. T, C. The two strands join together to form a double helix. Bases of two nucleotides form hydrogen bonds i.e A combines with T by two hydrogen bonds (A T) and G combines with C by three hydrogen bonds (G C). However, the sequence of bonding is such that for every A.T.G.C. on one strand there would be T, A, C, G on the other strand. Therefore, the two chains are nucleotides on one chain are the photocopy of the sequence of nucleotides on the other chain.

The two strands of double helix run in antiparallel direction i.e. they have opposite polarity. The left-hand strand has 5'-3' polarity, whereas the right hand has 3'-5' polarity to the first one. The polarity is due to the direction of phosphodiester linkage complementary to each other i.e. sequence of as compared to the first one. The DNA model also suggested a copying mechanism of the genetic material. DNA replication is the fundamental and unique event underlying growth and reproduction in all living organisms ranging from the smallest viruses to the most complex of all creatures including man. DNA replicates by a semiconservative mechanism that was experimentally proved by Mathew, Meselson and Frank W. Stahlin in 1958. If changes occur in the sequence or composition of base pairs of DNA, mutation takes place. Though the presence of adenine, guanine, thymine and cytosine is a universal phenomenon, yet unusual bases in DNA molecules also occur. In some bacteriophages, 5- hydroxymethyl- cytosine (HMC) replaces cytosine of the DNA molecule when methylation of adenine, guanine and cytosine occurs. This results in changes in these bases: figure 2.3.

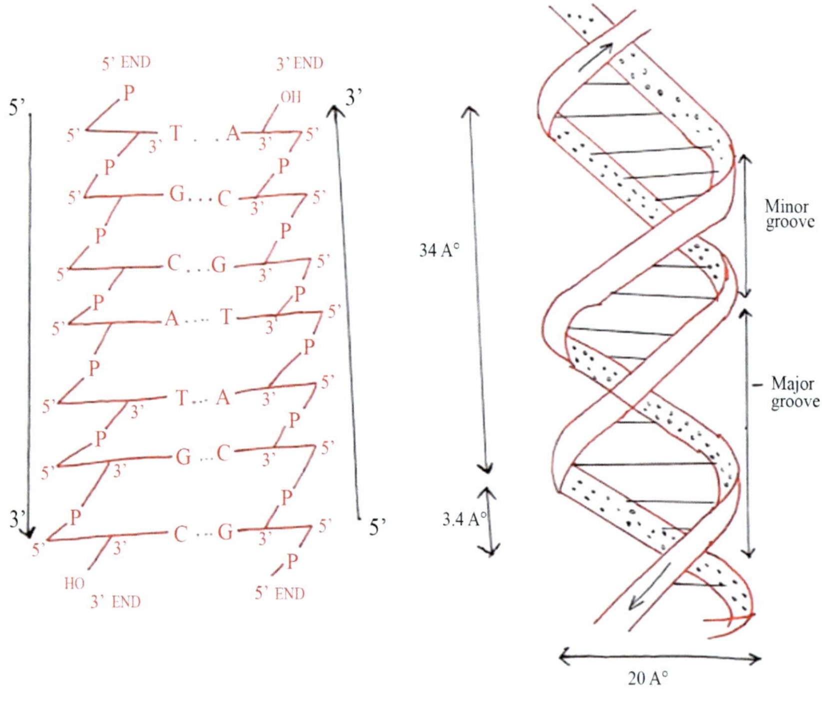

Figure 2.3: Watson and Crick's Model of DNA.

Three major forms

- A-DNA
- B-DNA
- Z-DNA

B-DNA is biologically the most common

- Right-handed (20 angstroms (A°) diameter)
- Complementary base-pairing (Watson-crick)
  - A-T = G-C
- Each base pair has the same width

- 10.85 a from c1’ to c1’
- A-T and G-C pairs are interchangeable.

“Pseudo-dyad” axis of Symmetry

Ideal **B-DNA** has 10 base pairs per turn.

Base thickness. Aromatic rings with 3.4 A° **thickness to rings.**

- Pitch = 10 x 3.4 = 34 A° **per** complete turn.
- Axis passes through the middle of each bp
- Minor groove is narrow.
- Major groove is wide
- Class exercise: explore the structure of B-DNA. Pay special attention to the major, minor grooves.
- Minor: exposes edge from which c1’ atoms extend.
- Major: exposes the opposite edge of the base pair.
- The pattern of H-bond possibilities is more specific and more discriminating in the major groove.

A-DNA

It is a right-handed helix. It is wider and flatter than B-DNA.11.6 bp per turn. The Pitch is 34 A°. An axial hole. The base planes are tilted 20 degrees with respect to the helical axis. The helix axis passes “above” the major groove. The deep major and shallow minor groove. It was observed under dehydrating conditions. When relative humidity is ~ 75%. B-DNA → A-DNA (reversible). Most self-complementary oligonucleotides of < 10 bp crystallize in A-DNA conf. A-DNA has been observed in 2 contexts:

- At the active site of DNA polymerase (~ 3 bp)
- Gram (+) bacteria undergoing sporulation

It induces B-DNA to → A-DNA for resistance to UV-induced damage and cross-linking of pyrimidine bases

Z-DNA

It is a left-handed helix. It is seen in conditions of high salt concentrations. It reduces repulsions between closest phosphate groups on opposite strands (B-DNA) in complementary polynucleotides with alternating purines and pyrimidines.

- poly d(GC) · poly d(GC)
- poly d(AC) · poly d(GT)

It might also be seen in DNA segments with the above characteristics with 12 base pairs per turn. A pitch of 44 degrees. A deep minor groove. No discernible major groove. The reversible change from B-DNA to Z-DNA in localized regions may act as a "switch" to regulate gene expression. Transient formation behind actively transcribing RNA polymerase.

Palindromes in DNA Sequences

The genetic palindromes are similar to verbal palindromes. A palindromic sequence in DNA is one in which the 5' to 3' base pair sequence is identical on both strands (the 5' and 3' ends refer to the chemical structure of the DNA).

RNA is single-stranded and does not have to stay in the nucleus. RNA is not found in chromosomes because it does not carry the genetic code; however, it can read the DNA code and take the information out of the nucleus.

- RNA's main work is to build proteins.
- The building blocks of RNA are Nucleotides, just like DNA.

A Nucleotide in RNA is still made of 3 important things:

1. 5-Carbon Sugar - **Ribose** (instead of Deoxyribose)
2. Phosphate
3. Nitrogen base

There are 4 nitrogen bases in RNA, A,G,C, and **U** that pair together)

A →**U** C →G

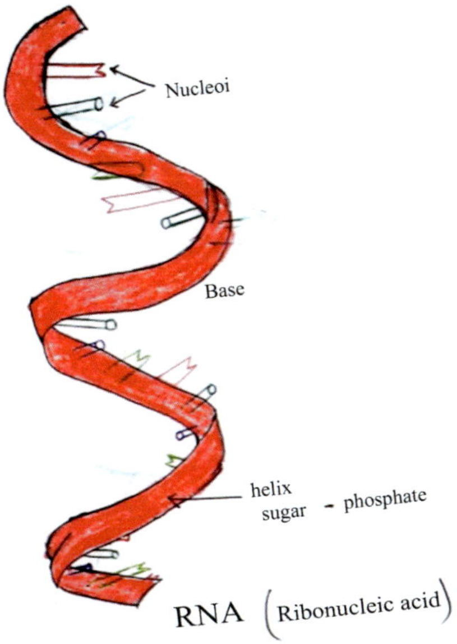

Three Main Types of RNA

Messenger RNA (mRNA): It carries copies of instructions for the assembly of amino acids into proteins from DNA to the rest of the cell (serve as a "messenger").

Ribosomal RNA (rRNA): It makes up the major part of ribosomes, which is where proteins are made.

Transfer RNA (tRNA): The transfers amino acids to ribosomes during protein synthesis.

DNA Denaturation (Melting)

Strands are held together by weak, noncovalent bonds. The strands start to separate at a specific temperature.

- Within a few degrees, the process is complete.
- The solution contains single-stranded molecules.
- Higher single strand absorbance at 260 nm.
- Hydrophobic base interactions arereduced.
- Bases absorb photons more efficiently.

Melting temperature (T_m) is half denatured. T_m increases with G-C content (%G + %C). A-T-rich sections melt before G-C-rich segments.

DNA Renaturation

J. Marmur (Harvard, 1960) – first described Slowly cool heat-denatured DNA. The temperature quickly is brought to ~25°C below T_m & incubated a while complementary single-stranded DNAs can reassociate or reanneal

Renaturation very useful genome complexity: variety & copy number and Hybridization: molecular identification. Renaturation governed by the ionic strength of the solution**,** Temperature**,** Time**,** DNA concentration and size of the interacting molecules

Smaller Genome - Faster Renaturation

More copies of small genomes and increases the chance of collision between complementary fragments and Viral/bacterial genomes are symmetrical curves.

Circular and Superhelical DNA

Almost in all the prokaryotes and a few viruses, the DNA is organized in the form of closed circles. The two ends of the double helix get covalently sealed to form a closed circle. Thus, a closed circle contains two unbroken complementary strands. Sometimes one or more nicks or breaks may be present on one or both strands, for example, DNA of phage PM, Besides some exceptions, the covalently closed circles are twisted into superhelix or supercoils and is associated with basic proteins but not with histones found complexed with all eukaryotic DNA these histones like proteins appears to help the organization of bacterial chromatin structure with the result of nucleosome like structure, folding and supercoiling of DNA and association of DNA polymerase with nucleoids. These nucleoid-associated proteins include HU proteins, IHF, proteins HI, Fir A, H-NS and Fis. In archaeobacteria (e.g. Archaea) the chromosomal DNA exists in protein associated form. Histone-like proteins have been isolated from nucleoprotein complexes in *Thermoplasma acidophilum* and *Halobacterium salinarum*.

Thus, the protein associated DNA and nucleosome like structures are defected in helix coils clockwise from the axis the coiling is termed as positive contrast, if the path of coiling is anticlockwise, the coil is called left-handed or negative coil. The two ends of a linear DNA helix can be joined to make each strand continuous. However, if one end rotates at 360° with respect to the other. These histones like proteins are appearing to help the organization of bacterial DNA into a coiled variety of bacteria. The unwinding of the double joined resulting in the formation of a twisted circle in opposite sense i.e. opposite helix; the ends are to unwinding direction.

Such a twisted circle appears as an 8 it has one node or crossing over point. If it is twisted at 720° before joining, the resulting superhelix will contain two nodes. The enzyme topoisomerases alter the topological form i.e. supercoiling of a circular DNA molecule. Type I topoisomerases (e.g. *E. coli* top A) relax the negatively supercoiled DNA by breaking the 5'-phosphoryl end and then resealing the nicked phosphodiester backbone (Moat and Foster, 1995). Type II Topoisomerases need energy to unwind the DNA molecules resulting in the introduction of supercoils. One type of type II isomerase, DNA gyrase is apparently responsible for the negatively supercoiled state of the bacterial chromosome. Super coiling is essential for efficient replication and transcription of prokaryotic DNA. The bacterial chromosome is believed to contain about 50 negatively supercoiled loops topological unit, the boundaries of which may be defined by the sites on DNA that limit its rotation (Wang, 1982; Mout and Foster, 1995). One of the phosphodiester bonds in dsDNA allows the 3'-OH end to swivel around or domains.

Organization of DNA

In addition to the organization of DNA in prokaryotes and lower eukaryotes as discussed earlier in eukaryotes, the DNA helix is highly organized into the well-defined DNA-protein complex termed as nucleosomes. Among the proteins, the most prominent are histones. The histones are small and basic proteins rich in amino acids such as cells there are five types of histones e.g. H, H, A, H, B, H, and H. Eight histone molecules (two each of H, A, H, B, H, and H) form an octamer ellipsoidal structure of about 1 nm to 7 nm in diameter. DNA coils around the surface of the ellipsoidal structure of histones 166 base pairs (about 7/4 turns) before proceeding onto the next and form. Thus, a nucleosome is an octamer or four histone proteins complexed with DNA lysine and/or arginine. Almost in all eukaryotic long and 6.5- complex structure, the nucleosome. The histones play important role in determining eukaryotic chromosomes by determining the conformation known as chromatin. The nucleosomes are the repeating units of DNA organization which are often termed as beads. The DNA isolated from chromatin looks like a string of 146 base pairs of DNA lying in the helical path and the histone. DNA assembly is known as the nucleosome core. The stretch of DNA between the nucleosome is known as linker' which varies in length from 14 to over 100 base pairs. The H_1 is associated with the linker region and helps the folding of DNA into a complex structure called chromatin or beads. As a result of m chromatin becomes visible as chromosomes during cell division.

Structure of Ribonucleic Acid (RNA)

The RN TMV, yellow mosaic virus, influenza virus, foot and mouth disease is usually single-stranded except viruses which have dsDNA. The retrovirus is entirely single strand of the RNA which folded either at certain regions or to form a hairpin-shaped structure. In the hairpin-shaped structures, the complementary bases are linked by hydrogen bonds which give stability to the molecules. However, no complementary bases are found in the unfolded region. The RNA does not possess an equal purine-pyrimidine ratio, as it is found in the DNA. Like DNA, the RNA is also the polymer of four nucleotides each one containing D-ribose, phosphoric acid and a nitrogenous base. The bases are two purines (A, G) and two are not found in RNA. A pairing between bases occurs as A-U and G-C pyrimidines (C, U). Thymine is not found in RNA. The pairing is between bases occurs as A-U and G-C. The nucleotides formed by the four bases are adenosine monophosphate (AMP), guanosine monophosphate (GMP), cytosine monophosphate (CMP), and uridine monophosphate (UMP). These are found freely in nucleoplasm but in the form of triphosphates e.g. ATP, GTP, UTP and CTP. As a result of polymerization, the ribonucleotides form a polynucleotide chain of RNA. If the RNA is involved in genetic mechanisms, it is called genetic RNA as found in plants, animals and bacterial viruses. The DNA acts as genetic material and RNA follows the order of DNA. In such cells, the RNA does not have a genetic role. Therefore, it is called non-genetic RNA. The non-genetic RNA is of three types: (i) ribosomal RNA (rRNA), (ii) transfer RNA (tRNA) or soluble RNA (sRNA), and (iii) messenger RNA (mRNA) or template RNA. These three types of RNA differ from each other in structure, site of synthesis in eukaryotic cell and function.

DNA Replication

The transmission of chromosomal DNA from generation to generation is crucial to cell propagation. This can only be achieved when chromosomal DNA is accurately replicated, providing two copies of the entire genome for faithful distribution into each daughter cell. Cells need to make a copy of DNA before dividing so each daughter cell has a complete copy of genetic information.

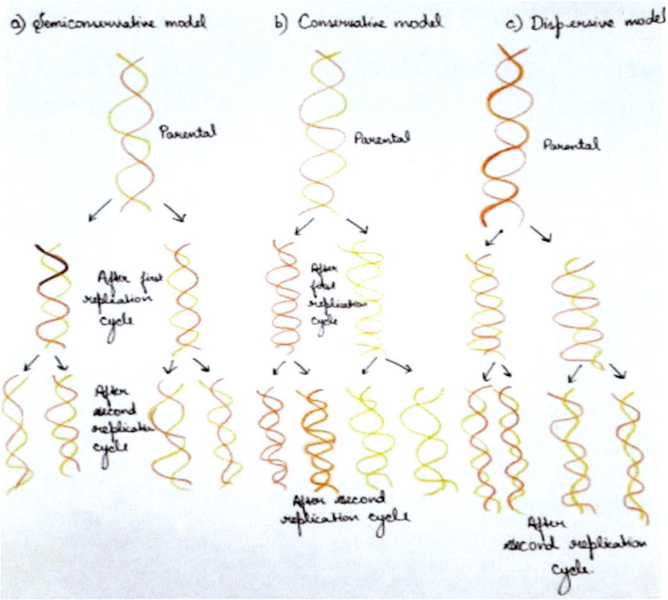

Replication of DNA

Base pairing allows each strand to serve as a template for a new strand. The new strand is 1/2 parent template & 1/2 new DNA. It has anti-parallel strands. The nucleotides in the DNA backbone are bonded from phosphate to sugar between 3′& 5′ carbons.

- DNA molecule has "direction"
- Complementary strand runs in the opposite direction

Meselson and Stahl's Experiment

Two isotopes of nitrogen:

- ^{14}N common form; ^{15}N rare heavy form.
- *E. coli* were grown in a ^{15}N media first, and then transferred to ^{14}N media.
- Cultured *E. coli* were subjected to equilibrium density gradient centrifugation.

Modes of Replication

Replicons: units of replication. Theta replication: circular DNA, *E. coli*; the single origin of replication forming a replication fork, usually a bidirectional replication. Rolling-circle replication: virus, F factor of *E. coli* and a single origin of replication.

Linear Eukaryotic Replication

Eukaryotic cells; thousands of origins; a typical replicon: 200,000 ~ 300,000 bp in length. Linear eukaryotic replication.

Direction of Replication

DNA polymerase adds nucleotides only to the 3′ end of a growing strand. The replication can only go 5′→3′.

Bonding in DNA

Hydrogen bonds and covalent phosphodiester bonds.

Replication: 1st Step

Initiation: 245 bp in the *oriC* (single origin replicon); an initiation protein. The unwinding of DNA is performed by **Helicase**. **Gyrase** removes supercoiling ahead of the replication fork. Single-stranded DNA is prevented from annealing by single-stranded binding proteins. Primers: an existing group of RNA nucleotides with a 3′-OH group to which a new nucleotide can be added; usually 10 ~ 12 nucleotides long.

Primase: RNA Polymerase

Unwind DNA

Helicase enzyme unwinds part of the DNA helix. Stabilized by single-stranded binding proteins.

Prevent the DNA Molecule from Closing

DNA gyrase is an enzyme that prevents tangling upstream from the replication fork.

Replication: 2nd step

RNA Primase. Adds a small section of RNA (RNA primer) to the 3' end of template DNA. DNA polymerase 3 (the enzyme that builds a new DNA strand) can only add nucleotides to existing strands of DNA.

DNA Synthesis Requires a RNA Printer

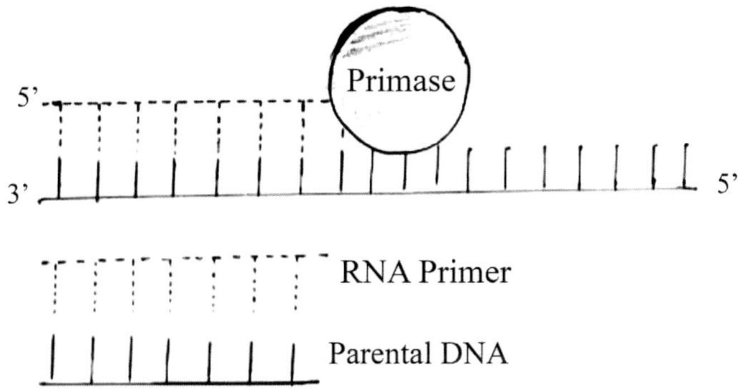

Elongation: It is carried out by DNA polymerase III. The removing RNA primer: DNA polymerase I. DNA ligase: connecting nicks after RNA primers are removed. **Termination:** when a replication fork meets or by a termination protein.

Replication: 3rd step

Build the daughter DNA strand. Add new complementary bases. With the help of the enzyme DNA polymerase III.

Replication: 4th step

Replacement of RNA primer by DNA is done by DNA polymerase I.

Direction of replication

Leading strand: It undergoes continuous replication. **Lagging strand:** It undergoes discontinuous replication. **Okazaki fragments:** The discontinuously synthesized short DNA fragments form the lagging strand.

Leading &Lagging Strands

Limits of DNA polymerase III. It can only build onto 3′ end of an existing DNA strand.

Lagging Strand

Okazaki fragments. It is joined by ligase and the "spot welder" enzyme.

Leading Strand

It is produced by continuous synthesis.

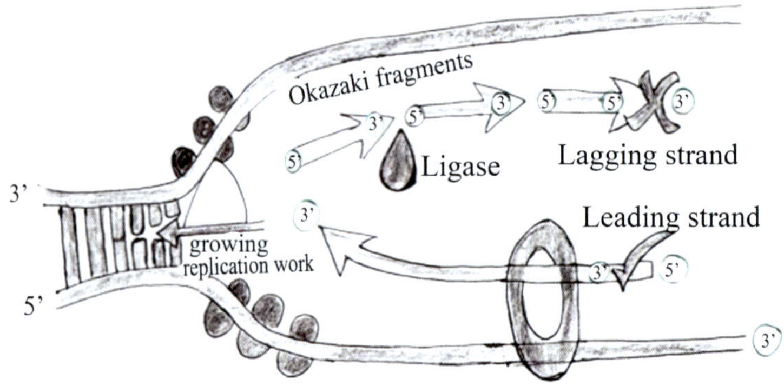

DNA Replication on the Lagging Strand

RNA primer is added. It is built by primase. It serves as a starter sequence for DNA polymerase III. However, short segments called Okazaki fragments are made because they can only go in a 5' →3' direction. Replacing RNA primers with DNA

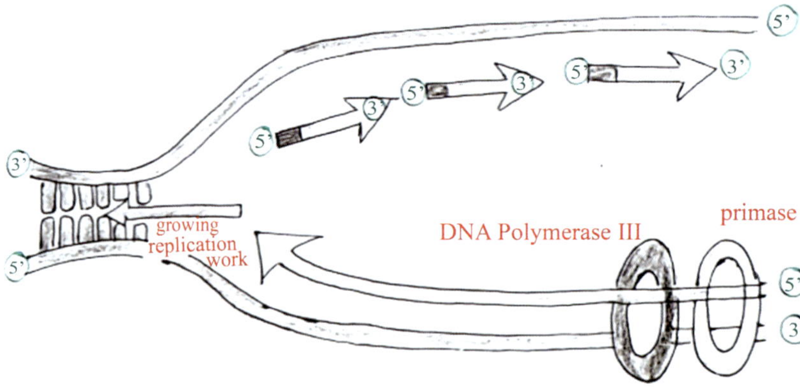

DNA Polymerase I

It removes sections of the RNA primer and replaces them with DNA nucleotides.

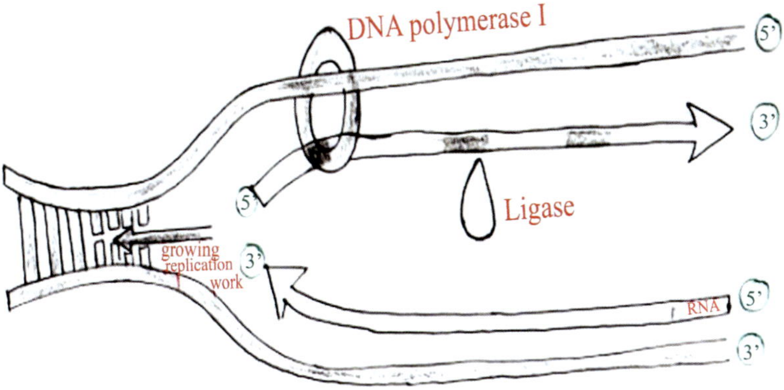

Replication Fork

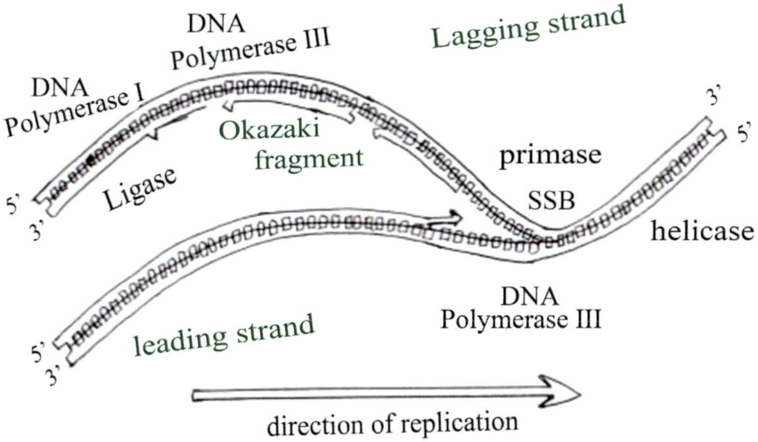

SSB = single-stranded binding proteins.

Stabilization of the Replication Machine

DNA polymerase delta and epsilon are attached to the DNA and held in place by other proteins. Sliding-clamp proteins (proliferating cell nuclear antigen - PCNA), allow the stable binding of DNA polymerase and strand synthesis. Clamp-loading proteins (replication factor C - RFC), aid in attaching the sliding-clamp proteins. The DNA is copied by a replication machine that travels along each replication fork. One of the most important members of this complex is DNA polymerase. It is able to add DNA subunits and make new DNA.

DNA Polymerase

It was discovered in 1956 by Kornberg first time from *E. coli* bacteria.

Bacteria have 3 types

DNA Pol I, II, and III. DNA Pol III is involved in the replication of DNA. DNA Pol I involved in repair. Humans have 4 types of polymerase. DNA Pol alpha, beta, delta - nuclear DNA. DNA Pol gamma - mitochondrial DNA. All DNA polymerases have 2 properties. Only synthesize DNA in one direction 5' to 3'. Only add to the end of existing double-stranded DNA. Therefore, they cannot start the synthesis of DNA from scratch. RNA polymerases can, but not DNA polymerases.

DNA Polymerase

It has a proofreading mechanism built. Checks for base matching. Removes mismatched bases by going backward. The reason why it is not able to build DNA in the 3' to 5' direction. No energy from ATP hydrolysis. Makes just one error in 10E8 or 10E9 (billion) bases added. The unwinding of DNA strands by helicases. Require ATP. Single-stranded DNA-binding proteins. Bind to single-stranded DNA to stabilize the structure. RPA (replication protein A - in eukaryotes). Topoisomerase - helps with the prevention of DNA strand twisting - 'swivels'. **Two types**

Type I - Break one strand only and then rejoin.

Type II - Break both strands and then rejoin.

Telomerase

It is an enzyme with a role in reverse transcriptase. It acts without a DNA template, as it has its own RNA template stored inside.

The Fidelity of DNA Replication

Proofreading: DNA polymerase I: $3'\rightarrow5'$ exonuclease activity removes the incorrectly paired nucleotide.

Mismatch repair: correcting errors after replication is complete.

Eukaryotic DNA Replication

Eukaryotic DNA polymerase. DNA polymerase α- acts like Primase to initiate. DNA polymerase δ- replicates the lagging strand. DNA polymerase e- replicates the leading strand.

Replication at the ends of Chromosomes

Telomeres and telomerase. DNA repair mechanism. Damaged DNA must be repaired. If the damage is passed on to subsequent generations, then we use the evolutionary term - mutation. It must take place in the germ cells - the gametes -eggs and sperm. If the damage is to somatic cells (all other cells of the body bar germ cells) then just that one individual is affected.

Damage from where Consequences of DNA replication errors. Chemical agents acting on the DNA.UV light imparts energy into a DNA molecule. Spontaneous changes to the DNA.

Repair Mechanisms

1. Reversal of damage
2. Excision repair
3. Mismatch repair
4. Recombination repair
5. Error-prone repair
6. Restriction-modification systems

1. Reversal of damage

Enzymatically undo the Damage

- Photoreactivation
- Removal of methyl groups

Photolyase Breaks apart Pyrimidine Dimers

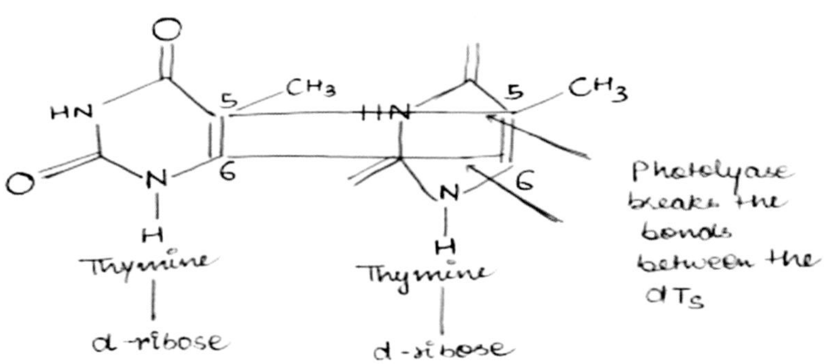

1a. Photoreactivation

Photolyase: It binds pyrimidine dimers and catalyzes a photochemical reaction. It breaks the cyclobutane ring and reforms two adjacent T's. 2 subunits, encoded by *phrA* and *phrB*.

Conversion of 6-O-methyl-dG back to dG

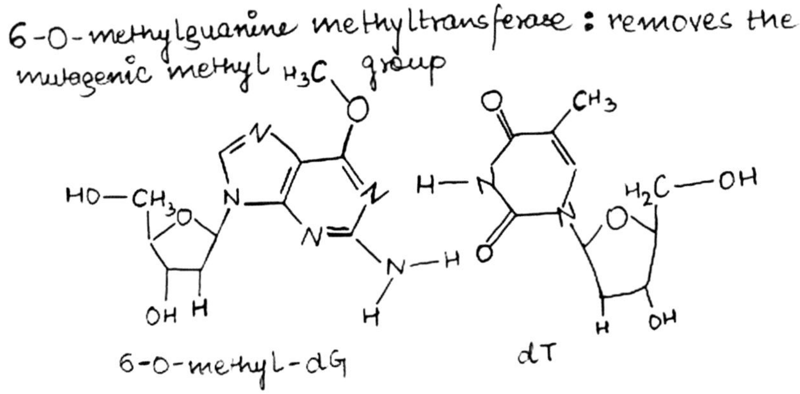

1b. Removal of Methyl Groups

6-O-methylguanine methyltransferase. It recognizes 6-O-methylguanine in DNA, removes the methyl group. Transfers the methyl group to an amino acid of the enzyme. "Suicide" mechanism. It is encoded by the ada gene in *E. coli*.

Excision Repair

It removes the damage (base or DNA backbone) and ss nick/gap provides the 3'OH for DNA Pol I initiation. DNA ligase seals nick.

Nucleotide Excision Repair

It cut out a segment of DNA around a damaged base.

Base Excision Repair

It cut out the base, then cut next to the apurinic/apyrimidinic site, and let DNA Pol I repair.

Mismatch Repair

The action of DNA polymerase III (including proofreading exonuclease) results in 1 misincorporation per 10^8 bases synthesized. Mismatch repair reduces this rate to 1 change in every 10^{10} or 10^{11} bases. Recognize mispaired bases in DNA, e.g. G-T or A-C base pairs. These do not cause large distortions in the helix: the mismatch repair system apparently reads the sequence of bases in the DNA.

Role of Methylation in Discriminating Parental and Progeny Strands

dam methylase acts on the A of GATC (note that this sequence is symmetrical or pseudopalindromic). Methylation is delayed for several minutes after replication. Mismatch repair works on the un-methylated strand (which is newly replicated) so that replication errors are removed preferentially.

Action of MutS, MutL, MutH

MutS: recognizes the mismatch (heteroduplex). **MutL**: a dimer; in presence of ATP, binds to MutS-heteroduplex complex to activate MutH. **MutH**: endonuclease that cleaves 5' to the G in an unmethylated GATC, leaves a nick.

Eukaryotic Homologs in Mismatch Repair

Human homologs to *mutL* (hMLH1) and *mutS* (hMSH2, hMSH1) have been discovered. Mutations in them can cause one of the most common hereditary cancers, hereditary nonpolyposis colon cancer (HNPCC).

Role of *umuC* and *umuD* Genes in Error-prone Repair

It is named so for the UV **non**-mutable phenotype of mutants with defects in these genes. It is needed for bypass synthesis; the mechanism is under investigation e.g. these proteins may reduce the template required for the polymerase. UmuD protein is proteolytically activated by LexA.

Evolution acts on Mutations

If we did not have a mutation, then we would all be the same. Any changes in the environment would be deleterious to all members of the population equally. There would be no evolution. But mutations do exist and are supported by comparison of related organisms.

Introduction

DNA is a relatively stable molecule, but the earth's natural environment is quite toxic, and damage to DNA is inevitable. DNA can also be altered by mistakes made during its own replication or recombination. Damage and sequence alterations to DNA are often quickly repaired, but when they are not, the DNA becomes permanently altered and harbors a mutation. Mutations are changes in the DNA sequence, and when mutations occur in germ-line cells, these changes are inheritable. A cancer cell has mutations that prevent cell death, resulting in loss of cell cycle control and unregulated cell division, which leads to malignant tumors that can end the life of the entire organism. The cell has a limited amount of time to fix the initial alteration and restore the DNA to its normal sequence before replication converts the alteration into a mutation that will be passed on to the next generation.

Mutation

Somatic mutations: It occurs in somatic cells and only affects the individual in which the mutation arises. **Germ-line mutations:** They alter gametes and passed to the next generation.

Mutations are quantified in two ways

- Mutation rate = probability of a particular type of mutation per unit time (or generation).
- Mutation frequency = number of times a particular mutation occurs in a population of cells or individuals.

Types of DNA Mutations

A mutation is a change in a DNA sequence that is propagated through cellular generations. Mutations can be as small as a single base pair or can range from a few base pairs to thousands. Mutations of one or a few base pairs usually result from errors in replication or damaged nucleotides. Mutation can have different effects on gene function.

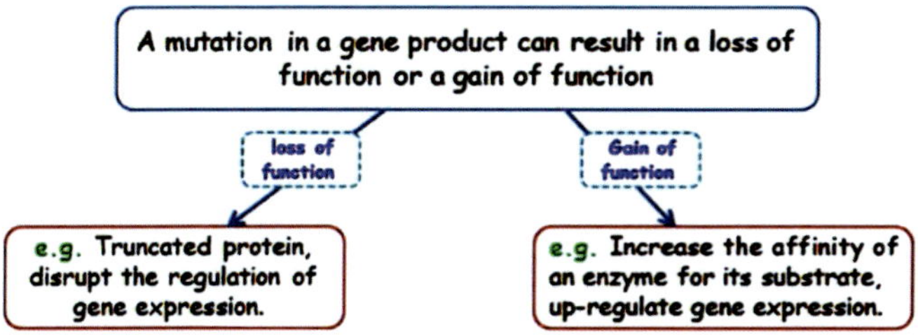

Studies in *Drosophila melanogaster* suggest that mutations that alter the protein sequence are most likely to be harmful; about 70% have a negative effect, and the rest either are neutral or have a weak beneficial effect.

A Point Mutation can alter one Amino Acid

A change in a single base pair is often referred to as a point mutation. Point mutations fall into two categories. A transition mutation is the exchange of a purine-pyrimidine base pair for the other purine-pyrimidine base pair: C≡G becomes T=A, or T=A becomes C≡G. A transversion mutation is the replacement of a purine-pyrimidine base pair with a pyrimidine-purine base pair or vice versa.

For example: C≡G becomes either G≡C or A=T.

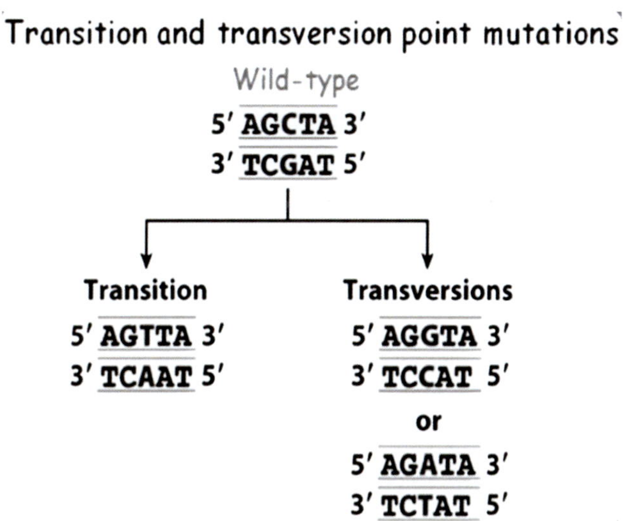

Transition mutations are nearly 10 times more frequent than transversions. A point mutation in the protein-coding region of a gene can result in an altered protein with partial or complete loss of function. If the protein is central to cell viability, the cell could die. Point mutations are known to cause a wide variety of human diseases. One example is sickle-cell anemia, resulting from a transversion that produces an amino acid change in hemoglobin: a GAG encoding glutamate (E) at residue 6 changes to GTG, encoding valine (V) (E6V). The most harmful mutations are those occurring in the genes involved in DNA repair because these often result in cancer. Many mutations in one cell can result in cancer because a mutation will occur in a gene (or genes) that encodes a protein needed to control cell division. In normal cells, oncogenes encode proteins that drive the cell division cycle forward, and tumor suppressor genes encode proteins that suppress cell division. Many tumor suppressors are transcription factors that regulate the expression of genes that drive the cell cycle. The transcription factor p53 and the retinoblastoma protein are examples of tumor suppressors that are mutated in many types of cancer. The DNA sequence encoding a protein is read in codons. Each codon corresponds to an amino acid. A silent mutation is a nucleotide change that produces a codon for the same amino acid. For example, GAA and GAG both code for glutamate. A missense mutation is a nucleotide change that results in a different amino acid, such as a change from glutamate (GAA) to glutamine (CAA). A nonsense mutation changes the nucleotide sequence so that instead of encoding an amino acid, the triplet functions as a stop codon, terminating the protein. Another type of mutation is the gain or loss of one or more base pairs. Insertion mutations occur when one or more base pairs are added to the wild-type sequence. Deletion mutations are due to the loss of one or more base pairs. Insertion and deletion mutations are collectively referred to as indels. The DNA sequence from the start codon to the stop codon is referred to as a reading frame. Because nucleotides are decoded in triplets, an indel mutation of only one or two base pairs in the coding sequence of a protein throws off the reading frame after the mutation, resulting in a frameshift mutation.

Mechanisms of DNA Repair

The most prevalent means that cells use to repair damaged DNA is excision repair, of which there are two types: base excision and nucleotide excision repair. Base excision repair (BER) functions at the level of a single damaged nucleotide that distorts DNA very little. It is also the main pathway for the repair of single-strand DNA breaks that lack a ligatable junction and therefore require "cleaning" of the 3′ or 5′ terminus for ligation. Nucleotide excision repair (NER) targets large, bulky lesions and removes DNA on either side of them. In contrast to base excision repair, NER does not require specific recognition of a damaged nucleotide and thus it can remove DNA lesions.

Base Excision Repair

(a) In bacteria, a glycosylase excises a damaged nucleotide base, and then an AP endonuclease nicks the backbone at the abasic site. Nick translation by Pol I excise the 5′ deoxyribose phosphate (5′-dRP) and some dNMPs, and synthesize a new strand. Ligase seals the gap.

(b) Eukaryotic BER, after the first two steps, can take either of two paths. In long patch repair, a DNA polymerase extends the DNA strand from the 3′ terminus, displacing the 5′ single-stranded DNA; this is followed by cleavage by a flap endonuclease and ligation. In short patch repair, only one nucleotide is inserted (by Pol β) prior to ligation.

The NER pathway uses several proteins, including UvrA, UvrB and UvrC, which recognize the lesion and make incisions on either side, allowing UvrD (helicase II) to displace a section of lesion-containing DNA. The single-strand gap is filled in by Pol I, and the DNA is sealed by ligase. A transcription-coupled repair (TCR) path can also be taken in which RNA polymerase stalls at the lesion on the coding strand. After the RNA polymerase is displaced, the reaction proceeds, using UvrA through UvrD, Pol I, and ligase.

3. REGULATION OF GENE EXPRESSION: BASIC PRINCIPLES OF TRANSCRIPTION, TRANSCRIPTION IN PROKARYOTES, TRANSCRIPTION IN EUKARYOTES, ITS REGULATION AND RNA PROCESSING

Introduction

It is the process of the formation of the RNA. It takes place through the usual process of complementary base pairing, catalyzed and scrutinized by the enzyme RNA polymerase. It occurs unidirectionally in which RNA chain (transcript) is synthesized from 5’ to 3 directions. It is chemically and enzymatically very similar to DNA replication. However, there are the following important that differ from that of replication. The new strand is made from ribonucleotides (NTPs) rather than deoxyribonucleotides (NTPs). It is primer-independent. RNA polymerase can initiate a new RNA chain on a DNA template without the need for a primer. It is a selective process i.e transcription selectively copies only certain parts of the genome called a transcription unit. In contrast, during replication entire genome replicates. Only one of the DNA strands acts as a template on which the RNA strand is built. It is unidirectional.

Prokaryotes

Promoter

Pribnow Box (also called the -10 element) – TATAAT. 35 element- TTGACA.

Eukaryotes

Promoter – (asymmetrical sequence). Basic core promoter –TATA box (TATAAA(A)); within 50bp upstream of the start site; found in unicellular eukaryotes. The core promoter plus has downstream promoters and proximal promoter elements.

Initiator Sequences

Regulatory Elements/Response Element: Response elements are the recognition sites of certain transcription factors. Most of them are located within 1 kb from the transcriptional start site. Enhancer elements -upon binding with transcription factors (activators), can enhance transcription; located either upstream or downstream of the transcriptional initiation site.

- Upstream enhancer elements
- Downstream enhancers
- Distal enhancer elements

Silencers - upon binding with transcription factors (repressors), can repress transcription.

Proteins involved in Transcription

RNA polymerase, general (or basal) transcription factors are TFIIA, TFIIB, TFIID, TFIIE, TFIIF, TFIIH. transcription factors that bind to regulatory elements are holoenzyme or initiation complex and transcription factors have common DNA binding motifs.

- Zinc finger
- Helix-turn-helix
- Leucine zipper

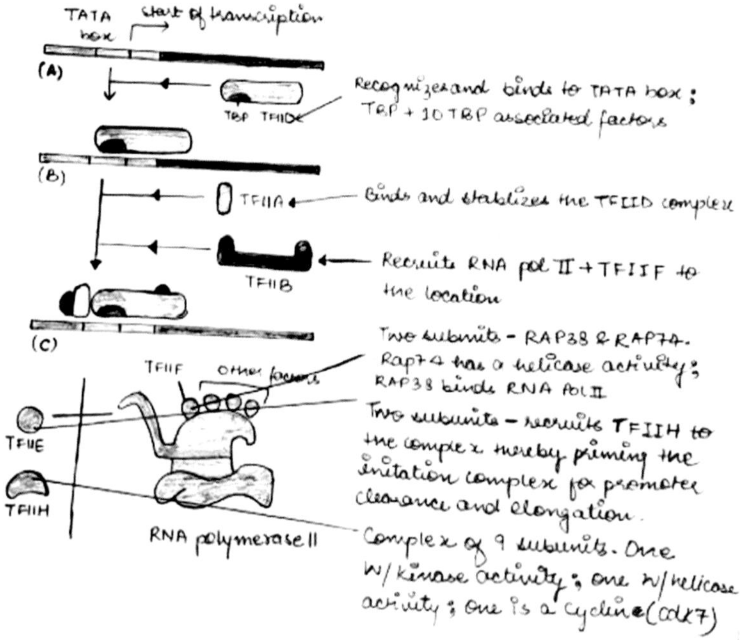

Synthesis of enzymes comes into the control of genetic material i.e., DNA in living organisms. Enzymes synthesis occurs by two processes viz. translation and transcription. Translation refers to the synthesis of mRNA. It is controlled by the promoter region of the gene.

Pathway for Gene Expression

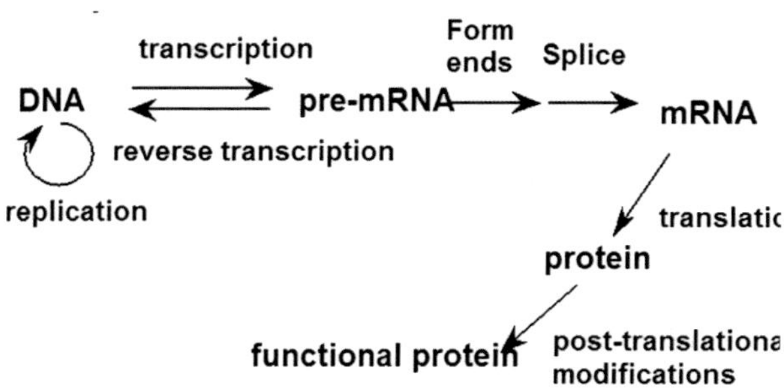

The Reaction Catalyzed by RNA Polymerase

It catalyzes the synthesis of RNA directed by DNA as a template = **transcription.** Makes an RNA chain with a sequence complementary to the template strand of DNA. Does not require a primer; can start RNA synthesis at a site on the DNA template *E. coli* RNA polymerase. Synthesizes all classes of RNA.

- mRNA
- rRNA
- tRNA

Core= α_2**ßß' catalyzes the elongation of an RNA chain.** Holoenzyme = α_2**ßß's catalyzes initiation or RNA synthesis specifically at a promoter.**

Promoter

The sequence of DNA is required for accurate, specific initiation of transcription. The sequence of DNA to which RNA polymerase binds to accurately initiate transcription. In most cases, RNA polymerase binds to a DNA sequence including the initiation site, but it can be directed there by sequences flanking the initiation site.

RNA Polymerase at a Promoter

- **α**: assembly and binds to UP.
- **ß+ß'**: form catalytic center.
- 6: binds -10 and -35 of the promoter to confer specificity during initiation.

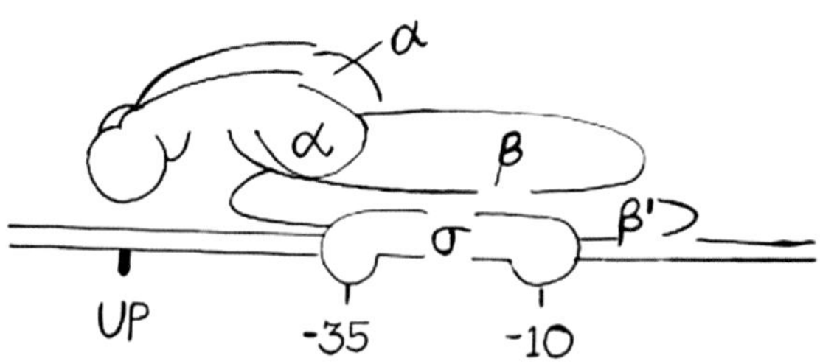

Mode of Action of 6 Factors

The 6 factor causes RNA polymerase to be selective in its choice of initiation sites by affecting the dissociation rate of polymerase from DNA. Core dissociates from general DNA with a half-time of 60 min; use in elongation. Holoenzyme dissociates from general DNA with a half-time of 1 sec. Holoenzyme dissociates from promoter DNA with a half-time of hours (in the absence of rNTPs).

Transcription Cycle

Initiation: Holoenzyme binds to the promoter, unwinds DNA, and forms phosphodiester bonds between 7 to 12 nucleotides.

Elongation: Core elongates RNA with high processivity. May use NusA.

Termination: Polymerase dissociates from template DNA and releases new RNA.

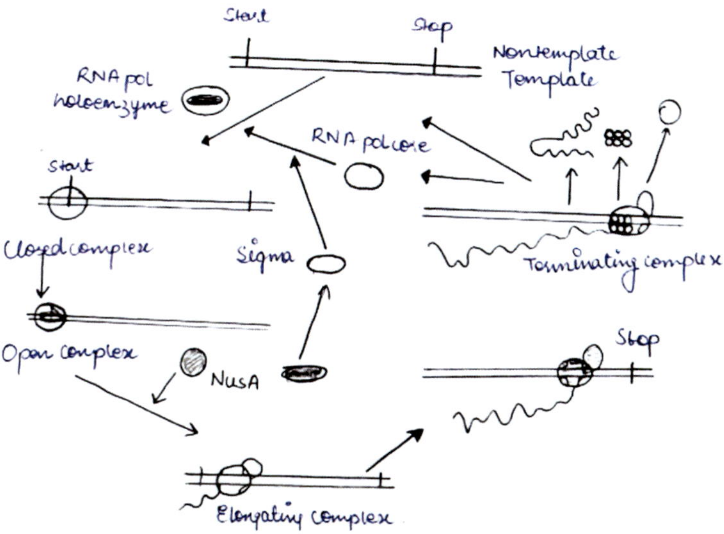

Sites on RNA Polymerase Core

Enzyme covers about 60 bp of DNA, with about 17 bp unwound = transcription bubble. The bubble must contact the active site for polymerization. At the beginning of the bubble, the DNA is unwound, implicating a helicase activity. At the end of the bubble, the DNA is rewound. Effects of transcription on supercoiling of the template. The unwinding (untwisting) and rewinding of the DNA template introduce positive supercoils ahead of the polymerase and negative supercoils behind it. Unwinding will decrease T by 1 for every 10 bp unwound. Since DL=0, DW=-DT, and DW=+1 for every 10 bp unwound. The opposite occurs upon rewinding. DT=+1 for every 10 bp rewound, and thus DW=-1 for every 10 bp rewound. Topoisomerases aid transcription *in vivo*.

RNA Polymerases in Eukaryotes

Name	Makes	α-amanitin
RNA Polymerase I	pre-rRNA	insensitive
RNA Polymerase II	pre-mRNA some snRNAs	very sensitive
RNA Polymerase III	pre-tRNA other small RNAs some snRNAs	less sensitive

Subunit Structure of Eukaryotic RNA Polymerases

All 3 have multiple subunits (8 to 14). The molecular weight of each polymerase is about 500,000. Some subunits are common to all 3 RNA polymerases. All 3 RNA polymerases have subunits that are homologous to the bacterial ß, ß' and α subunits. The transcription and translation take place in different cellular comparts: transcription takes place in the nucleus and translation takes place in the cytosol. In prokaryotes, transcription of mRNA and translation occur simultaneously. Thus, mRNA molecules undergo little or no modification after synthesis by RNA polymerase in prokaryotes. In contrast, pre-tRNA and pre-rRNA undergo processing like cleavage, the addition of nucleotides and chemical modification after synthesis. Although both prokaryotes and eukaryotes modify pre-tRNA and pre-rRNA, eukaryotes very extensively process pre-mRNA destined to become mRNA. The primary transcript of an RNA polymerase is referred to as pre-RNA. Processing of eukaryotic pre-mRNA involves 5' capping, 3' polyadenylation, splicing and RNA editing before being transported to the cytosol, where they are translated by ribosomes. The final product of gene expression is a polypeptide chain of amino acids whose sequence is dictated by the genetic code.

- mRNA transcribed from genomic DNA.
- Transfer RNA (tRNA) to transport amino acids.
- Ribosome to "read" mRNA, align amino acids attached to tRNA and create the peptide bonds between adjacent amino acids.

Ribosome

Two subunits composed of protein and ribosomal RNA (rRNA). The rRNA is a structural component of the ribosome subunits.

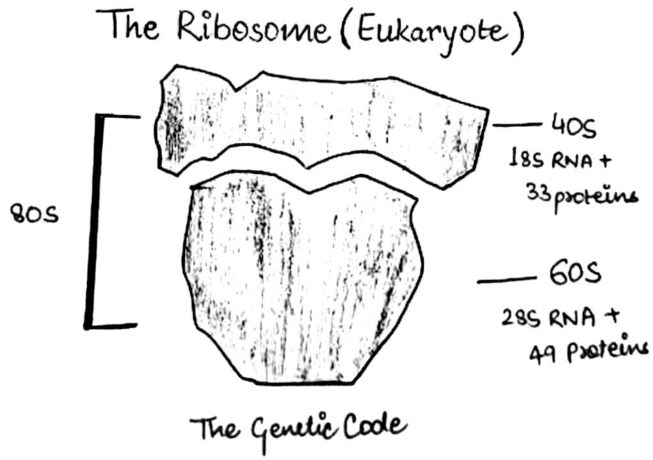

tRNA

tRNAs are small, extremely stable RNA structures shaped like a cloverleaf due to internal base pairing. They are almost identical in both prokaryotes and eukaryotes. They contain a number of post-transcriptional modifications, including non-traditional bases (other than (A, U, G and C). The amino acid attaches to the 3' end of the tRNA. tRNA can base pair with the codon in mRNA at the anticodon on the tRNA.

Charging tRNA with an Amino Acid

An individual tRNA is specific for a single amino acid. (There are at least 20 different tRNAs). Charging is the process of covalently attaching an amino acid to the tRNA. Charging is accomplished by aminoacyl tRNA synthetases, each of which is specific for an amino acid.

- Initiation—the two subunits of the ribosome come together and the start codon on the mRNA in the ribosome is aligned to set the reading frame.
- Elongation—charged tRNAs attach and peptide bonds form between the amino acids.

- Termination

Prokaryotes

Translation takes place directly after transcription. mRNA is not modified. Transcription and translation take place in the same area.

Eukaryotes

The transcript is modified before leaving the nucleus (5' cap and 3' poly-A tail).

- Modifications increase the translation efficiency and lifespan of the mRNA.
- Translation takes place on ribosomes located in the rough endoplasmic reticulum (translation is physically separated from transcription).
- Influence on gene expression forms the basis of function in many antibiotics.
- The purpose of antibiotics is to impair function in the prokaryote without disrupting function in eukaryotes.

Antibiotics

- Tetracycline: Blocks attachment of the AA-tRNA to the A site in the ribosome.
- Erythromycin: Blocks the movement of the ribosome down the mRNA.
- Streptomycin: Blocks the formation of the initiation complex and causes miscoding.
- Rifamycin: Blocks transcription by blocking the binding of RNA polymerase.

Genetic Code

It consists of 64 triplet codons (A, G, C, U) $4^3 = 64$. All codons are used in protein synthesis. It has 20 amino acids. It has 3 termination (stop) codons: UAA, UAG, and UGA. So, AUG (methionine) is the start codon (also used internally). It has multiple codons for a single amino acid = degeneracy. It has 5 amino acids are specified by the first two nucleotides only.

The Genetic Code

UUU	Phe	UCU	Ser	UAU	Tyr	UGU	Cys
UUC		UCC		UAC		UGC	
UUA	Leu	UCA		UAA	Stop	UGA	Stop
UUG		UCG		UAG		UGG	Trp
CUU	Leu	CCU	Pro	CAU	His	CGU	Arg
CUC		CCC		CAC		CGC	
CUA		CCA		CAA	Gln	CGA	
CUG		CCG		CAG		CGG	
AUC	IIe	ACU	Thr	AAU	Asn	AGU	Ser
AUC		ACC		AAC		AGC	
AUA		ACA		AAA	Lys	AGA	Arg
AUG	Met	ACG		AAG		AGG	
GUU	Val	GCU	Ala	GAU	Asp	GGU	Gly
GUC		GCC		GAC		GGC	
GUA		GCA		GAA	Glu	GGA	
GUG		GCG		GAG		GGG	

Codon-anticodon Interactions

The codon-anticodon base-pairing is antiparallel. The third position in the codon is frequently degenerate. One tRNA can interact with more than one codon (therefore 50 tRNAs). The Wobble rules.

- C with G or I (inosine).
- A with U or I.
- G with C or U.
- U with A, G, or I.
- I with C, U, or A.

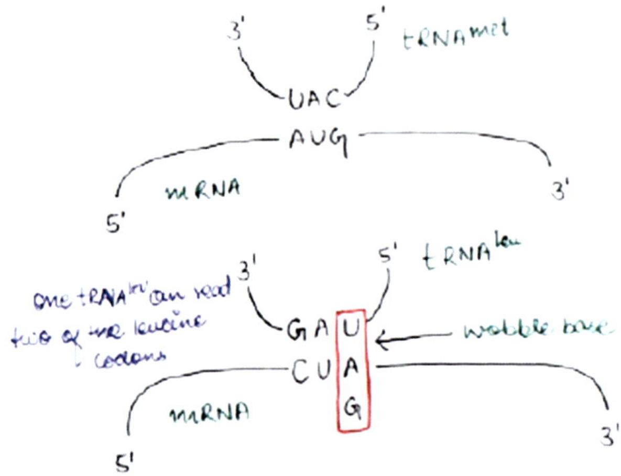

Termination Point

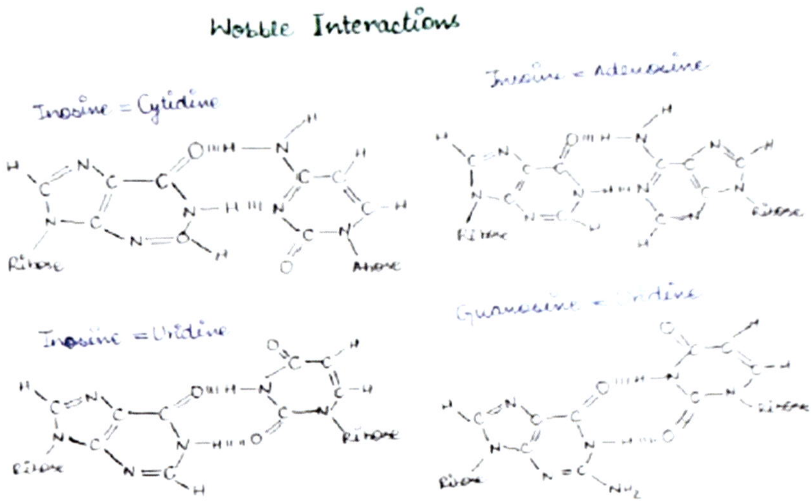

Initiation in Prokaryotes and Eukaryotes

Initiation can occur at internal AUG codons in prokaryotic mRNA. Initiation in eukaryotes occurs only at the first AUG codon. Lac operon in *E. coli* is transcribed as a polycistronic mRNA with multiple AUG codons.

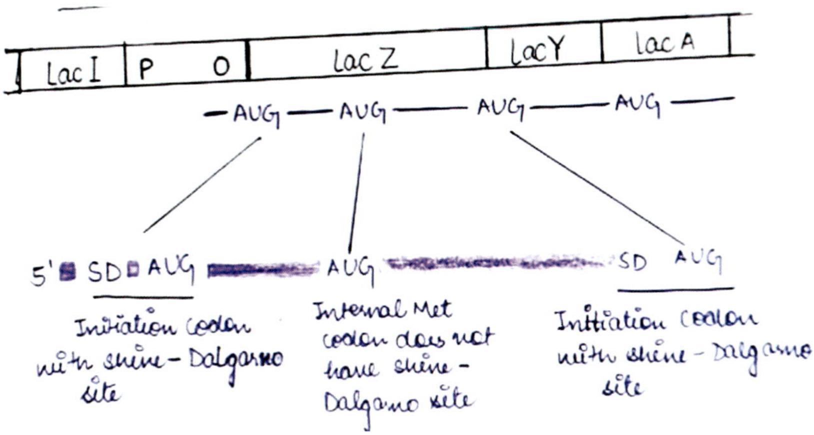

Reading Frame

The reading frame is determined by the AUG initiation codon. Every subsequent triplet is read as a codon until reaching a stop codon.

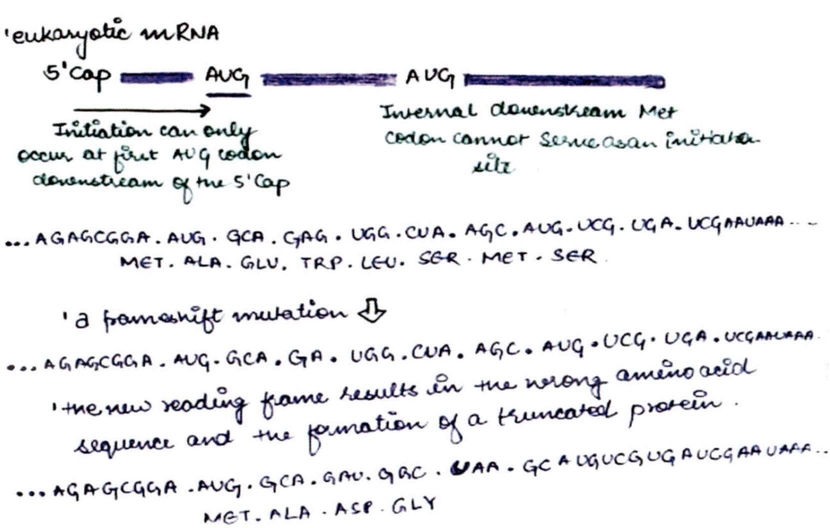

Mutations affecting Translation

Hemoglobin wayne (3’ terminal frameshift mutation). Missense mutations are (e.g., AGC Ser to AGA Arg). Nonsense mutations are (e.g., UGG Trp to UGA Stop). Read through, reverse terminator, or sense mutations (e.g., UAA Stop to CAA Gln) as in hemoglobin constant spring. Silent mutations are (e.g., CUA Leu to CUG Leu) do not affect translation.

Normal α-globin •ACG•UCU•AAA•UAC•CGU•UAA•GCU GGA GCC
UCG GUA •THR•SER•LYS•TYR•ARG

Wayne α-globin •ACG•UCA•AAU•ACC•GUU•AAG•CUG•GAG
CCU CGG UAG THR•SER•ASN•THR
VAL•LYS•LEU•GLU•PPO•ARG

mutated region

Post-translational Modifications

- Cleavage of signal peptides
- Phosphorylation
- Amidation
- Glycosylation
- Hydroxylation
- Ubiquitination
- Addition of prosthetic groups
- Iodination
- Adenylation
- Sulfonation
- Prenylation
- Myristoylation
- Acylation
- Acetylation
- Methylation
- Oxidative crosslinking
- N-Glutamyl cyclization
- Carboxylation

N-terminal Modifications

- N-formyl Met (a by-product of the translation process and incomplete cleavage by deformylase).

- Aminopeptidase cleavage of Met.
- N-terminal acetylation.
- N-terminal myristoylation.
- N-terminal glutamyl cyclization.

C-terminal Modification

- C-terminal prenylation
  - Farnesylation
  - Geranylation/geranyl-geranylation
- C-terminal amidation

Phosphorylation

- Hydroxyl groups - Ser, Thr, and Tyr.
- Amide groups of His, Lys & Carboxy group Asp.
- It often causes changes in protein conformation and/or protein-protein interactions.

C-terminal Amidation

- Neuropeptides and hormones
- C-terminal glycine is hydroxylated alpha-hydroxy-glycine.
- Glyoxylate. C-terminal amidation is essential to the biological activity of many neuropeptides and hormones.

Hydroxylation and Carboxylation

- Hydroxylation (previous lecture).
  - Lys and Pro (e.g. collagen).
  - Requires vitamin C.
- Carboxylation

  Requires vitamin K.

Iodination

- Special case: Tyr → thyroxin.

Protein Acylation

- Methylation and acetylation.
- Acetylation of histone Lys.
- S-acylation (Cys).
- Prenylation (C-terminus).
- Myristoylation (N-terminus).

Acetylation

N-terminal Acetylation

- No simple sequence consensus.
- It usually accompanied by cleavage of the N-terminal Met.

Methylation

- It can occur at carboxyls, amides, sulfhydryl, etc.
- S-Adenosyl Methionine (S-AM) is the activated carbon source.

Histones

- DNA binding proteins.
- The chromatin structure.
- It affects gene expression.

Operon

A unit of bacterial gene expression and regulation, including structural genes and control elements in DNA recognized by regulator gene product(s). Trans-acting – A product that can function on any copy of its target DNA. This implies that it is a diffusible protein or RNA. Cis-acting – A site that affects the activity only of sequences on its own molecule of DNA (or RNA); this property usually implies that the site does not code for protein. Regulator gene – A gene that codes for a product (typically a protein) that controls the expression of other genes (usually at the level of transcription). Structural gene – A gene that codes for any RNA or protein product other than a regulator. In negative regulation, a repressor protein binds to an operator to prevent a gene from being expressed. In positive regulation, a transcription factor is required to bind at the promoter in order to enable RNA polymerase to initiate transcription.

- In **inducible regulation**, the gene is regulated by the presence of its substrate (the **inducer**).
- In **repressible regulation**, the gene is regulated by the product of its enzyme pathway (the **corepressor**).

Gene regulation *in vivo* can utilize any of these mechanisms, resulting in all four combinations: **negative inducible**, **negative repressible**, **positive inducible** and **positive repressible**.

Structural Gene Clusters are Coordinately Controlled

Genes coding for proteins that function in the same pathway may be located adjacent to one another and controlled as a single unit that is transcribed into a polycistronic mRNA.

The *lac* operon is Negative Inducible

Transcription of the *lacZYA* operon is controlled by a repressor protein (the *lac* repressor) that binds to an operator that overlaps the promoter at the start of the cluster. It shows constitutive expression-A state in which a gene is expressed continuously. In the absence of β-galactosides, the *lac* operon is expressed only at a very low (basal) level. The repressor protein is a tetramer of identical subunits coded by the *lacI* gene. **β**-galactoside sugars, the substrates of the *lac* operon, are its inducer. The addition of specific β-galactosides induces transcription of all three genes of the *lac* operon. The *lac* mRNA is extremely unstable; as a result, induction can be rapidly reversed.

Lac Repressor is Controlled by a Small-molecule Inducer

An inducer functions by converting the repressor protein into a form with a lower operator affinity. Repressor has two binding sites, one for the operator DNA and another for the inducer. Gratuitous inducer – Inducers that resemble authentic inducers of transcription, but are not substrates for the induced enzymes.

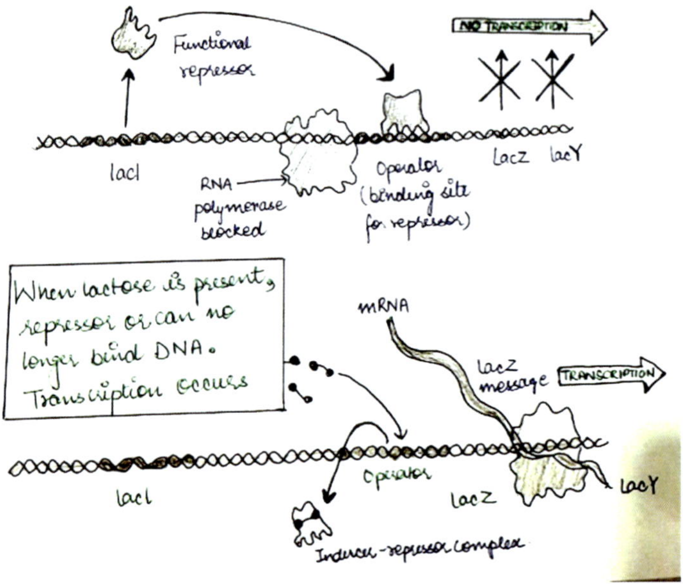

Lac Operon

The repressor is inactivated by an allosteric interaction in which binding of the inducer at its site changes the properties of the DNA-binding site (allosteric control). The true inducer is allolactose, not the actual substrate of β-galactosidase.

cis-Acting Constitutive Mutations Identify the Operator

Mutations in the operator cause constitutive expression of all three *lac* structural genes. These mutations are *cis*-acting and affect only those genes on the contiguous stretch of DNA. Mutations in the promoter prevent the expression of *lacZYA* and are uninducible and *cis*-acting. Cis-dominant – A site or mutation that affects the properties only of its own molecule of DNA, often indicating that a site does not code for a diffusible product.

Trans-acting Mutations Identify the Regulator Gene

Mutations in the *lacI* gene are Trans-acting and affect the expression of all *lacZYA* clusters in the bacterium. Mutations that eliminate *lacI* function cause constitutive expression and are recessive (*lacI*$^{-}$). Mutations in the DNA-binding site of the repressor are constitutive because the repressor cannot bind the operator. Mutations in the inducer-binding site of the repressor prevent it from being inactivated and cause uninducibility. When mutant and wild-type subunits are present, a single *lacI*$^{-d}$ mutant subunit can inactivate a tetramer whose other subunits are wild-type. It is dominant negative. Interallelic complementation – The change in the properties of a heteromultimeric protein brought about by the interaction of subunits coded by two different mutant alleles. The mixed protein may be more or less active than the protein consisting of subunits of only one or the other type. Negative complementation – This occurs when interallelic complementation allows a mutant subunit to suppress the activity of a wild-type subunit in a multimeric protein. *lacI^-d^* mutations occur in the DNA-binding site. Their effect is explained by the fact that repressor activity requires all DNA-binding sites in the tetramer to be active.

Positive Gene Regulation

If glucose levels are low (along with overall energy levels), then cyclic AMP (cAMP) binds to cAMP receptor protein (CRP) which activates transcription. If glucose levels are sufficient and cAMP levels are low (lots of ATP), then the CRP protein has an inactive shape and cannot bind upstream of the *lac* promoter.

Points of Control of Gene Regulation

The control of gene expression can occur at any step in the pathway from a gene to a functional protein.

- Packing/unpacking DNA
- Transcription
- mRNA processing
- mRNA transport
- Translation
- Protein processing
- Protein degradation

DNA sequence that can move in the genome. It is also called a mobile DNA element or transposable element. It has “selfish DNA” that exist only to maintain themselves. **Transposition:** The process by which these sequences are copied and inserted into a new site in the genome. Probably had a significant influence on evolution. In the 1940s, Barbara McClintock discovered the first transposable element in maize, earned a Nobel prize in 1983. In the late 1960s, transposition was also found in bacteria.

Retrotransposons

Most mobile elements in bacteria are DNA transposons. In contrast, most mobile elements in eukaryotes are retrotransposons, but eukaryotic DNA transposons also occur.

DNA Transposons

- Bacterial insertion sequences (IS element).
- P element in *Drosophila.*

Retrotransposons

- LTR retrotransposons.
- Non-LTR retrotransposons: the most common type of transposons in mammals.

Non-LTR Retrotransposons

- Long interspersed elements (LINEs) ≈6 kb in humans account for 21% of the genome.
- Short interspersed elements (SINEs) ≈300 bp in humans account for 13% of the genome.

Transposable Elements

Transposable elements are called mobile genetic elements. It comprises 45% of human chromosomal DNA "middle repetitive DNA". It contributes to spontaneous mutation, genetic rearrangements, and horizontal transfer of genetic material. The aid speciation and genomic change (in bacteria transposons are often associated with antibiotic resistance genes). The cells must depress transposition to insure genetic stability.

Types of Transposable Elements

It has repeated ends, usually inverted, sometimes direct. The repeated ends themselves are IS elements and can independently transpose. The ends mobilize all intervening DNA. It often has antibiotic resistance genes (examples Tn3 (ampicillin), Tn5 (kanamycin), Tn10 (tetracycline). It often resides on plasmids.

Types of DNA Transposons

Excisive mechanism examples: Tn5, Tn10, P elements.

Replicative mechanism examples: Tn3, bacteriophage Mu.

Excisive Transposons

Cut-and-paste mechanisms are called excisive transposons. It has cut them out of the original site, producing a double-strand break. Cut target site and ligate to element ends, thereby inserting at the new site. The original site breaks repaired usually with sister chromosome, restoring transposon at the original site and sometimes end healed without transposon, can also be associated with deletion at the excision site.

Replicative Transposons

The original cut of the transposon is only nick and only one strand is initially ligated. Element replicates through itself. Produces as intermediate a "co-integrate" structure. Co-integrate is resolved by resolvase (as TnpR of Tn3) and at a specific site (as *res* of Tn3).

"Degenerate" Transposons

Many naturally occurring transposable elements have suffered from mutations and are no longer active. Some of these may have cis-acting end mutations and cannot be mobilized. Others may have intact ends but no transposase: these can be mobilized by an element that is tnp+ ("autonomous" element). Ac Ds system is an example of the latter: Ac (activator) can mobilize Ds (dissociator).

- MITEs (minature inverted repeat transposable elements) are nonautonomous DNA elements.
- SINEs are retrotransposon version (LINEs).

Classification of Retroelements

It has an obligate RNA intermediate, use reverse transcriptase (RT, RNA-dependent DNA polymerase).

- LTR-retroelements: long terminal repeatsTy1/copia, Ty3/gypsy and retroviruses.
- Non-LTR-retroelements "retroposons" LINES.

Characteristics of LTR Retroelements

Long terminal repeats are required for the replication cycle. **Genes:** gag, pol, (viruses also have env). Pol is a polyprotein that gives rise to RT (reverse transcriptase), IN (integrase) RH (RNase H), PR (protease). Forms VLPs virus-like particles. Integrase is functionally and structurally similar to transposase of DNA transposons, DDE motif. The integration gives characteristic TSD.

Characteristics of Non-LTR Retroelements

It contains 2 ORFs, orf1, orf2. The variable TSD. ORF2 gives rise to EN, endonuclease, (similar to APE) and RT. It uses target primed reverse transcription TPRT. It can transduce 3' downstream non-element segments.

Human L1 (LINE-1) Retroelement

It is present in 15% of human DNA. It contains 520,000 copies, only 3-5,000 are full-length. It is associated with human disease loci. Transpose specifically integrates it in the germline.

Subjective Questions

1. Define replication. Explain the process of replication?
2. Define transcription. Explain the process of transcription.
3. Write about Inhibitors of Transcription?
4. Write a note on post-transcriptional modifications?
5. Define genes and write on genetic code?
6. Define translation. Describe translation?
7. Explain DNA damage and DNA repair mechanisms?
8. Write very briefly about diseases of DNA repair?
9. Explain salient features of the Lac operon model?
10. Define recombinant DNA. How it is obtained? Write applications of this technology.
11. Write about PCR?
12. Write about Okazaki fragments?
13. Write about transcription factors?
14. Write a note on genetic engineering?
15. Write about factors of translation?
16. Write about the formation and functions of RNA primers.
17. Write about genes of lac operon model?

18. Write a note on the replication fork?
19. Write a note on RNA polymerases?
20. What are transposons?
21. Explain DNA transposons?

Objective Questions

1. A nucleoside consists of

(A) Nitrogenous base
(B) Purine or pyrimidine base + sugar
(C) Purine or pyrimidine base + phosphorous
(D) Purine + pyrimidine base + sugar + phosphorous

Answer: B

2. A nucleotide consists of

(A) A nitrogenous base like choline
(B) Purine + pyrimidine base + sugar + phosphorous
(C) Purine or pyrimidine base + sugar
(D) Purine or pyrimidine base + phosphorous

Answer: B

3. A purine nucleotide is

(A) AMP
(B) UMP
(C) CMP
(D) TMP

Answer: A

4. A pyrimidine nucleotide is

(A) GMP
(B) AMP
(C) CMP
(D) IMP

Answer: C

5. Adenine is

(A) 6-Amino purine
(B) 2-Amino-6-oxypurine

(C) 2-Oxy-4-aminopyrimidine
(D) 2, 4-Dioxypyrimidine

Answer: A

6. **2, 4-Dioxypyrimidine is**
(A) Thymine
(B) Cytosine
(C) Uracil
(D) Guanine

Answer: C

7. **The chemical name of guanine is**
(A) 2,4-Dioxy-5-methylpyrimidine
(B) 2-Amino-6-oxypurine
(C) 2-Oxy-4-aminopyrimidine
(D) 2, 4-Dioxypyrimidine

Answer: B

8. **Nucleotides and nucleic acids concentration are often also expressed in terms of**
(A) ng
(B) mg
(C) meq
(D) OD at 260 nm

Answer: D

9. **The pyrimidine nucleotide acting as the high energy intermediate is**
(A) ATP
(B) UTP
(C) UDPG
(D) CMP

Answer: C

10. **The carbon of the pentose in ester linkage with the phosphate in a nucleotide structure is**
(A) C1
(B) C3

(C) C4
(D) C5

Answer: D

11. Uracil and ribose form

(A) Uridine
(B) Cytidine
(C) Guanosine
(D) Adenosine

Answer: A

12. The most abundant free nucleotide in mammalian cells is

(A) ATP
(B) NAD
(C) GTP
(D) FAD

Answer: A

13. The mean intracellular concentration of ATP in mammalian cell is about

(A) 1 mM
(B) 2 mM
(C) 0.1 mM
(D) 0.2 mM

Answer: A

14. The nucleic acid base found in mRNA but not in DNA is

(A) Adenine
(B) Cytosine
(C) Guanine
(D) Uracil

Answer: D

15. In RNA molecules, 'Caps'

(A) Allow tRNA to be processed
(B) Are unique to eukaryotic mRNA

(C) Occur at the 3' end of tRNA

(D) Allow correct translation of prokaryotic mRNA

Answer: B

16. In contrast to eukaryotic mRNA, prokaryotic mRNA

(A) Can be polycistronic

(B) Is synthesized with introns

(C) Can only be monocistronic

(D) Has a poly-A tail

Answer: A

17. The size of small stable RNA ranges from

(A) 0–40 nucleotides

(B) 40–80 nucleotides

(C) 90–300 nucleotides

(D) More than 320 nucleotides

Answer: C

18. The number of small stable RNAs per cell ranges from

(A) 10–50,000

(B) 50,000–1,00,000

(C) 1,00,000–10,00,000

(D) More than 10 lakhs

Answer: C

19. Molecular weight of heterogeneous nuclear RNA (hnRNA) is

(A) More than 107

(B) 105 to 106

(C) 104 to 105

(D) Less than 104

Answer: A

20. In RNA molecule guanine content does not necessarily equal its cytosine content nor does its adenine content necessarily equal its uracil content since it is a

(A) Single strand molecule

(B) Double-stranded molecule

(C) Double-stranded helical molecule
(D) Polymer of purine and pyrimidine ribonucleotides.

Answer: A

21. The nitrogenous base present in the RNA molecule is

(A) Thymine
(B) Uracil
(C) Xanthine
(D) Hypoxanthine

Answer: B

22. RNA does not contain

(A) Uracil
(B) Adenine
(C) Thymine
(D) Ribose

Answer: C

23. The sugar moiety present in RNA is

(A) Ribulose
(B) Arabinose
(C) Ribose
(D) Deoxyribose

Answer: C

24. In RNA molecule

(A) Guanine content equals cytosine
(B) Adenine content equals uracil
(C) Adenine content equals guanine
(D) Guanine content does not necessarily equal its cytosine content.

Answer: D

25. Methylated purines and pyrimidines are characteristically present in

(A) mRNA
(B) hnRNA
(C) tRNA
(D) rRNA

Answer: C

26. Thymine is present in

(A) tRNA

(B) Ribosomal RNA

(C) Mammalian mRNA

(D) Prokaryotic mRNA

Answer: A

27. The approximate number of nucleotides in tRNA molecule is

(A) 25

(B) 50

(C) 75

(D) 100

Answer: C

28. In every cell, the number of tRNA molecules is at least

(A) 10

(B) 20

(C) 30

(D) 40

Answer: B

29. The structure of tRNA appears like a

(A) Helix

(B) Hairpin

(C) Cloverleaf

(D) Coil

Answer: C

30. Although each specific tRNA differs from the others in its sequence of nucleotides, all tRNA molecules contain a base -paired stem that terminates in the sequence CCA at

(A) 3′ Termini

(B) 5′ Termini

(C) Anticodon arm

(D) 53 ″ -Termini

Answer: A

31. Transfer RNAs are classified on the basis of the number of base pairs in

(A) Acceptor arm
(B) Anticodon arm
(C) D arm
(D) Extra arm

Answer: D

32. In tRNA molecule D arm is named for the presence of the base:

(A) Uridine
(B) Pseudouridine
(C) Dihydrouridine
(D) Thymidine

Answer: A

33. The acceptor arm in the tRNA molecule has

(A) 5 Base pairs
(B) 7 Base pairs
(C) 10 Base pairs
(D) 20 Base pairs

Answer: B

34. In tRNA molecule, the anticodon arm possesses

(A) 5 Base pairs
(B) 7 Base pairs
(C) 8 Base pairs
(D) 10 Base pairs

Answer: A

35. The T ψ C arm in the tRNA molecule possesses the sequence

(A) T, pseudouridine and C
(B) T, uridine and C
(C) T, dihydrouridine and C
(D) T, adenine and C

Answer: A

36. Double helical structure model of the DNA was proposed by

(A) Pauling and Corey
(B) Peter Mitchell
(C) Watson and Crick
(D) King and Wooten

Answer: C

37. DNA does not contain

(A) Thymine
(B) Adenine
(C) Uracil
(D) Deoxyribose

Answer: C

38. The sugar moiety present in DNA is

(A) Deoxyribose
(B) Ribose
(C) Lyxose
(D) Ribulose

Answer: A

39. DNA rich in A-T pairs have

(A) 1 Hydrogen bond
(B) 2 Hydrogen bonds
(C) 3 Hydrogen bonds
(D) 4 Hydrogen bonds

Answer: B

40. In a DNA molecule

(A) Guanine content does not equal cytosine content
(B) Adenine content does not equal thymine content
(C) Adenine content equals uracil content
(D) Guanine content equals cytosine content

Answer: D

41. DNA rich in G-C pairs have

(A) 1 Hydrogen bond
(B) 2 Hydrogen bonds

(C) 3 Hydrogen bonds
(D) 4 Hydrogen bonds

Answer: C

42. The fact that DNA bears the genetic information of an organism implies that

(A) Base composition should be identical from species to species
(B) DNA base composition should change with age
(C) DNA from different tissues in the same organism should usually have the same base composition (D) DNA base composition is altered with the nutritional state of an organism

Answer: C

43. The width (helical diameter) of the double helix in B-form DNA in nm is

(A) 1
(B) 2
(C) 3
(D) 4

Answer: B

44. The number of base pairs in a single turn of B-form DNA about the axis of the molecule is

(A) 4
(B) 8
(C) 10
(D) 12

Answer: C

45. The distance spanned by one turn of B form DNA is

(A) 1.0 nm
(B) 2.0 nm
(C) 3.0 nm
(D) 3.4 nm

Answer: D

46. In a DNA molecule the thymine concentration is 30%, the guanosine concentration will be

(A) 10%

(B) 20%
(C) 30%
(D) 40%

Answer: B

47. IN a DNA molecule, the guanosine content is 40%, the adenine content will be

(A) 10%
(B) 20%
(C) 30%
(D) 40%

Answer: A

48. An increased melting temperature of duplex DNA results from a high content of

(A) Adenine + Guanine
(B) Thymine + Cytosine
(C) Cytosine + Guanine
(D) Cytosine + Adenine

Answer: C

49. A synthetic nucleotide analog, 4-hydroxypyrazolopyrimidine is used in the treatment of

(A) Acute nephritis
(B) Gout
(C) Cystic fibrosis of lung
(D) Multiple myeloma

Answer: B

50. A synthetic nucleotide analog, used in the chemotherapy of cancer and viral infections is

(A) Arabinosyl cytosine
(B) 4-Hydroxypyrazolopyrimidine
(C) 6-Mercaptopurine
(D) 6-Thioguanine.

Answer: A

51. Genetic information flows from

(A) DNA to DNA
(B) DNA to RNA
(C) RNA to cellular proteins
(D) DNA to cellular proteins

Answer: B

52. Genetic code is

(A) Collection of codon
(B) Collection of amino acids
(C) Collection of purine nucleotide
(D) Collection of pyrimidine nucleotide

Answer: A

53. Degeneracy of genetic code implies that

(A) Codons do not code for specific amino acid
(B) Multiple codons must decode the same amino acids
(C) No anticodon on tRNA molecule
(D) Specific codon decodes many amino acids

Answer: B

54. Genetic code is

(A) Overlapping
(B) Non-overlapping
(C) Not universal
(D) Ambiguous

Answer: B

55. mRNA is complementary to the nucleotide sequence of

(A) Coding strand
(B) Ribosomal RNA
(C) tRNA
(D) Template strand

Answer: A

56. In DNA replication the enzyme required in the first step is

(A) DNA directed polymerase

(B) Unwinding proteins

(C) DNA polymerase

(D) DNA ligase

Answer: B

57. The smallest unit of DNA capable of coding for the synthesis of a polypeptide is

(A) Operon

(B) Repressor gene

(C) Cistron

(D) Replicon

Answer: C

58. Termination of the synthesis of the RNA molecule is signaled by a sequence in the template strand of the DNA molecule, a signal that is recognized by a termination protein, the

(A) Rho (ρ) factor

(B) σ factor

(C) δ factor

(D) ε factor.

Answer: A

59. All Pribnow boxes are variants of the sequence:

(A) 5′–TATAAT –3′

(B) 5′–GAGCCA –3′

(C) 5′–UAACAA –3′

(D) 5′–TCCTAG –3′

Answer: A

60. 5'-Terminus of mRNA molecule is capped with

(A) Guanosine triphosphate

(B) 7-Methylguanosine triphophate

(C) Adenosine triphosphate

(D) Adenosine diphosphate

Answer: B

61. The first codon to be translated on mRNA is

(A) AUG

(B) GGU

(C) GGA

(D) AAA

Answer: A

62. AUG, the only identified codon for methionine is important as

(A) A releasing factor for peptide chains

(B) A chain terminating codon

(C) Recognition site on tRNA

(D) A chain initiating codon

Answer: D

63. In the biosynthesis of proteins the chain terminating codons are

(A) UAA, UAG and UGA

(B) UGG, UGU and AGU

(C) AAU, AAG and GAU

(D) GCG, GCA and GCU

Answer: A

64. The formation of initiation complex during protein synthesis requires a factor:

(A) IF-III

(B) EF-I

(C) EF-II

(D) IF-I

Answer: A

65. The amino-terminal of all polypeptide chains at the time of synthesis in *E. coli* is tagged to the amino acid residue:

(A) Methionine

(B) Serine

(C) N-formyl methionine

(D) N-formal serine

Answer: C

66. Initiation of protein synthesis begins with the binding of

(A) 40S ribosomal unit on mRNA

(B) 60S ribosomal unit

(C) Charging of tRNA with specific amino acid

(D) Attachment of aminoacyl tRNA on mRNA

Answer: A

67. Initiation of protein synthesis requires

(A) ATP

(B) AMP

(C) GDP

(D) GTP

Answer: D

68. The enzyme aminoacyl tRNA synthetase is involved in

(A) Dissociation of discharged tRNA from 80S ribosome

(B) Charging of tRNA with specific amino acids

(C) Termination of protein synthesis

(D) Nucleophilic attack on esterified carboxyl group of peptidyl tRNA

Answer: B

69. In the process of activation of amino acids for protein synthesis, the number of high energy phosphate bond equivalent utilized is

(A) 0

(B) 1

(C) 2

(D) 4

Answer: C

70. Translation results in a product known as

(A) Protein

(B) tRNA

(C) mRNA

(D) rRNA

Answer: A

71. The enzyme DNA ligase

(A) Introduces superhelical twists
(B) Connects the end of two DNA chains
(C) Unwinds the double helix
(D) Synthesises RNA primers

Answer: B

72. Restriction endonucleases

(A) Cut RNA chains at specific locations
(B) Excise introns from hnRNA
(C) Remove Okazaki fragments
(D) Act as defensive enzymes to protect the host bacterial DNA from DNA of foreign organisms

Answer: D

73. The complementary base of adenine in RNA is

(A) Thymine
(B) Cytosine
(C) Guanine
(D) Uracil

Answer: D

74. Extranuclear DNA is present in

(A) Ribosomes
(B) Endoplasmic reticulum
(C) Lysosomes
(D) Mitochondria

Answer: D

75. Mitochondrial DNA is present in

(A) Bacteria
(B) Viruses
(C) Eukaryotes
(D) All of these

Answer: C

76. Ribothymidine is present in

(A) DNA

(B) tRNA

(C) rRNA

(D) hnRNA

Answer: B

77. Ten base pairs are present in one turn of the helix in

(A) A-DNA

(B) B-DNA

(C) C-DNA

(D) Z-DNA

Answer: B

78. Transfer RNA transfers

(A) Information from DNA to ribosomes

(B) Information from mRNA to cytosol

(C) Amino acids from cytosol to ribosomes

(D) Proteins from ribosomes to cytosol

Answer: C

79. Synthesis of DNA is also known as

(A) Duplication

(B) Replication

(C) Transcription

(D) Translation

Answer: B

80. Replication of DNA is

(A) Conservative

(B) Semi-conservative

(C) Non-conservative

(D) None of these

Answer: B

81. Direction of DNA synthesis is

(A) 5' → 3'

(B) 3' → 5'

(C) Both (A) and (B)

(D) None of these

Answer: A

82. Formation of RNA primer:

(A) Precedes replication

(B) Follows replication

(C) Precedes transcription

(D) Follows transcription

Answer: A

83. Okazaki pieces are made up of

(A) RNA

(B) DNA

(C) RNA and DNA

(D) RNA and proteins

Answer: C

84. Okazaki pieces are formed during the synthesis of

(A) mRNA

(B) tRNA

(C) rRNA

(D) DNA

Answer: D

85. After formation of replication fork

(A) Both the new strands are synthesized discontinuously

(B) One strand is synthesized continuously and the other discontinuously

(C) Both the new strands are synthesized continuously

(D) RNA primer is required only for the synthesis of one new strand

Answer: B

86. An Okazaki fragment contains about

(A) 10 Nucleotides
(B) 100 Nucleotides
(C) 1,000 Nucleotides
(D) 10,000 Nucleotides

Answer: C

87. RNA primer is formed by the enzyme:

(A) Ribonuclease
(B) Primase
(C) DNA polymerase I
(D) DNA polymerase III

Answer: B

88. In RNA, the complementary base of adenine is

(A) Cytosine
(B) Guanine
(C) Thymine
(D) Uracil

Answer: D

89. Another name for reverse transcriptase is

(A) DNA dependent DNA polymerase
(B) DNA dependent RNA polymerase
(C) RNA dependent DNA polymerase
(D) RNA dependent RNA polymerase

Answer: C

90. mRNA is a complementary copy of

(A) 5′-3′ strand of DNA+
(B) 3′-5′ strand of DNA
(C) Antisense strand of DNA
(D) tRNA

Answer: B

CHAPTER 6

CELL GENETICS

1. HISTORICAL DEVELOPMENT IN GENETICS AND MENDEL'S LAW'S OF HEREDITY

Gregor Mendel

He was born in 1822 in Moravia (now part of the Czech Republic). The son of a tenant farmer; joined a monastery to get an education. He was deeply interested in science, particularly heredity. At the monastery in Brno, Moravia, Mendel received the support of Abbot Napp. From "1851-1855, studied at the University of Vienna, but did not receive a degree. Heredity appeared random and unpredictable. Many traits seemed to blend in the offspring, suggesting a liquid factor controlled heredity. Yet some traits, such as red hair, did not blend away. With Abbot Napp's encouragement, Mendel studied heredity in peas, carefully choosing traits that did not appear to blend. Collected data from 1856 - 1865. Mendel's creative contribution: he was the first to follow single traits from generation to generation instead of trying to document and follow every trait in the plants. Mendel presented his findings to the Association of Natural Research in Brno in 1865. Few people recognized the significance of Mendel's research. His quantitative methods were uncommon at the time, and the "blending" theory was widely accepted. In 1868, Mendel became abbot of his monastery. His religious work left little time for research, which he set aside, though he was always convinced he had made a valuable contribution to science. Mendel died in 1884. Sixteen years later, in 1900" his work was rediscovered by Hugo de Vries and others looking for clues into the puzzle of heredity. Though criticized in some details, the main body of Mendel's work still stands.

Gametes: reproductive cells produced by sexually reproducing organisms.

Male gametes = sperm
In plants: contained in pollen
Female gametes = eggs
In plants: contained in ovules
Ovules contained in carpels

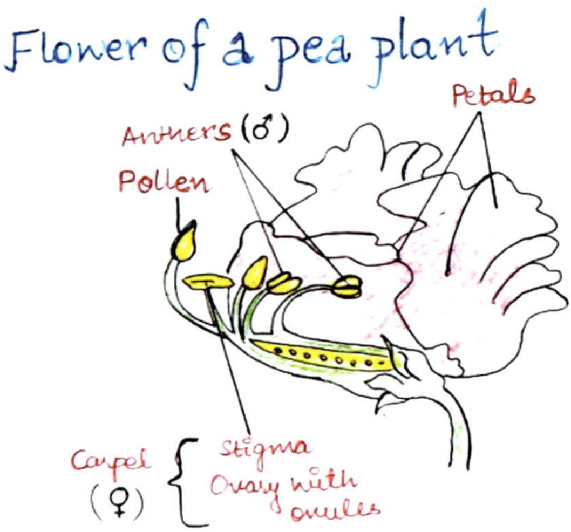

Fertilization

Fertilization: The Fusion of egg and sperm is called fertilization.

Self-fertilized: Fusion of sperm and egg from the same plant is called self-fertilization.

Cross fertilized: Fusion of egg and sperm from two different plants figure 1.1.

Produced hybrids

F_1: First generation

F_2: Second generation

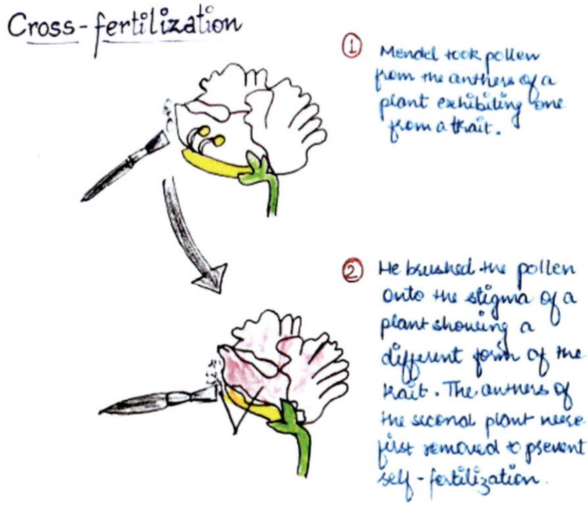

Figure 1.1: Cross-fertilization.

Mendel's Laws

A scientific law is an evidence-based description of a natural phenomenon in a given set of circumstances. Mendel's three Laws of Heredity describe what Mendel observed in patterns of inherited traits.

Laws of Heredity

- Law of Dominance
- Law of Segregation
- Law of Independent Assortment

Law of Dominance

Traits are controlled by two factors that can be called "dominant" or "recessive." A "dominant" trait shows if the offspring inherits at least one dominant factor from one parent. A "recessive" trait shows only if the offspring inherits two recessive factors, one from each parent. "In this cross between two purple-flowered pea plants, one-quarter of the offspring have white flowers. Based just on this information, which is dominant? white or purple flowers. Hint: "Dominance" is not based on the numbers of individuals with the trait. It is based on the number of copies of the allele that must be inherited to show the trait.

Mendel's Interpretations

Genes are the hereditary information that determines a single trait. Alleles are alternate forms of a gene. When an organism inherits two identical alleles for a trait, it is said to be homozygous for the trait". When an organism inherits different alleles for one trait, the organism is called heterozygous for the trait figure 1.2.

Figure 1.2: Heterozygous.

The offspring of a purple-flowered pea plant and a white-flowered pea plant all have purple flowers. The purple trait is dominant figure 1.3.

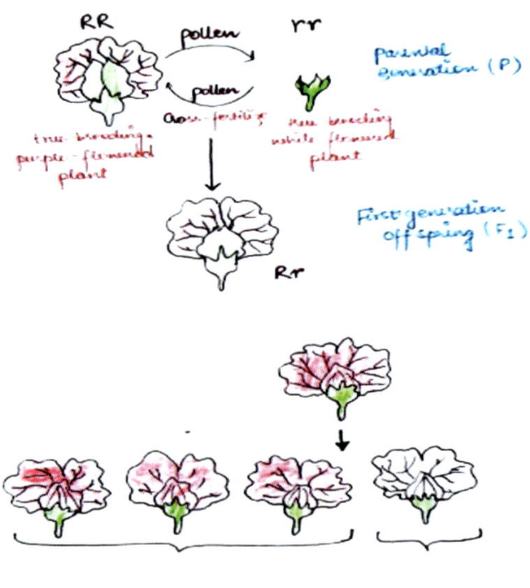

Figure 1.3: Law of dominance.

Offspring of the F1 generation (the hybrids) may be purple-flowered if they inherit at least one factor for purple flowers, or may be white flowered if they inherit the white factor from both parents.

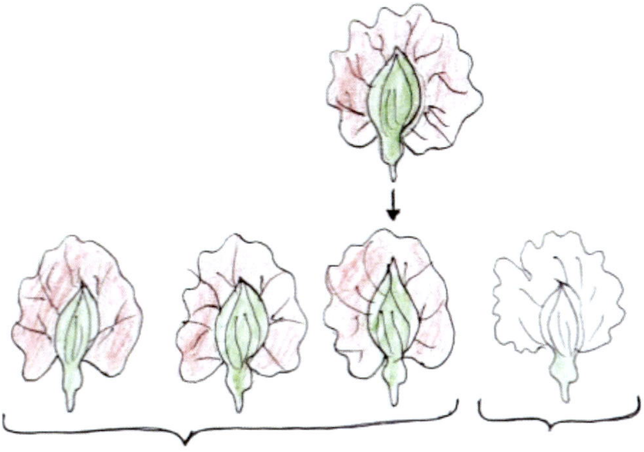

The "purple-flowered trait is dominant because each individual that inherits at least one copy of the purple allele (R) shows the purple phenotype. The white-flowered trait is recessive because an individual must inherit two copies of the white allele (r) to show the white phenotype".

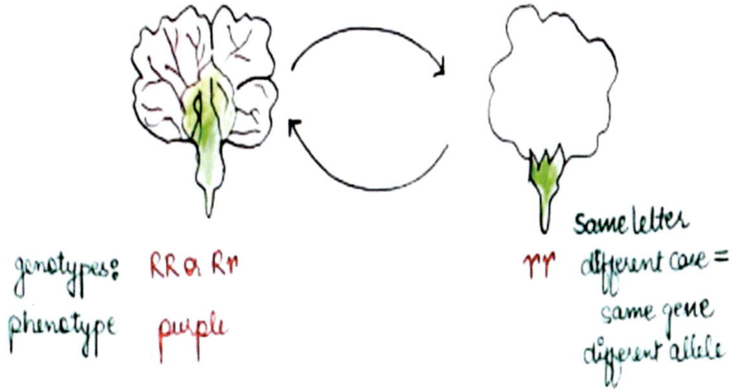

Law of Segregation

Each individual has a pair of factors controlling each trait, one inherited from each biological parent. These two factors separate are during the formation of gametes (sex cells). Only one ends up in each sex cell. The homozygous parents in the P generation can pass one kind of allele to their offspring figure 1.4.

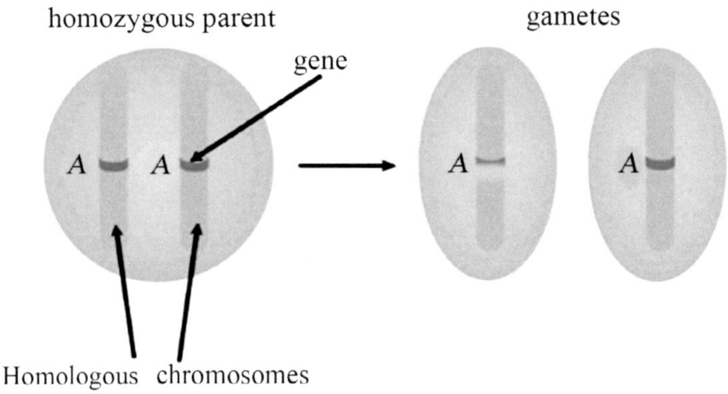

Figure 1.4: Law of segregation.

The heterozygous parents of the F1 generation have two alleles for the gene in question, and can pass one or the other, but not both, to their offspring figure 1.5.

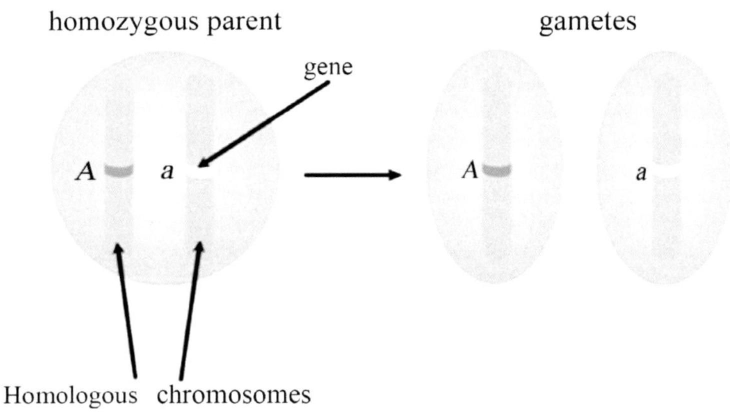

Figure 1.5: Heterozygous parents produce gametes.

The genotypes can be represented with letters, which symbolize the alleles: capital for dominant alleles, the small case for recessive. When the gametes join to produce the F1 generation, all offspring of homozygous dominant and homozygous recessive parents are heterozygous figure 1.6.

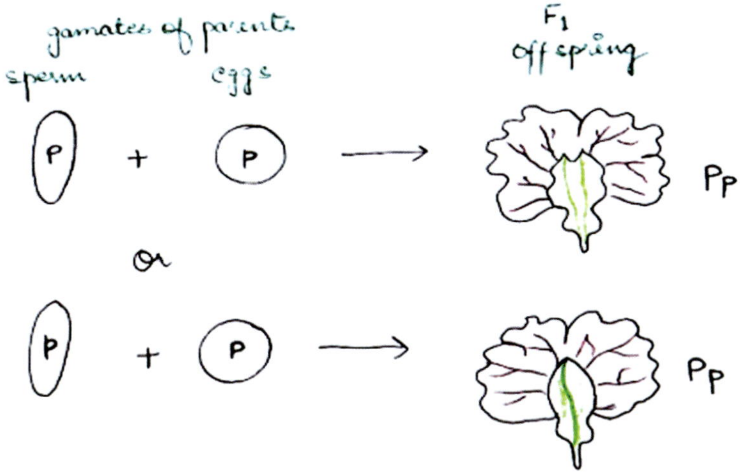

Figure 1.6: Gametes join to produce the F1 generation.

The heterozygous F1 individuals can transfer either a dominant OR a recessive allele in each of their gametes figure 1.7.

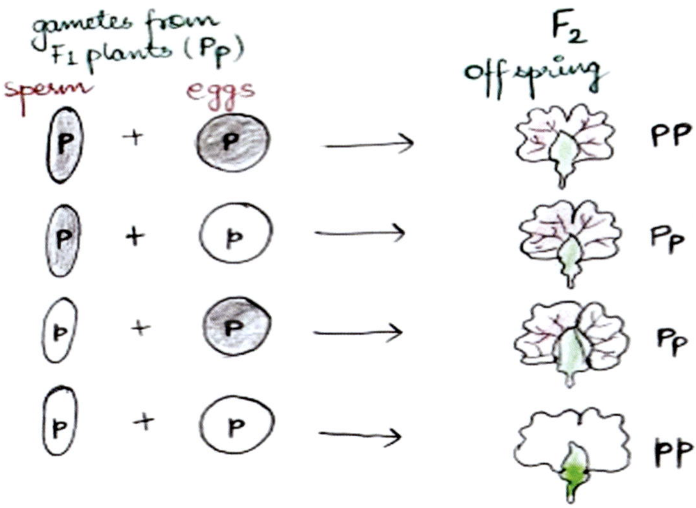

Figure 1.7: Heterozygous F1 individuals can transfer either a dominant or a recessive allele.

A punnet's square is one way to predict the outcome of a cross by showing all the possible combinations of all the possible gametes.

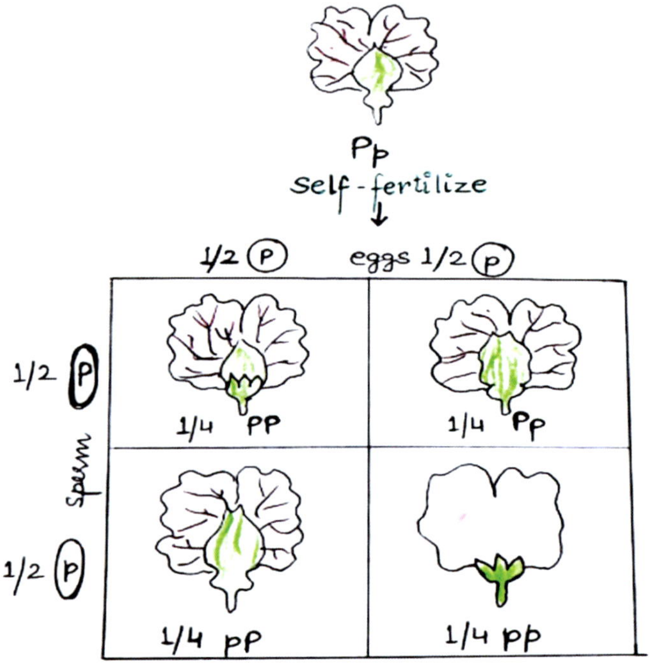

Figure 1.8: Punnet's square.

Law of Independent Assortment

The segregate independently of one another when genetic factors segregate in the gametes. "A dominant allele for one trait does not guarantee inheritance of a dominant allele for a different trait. All organisms have multiple inheritable traits controlled by genes. Each trait is inherited independently of the others. A pea plant may, for example, have yellow seeds "dominant" but white flowers "recessive" figure 1.9.

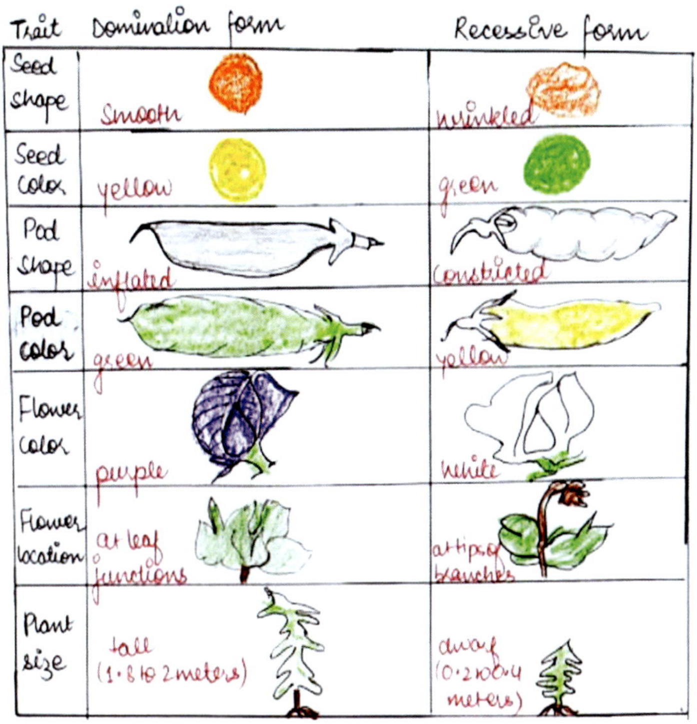

Figure 1.9: Law of independent assortment.

Laws: "Proven" Forever

Mendel's Laws were good descriptions of what he observed in the peas and other plants he worked with. New knowledge accumulated since Mendel's time has refined his ideas. "While his laws still hold true in some instances, there are many exceptions that we will explore in the next presentations. Genes may have multiple alleles, such as dominant and recessive alleles. Chromosomes, which carry genes, separate from one another during gamete formation. Chromosomes sort independently of one another during gamete formation, but each gamete gets one of each kind of chromosome".

Mutations

The change in a DNA sequence results in the appearance of a new allele. It can occur at any time, but not always beneficial. It can be harmful and result in death. It can result in the organism having traits that make them better suited to the environment.

Incomplete Dominance

A condition where all three genotypes are expressed and phenotypic level is contradicting Mendel's conclusions. The genotypic level is consistent with Mendel's laws figure 1.10.

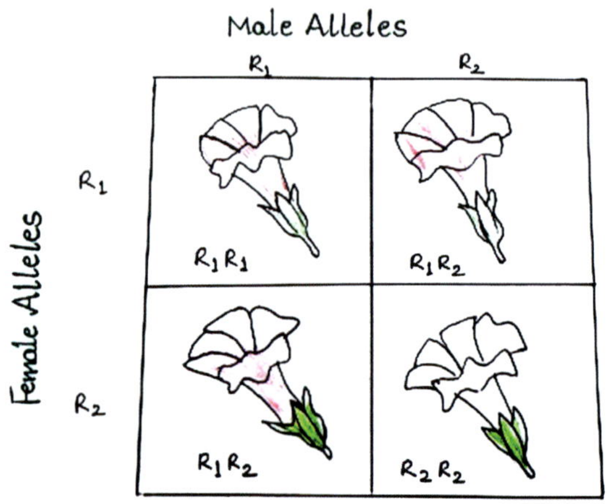

Figure 1.10: Incomplete dominance.

Lethality

A condition in which the inheritances of a lethal combination of alleles results in death of the organism; an example is Huntington's disease. Lucien Cuenot studied the inheritance of coat color in mice figure 1.11.

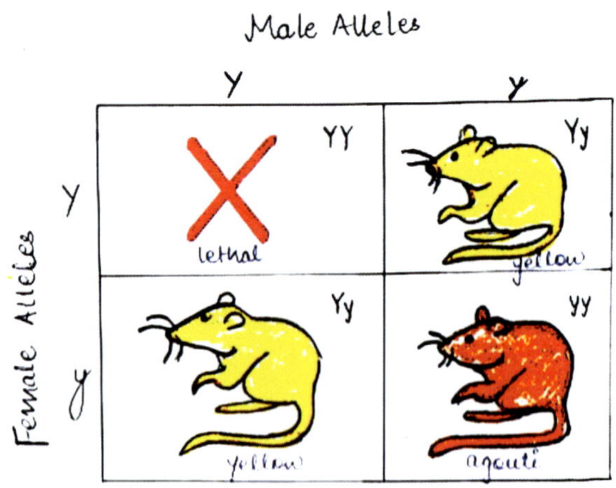

Figure 1.11: Lethality.

Pleiotropy

A single gene affects two or more traits examples: One gene affects whether seed coat is round/wrinkled and Cat fur figure 1.12.

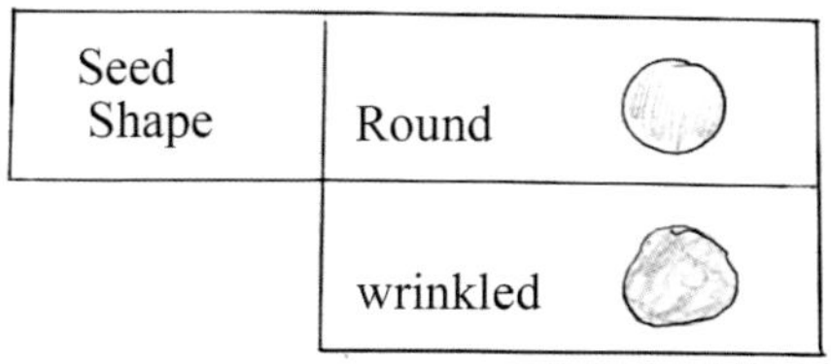

Figure 1.12: Pleiotropy.

Polygenic Inheritance

A trait affected by many genes **examples:** Height, weight, skin color.

Monogenic Inheritance

Traits determined by a single gene with two alleles **examples:** Flower color in four-o'clock plants. The cellular structure holds genetic information.

Mitosis

It is a form of asexual reproduction. It occurs when an organism grows or replaces damaged cells. Prior to mitosis, cells undergo replication. The process in which chromatin is copied. It produces diploid cells.

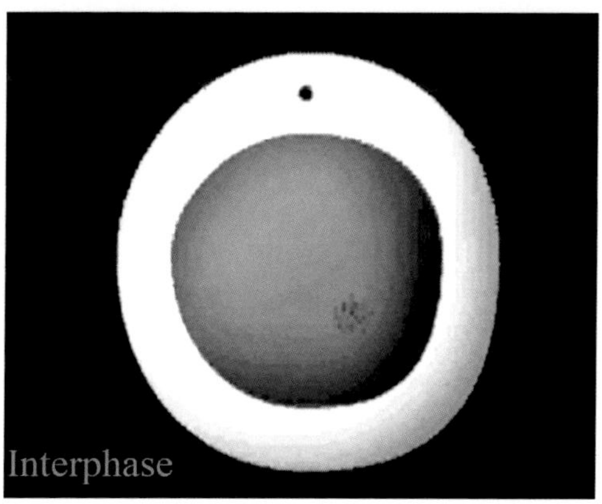

Figure 1.13: Stage of interphase.

Prophase

It is the start of mitosis. The chromatin condenses into rod-like chromosomes. Each chromosome consists of sister chromatids, connected at the centromere. The nuclear membrane disappears figure 1.14.

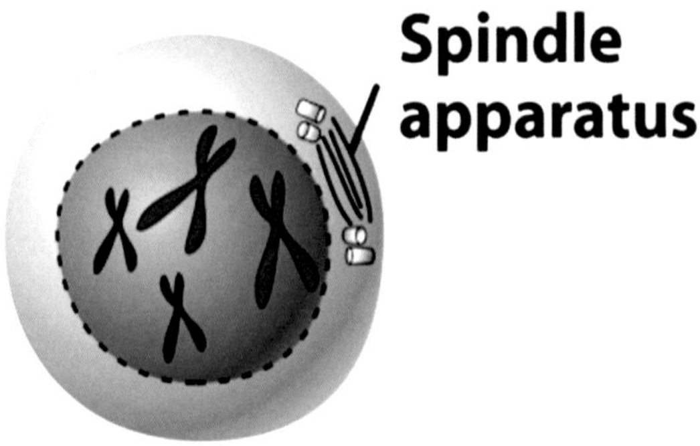

Figure 1.14: Prophase.

Metaphase

Chromosomes align themselves in the flat plane at the cell equator figure 1.15.

Figure 1.15: Metaphase.

Anaphase

Centromeres are split in anaphase and sister chromatids-now chromosomes-are pulled to opposite poles of the cell.

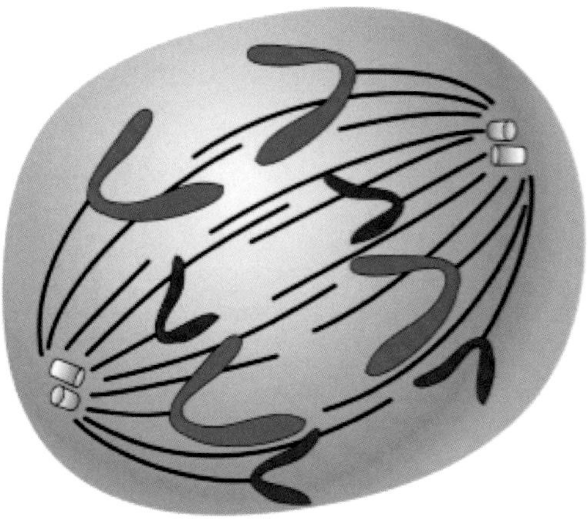

Figure 1.16: Anaphase.

Telophase

Chromosomes are unraveled, returning the chromatin to its non-dividing threadlike state and the nuclear membrane is assembled figure 1.17.

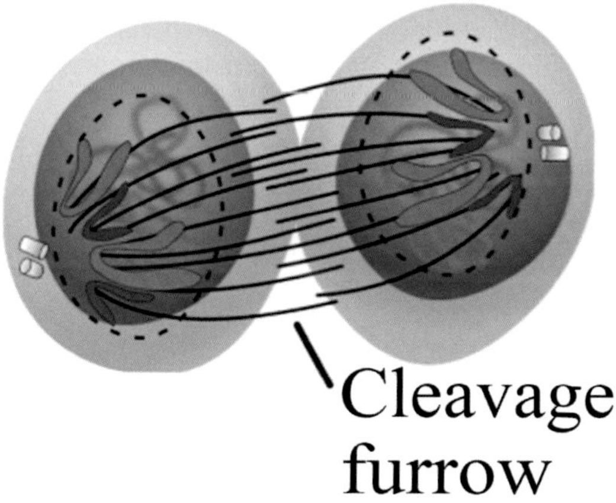

Figure 1.17: Telophase.

Cytokinesis

The division of the cytoplasm is in cytokinesis. It begins during anaphase and telophase. It differs in animals and plant cells. Plant cells form a cell plate. Membranous vesicles congregate at the center of the cell. The vesicles are containing cell wall material. Animal cells form a cleavage furrow. It forms around the periphery of the dividing cell. The furrow becomes deeper and deeper until the membrane pinches off forming two cells figure 1.18.

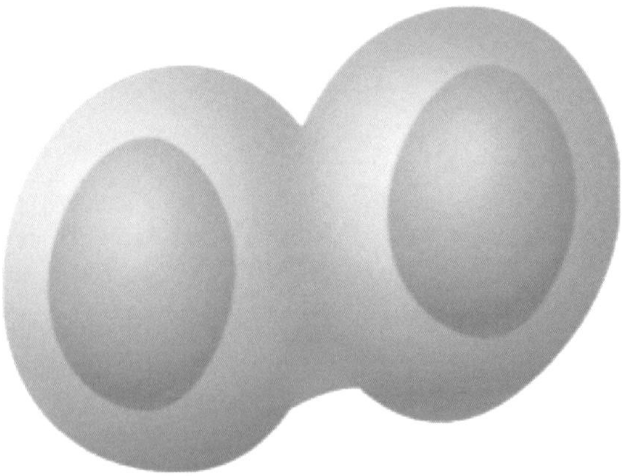

Figure 1.18: Cytokinesis.

Chromosomes Pairs

Homologous pairs: chromosomes that are closely matched in size and shape and determine the same traits. Sex chromosomes are determining the gender of the organism.

Chromosomal Theory of Inheritance

The two members of each pair of a homologous pair of chromosomes carry alleles for the same genes and, therefore, affect the same traits. It was proposed in 1903.

Mapping Genes

The location of a gene on a chromosome is currently trying to map all human genes to the appropriate chromosome **example:** the gene responsible for sickle cell anemia is located on chromosome 11.

Sexual Reproduction

It is a type of reproduction in which genetic information from the female combines with the male. It requires fertilization. The fusion of egg and sperm is called gametes. These cells need to have half the amount of genetic information.

Meiosis

It produces haploid cells (gametes-eggs and sperm). It possesses only one member of each pair of homologous chromosomes. Chromosomes replicate before meiosis. It requires two rounds of division.

Prophase I

The chromatin is condensing into compact chromosomes. The nuclear envelope is disappearing. The synapsis is occurring. The homologous pairs of chromosomes are closely aligned allowing exchange of chromosome segments.

Metaphase I

The aligned pairs of replicated chromosomes move to the equator of the dividing cell figure 1.19.

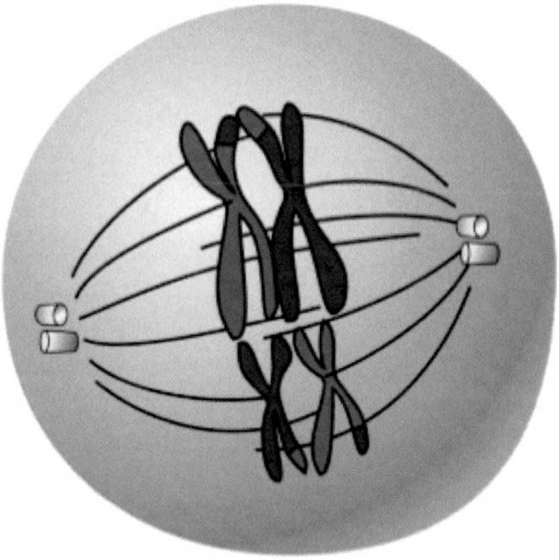

Figure 1.19: Metaphase I.

Anaphase I

The members of homologous pairs of chromosomes separate from each other. They move to the opposite poles of the cell. The chromosomes experience independent assortment figure 1.20.

Figure 1.20: Anaphase I.

Telophase I

The chromosomes cluster at the opposite poles of the cell and begin to decondense in telophase I. The nuclear envelope may reform and cytokinesis occurs.

Interkinesis: phase between meiosis I and II. The daughter cells are now haploid figure 1.21.

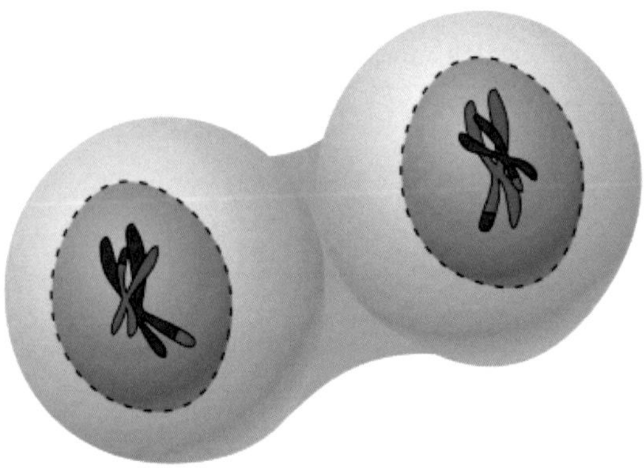

Figure 1.21: Interkinesis.

Prophase II

The partially unraveled chromosomes are condensing again figure 1.22.

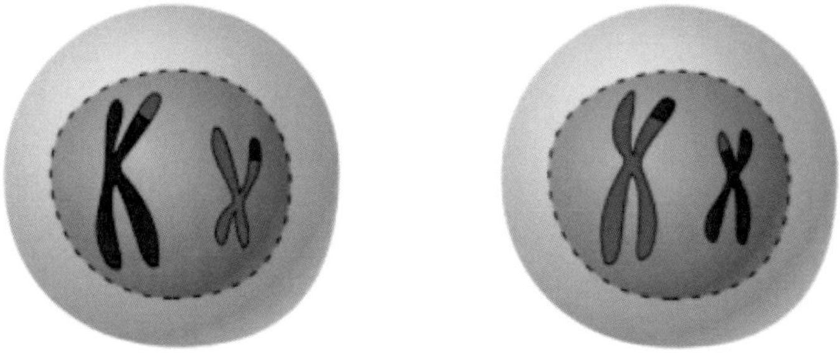

Figure 1.22: Prophase II.

Metaphase II

The chromosomes move to the cell equator. It has no longer homologous pairs, so chromosomes line up singly at the middle of the cell figure 1.23.

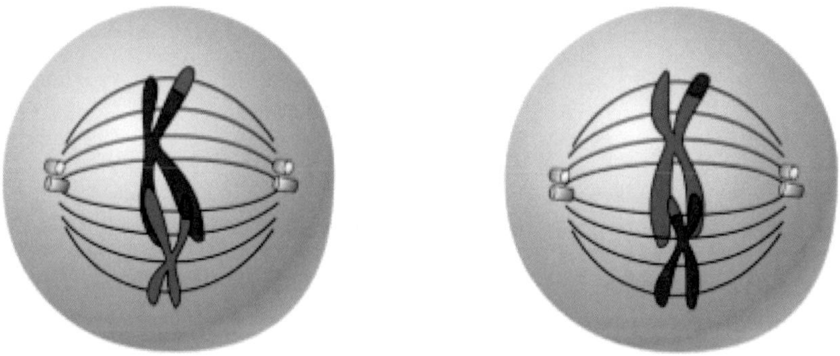

Figure 1.23: Metaphase II.

Anaphase II

The centromeres are dividing and chromosomes are separated. It moves to the opposite poles of the cell figure 1.24.

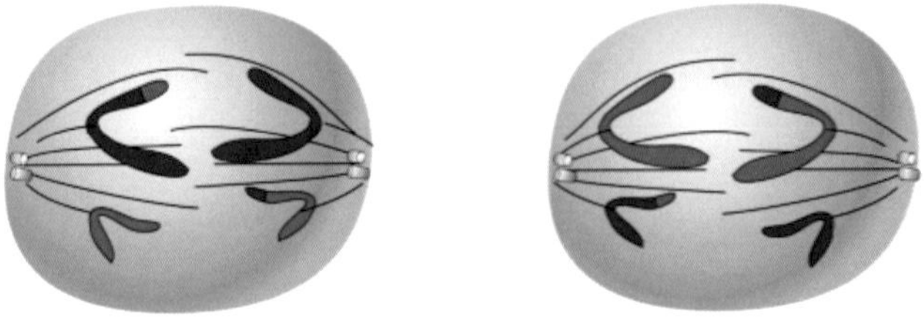

Figure 1.24: Anaphase II.

Telophase II

The clustered chromosomes at the cell pole begin to decondense and the nuclear membrane is developing. The cytokinesis is occurring and produces four non-identical haploid cells figure 1.25.

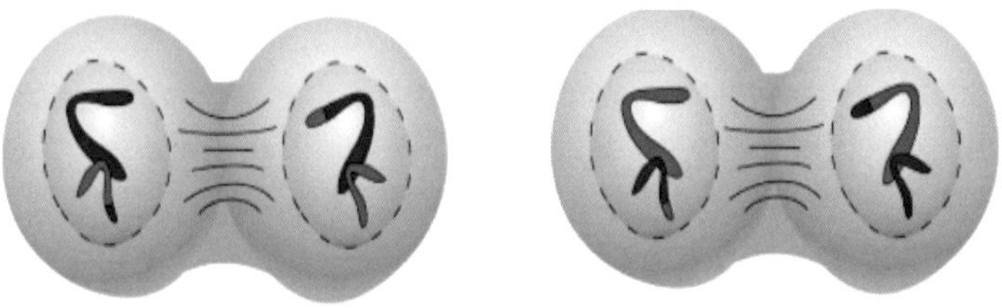

Figure 1.25: Telophase II.

Cell Cycle

It is repetitive sequence of events that characterizes the life of a cell and consists of two main phases are **figure 1.26.**

Interphase

Cells are not dividing and 90% of the cell cycle is this phase.

M phase: It includes mitosis and cytokinesis.

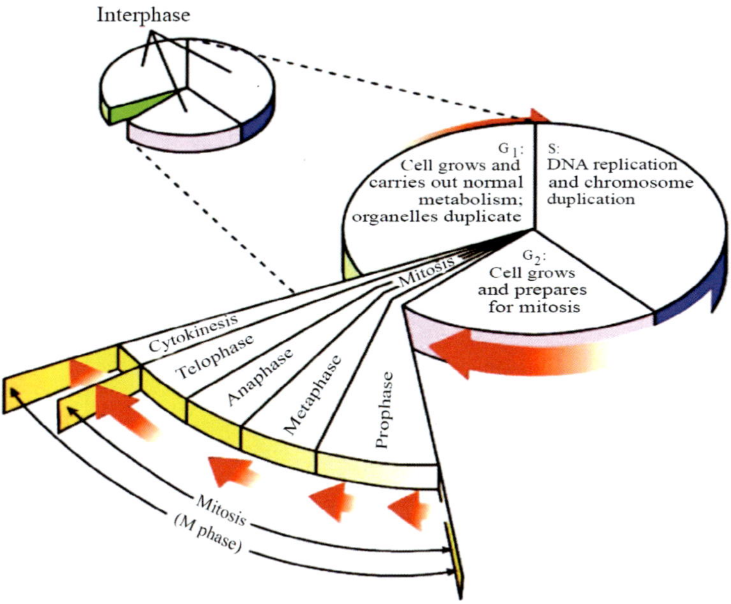

Figure 1.26: Cell cycle.

Interphase has Three Subphases

- **G_1 (first gap):** Cell makes copies of organelles and grows larger
- **S (synthesis):** Genetic material is copied
- **End of this phase, cells chromosomes are doubled**
- **Copies are attached:** thus total number of chromosomes remains the same
- **G_2 (second gap):** Cell prepares upcoming M phase

Regulation of Cell Cycle

The cell cycle must be regulated or can result in cancer and uncontrolled cell growth.

Metastasize: A cell breaks free from the original cancerous mass and resides in a new area in the body

- Two checkpoints
- First between G_1 and S
- Second between G_2 and M

- To pass checkpoints, cell must possess appropriate amounts of protein in the cytoplasm.
- These proteins activate other proteins necessary for the production of genetic material and mitosis figure 1.27.

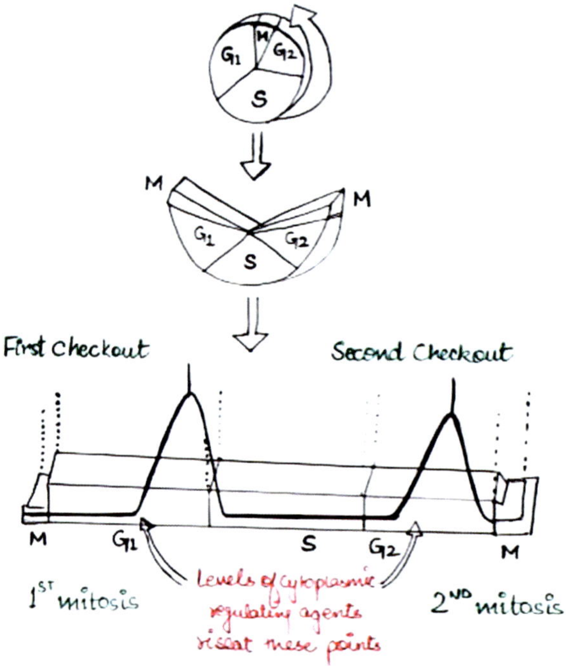

Figure 1.27: Check points in cell cycle.

To pass checkpoints, a cell must possess appropriate amounts of regulating proteins in the cytoplasm when regulator concentration is high, the cell cycle progresses. The cell cycle is suspended at that stage when genetic factors segregate in the gametes. External and internal regulatory agents also influence passage through checkpoints. Chromosomes contain genes. Chromosomes follow the law of independent assortment, not genes. If two genes are on the same chromosome, the two genes are inherited together or are said to be linked or in linkage groups. **example: sex linkage**

Sex Chromosomes

It contains other genes aside from those to determine gender **example:** eye color and gender in fruit flies figure 1.28.

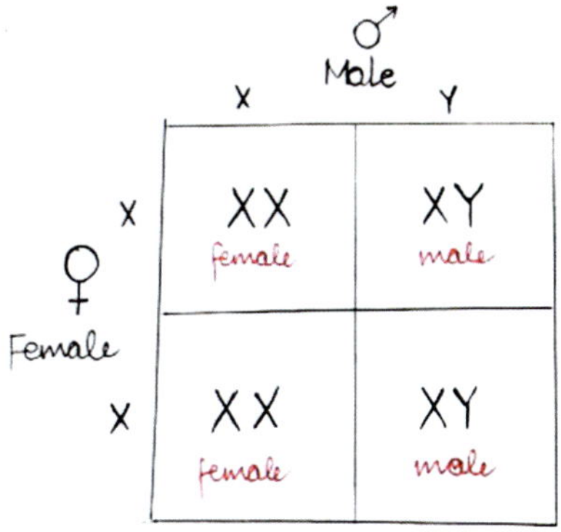

Figure 1.28: Sex chromosomes.

Crossing Over

The chromosomes can exchange segments during meiosis. The exchanges of the genetic material between chromatids of homologous chromosomes occurs at the chiasmata. It is an important mechanism for creating new combinations of genes and disrupts linkage groups figure 1.29.

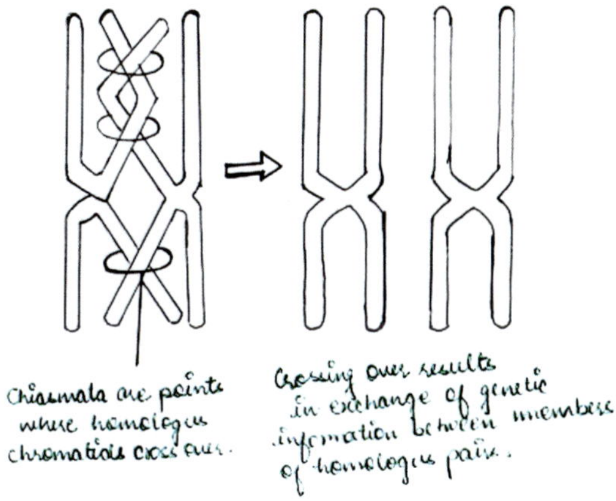

Figure 1.29: Crossing over.

DNA is the Genetic Material

The first genetic material is generally believed to have been RNA, initially manifested by self-replicating RNA molecules floating on bodies of water. This hypothetical period in the evolution of cellular life is known as the RNA world. Frederich Meisner studied the fundamental constituents of life in the 1860s. He discovered an unknown substance containing carbon, nitrogen, oxygen, and phosphorus. It came from the nucleus of the cell and was called nucelin. His students renamed the substance nucleic acid after finding it was acidic in 1881. He discovered that nucleic acids were contained in chromatin.

Fred Griffiths, medical officer for the British Ministry of Health in 1928 studied *Pneumoncoccus* bacteria. Two kinds: smooth and rough and experimented by injecting the two types in mice.

- Smooth: Virulent form that appears smooth and shiny when grown on an agar plate
- Rough: harmless form that appears rough when grown on an agar plate.

Grifth performed transformation experiements with two different strains of the bacterium Diplococcus pneumonia (virulent strains), which cause pneumonia in certain vertebrates (such as human and mice), and avirulent strains which does not cause pneumonia. The difference in virulence is related to the polysaccharide capsule of the bacterium. Virulent strains have this capsule, whereas avirulent strains do not. The non-capsulated bacteria are readily engulfed and destroyed by phagocytic cells. The virulent strain bacteria is posse's polysaccharide capsule is not easily engulfed. So they are able to multiply and cause pneumonia figure 1.30 (Further readings are suggested in the Pathfinder Life sciences book-Parnav Kumar and Usha Mina).

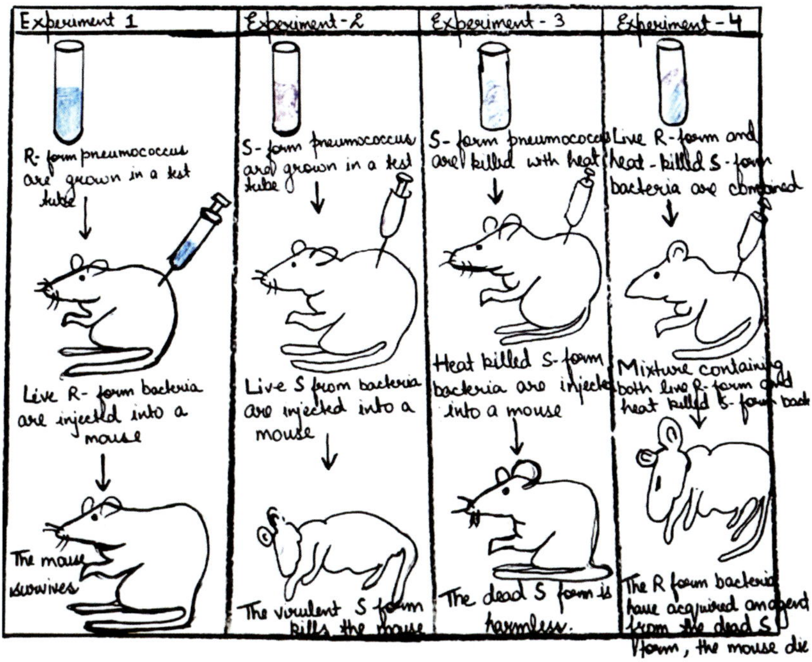

Figure 1.30: Griffith transformation experiments.

Griffiths identified the material as the transforming principle. Avery, MacLeod and McCartney studied the transforming principle for 20 years. They determined that the transforming agent was DNA. Alfred Hershey and Martha Chase studied viruses that infect bacteria. Viruses are called bacteriophages. Viruses are made of Protein coat and nucleic acid. Viruses mix their genes with host genes to hijack cell machinery and use it to produce new viruses. Usually kills a host cell. Hershey and Chase labeled protein and DNA differently with isotopes. Isotopes are variants of elements that share the same chemical properties but differ in the number of neutrons. It labeled DNA with ^{32}P and protein with ^{35}S.

Eukaryotic Chromosome Structure

DNA from all 46 chromosomes in a human cell nucleus was laid out end to end, it would measure approximately two meters. However, the diameter would be only 2 nm. Considering that the size of a typical human cell is about 10 μm (100,000 cells lined up to equal one meter), DNA must be tightly packaged to fit in the cell's nucleus. At the same time, it must also be readily accessible for the genes to be expressed. During some stages of the cell cycle, the long strands of DNA are condensed into compact chromosomes. There are a number of ways that chromosomes are compacted to fit in the cell's nucleus and be accessible for gene expression. In the first level of compaction, short stretches of the DNA double helix wrap around a core of eight histone proteins at regular intervals along the entire length of the chromosome. The DNA-histone complex is called chromatin. The beadlike, histone DNA complex is called a nucleosome. DNA connecting the nucleosomes is called linker DNA. A DNA molecule in this form is about seven times shorter than the double helix without the histones. The beads are about 10 nm in diameter, in contrast with the 2-nm diameter of a DNA double helix. The next level of compaction occurs as the nucleosomes and the linker DNA between them are coiled into a 30-nm chromatin fiber. This coiling further shortens the chromosome so that it is now about 50 times shorter than the extended form. In the third level of packing, a variety of fibrous proteins is used to pack the chromatin. These fibrous proteins also ensure that each chromosome in a non-dividing cell occupies a particular area of the nucleus that does not overlap with that of any other chromosome figure 1.31 and 1.32.

Chromatin: the complex of DNA and chromosomal proteins

- It is twice as much protein as DNA

Two Major Types of Proteins

- **Histones:** Abundant, basic proteins with a positive charge that bind to DNA 5 main types: H1, H2A, H2B, H3, H4, equal in mass to DNA, evolutionarily conserved
- **Non-histones:** All the other proteins associated with DNA differ markedly in type and structure amounts vary widely, >> 100% DNA mass and << 50% DNA mass

Histones (Components of the nucleosome)

The major protein component of chromatin

↳ small, very basic (lys, arg) proteins (to neutralize the highly charged DNA)

↳ assembly of histones with DNA requires, a molecular chaperone, nucleoplasmin

Five Types of Histones:

H1 Clamp for DNA wrapped around nucleosome

"Core" Histones: H2A, H2B, H3, H4 } Found in 2 copies of each in the nucleosome. The DNA is then wrapped around the protein core made of these proteins – together they form an octamer

H4: Is very highly conserved between species (98% identical between cows and peas) ~ 1% change in 600 million years

H3: Is also very conserved (97% identical)

Nucleosomal packaging of DNA

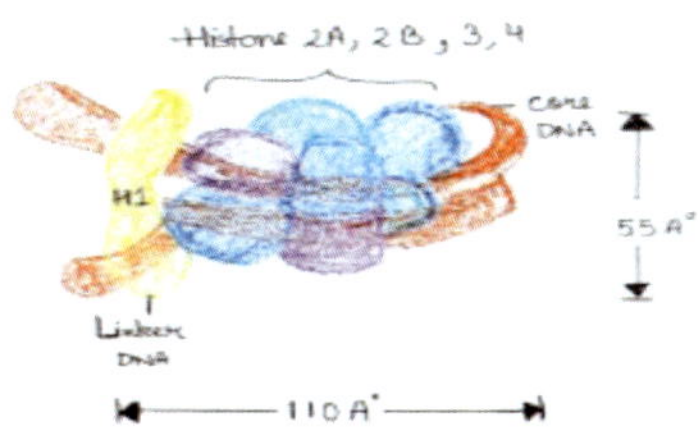

In the presence of Histone H1 175-200 bp DNA is associated with the nucleosome – but, only 146 bp is wrapped around the octamer (ie if H1 is removed ~ 146 bp is observed in nuclease digests)

Histone H1 – helps "clamp" the DNA onto the nucleosome and participates in higher-order chromatin folding.

Histone modifications

Histones are also modified in order to regulate

1. chromosome structure; compaction
2. gene activity

The main types of modifications are

a) Phosphorylation of serines
b) Methylation of lysines
c) Acetylation of lysines: neutralizes + charges

$$--NH_3^+ \rightarrow --NH-\overset{\overset{\displaystyle O}{\|}}{C}-CH_3$$

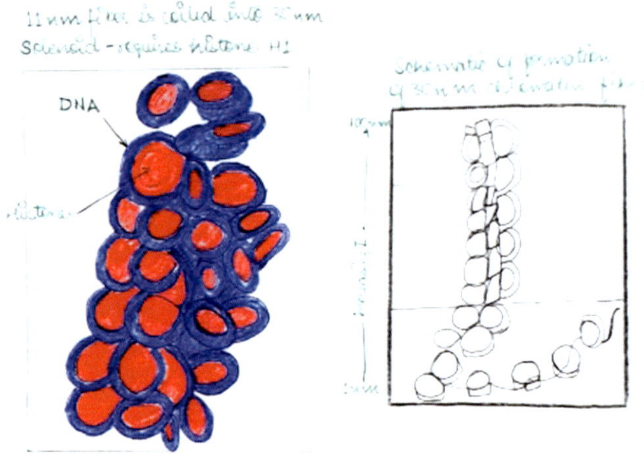

Figure 1.31: Nucleosomes.

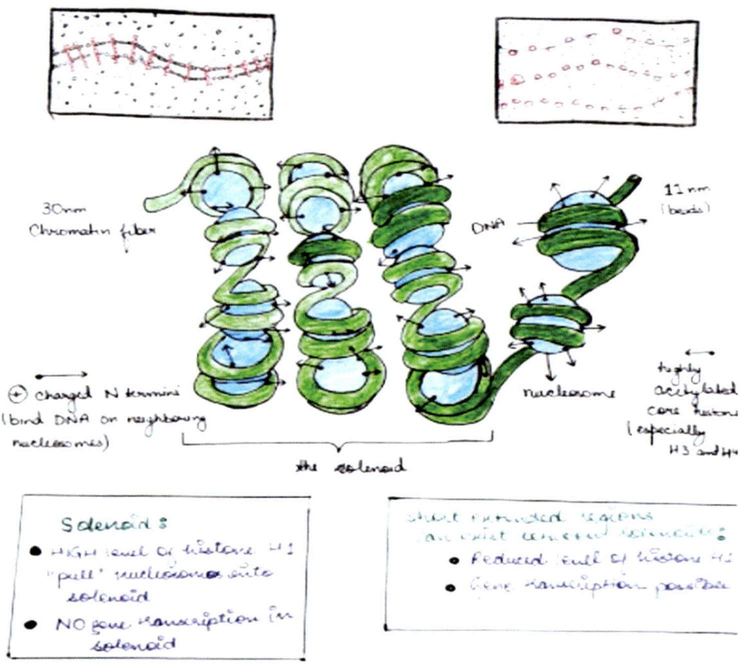

Figure 1.32: DNA binds with histones proteins.

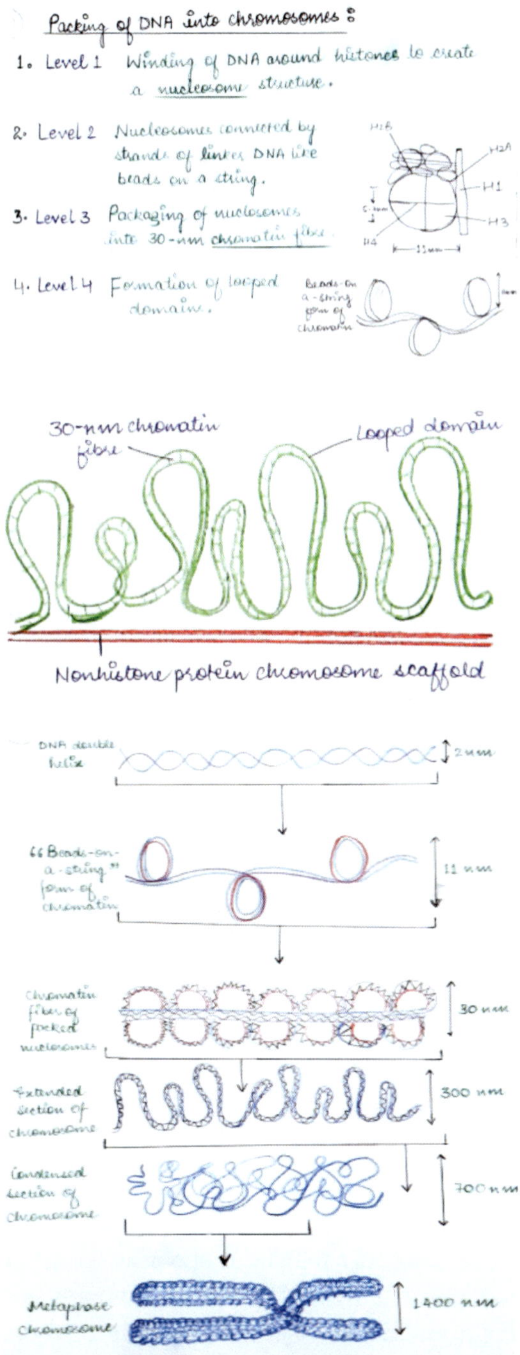

Figure 1.33: Non-histone protein chromosomes scaffold.

Centromeric DNA (CEN): Specialized sequences that function with microtubules.

Telomeric DNA: At the extreme ends of the chromosome, maintain stability, and consist of tandem repeats. Demonstrated to play a role in DNA replication and the stability of DNA.

Unique-sequence DNA: Often referred to as single-copy and usually code for genes.

Repetitive-sequence DNA: May be interspersed or clustered and vary in size.

- **SINEs:** Short interspersed repeated sequences (100-500 bp).
- **LINEs:** Long interspersed repeated sequences (>5,000 bp).
- **Microsatellites:** short tandem repeats (e.g., TTATTATTATTA).

Chromatin is divided into Euchromatin and Heterochromatin

It can be seen only during mitosis in individual chromosomes. The general mass of chromatin is in the form of euchromatin during interphase. Euchromatin is less tightly packed than mitotic chromosomes. The regions of heterochromatin remain densely packed throughout the interphase.

Chromosomes have Banding Patterns

Certain staining techniques cause the chromosomes to have the appearance of a series of striations called G-bands. The bands are lower in G • C content than the interbands. Genes are concentrated in the G • C-rich interbands.

Eukaryotic DNA has Loops and Domains Attached to a Scaffold

The DNA of interphase chromatin is negatively supercoiled into independent domains of ~85 kb. Metaphase chromosomes have a protein scaffold to which the loops of supercoiled DNA are attached.

Specific Sequences Attach DNA to an Interphase Matrix

DNA is attached to the nuclear matrix at specific sequences called MARs or SARs. The MARs are A • T-rich but do not have any specific consensus sequence. The centromere is essential for segregation. A eukaryotic chromosome is held on the mitotic spindle by the attachment of microtubules to the kinetochore that forms in its centromeric region. Centromeres often have heterochromatin that is rich in satellite DNA sequences.

Centromeres have short DNA Sequences in *S. cerevisiae*

CEN elements are identified in *S. cerevisiae* by the ability to allow a plasmid to segregate accurately at mitosis. CEN elements consist of short conserved sequences CDE-I and CDE-III that flank the A • T-rich region CDE-II.

The Centromere Binds a Protein Complex

A specialized protein complex that is an alternative to the usual chromatin structure is formed at CDE-II. The CBF3 protein complex that binds to "CDE-III is essential for centromeric function. The proteins that connect these two complexes may provide the connection to microtubules. Centromeres may contain repetitious DNA. Centromeres in higher eukaryotic chromosomes contain large amounts of repetitious DNA. The function of the repetitious DNA is not known.

Telomeres are Replicated by a Special Mechanism

The telomere is required for the stability of the chromosome end. A telomere consists of a simple repeat where a C+A-rich strand has the sequence $C_{>1}(A/T)_{1\text{-}4}$.

Telomeres Seal the Chromosome Ends

The protein TRF2 catalyzes a reaction in which 3 repeating unit of the G+T-rich strand forms a loop by displacing its homolog in an upstream region of the telomere.

Lampbrush Chromosomes are Extended

Sites of gene expression on lampbrush chromosomes show loops that are extended from the chromosomal axis.

Polytene Chromosomes form Bands

Polytene chromosomes of Dipterans have a series of bands that can be used as a cytological map.

Polytene Chromosomes expand at sites of Gene Expression

Bands that are sites of gene expression on polytene chromosomes expand to give "puffs." The nucleosome is the subunit of all chromatin. Micrococcal nuclease releases individual nucleosomes from chromatin as 11S particles. A nucleosome contains ~200 bp of DNA. Two copies of each core histone (H2A, H2B, H3, and H4). One copy of H1. DNA is wrapped around the outside surface of the protein octamer.

DNA is coiled in arrays of Nucleosomes

Greater than 95% of the DNA is recovered in nucleosomes or multimers when micrococcal nuclease cleaves DNA of chromatin. The length of DNA per nucleosome varies for individual tissues in a range from 154-260 bp.

Nucleosomes have a Common Structure

Nucleosomal DNA is divided into the core DNA and linker DNA depending on its susceptibility to micrococcal nuclease. The core DNA is the length of 146 bp that is found on the core particles produced by prolonged digestion with micrococcal nuclease. Linker DNA is the region of 8-114 bp that is susceptible to early cleavage by the enzyme. Changes in the length of linker DNA account for the variation in the total length of nucleosomal DNA. H1 is associated with linker DNA and may lie at the point where DNA enters and leaves the nucleosome.

Polytene Chromosome and Lambrush Chromosome

Giant chromosomes were first observed by E. G. Balbiani in 1881 in salivary gland cells of Dipteran species. "Polytene chromosomes are giant chromosomes found in the salivary glands, malpighian tubules, gut lining and adipose tissue cells of some Dipteran larvae. These chromosomes may reach a size up to 200 times (or more) the size of corresponding chromosomes at meiosis or in nuclei of normal mitotic cells. These giant chromosomes are formed by repeated cycles of endo-reduplication of single chromatids. This means that the chromatids replicate without cell division, as a result, the number of chromatids keeps increasing. This process is called Polyteny and therefore, these chromosomes are called polytene chromosomes. Polytene chromosomes show another characteristic feature i.e., somatic pairing. As a result of this, chromosomes always appear to be half of that in normal somatic cells. Lampbrush chromosomes were first seen in salamander oocytes by Fleming in 1882. The name lampbrush was given by Ruckert in 1892. The nomenclature comes from its similarity in appearance to the brushes used to clean lamp chimneys in old days. These chromosomes are transitory structures that exist during the extended Diplotene in oocytes of most invertebrates and vertebrates except mammals. The loops in these chromosomes are sites of RNA transcriptions". The condensed chromatin in chromomeres is generally not transcribed. The position of each loop is fixed. A given loop always contains the same DNA sequence. They have been widely used in studies of chromosome organization, gene expression, molecular morphology of RNA transcription, etc.

Chromosome Banding

Chromosomes are stained with various dyes enabling the chromosome segments to be identified. Most methods can distinguish 550 bands/ haploid set. High-resolution methods can distinguish up to 850 bands/ haploid set that can allow the identification of small interstitial deletions.

G-Banding

Dye gives chromosomes a striped appearance because it stains the regions of DNA that are rich in adenine (A) and thymine (T) base pairs. Regions that stain as dark G bands replicate late in the S phase of the cell cycle and contain more condensed chromatin. \Light G bands generally replicate early in the S phase, and have less condensed chromatin.

Numerical Changes

Mutations and Polymorphisms

Mutation: a permanent transmissible change in the genetic material, usually in a single gene. Polymorphism: two or more genetically determined, proportionally represented phenotypes in the same population.

Types of Mutations

- Genomic: abnormal chromosome number (monosomy, polysomy, aneuploidy).
- Chromosomal: abnormal chromosome structure.
- Gene: DNA sequence changes in specific genes.

Chromosome Morphology

Telomere: chromosome ends. Centromere: site of spindle attachment. Constriction of the metaphase chromosome at the centromere defines two arms. Nucleosome: DNA double helix wrapped around histone proteins.

Chromosome Morphology Changes during the Cell Division Cycle

DNA double helix: 2nm diameter Interphase (G1, S, G2). Chromatin "beads on a string:" 11 nm. Chromatin in nucleosomes: 30 nm in Metaphase (Mitosis). Extended metaphase chromosomes: 300 nm. Condensed metaphase chromosomes: 700 nm

Linkage and Crossing Over

Each species of an organism must contain hundreds to thousands of genes. "Yet most species have at most a few dozen chromosomes. Therefore, each chromosome is likely to carry many hundred or even thousands of different genes. The transmission of such genes will violate Mendel's law of independent assortment. In eukaryotic species, each linear chromosome contains a long piece of DNA. A typical chromosome contains many hundred or even a few thousand different genes. The term linkage has two related meanings

1. Two or more genes can be located on the same chromosome
2. Genes that are close together tend to be transmitted as a unit

Chromosomes are called linkage groups. They contain a group of genes that are linked together. The number of linkage groups is the number of types of chromosomes of the species.

For example, in Humans

- 22 autosomal linkage groups
- An X chromosome linkage group
- A Y chromosome linkage group

Genes that are far apart on the same chromosome may independently assort from each other. This is due to crossing-over.

Crossing over may produce Recombinant Phenotypes

In diploid eukaryotic species, "linkage can be altered during meiosis as a result of crossing over. It occurs during prophase I of meiosis at the bivalent stage. Non-sister chromatids of homologous chromosomes exchange DNA segments. Bateson and Punnett discovered two traits that did not assort independently. In 1905, William Bateson and Reginald Punnett conducted a cross in sweet pea involving two different traits. Flower color and pollen shape. This is a dihybrid cross that is expected to yield a 9:3:3:1 phenotypic ratio in the F_2 generation. However, Bateson and Punnett obtained surprising results. They suggested that the transmission of the two traits from the parents was somehow coupled. The two traits are not easily assorted in an independent manner". However, they did not realize that the coupling is due to the linkage of the two genes on the same chromosome.

2. MUTATION-SPONTANEOUS AND INDUCED, CHROMOSOMAL CHANGES-SEX CHROMOSOMES AND THEIR DETERMINATION IN MAN AND DROSOPHILA, SEX LINKED CHARACTERS, CYTOPLASMIC INHERITANCE-PLASMA AND NUCLEAR

Spontaneous Mutations

Spontaneous mutations can be produced by different processes. Spontaneous lesions and replication errors generate most of the base-substitution and frameshift mutations. Replication errors may also cause some deletions that occur in the absence of mutagenic treatment.

Extrachromosomal inheritance: The phenotype of maternal parent is solely responsible for the phenotype of offspring - an example of maternal inheritance of chloroplasts.

Mitochondrial heredity

About 30,000 genes on 23 human chromosomes (3.3 billion base pairs/ haploid cell). The mitochondrial Genome contains 37 genes.

- 13 code for some of the proteins involved in oxidative respiration
- 22 tRNA genes

- 2 rRNA genes
- some proteins involved in mt protein synthesis
- 16,569 base pairs, circular

Human Mitochondrial Diseases

It is maternally inherited: only the offspring of affected mothers are affected. Show deficiency in mitochondrial function. It is caused by a mutation in a mitochondrial gene.

Example: myoclonic epilepsy and ragged red fiber disease (MERRF)

- Deafness, dementia, seizures
- Point Mutation in a mitochondrial tRNA

Human Mitochondrial Diseases

- Leber's Hereditary Optic Neuropathy (LHON)
  - Sudden bilateral blindness.
  - Point mutation in small subunit of NADH dehydrogenase.
  - Kearns-Sayre Syndrome (KSS).
  - Symptoms in eyes, muscles, heart, brain.
  - Deletion mutation in mtDNA.

Maternal Effect

Pattern offspring is determined by the genotype of the mother, regardless of phenotype.

Maternal Effect Genes in *Drosophila* Development

It was discovered by **Edward Lewis, Christiane Nusslein-Volhard and Eric Weischaus** (Nobel Prize in Physiology or Medicine, 1995). The gene products are deposited in the egg by cells of the mother's ovary. These genes set up the first embryonic axes: dorsal-ventral and anterior-posterior and control later zygotic genes.

Sex-linked Chromosome

A trait whose gene is located on a sex chromosome (almost always the X-chromosome) and common sex-linked diseases are

- Colorblindness

- Hemophilia
- Duchenne muscular dystrophy

Mutation

Any sudden change occurring in the hereditary material is called a mutation. They may be harmful, beneficial or neutral. In multicellular organisms, there are two broad categories of mutations: Somatic mutations & germ line mutations.

Somatic Mutations

Arise in the somatic cells. Passed on to other cells through the process of mitosis. The effect of these mutations depends on the type of the cell in which they occur & the developmental stage of the organism. If occurs early in development, the larger the clone of the mutated cells.

Germ Line Mutations

They occur in the cells that produce gametes. Passed on to future generations. In multicellular organisms, the term mutation is generally used for germ line mutations. The term mutation was given by **Devries** in 1901 while studying evening primrose *Oenothera lamarckiana*. Most of these were chromosomal variations. Some were point variations. Originally the term mutation was given to both chromosomal as well as point mutations. Recently chromosomal mutations have been studied separately. The term mutation is now given only to point mutations.

Types of Gene Mutation

A number of ways to classify gene mutations. On the basis of the molecular nature of the defect. On the nature of the phenotypic effect-- the amino acid sequence of the protein is altered or not. On the basis of the causative agent of the mutation.

Mutations on the basis of the molecular nature of the defect

- Base substitution.
- Insertions & deletions.

Base Substitution

The simplest type of gene mutation. Involves the alteration of a single nucleotide in the DNA. A base substitution usually leads to base pair substitution. Base substitution is of two types.

Transition

- Purine is replaced with a purine
- A into G
- G into A
- Pyrimidine is replaced with a pyrimidine
- C into T
- T into C

Transversions

- A purine is replaced by a pyrimidine
- A into C
- A into T
- G into T
- G into C
- **or a pyrimidine is replaced by a purine**
- C into G
- C into A
- T into G
- T into A

Insertions & Deletions

Addition or the removal, respectively, of one or more nucleotide pairs. Usually changes the reading frame, altering all amino acids encoded by codons following the mutation Also called as frame shift mutations. Additions or deletions in the multiples of three nucleotides will lead to the addition or deletion of one or more amino acids. These mutations are called in-frame insertions and deletions, respectively.

Mutations on the basis of the Phenotypic effects of Mutations

The most common phenotype in natural populations of the organism is called as wild-type phenotype. The effect of mutation is considered with reference to the wild-type phenotype.

Forward Mutation

A mutation that alters the wild-type phenotype Reverse mutation (reversion). A mutation that changes a mutant phenotype back into the wild-type.

Missense Mutation

A base is substituted that alters a codon in the mRNA resulting in a different amino acid in the protein product.

Nonsense Mutation

It changes a sense codon into a nonsense codon. Nonsense mutation early in the mRNA sequence produces a greatly shortened & usually nonfunctional protein. Silent mutation: alters a codon but due to degeneracy of the codon, the same amino acid is specified.

Neutral Mutation

A mutation that alters the amino acid sequence of the protein but does not change its function as the replaced amino acid is chemically similar or the affected amino acid has little influence on protein function.

Loss of function mutations

Complete or partial loss of the normal function. The structure of protein is so altered that it no longer works correctly. Mutations can occur in the regulatory region that affects transcription, translation or splicing of the protein. Frequently recessive.

Gain of Function Mutations

It produces an entirely new trait. It causes a trait to appear in inappropriate tissues or at inappropriate times in development. It is frequently dominant.

Conditional Mutations

It is expressed only under certain conditions.

Lethal Mutations

It causes the death of the organism.

Suppressor Mutation

Suppresses the effect of other Mutation

It occurs at a site different from the site of the original mutation. The organism with a suppressor mutation is a double mutant but exhibits the phenotype of unmutated wild type. The difference from a reverse mutation is that the mutated site is reverted back into the wild type sequence.

On the basis of the Causative Agent of Mutation

Spontaneous

Mutations that result from natural changes in DNA.

Induced

The results from changes caused by environmental chemicals & radiations. Any environmental agent that increases the rate of mutation above the spontaneous is called a mutagen such as chemicals & radiations.

Chemical Mutagens

The first discovery of a chemical mutagen was made by **Charlotte Auerbach**.

Base Analogs

Chemicals with structures similar to that of any of the four standard bases of DNA. DNA polymerases cannot distinguish these analogs. They may be incorporated into newly synthesized DNA molecules.

2-amino purine (P)

Base analog of adenine. Normally pairs with thymine. May mispair with cytosine. Causes a transition mutation. Both base analogs produce transition mutations. Mutations by base analogs can be reversed by the treatment with the same analog or different analog.

Alkylating Agent

Chemicals that donate alkyl groups e.g. ethyl methanesulfonate (EMS). It adds an ethyl group to guanine and produces 6-ethylguanine, which pairs with thymine and leads to CG: TA transitions. Also adds an ethyl group to thymine to produce 4-ethylthymine, which then pairs with guanine, leading to a TA: CG transition. Mutations produced by EMS can be reversed by additional treatment with EMS. Mustard gas is another alkylating agent. Nitrous acid produces exclusively transition mutations. Both C.G T.A & T.A C.G transitions are produced. Thus mutations can be reversed with nitrous acid.

Hydroxyl Amine

Specific base modifying mutagen which adds a hydroxyl group to cytosine producing hydroxlamine cytosine which pairs with adenine instead of guanine. This Leads to C.G: T.A transitions. It acts only on cytosine thus can not revert the mutation produced.

Oxidative Reactions

Reactive forms of oxygen like superoxide radicals, hydrogen peroxide and hydroxyl radicals produced in the course of normal aerobic metabolism or by radiation, ozone, peroxides, and certain drugs cause damage to DNA & induce mutations by chemical changes. Oxidation converts guanine into 8-oxy-7,8-dihydrodeoxyguanine, which mispairs with adenine leading to G.C and T.A transversion.

Intercalating Agents

Proflavin, acridine orange, ethidium bromide, and dioxin. They are about the same size as a nucleotide. They produce mutations by sandwiching themselves (intercalating) between adjacent bases in DNA. They distort the three-dimensional structure of the helix and cause single-nucleotide insertions and deletions in replication. These insertions and deletions frequently produce frameshift mutations.

Radiations

Ionizing radiations. In 1927, **Herman Muller** demonstrated that mutations could be induced by X-rays. X-rays, gamma rays, and cosmic rays are all capable of penetrating tissues and damaging DNA. They remove electrons from the atoms that they encounter, changing stable molecules into free radicals and reactive ions which then alter the structures of bases and break phosphodiester bonds in DNA. Ionizing radiation also frequently results in double-strand breaks in DNA.

Mutation Rates

The frequency with which a gene changes from the wild type to a mutant is referred to as the mutation rate. Expressed as the number of mutations per biological unit i.e. mutations per cell division, per gamete per round of replication e.g. the mutation rate for achondroplasia (hereditary dwarfism) is about 4 mutations per 100,000 gametes.

Mutation Frequency

Incidence of a specific type of mutation within a group of individual organisms e.g. for achondroplasia, the mutation frequency in the United States is about $2x10^{-4}$.

Sex chromosomes and their determination in man and Drosophila

Chromosomes are thread-like molecules that carry hereditary information for everything from height to eye color. They are made of protein and one molecule of DNA, which contains an organism's genetic instructions, passed down from parents. In humans, animals, and plants, most chromosomes are arranged in pairs within the nucleus of a cell. Humans have 22 of these chromosome pairs, called autosomes.

Humans have an additional pair of sex chromosomes for a total of 46 chromosomes. The sex chromosomes are referred to as X and Y, and their combination determines a person's sex. Typically, human females have two X chromosomes while males possess an XY pairing. This XY sex-determination system is found in most mammals as well as some reptiles and plants.

Whether a person has XX or XY chromosomes is determined when a sperm fertilizes an egg. Unlike the body's other cells, the cells in the egg and sperm — called gametes or sex cells — possess only one chromosome. Gametes are produced by meiosis cell division, which results in the divided cells having half the number of chromosomes as the parent, or progenitor, cells. In the case of humans, this means that parent cells have two chromosomes and gametes have one.

All of the gametes in the mother's eggs possess X chromosomes. The father's sperm contains about half X and half Y chromosomes. The sperms are the variable factor in determining the sex of the baby. If the sperm carries an X chromosome, it will combine with the egg's X chromosome to form a female zygote. If the sperm carries a Y chromosome, it will result in a male.

During fertilization, gametes from the sperm combine with gametes from the egg to form a zygote. The zygote contains two sets of 23 chromosomes, for the required 46. Most women are 46XX and most men are 46XY, according to the World Health Organization.

There are some variations, though. Recent research has found that a person can have a variety of different combinations of sex chromosomes and genes, particularly those who identify as LGBT. For example, a certain X chromosome called Xq28 and a gene on chromosome 8 seem to be found in higher prevalence in men who are gay, according to a 2014 study in the journal Psychological Medicine.

A few births out of a thousand of babies are born with a single sex chromosome (45X or 45Y) and are referred to as sex monosomies. Others are born with three or more sex chromosomes (47XXX, 47XYY or 47XXY, etc.) and are called sex polysomies. "In addition, some males are born 46XX due to the translocation of a tiny section of the sex determining region of the Y chromosome," said WHO. "Similarly some females are also born 46XY due to mutations in the Y chromosome. Clearly, there are not only females who are XX and males who are XY, but rather, there is a range of chromosome complements, hormone balances, and phenotypic variations that determine sex." It is important to remember that sex and gender have two separate definitions and many cultures include more labels than simply "male" and "female" to identify others. "While the chromosomes for other parts of the body are the same size and shape — forming an identical pairing the X and Y chromosomes have different structures. The X chromosome is significantly longer than the Y chromosome and contains hundreds more genes. Because the additional genes in the X chromosome have no counterpart in the Y chromosome, the X genes are dominant. This means that almost any gene on the X, even if it is recessive in the female, will be expressed in males. These are referred to as X-linked genes. Genes found only on the Y chromosome are referred to as Y-linked genes, and expressed only in males. Genes on either sex chromosome can be called sex-linked genes".

In Man: Females have two X chromosomes in their cells, while males have both X and a Y chromosome in their cells. Egg cells all contain an X chromosome, while sperm cells contain an X or Y chromosome. This arrangement means that it is the male that determines the sex of the offspring when fertilization occurs.

In Drosophila: sex determination is achieved by a balance of female determinants on the X chromosome and male determinants on the autosomes. Normally, flies have either one or two X chromosomes and two sets of autosomes. If there is but one X chromosome in a diploid cell (1X:2A), the fly is male.

Sex Linked Characters and Cytoplasmic Inheritance-plasma and Nuclear

The existence of genes as segments of nucleic acid molecules, located in the chromosome of the nucleus, has been demonstrated by several experiments. "The nuclear genes control the phenotypes of the organisms and are concerned with the transmission of hereditary character from one generation to next generation is known and predictable Mendelian fashion. The inheritance of genes of nuclear chromosomes is characterized by the fact that the genes from male and female parents contribute equally to the genetic constitution of the offspring. Therefore, in it the reciprocal crosses between parents of different homozygous genotypes will produce offspring of identical phenotypes except for sex-linked genes. However, in certain cases, although male and female parents contribute equally their nuclear genes to the offspring, the results show a non-Mendelian inheritance pattern and the result of reciprocal crosses varies. These variations suggest that the genes for the inheritance of certain characters do not occur within the nucleus, but they are present in the cytoplasm and play an important role in the transmission of certain specific traits, which are not controlled by nuclear genes. Therefore, it builds up the concept of cytoplasmic inheritance. The genes for cytoplasmic inheritance are independent, self-replicating nucleic acids. Evidence for cytoplasmic inheritance was first reported by Correns in *Mirabilis jalapa* and by Bar in *Pelargonium zonule* in 1908. Rhoades described cytoplasmic male sterility in maize in 1933. In 1943, Sonneborn discovered kappa particles in Paramoecium and described its cytoplasmic inheritance. The presence of DNA in chloroplasts was first demonstrated by Ris in plant cells.

In 1963, Nass and his co-workers proved the existence of DNA in mitochondria. Subsequently, from time to time, observations by several scientists have been reported the important role of cytoplasm in genetics. Thus, on the basis of observations made on the cytoplasmic inheritance of some specific traits, it has been suggested that cytoplasm is also genetically active. Extra-chromosomal inheritance, extra-nuclear inheritance, somal inheritance and maternal inheritance are all synonyms. All these terms can be defined as the inheritance of characteristics of only one of the two parents, usually the female parent to the progeny. The reciprocal crosses show consistent differences as well as there is a lack of segregation in F_2 and subsequent generations".

Cytoplasmic Inheritances

They show the following characteristic features: Hereditary traits which are transmitted by cytoplasm do not show Mendelian segregation in crosses and in reciprocal crosses with respect to a particular set of characteristics controlled by a set of cytoplasmic genes produce dissimilar hybrids. Most of the recorded cytoplasmically inherited characteristics would follow the maternal line, i.e., uniparental mode of transmission. In higher plants and animals, ovum or egg cell is comparatively large and contains a large amount of cytoplasm. But male gametes or sperms have a very little amount of cytoplasm. So, under this situation, most cytoplasmic factors are transmitted to the progeny through the ovum of the mother. "It is known as maternal inheritance or trans-ovarian transmission. In this mode of transmission, all the offspring of the parents have the maternal condition and only female progeny can transmit the cytoplasmic characteristics to the succeeding generations. Hence the reciprocal crosses yield different or non-Mendelian results. The true cytoplasmic inheritance is one that involves plastids (chloroplasts) and mitochondria. Thus, cytoplasmic inheritance is again of two types, viz., 1. plastid inheritance and 2. mitochondrial inheritance. The former is applicable to plants only because plastids are found only in plants. The mitochondrial inheritance is common for both plants and animals. The cytoplasmic inheritance is governed by genes that are found in chloroplasts and mitochondria. The genes which govern cytoplasmic inheritance are called plasma genes or cytoplasmic genes or cytogenes or extra nuclear genes". These genes are made of DNA found in chloroplasts (cp-DNA) and mitochondria (mt- DNA).

The important features of cytoplasmic DNA in comparison to nuclear DNA are briefly described below:

1. The cytoplasmic DNA is of two types, viz., chloroplast DNA (cp-DNA) and mitochondrial DNA (mt-DNA).
2. The cytoplasmic DNA is usually a circular molecule except in ciliate protozoa where it is a linear molecule. Nuclear DNA on the other hand is usually linear in eukaryotes and circular in prokaryotes.
3. The synthesis of chloroplast DNA and mitochondrial DNA continues throughout the cell cycle, whereas synthesis of nuclear DNA occurs only during the interphase (S) of the cell cycle.
4. Cytoplasmic DNA replicates both in chloroplasts and mitochondria in a semiconservative fashion, whereas nuclear DNA replicates in the chromosomes in a semiconservative manner.
5. Both cytoplasmic and nuclear DNAs are liable to mutation. Mutation of cytoplasmic DNA leads to change in the behavior of plasma genes and mutation of nuclear DNA leads to alteration in nuclear genes.
6. Cytoplasmic DNA is capable of coding for RNA and protein synthesis like nuclear DNA.
7. In vitro synthesis of cytoplasmic DNA is possible in the respective organelle. In vitro synthesis of nuclear DNA is also possible.
8. The synthesis of cytoplasmic DNA is blocked when treated with ethidium bromide, acriflavin and actinomycin-D. Similar is the case with nuclear DNA. Thus, cytoplasmic DNA differs from nuclear DNA in four main aspects.

Subjective Questions

1. Define gene structure and function?
2. What are chromosomes and cell division?
3. Explain the common chromosome disorders?
4. What is single-gene (Mendelian) inheritance?
5. Define the identifying genes for Mendelian diseases?
6. What are polygenic inheritance and complex diseases?
7. Differentiate between genes and populations?

8. What are genes and hemoglobin?
9. Define genes and development?
10. What are genes and explain about cancer with example?
11. What are genes and biochemistry?
12. Describe the genes and drugs and treatment?
13. What is Mendel's law of heredity?
14. Explain the cell division?
15. Define briefly meiosis and mitosis?
16. What is Linkage and crossing over?
17. Explain the sex chromosomes?
18. Describe the sex-linked characters?
19. Define cytoplasmic inheritance plasma and nuclear?
20. Explain the mutation?
21. Explain the Law of segregation?
22. What is the law of dominance? Give example?
23. Explain about the deletion, insertion of a gene?
24. Briefly explain the non-sense mutation with an example?
25. Describe the law of independent assortment?

Objective Questions

1. The tendency of an offspring to resemble its parent is known as

a. Variation

b. Heredity

c. Resemblance

d. Inheritance

Answer: B

2. Who is known as the "Father of Genetics"?

a. Morgan

b. Mendel

c. Watson

d. Bateson

Answer: B

3. The alternate form of a gene is

a. Alternate type

b. Recessive character

c. Dominant character

d. Allele

Answer: D

4. The genotypic ratio of a monohybrid cross is

a. 1:2:1

b. 3:1

c. 2:1:1

d 9:3:3:1

Answer: A

5. The crossing of F1 to either of the parents is known as

a. Test cross

b. Back cross

c. F1 cross

d. All of the above

Answer: B

6. Which of the following statements is true regarding the "law of segregation"?

a. Law of segregation is the law of purity of genes

b. Alleles separate from each other during gametogenesis

c. Segregation of factors is due to the segregation of chromosomes during meiosis

d. All of the above

Answer: D

7. Homozygosity and heterozygosity of an individual can be determined by

a. Back cross

b. Self-fertilization

c. Test cross

d. All of the above

Answer: C

8. An exception to Mendel's law is

a. Independent assortment

b. Linkage

c. Dominance

d. Purity of gametes

Answer: B

9. Pea plants were used in Mendel's experiments because

a. They were cheap

b. They had contrasting characters

c. They were available easily

d. All of the above

Answer: B

10. The smallest unit of genetic material which produces a phenotypic effect on mutation is

a. Muton

b. Gene

c. Recon

d. Nucleic acid

Answer: A

11. Mendel's findings were rediscovered by

a. Correns

b. De Vries

c. Tschermark

d. All

Answer: D

12. Alleles are

a. Alternate forms of genes

b. Linked genes

c. Chromosomes that have crossed over

d. Homologous chromosomes

Answer: A

13. When the activity of one gene is suppressed by the activity of a non-allelic gene, it is known as

a. Pseudo-dominance

b. Hypostasis

c. Epistasis
d. Incomplete dominance

Answer: C

14. Cystic fibrosis is

a. Sex-linked recessive disorder
b. Autosomal dominant disorder
c. Autosomal recessive disorder
d. Sex-linked dominant disorder

Answer: C

15. 9:7 ratios in the F2 generation represents

a. Incomplete dominance
b. Co-dominance
c. Epistasis
d. Complementary interaction

Answer: C

16. A small amount of lethal mutation is always present in the population due to

a. Positive selection
b. Negative selection
c. Frequency-dependent selection
d. Mutation-selection balance

Answer: D

17. If a plant with genotype AaBb is self-fertilized, the probability of getting AABB genotype will be (A and B are not linked)

a. ½
b. ¼
c. ⅛
d. 1/16

Answer: D

18. How many phenotypes can occur in the human blood group ABO with alleles $I^A I^B i$?

a. 2
b. 3

c. 4

d. 1

Answer: C

19. The geometrical device that helps to find out all the possible combinations of male and female gametes is known as

a. Bateson Square

b. Mendel Square

c. Punnett Square

d. Mendel's Cube

Answer: C

20. Which term represents a pair of contrasting characters?

a. Heterozygous

b. Homozygous

c. Codominant genes

d. Allelomorphs

Answer: D

21. DNA replication can be described as

a. Semiconservative

b. Conservative

c. Degenerative

d. Dispersive e. radical.

Answer: A

22. If in the F1 and F2 generations the same characteristics appeared in both males and females, it would be safe to assume that these traits for eye color and wing length

A. Are sex-linked

B. Vary in dominance according to sex

C. Are sex-influenced characteristics

D. Are autosomal characteristics

E. Follow the Mendelian rule of independent assortment

Answer: D

23. In the F2 generation, the results are best explained by the fact that

a. The test cross with the F1 flies resulted in sterile offspring

b. These genes for eye color and wing shape do not pass through the F1 generation

c. These genes for eye color and wing shape are found on the same chromosome

d. Crossing over decreases variability e. the genes are sex-linked

Answer: C

24. If a single locus controls wing shape, then the alleles for this gene act as

a. Dominant-recessive alleles

b. Incomplete-dominance alleles

c. Codominant alleles

d. Multiple alleles

e. Variable alleles

Answer: A

25. The electrophoretic separation of the pieces of DNA in each of the four samples was achieved because of the differential migration of the DNA fragments in an electric field. This differential migration was caused by the

a. Relative amounts of radioactivity in the DNA

b. Number of cleavage points per fragment

c. Size of each fragment

d. Overall positive charge of each fragment

e. Solubility of each fragment

Answer: C

CHAPTER 7

CELL MICROBIOLOGY

1. HISTORY OF MICROBIOLOGY: SPONTANEOUS GENERATION THEORY, THE ROLE OF MICROBES IN FERMENTATION AND GERM THEORY OF DISEASE

Introduction

The study of microbes or microorganisms is called microbiology. The microbes, or microorganisms are minute living things that are usually unable to be viewed with the naked eye. Some examples of microbes are as follows: Bacteria, fungi, protozoa, algae, viruses, and some are parasites (helminths). Some are pathogenic and some are not. Most bacterial species cannot cause disease

Beneficial Roles of Microbes

- Production of antibiotics and foodstuffs.
- Decompose organic waste.
- Produce industrial chemicals such as ethyl alcohol and acetone.
- Produce fermented foods such as vinegar, cheese, and bread.

Ancient Microbiology

Ancient people recognized many factors involved in diseases. Most ancient people documented that some diseases are communicable, example: When the black death struck Europe entire villages were abandoned in an effort to escape the highly infectious plague (1347 A.D). No medical knowledge existed in Medieval England to cope with the disease. The Romans were a hygienic bunch and were concerned with health and cleanliness (waste and sewage). The Romans invented the first underground sewage system. The Romans understood that sewage could cause disease, therefore, decided to build an underground sewage system which is an idea we still use today. A network of pipes brought clean water into the city of Rome and removed waste. Waste flushed from the latrines flowed through a central channel into the main sewage system and thence into a nearby river or stream. The public bathhouses were the place where people went to socialize and do business as well as get clean.

History of Microbiology

First microbes were observed in 1673. Robert Hooke- in 1665 reported that living things were composed of cells. He devised the compound microscope and illumination system.

Antoni van Leeuwenhoek- (1673-1723): He is considered the father of bacteriology. He made simple microscopes and began observing with them (1674). He discovered bacteria (he called them animalcules). Leeuwenhoek's microscope:

- A screw for adjusting the height of the object being examined
- A metal plate serving as the body
- A skewer to impale the object and rotate it
- The lens itself, which was spherical

Spontaneous Generation

Aristotle synthesized the hypothesis which stated that some vital force contained in given organic matter can create living organisms from inanimate objects. In basic terms, spontaneous generation stated that living organisms arise from nonliving matter. Spontaneous generation was disproved in 1668 by Italian Scientist, **Francesco Redi**.

Redi's Experiment: Where do maggots come from?

Hypothesized: Redi suspected that flies landing on the meat laid eggs that eventually grew into maggots

Experiment: Placed meat in three separate jars

Disproving Spontaneous Generation: Redi's Experiment

Jar #1	**Jar #2**	**Jar #3**
Left open Maggots developed, Flies were observed laying eggs on the meat in the open jar	**Covered with netting** Maggots appeared on the netting and Flies were observed laying eggs on the netting	**Sealed** No maggots developed

Spontaneous Generation

John Needham- revived the theory of spontaneous generation in 1745. Lazzaro Spallanzani (1729-1799). In 1776 he demonstrated that microorganisms were already in the solution, the container, or the air. He took solutions that he knew would "breed" organisms and boiled them for up to an hour. The flasks were hermetically sealed to keep out contaminated air.

Biogenesis: In 1858 German scientist, Rudolf Virchow challenged spontaneous generation with his concept of biogenesis. Living organisms arise from pre-existing life. Virchow presented his idea to the scientific community, but could not back it up with a convincing experiment.

In 1861, a French scientist by the name of Louis Pasteur demonstrated where microorganisms came from

Father of medical microbiology: he demonstrated that microorganisms exist in the air and could contaminate sterile solutions by passing air through cotton filters. The filter trapped tiny particles floating in the air

Louis Pasteur

Louis Pasteur performed numerous experiments to discover why wine and dairy products became sour. He found that bacteria were to blame (lactic acid fermentation). Pasteur called attention to the importance of microorganisms in everyday life and stirred scientists to think that if bacteria could make the wine "sick," then perhaps they could cause human illness.

The Golden Age of Microbiology 1857- 1914

Beginning with Pasteur's work, discoveries included the relationship between microbes and disease, immunity and antimicrobial medicine

The Germ Theory of Disease

Germ theory states that specific microscopic organisms are the cause of specific diseases.

Pasteur and Germ Theory of Disease

Pasteur showed that microbes are responsible for fermentation. Microbial growth is also responsible for spoilage of food. Pasteur demonstrated that spoilage bacteria could be killed with heat **(pasteurization)**. It was discovered that weak forms of disease could be used as an immunization against stronger forms and that rabies was transmitted by viruses too small to be seen under the microscopes of the time. He developed vaccines for anthrax (1881) and rabies (1885).

History of Vaccination

A precursor of smallpox **vaccination** was variolation. The Germ theory of disease was not known at this time. An early Asian method introduced dried scabs of smallpox patients and was later modified in Europe. The modification consisted of injecting infectious material under the skin. First tested among abandoned children and prisoners when it was declared safe, members of the English royal family were inoculated.

Smallpox Vaccine

In 1796: Edward Jenner is credited with the development of the smallpox vaccine. Folk wisdom suggested that dairy maids who had contracted cowpox seemed to be immune to smallpox. Infection with the cowpox virus produced a much less severe form of disease than smallpox. Jenner conducted an experiment in which he used scabs from the cowpox lesions on the arm of a dairy maid, Sarah Nelmes to create a small pox vaccine. He then used the material to vaccinate an 8-year-old boy, James Phipps. After being vaccinated Phipps appeared to develop immunity to the smallpox virus. Later Jenner vaccinated his own son and several other children. He obtained similar results. Worldwide elimination of smallpox was achieved in 1978. He called vaccination from vacca for cow. The protection is called immunity

Discovery of Penicillin

In 1928 Alexander Fleming discovered the first antibiotic. He observed that Penicillium fungus made an antibiotic, penicillin, that killed S. aureus. In the 1940s Penicillin was tested clinically and mass produced and was available towards the end of World War II.

Development of Agar

Angelina Hesse developed the use of agar to grow microorganisms. She was the wife of Walter Hesse who worked in Koch's laboratory. The benefit of agar was it was not attacked by most bacteria. Agar is better than gelatin because of its higher melting point (96°c) and solidifying (40–45°c) points.

The fermentation industry is composed of five major bio-ingredient categories. They are proteins & amino acids, organic acids, antibiotics, enzymes and vitamins & hormones.

Fermentation Industry

- The cost and availability of feed-stocks are derived.
- The efficiency of industrial microorganisms.
- Fermentation condition and optimization.
- Downstream process and end-product recovery efficiency.
- Fermentation by-product utilization.
- Utility consumption and labor cost.

Fermentation Media

Optimum balance of the media is mandatory for cells propagation and for the maximum production of the target metabolite.

Media Compositions

- Carbon source
- Nitrogen source
- Minerals
- Growth factors
- Mutants

Industrial Microorganisms

- Microbial screening (Wild strains).
- Microbial yield improvement (Mutation).
- Recombinant DNA (Genetically engineered).
- Microbial selection.
- Industrial microorganism.

Types of Fermentation

- Solid state fermentation
- Liquid state fermentation surface culture & submerged culture

Solid State Fermentation (SSF)

- Solid state fermentation process can be defined as microbial growth on particles without the presence of free water.
- Particles are a solid culture substrate such as rice or wheat bran saturated with water and inoculated with (mold, yeast, bacteria) in controlled room temperature.
- It is ideal for growing filamentous fungi.
- It has been used in Asia and developing nations.
- It is more cost-effective (smaller vessels lower water consumption, reduced wastewater treatment costs, lower energy consumption, and less contamination problems).

Applications

Potentially many high-value products such as extra-cellular enzymes, primary metabolites, and antibiotics could be produced in solid state fermentation. It is estimated that nearly a third of industrial enzyme produced in Japan is made by the SSF process. Production of organic and ethanol from starch substrates. The digestibility of fibers and lignocelluloses materials for both human and animal consumption.

Liquid State Fermentation: Submerged Culture

The submerged culture is performed in tanks that can reach in size for over 100,000 gallons. It is ideal for growing unicellular organisms such as bacteria and yeast.

Liquid State Fermentation Methods

1. Batch fermentation.
2. Fed-batch fermentation.
3. Continuous fermentation.
4. Semi-continuous fermentation.

Batch Fermentation

It is considered to be a closed system. The sterilized media in the fermenter is inoculated with the microorganism. Incubation is allowed under the optimum conditions (aeration, agitation and temperature). The during entire fermentation nothing is added except air, antifoam and acid/base.

Fed-batch Fermentation

It is an enhancement of batch fermentation and continues adding the nutrients (feeding) in a small dose during the fermentation. The method in controlling the nutrients feeding process is by measuring methods. The main advantage of fed-batch fermentation is the elimination of catabolite repression (feedback inhibition).

Continuous Fermentation

It is an open system and continuously sterile nutrient is added and the converted nutrient is taken out from the fermentor. In continuous process cell loss as a result of outflow must be balanced by the growth of the microorganism.

Important Factors for Continuous Fermentation

The system must be stable for at least 500 hours and maintain sterile conditions for all periods of fermentation time. The composition of nutrients must be constant all the time. It is maintaining the strain stability for constant high production yield.

Semi-continuous Fermentation

Semi-continuous fermentations, in which a fraction of fermentation is replaced with fresh media at regular intervals.

Enzymes

Enzymes are active biological molecules responsible for thousands of metabolic processes that sustain life. Most enzymes are proteins, although some catalytic enzymes are RNA molecules. In enzyme reactions, the molecules at the beginning of the process, called **substrates** and converted into different molecules that are called **end-products.** Enzymes are very specific as to which reaction they catalyze and the substrate involved in the reaction. It depends on enzyme activity, the bioconversion to end-products can be faster and reach the equilibrium state rapidly.

Factors of Enzyme

1. **Temperature:** The optimum is generally 40-60^{0}C. Some enzymes exhibit an optimum at almost 100^{0}C.
2. **Value of pH:** The optimum generally in the range from 5-7. The extreme values of 1.5-10.5 have been found.
3. **Activation:** Many chemicals activates the activity of the catalytic enzyme, such as inorganic ions.
4. **Inhibitors:** Many chemicals inhibits the activity of the catalytic enzyme.
5. There are two types of enzymes inhibition: Irreversible inhibitors (competitive inhibition) and reversible inhibitors (uncompetitive inhibition).
6. **Substrate inhibition:** High concentration of substrate may inhibit the catalytic activity of an enzyme.
7. **End-product inhibition**: In the case of multi-enzyme system (catalytic inhibition).

Activators

Cofactors and Coenzymes

- Some enzymes do not need any additional components to show full activity.
- Cofactors can be either inorganic (metals) or organic compounds (flavin and heme).
- Coenzymes include NAD^+, $NADP^+$ and ATP.
- These coenzymes transfer chemical group between enzymes.
- The chemical groups carried by the hydride ion (H^+) of NADH or NADPH.

$NAD^+ + 2H^+ + 2e^-$ NADH / $NADP^+ + 2O^-_2 + 2H^+$ NADPH $+ 2O_2$

Or phosphate groups carried by ATP.

ATP $+H_2o$ ADP$+P_1$ (-7.3kcal/ mole) / ATP $+H_2o$ AMP + PP_1 (-14kcal/mole)

Classification of Enzymes

The International of Biochemistry and Molecular Biology developed a nomenclature for enzymes (EC number). Each enzyme is classified by a sequence of four numbers led by EC. (E.C. 5.3.1.18 Glucose isomerase)

Classification

1. **EC1** *Oxidoreductases* (catalyze oxidation/reduction reactions).
2. **EC2** *Tranferases* (transfer a functional group).
3. **EC3** *Hydrolases* (catalyze the hydrolysis of various bonds).
4. **EC4** *Lyases* (cleave bonds by mean of hydrolysis /oxidation).
5. **EC5** *Isomerases* (isomerization within the same molecule).
6. **EC6** *Ligase* (join two molecules with covalent bonds).

Enzymes Production

Constitutive Enzymes: The microorganism produces the enzyme in minimal fermentation media.

Inducible Enzymes: The microorganism requires adding inducible agents in the media to produce the target enzyme.

Extracellular Enzymes: The microorganism secretes the enzyme in the fermentation media.

Intracellular Enzymes: The microorganism produces the enzyme inside the cell.

Enzymes are usually sold based on the activity (u/ml or u/gm). If the efficiency of enzymes is considered, their cost, is based on active enzyme protein u/mg protein (specific activity). The commercial exploitation of enzymes ranges from high-volume but low cost (industrial enzymes) to low volume, but high cost (enzymes for medical, scientific and analytical use). Workers handling industrial enzymes should use protective clothing and eye protection. Food enzymes in foods processing have been show to be safe through many years of the manufacturing practice.

Robert Koch in 1867 provided proof that bacterium causes anthrax and provided the experimental steps, Koch's postulates, used to prove that a specific microbe causes a specific disease. The pathogen must be present in all cases of the disease. The pathogen must be isolated and grown in lab in pure culture. Pathogen from pure cultures must cause disease when inoculated into healthy, susceptible lab animals. The same pathogen must be isolated from the diseased lab animal. Trillion microbes in/on the human body such as indigenous microflora, opportunistic pathogens, producers of vitamins, enzymes, infectious disease (caused by microbes) and microbial intoxication (caused by toxins of microbes).

Agostino Bassi in 1835 showed a silkworm diseases were caused by a fungus.

Pasteur in 1865 believed that another silkworm disease was caused by a protozoan.

Ignaz Semmelweis in 1840 advocated hand washing to prevent transmission of fever from one OB patient to another.

Joseph Lister in 1860, **was** the father of antiseptic surgery. He used a chemical disinfectant (carbolic acid) to prevent surgical wound infections after reading Pasteur's work showing microbes are in the air, can spoil food, and cause animal disease.

Microbes in the Environment

- Saprophytes: decompose materials
- Genetically engineered bioremediators
- Microbial ecology

- Food chains plankton

Phytoplankton

Zooplankton

- Biotech in food/beverages
- Antibiotics
- Photosynthesis: production of oxygen

Only a few hundred years old as a science. *Mycobacterium tuberculosis* has been found in the DNA of 3000-year-old mummies. The causative agent of tuberculosis.

Anton van Leeuwenhoek (1632-1723): Animalcules (1678); live microbes and 3 basic shapes of bacteria

- Cocci
- Bacilli
- Spirilla

Anton van Leeuwenhoek Father of Microbiology, bacteriology, protozoology, microscopy and no disease connection suspected

Robert Hooke (1635-1703)

He observed slices of cork in 1665, England and defined about the "Cells" no stains. The beginning of the "cell theory" and all living things are composed of cells.

Spontaneous Generation

The theory held by most until the late 1800's, Abiogenesis, disproved and led to theory, two scientists paved the way.

- Francesco Redi
- Lazzaro Spallanzani

Redi's experiment in 1668 meat in an open jar = flies and meat in a sealed jar or one with a cloth over it = no flies.

Spallanzani's Experiment – 1768

Sterilize solutions using heat. Seal one, leave the other open. The growth/ turbidity in open flask only.

Edward Jenner (1749-1823)

- Smallpox outbreaks
- Cowpox
- Cowpox vaccine
- Inoculation

Louis Pasteur (1822-1895)

- Pasteur institute
- French biochemist
- Ultimately proved biogenesis
- Germ theory of fermentation
- Germ theory of disease
- Aerobes and anaerobes
- Pasteurization
- Father of bacteriology & immunology
- Vaccines, attenuation and immunity

Koch's Postulates – 1884

It was used anthrax in sheep to develop, pathogen present in each case, It must isolate from a host, grow in pure culture, reproduce disease from culture in others, repeat about Tuberculosis, cholera, typhoid.

Exceptions to Koch's Postulates

It is limited to pathogens that may be cultured *in vitro*, therefore excludes viruses and certain unusual bacteria, obligate intracellular pathogens and fastidious microorganisms. Some pathogens change when grown *in vitro* instead of *in vivo*. **Antibiotics** were discovered Alexander Fleming, 1928**.**

Polio Vaccines

Injection (Salk, 1955)

Oral (Sabin, 1957)

Chicken Pox Vaccine

It was developed in 1974 and available in 1995 such as in *Haemophilus influenza* vaccine (HiB): Meningitis and Hepatitis B vaccinations. Luc Montagnier (1982) was discovered HIV and Robert Gallo (1984) HIV causes AIDS.

Disease is a disturbance in the state of health. Microbes cause disease in the course of stealing space, nutrients, and/or living tissue from their symbiotic hosts. Microbes do most of the following: gain access to the host (contamination). Adhere to the host (adherence), replicate on the host (colonization), invade tissues (invasion), produce toxins or other agents that cause host harm (damage).

Characteristics

Infectivity – the ability to infect a host

Pathogenicity – the ability to cause disease in the host

Virulence – the ability to cause severe disease in the host

Immunogenicity –the ability to induce an immune response in the host

Infectious disease

The infectious dose is the number of organisms needed to successfully infect the latent period. The incubation period is the interval from exposure to clinical symptoms. The infectious period is the interval during which the host can transmit infection. Reproductive rate is the ability of an agent to spread in populations. The outbreak is limited spread. Endemic is usually present; steady prevalence. An epidemic is rapid spread. Pandemic is occurring across countries and in multiple populations.

Modes of Transmission

Direct transmission through droplet, aerosol and skin to skin.

Indirect transmission through fomites (clothes, blankets, door handles, etc), vectors (e.g. mosquitoes), food and water intermediate hosts (e.g. snails).

Clinical Responses

An inapparent infection has no clinical symptoms generated. The carrier state has usually no clinical symptoms but host can transmit infection for long periods. The clinical symptoms are mild disease, severe disease, residual impairment and death.

- Bacteria thrive on appropriate media, stain gram-positive or negative
- Viruses are obligate intracellular parasites that only replicate intracellularly (DNA and RNA).
- Fungi is non-motile filamentous and branching strands of connected cells.
- Metazoa is multicellular animals (e.g. parasites) with complicated life cycles often involving several hosts.
- Protozoa are single cell organisms with a well-defined nucleus.
- Rickettsia is a very small bacteria spread by ticks.
- Prions are unique proteins lacking genetic molecules.
- Chlamydia is bacteria lacking cell walls.

Bacteria are classified by their gram stain characteristics. Gram staining is the application of a crystal violet dye to a culture of bacteria. Bacteria that retain the color of the dye are called gram-positive; bacteria that don't are gram-negative. The gram stain attaches to peptidoglycan in the bacterial cell wall. In gram-negative bacteria, the peptidoglycan layer is protected by an outer membrane.

Viruses are acellular, obligate intracellular organisms. The complete infectious virus is termed a virion. The virion consists of the specific nucleic acid (DNA or RNA) surrounded by a protein coat (capsid). Some viruses are enveloped which means that they possess a lipoprotein coat that surrounds the capsid and is acquired from infected host cell membrane. Viruses that lack an envelope are "naked." Viruses are typically classified by: genetic material (DNA vs. RNA), strandedness (single vs. double) and size and shape of the capsid and whether its enveloped or non-enveloped.

Method of Replication

All fungi are chemoheterotrophs. Pathogenic fungi have two forms: yeasts (unicellular) and molds (multicellular). Some fungi are dimorphic (this is particularly true for the pathogenic fungi). Molds grow as filamentous, branching strands of connected cells known as hyphae.

Types of Parasites

Protozoa: single-celled, microscopic organisms that can perform all necessary functions of metabolism and reproduction. Some protozoa are free-living, while others parasitize other organisms for their nutrients and life cycle. The morphology of protozoa varies widely and includes oval, spherical and elongated cells that can range in size from 5-10 to 1-2 mm. Structurally, the protozoa resemble other eukaryotic cells and possess a cytoplasmic membrane that encloses cytoplasm containing membrane-bound nuclei, mitochondria, 80s ribosomes and a variety of organelles. Helminths: A large, multicellular organism (worm) that is generally visible to the naked eye in its adult stages. Helminths can be free-living or parasitic Nematodes: Roundworms, Trematodes: Flukes and Cestodes: Tapeworms. Prions are abnormal, transmissible agents that are able to induce abnormal folding of normal cellular prion proteins in the brain, leading to brain damage and the characteristics signs and symptoms of the disease. Prion diseases are usually rapidly progressive and always fatal.

Bacterial photosynthesis is a light-dependent, anaerobic mode of metabolism. Carbon dioxide is reduced to glucose, which is used for both biosynthesis and energy production. Depending on the hydrogen source used to reduce CO_2, both photolithotrophic and photoorganotrophic reactions exist in bacteria. Bacteria require organic molecules to provide their carbon and energy. The energy-yielding catabolic reactions can be of many different types, although they all involve electron-transfer reactions in which the movement of an electron from one molecule to another is coupled with an energy-trapping reaction that yields ATP.

- Phototroph uses light for energy.
- Chemotroph uses chemical reactions for energy.
- Autotroph takes in CO_2 for a carbon source.
- Heterotroph takes in organic compounds for a carbon source.
- Photoautotroph uses Energy from light / carbon from the atmosphere.
- Photoheterotroph uses Energy from light / carbon from organic compounds.
- Chemoautotroph uses Energy from chemical reactions (inorganic) / carbon from the atmosphere.
- Chemoheterotroph uses Energy from organic compounds/carbon

from organic compounds. It obtains energy from compounds - chemicals.

- Chemoheterotroph uses energy is used in anabolism. It obtains energy from chemicals. Carbon from the atmosphere not from compounds "eaten". Chemoautotroph, *Thiobacillus ferroxidans* and energy used in the Calvin-Benson cycle to fix CO_2.
- Phototrophs use light energy.
- Photoautotrophs can use the light energy in the Calvin-Benson cycle to fix CO_2, and make sugars to use for energy. Photoheterotrophs use light energy and organic compounds.

Nutritional type	Energy source	Carbon source	Example
Photoautotroph	Light	CO_2	**Oxygenic**: Cyanobacteria plants. **Anoxygenic**: Green, purple bacteria.
Photoheterotroph	Light	Organic compounds	Green, purple nonsulfur bacteria.
Chemoautotroph	Chemical	CO_2	Iron-oxidizing bacteria.
Chemoheterotroph	Chemical	Organic compounds	Fermentative bacteria. Animals, protozoa, fungi, bacteria.

Viruses

Viruses are both living or non-living. They have some properties of life but not others for example, viruses can be killed, even crystallized like table salt. However, they can't maintain a constant internal state (homeostasis). A virus is a noncellular particle made up of genetic material (DNA) and protein that can invade living cells.

Discovery of Viruses

Beijerinck (1897) was coined the Latin name "virus" meaning poison. He studied filtered plant juices & found they caused healthy plants to become sick.

Tobacco Mosaic Virus: Wendell Stanley in 1935 was crystallized sap from sick tobacco plants and He discovered viruses were made of nucleic acid and protein

Smallpox: Edward Jenner in 1796 was developed a smallpox vaccine using milder cowpox viruses, Deadly viruses are said to be virulent, smallpox had been eradicated in the world today, but is now coming back. Viruses are smaller than the smallest cell, measured in nanometers, Viruses couldn't be seen until the electron microscope was invented in the 20th century.

Characteristics

Nonliving structures, noncellular, have a protein coat called the capsid. It has a nucleic acid core containing DNA or RNA. It is capable of reproducing only when inside a host cell. Some viruses are enclosed in a protective envelope. Some viruses may have spikes to help attach to the host cell. Most viruses infect only specific host cells. The capsids (coats) are made of individual protein subunits called capsomeres. Viruses are inactive outside of host cells, lack ribosomes and enzymes needed for metabolism. Use the raw materials and enzymes of the host cell to be able to reproduce. Some viruses cause disease such as: smallpox, measles, mononucleosis, influenza, colds, warts, AIDS, Ebola. Some viruses may cause some cancers like leukemia. Virus-free cells are rare. Viruses come in a variety of shapes. Some may be in helical shape like the Ebola virus. Some may be polyhedral shapes like the influenza virus. Others have more complex shapes like bacteriophages are RNA or DNA virus do or do not have an envelope, capsid shape, host they infect. Viruses that attack bacteria are bacteriophage (phage) T-phages are a specific class of bacteriophages with icosahedral heads, double-stranded DNA, and tails. **T-phages** The most commonly studied T-phages are T4 and T7. They infect *E. coli*, intestinal bacteria. Six small spikes at the base of a contractile tail are used to attach to the host cell. Inject viral DNA into cell. T-4 Bacteriophage head with 20 triangular surfaces, capsid contains DNA, head & tail fibers made of protein

Retroviruses

It is containing RNA, not DNA. Family Retroviridae. Contain an enzyme called Reverse Transcriptase. When a retrovirus infects a cell, it injects its RNA and reverse transcriptase enzyme into the cytoplasm of that cell. The enzyme reverse transcriptase (or RTase), which causes the synthesis of a complementary DNA molecule (cDNA) using virus RNA as a template. HIV and AIDS virus is a retrovirus. Feline Leukemia virus is also a retrovirus.

Viral Replication

Once a viral genome has entered a cell, the cell begins to manufacture viral proteins. The virus makes use of host enzymes, ribosomes, tRNAs, amino acids, ATP, and other molecules. Viral nucleic acid molecules and capsomeres spontaneously self-assemble into new viruses. Phages are the best understood of all viruses. Phages have two reproductive mechanisms: the lytic cycle and the lysogenic cycle

Viral Attack

It is host specific. Humans rarely share viral diseases with other animals. Eukaryotic viruses usually have protective envelopes made from the host cell membrane.

Lytic Cycle

The lytic cycle is a phage replicative cycle that culminates in the death of the host cell. The lytic cycle produces new phages and lyses (breaks open) the host's cell wall, releasing the progeny viruses. A phage that reproduces only by the lytic cycle is called a virulent phage. Bacteria have defenses against phages, including restriction enzymes that recognize and cut up certain phage DNA.

Steps of Lytic Cycle

Attachment to the cell, penetration (injection) of viral DNA or RNA, replication (Biosynthesis) of new viral proteins and nucleic acids, assembly (Maturation) of the new viruses, release of the new viruses into the environment (cell lyses).

Lysogenic Cycle

The lysogenic cycle replicates the phage genome without destroying the host. The viral DNA molecule is incorporated into the host cell's chromosome. This integrated viral DNA is known as a prophage. Every time the host divides, it copies the phage DNA and passes the copies to daughter cells. An environmental signal can trigger the virus genome to exit the bacterial chromosome and switch to the lytic mode. Phages that use both the lytic and lysogenic cycles are called temperate phages.

Viral Latency

Some viruses have the ability to become dormant inside the cell called latent viruses. They may remain inactive for long periods of time (years). Later, they activate to produce new viruses in response to some external signal.

HIV and Herpes viruses are examples. Once a prophage cell is activated, the host cell enters the lytic cell. New viruses form & the cell lyses (bursts).

Latency in Eukaryotes

Some eukaryotic viruses remain dormant for many years in the nervous system tissues. Chickenpox (caused by the virus *Varicella zoster*) is a childhood infection. It can reappear later in life as shingles, a painful itching rash limited to small areas of the body. Herpes viruses also become latent in the nervous system. A herpes infection lasts for a person's lifetime such as Genital herpes (Herpes Simplex 2), Cold sores or fever blisters (Herpes Simplex1).

Treatment for Viral Disease

Vaccines

An attenuated virus is a weakened, less vigorous virus. Attenuated virus is capable of stimulating an immune response and creating immunity, but not causing illness.

Viral Treatments

Interferons are naturally occurring proteins made by cells to fight viruses. Antiviral drugs (AZT), Protease inhibitors – prevents capsid formation.

Viroids & Prions

Viroids

Small, circular RNA molecules without a protein coat, infect plants, Potato famine in Ireland, Resemble introns.

Prions

Prions are "infectious proteins". They are normal body proteins that get converted into an alternate configuration by contact with other prion proteins. They have no DNA or RNA. The main protein involved in human and mammalian prion diseases is called "PrP".

2. MICROBES IN SOIL

Introduction

It is causes neurons to rapidly degenerate examples: Mad cow disease (bovine spongiform encephalitis: BSE). People in New Guinea used to suffer from kuru, which they got from eating the brains of their enemies. Microorganisms are used routinely in engineered waste treatment systems such as sewage treatment plants. They are also of critical importance in the recovery process of natural environments degraded by human activities, such as in the self-purification of streams receiving sewage and runoff, and the natural attenuation of industrial contaminants leaked or spilled onto soil. On the other hand, microorganisms have the potential to create substantial environmental problems. For example, they may deplete oxygen, generate unpleasant tastes and odors, clog equipment, and corrode pipes.

Now we consider the prokaryotic groups, bacteria and archaea. We also examine the eukaryotic groups containing single- celled organisms: protozoans, algae, fungi and slime molds, even though they also include many multicellular, macroscopic species. There is a wide range of diversity within the world of microorganisms in terms of survival strategy: where they find energy, how they grow, and what environments they prefer.

Energy Sources

The two major sources of energy are chemical oxidation and photosynthesis.

Carbon Sources

Since it is a major constituent of cell materials, all organisms need a source of carbon. Heterotrophs (including fungi, protozoans, and most bacteria) require organic carbon, whereas autotrophs (algae and some bacteria) consume inorganic carbon (carbon dioxide and bicarbonate).

Microbial cells are also commonly classified on the basis of the environments they prefer. Several factors are generally considered, including the presence of oxygen, temperature, salt tolerance, and pH. Strict aerobes require oxygen; cells able to grow at very low oxygen levels may be referred to as microaerophilic. Facultative anaerobes, can grow with or without oxygen. Anaerobic metabolism may be respiratory (using a variety of inorganic terminal electron acceptors such as nitrate, nitrite, ferric iron, sulfate, or carbon dioxide) or fermentative (using an organic terminal electron acceptor). Anoxic in the absence of oxygen, and thus is equivalent to anaerobic. The ability to utilize nitrate and/or nitrite as alternative terminal electron acceptors (denitrification). Microbes thrive under cold temperature conditions, ranging from below 0°C to the mid-teens. Organisms that prefer moderate temperatures are referred to as m. Their temperature preferences range from the mid-teens to high-30s or low-40°C. A relatively few organisms, mainly bacteria, archaea, and fungi, prefer above 45 to 50°C and are called t. Some prokaryotic extremophiles are (temp. optimum above 80°C), a few even growing at above 100°C. Certain soil bacteria live symbiotically in the roots of legumes (N_2-fixing bacteria) -convert inert N_2 gas into amines. Mycorrhizal fungi -provide increased N, P, K, and antibiotics. Ruminant bacteria of cattle, sheep benefit these animals in metabolizing cellulose. Conversion of dinitrogen gas (N_2) to ammonia (NH_3). Availability of fixed N often factors most limiting to plant growth. N-fixation ability is limited to few bacteria, either as free-living organisms or in symbiosis with higher plants. First attempt to increase forest growth through N-fixation in Lithuania, 1894 (lupines in Scots pine). Rare, extremely energy-consuming conversion because of stability of triply bonded N_2. Produces fixed N which can be directly assimilated into N-containing biomolecules.

N-fixation requires energy input

Reduction reaction, e^- must be added (sensitive to O_2). Requires ~35 kJ of energy per mol of N fixed (theoretically). Actual cost: ~15-30g CH per g of NH_3 produced. Assimilation of NH_3 into organic form takes 3.1-3.6 g CH.

Rhizobia fix nitrogen in symbiotic association with leguminous plants. Rhizobia fix N for the plant and the plant provides Rhizobia with carbon substrates. All nitrogen-fixing systems appear to be identical. They require nitrogenase, a reductant (reduced ferredoxin), ATP, O-free conditions and regulatory controls (ADP inhibits and NH_4^+ inhibits expression of nif genes. Biological nitrogen fixation is the reduction of atmospheric nitrogen gas (N_2) to ammonium ions (NH_4^+) by the oxygen-sensitive enzyme, nitrogenase. Reducing power is provided by NADPH/ferredoxin, via an Fe/Mo center. Plant genomes lack any genes encoding this enzyme, which occurs only in prokaryotes (bacteria). Even within the bacteria, only certain free-living bacteria (Klebsiella, Azospirillum, Azotobacter), blue-green bacteria (Anabaena) and a few symbiotic. Rhizobial species are known nitrogen-fixers. Another nitrogen-fixing association exists between an Actinomycete (Frankia spp.) and alder (Alnus spp.). The enzyme nitrogenase catalyzes the conversion of atmospheric, gaseous dinitrogen (N_2) and dihydrogen (H_2) to ammonia (NH_3), as shown in the chemical equation below: $N_2 + 3\ H_2 \Rightarrow 2\ NH_3$. The above reaction seems simple enough and the atmosphere is 78% N_2. The incredibly strong (triple) bond in N_2 makes this reaction very difficult to carry out efficiently. In fact, nitrogenase consumes ~16 moles of ATP for every molecule of N_2 it reduces to NH_3, which makes it one of the most energy-expensive processes known in Nature.

Nitrogenase complex

Two protein components: nitrogenase reductase and nitrogenase: Nitrogenase reductase is a 60 kD homodimer with a single 4Fe-4S cluster. It is very oxygen-sensitive. It binds MgATP. The 4ATP required per pair of electrons transferred Reduction of N_2 to $2NH_3 + H_2$ requires 4 pairs of electrons, so 16 ATP are consumed per N_2. N_2 reduction to ammonia is thermodynamically favorable. However, the activation barrier for breaking the N-N triple bond is enormous and 16 ATP provides the needed activation energy. Each molecule of the enzyme contains 2 Mo, 32 Fe, 30 equivalents of acid-labile sulfide (FeS clusters, etc). Four 4Fe-4S clusters plus two FeMoCo, an iron-molybdenum cofactor. Nitrogenase is slow - 12 e^- pairs per second, i.e., only three molecules of N_2 per second.

Types of N-fixers important in soils
Cyanobacteria: Autotrophic N-fixers, protect nitrogenase with specialized *heterocyst* cells. Heterotrophic bacteria: Free-living or associative with rhizosphere. Use energy from decomposing organic matter to fix N, protect nitrogenase by rapidly converting O_2 to CO_2 through respiration. Symbiotic bacteria: Plants form nodules to house bacteria and provide C as an energy source (*Rhizobium*/*Bradyrhizobium* for legumes, *Frankia* for non-legumes). Nodules contain a form of hemoglobin that binds O_2, protecting the nitrogenase enzyme.
Nitrogen fixation in Klebsiella
Nif system is turned on when No fixed nitrogen Anaerobic Temperature below 30°C. Nitrogenase is made and converts N_2 to NH_3.

Bacterial associations with certain plant families, primarily legume species, make the largest single contribution to biological nitrogen fixation in the biosphere.

Composting is recycling naturally When we compost organic materials such as grass clippings, food and kitchen scraps, and leaves and brush, we are using the natural process of decay. Composting turns these wastes into a valuable humus-like product called compost. Composting is nothing more than speeding up the natural decay process. A compost pile is used to help us manage the "environmental factors that affect the rate of decay. These factors are air (more specifically oxygen) water, food, and temperature. The rate of composting (decay) and the quality of the resulting compost depend on how well you control these factors. Besides producing a valuable product, composting is recycling. It is one more way we can reduce the amount of waste we produce.

Pennsylvania is committed to recycling 35% of its municipal waste. Composting yard and kitchen wastes can help us maintain or exceed that recycling goal. Each one of us produces 650 lbs of waste each year that could be composted. Not too many people want another landfill next to their house, so why fill up the landfills we have with material that could be recycled. Not too many of us want higher bills for trash pick-up. By composting we can help keep down the cost of garbage handling. Over 60 percent of what we put in our landfills is organic waste that could be recycled by composting. Much of this can be recycled in your own backyard by composting.

Composting is a natural process and it is easy for you because you have a work crew that does almost all the work for you. Naturally occurring microorganisms (mainly bacteria and fungi) do the hard part of breaking down organic materials into compost. All you have to do to keep your decomposers working hard is provide them with what they need. The right kinds of food, and The right amount of air, water, and warmth. While any organic material can be composted, you should not try to compost all of them in your backyard compost pile. Things that are good to compost in your backyard pile include: Garden trimmings, kitchen scraps, grass clippings, leaves, potting soil, manure, sawdust, hair and straw.

Some organic wastes are better left out of your backyard compost pile because they can cause problems either during composting or when you use the compost. These forbidden materials include: meat fish and dairy products because of their tendency to attract pests and vermin (flies, rodents). Hard to kill weeds such as bindweed and quackgrass as well as weeds that have produced seed. A hot compost pile can kill these weeds and seeds. Most backyard compost piles do not reach high enough temperatures for long enough periods of time to kill these weeds and seeds. Cat and dog manures also tend to attract pests and could spread disease. Again, most backyard compost piles do not get hot enough for long enough to be certain of destroying plant diseases or insects". So to be sure you don't spread or reinfect your garden, do not put diseased or insect ridden plants into the compost pile.

Many people think it is necessary to add lime to the compost pile. "Finished compost is always near neutral in pH even if acidic materials were placed in the pile. For most garden uses, high pH compost is not desirable, and adding lime to the pile will often create an ammonia odor. Adding wood ash will add some potash to the pile which is good. But large amounts of wood ash will tend to increase the salt content of the finished compost. Many wood ashes are alkaline and will increase pile pH and lead to ammonia odor problems. Greens are nitrogen rich materials, those that have carbon to nitrogen ratios that are less than 25 to 1. Greens also tend to be more wet and soft than browns. They decompose very rapidly.

Mixing compost into heavy soil improves its tilth and friability making it easier to work and easier for roots and water to penetrate. This also allows water to drain more readily so you don't have saturated conditions in your soil. Compost helps to loosen heavy clay soils and keep them from getting compacted and compost helps to suppress, or combat some soil microbes that can attack plant roots. Compost cannot be considered a fertilizer, but it does contain important plant nutrients such as nitrogen and phosphorus. Compost will slowly release these nutrients to plants. The nutrients in compost are not likely to be leached and lost from the topsoil. Compost also contains many trace nutrients that are essential to plant growth. Before using your compost be sure it is fully mature (use the bag test or germination test). Immature compost could damage or kill plants, and could tie up soil nitrogen. When using compost as a soil amendment or soil conditioner, work no more than a 2-inch layer of compost into the upper 4 to 6 inches of soil. Compost used in this way will improve water and nutrient retention and will loosen heavy clay soils. Compost can be used as a surface mulch. Use a maximum depth of 3 inches for mulch. Never pile compost or any mulch around trunks of trees or shrubs. This makes an ideal environment for insects to attack the plant. Keep compost 3-4 inches away from the trunk and extend out to the dripline. Mulching protects the soil from temperature extremes and helps maintain soil moisture. Fully mature compost can be used as a topdressing on lawns. Used fine (screened) compost and apply about a ¼ inch layer to the lawn. The topdressing will be most effective if the lawn is cored before compost application. Rake the lawn after compost topdressing to move some compost into the core holes. Compost can replace peat in preparing potting mixes". The compost must also be very mature and fine (screened). Use no more than 1/3 compost by volume in your potting mix.

Various chemical and physical methods are available for preservation of food. But selection of appropriate methods is very important to protect the quality and nutritional value of the food. During preservation by these methods, following principles are involved.

Microbial Decomposition of Food

Preventing contamination of food by spoilage organisms. "Removing micro-organisms from contaminated food. For. eg. by washing. Slowing down the growth and activity of micro-organisms in food. For. eg. by freezing the food. Killing micro-organisms present in the food for eg. by heating the food or by adding some chemicals.

Self-decomposition of Food

Damaging self-enzyme of food that causes self-decomposition for eg. damaging of ripening enzyme of fruits by washing it in hot water. Preventing purely chemical reactions occurring in food that damage the food for eg. prevention of oxidative rancidity of lipid by addition of anti-oxidant. Prevention of mechanical damage of food caused by insects, birds, food handling devices, etc.

Methods for Prevention of Microbial Spoilage of Food

1. Preventing Food Contamination

The process of preventing contamination of food by spoilage micro-organisms is called asepsis. Inner tissue of healthy animal and plants do not contain micro-organisms and if present they usually do not spoil food. Most spoilage micro-organisms enter into the food by contamination from external sources like water, air, etc. Therefore, microbial decomposition of food can be prevented by preventing contamination of food by spoilage organisms. Several foods like egg contain natural covering around food that prevent the entry of spoilage organism into the food. Nowadays, a variety of such artificial barriers are made around the food to prevent contamination for eg. fruits and vegetables are stored for years by coating the wax around them. Some other artificial barriers that prevent contamination include wrappers, cartons and cans. Safe handling of food and good personal hygiene of food handlers are important methods to prevent or minimize contamination of food by spoilage organisms.

2. Removing microorganisms from contaminated food

It is the second level of preservation of microbial decomposition of food. If contamination of food by spoilage organisms is not prevented, spoilage of food still can be prevented by removing contaminated micro-organisms from the food. Some other techniques that preserve food by removing micro-organisms include:

Filtration

It is only one successful method for removing micro-organisms from contaminated food. Unfortunately, this method is applicable only for clear liquids like clear fruit juice, beverages, etc. To remove micro-organisms, liquid food is filtered through the sterile bacterial filter.

Centrifugation

Centrifugation removes some but not all micro-organisms from contaminated liquid food. When liquid food is centrifuged suspended particles, debris and some micro-organisms settle to the bottom which is later separated and removed from the food. This technique is commonly used for milk in dairy industries.

Washing

This method is commonly used for vegetables and fruits. Dusts and micro-organisms present on the surface of whole vegetables and fruits are removed by washing with sterile water. One limitation of washing is that it increases the water activity of some food and hence facilitates the growth of remaining spoilage organisms. Furthermore, if washing water is not sterile" it adds spoilage organisms in food.

Trimming

Trimming is the process of removing spoiled parts of food like vegetables and fruits by cutting. Trimming prevents the spread of micro-organisms from spoiled parts into inner healthy tissue.

3. Killing Microorganisms

Different methods that preserve food either by killing micro-organisms or by slowing down their growth and activity includes:

- Preservation of a high temperature
- Preservation by low temperature
- Preservation by drying
- Preservation by radiation (irradiation)
- Chemical agents used in the preservation of food

Preservation of Food by High Temperature

Heat is the most commonly used method of food preservation. Heat kills spoilage of "microorganisms by denaturation of cytoplasmic protein and enzyme. Two types of heat i.e. dry heat and moist heat are used in the preservation of food. Moist heat is more effective in killing micro-organisms than dry heat. Heat causes the denaturation of protein and enzymes by breaking down their bonds. Breaking of bond occurs by two mechanisms i.e. hydrolysis and oxidation. Hydrolysis requires water but oxidation does not. Hydrolysis requires less energy (heat) for the breaking of bond than oxidation. Therefore, if moisture is present in food, bonds of protein and enzymes are broken down by hydrolysis even at lower temperature. This is the reason why moist heat can kill micro-organisms completely at the condition of 121.5°C for 15 min but dry heat requires 160°C for 1-2 hrs to kill the same organisms. Various factors affect the killing of micro-organisms in food by heat.

Time-temperature relationship. "Density of micro-organisms in food, if the density is high, higher temperature is needed. Types of micro-organisms present in food, e.g. Spore are difficult to kill than vegetative cells. The growth stage of micro-organisms, e.g. micro-organisms present in log phase are easily killed but micro-organisms present in the late log phase and stationary phase are somewhat heat resistant. Micro-organisms in dry food are difficult to kill than in moist food. If food is acidic or alkaline, micro-organisms present in the log phase are easily killed by heat. On the other hand, if food is neutral, high temperature is needed to kill the same organisms. Penetration of heat into the inner layer of food also affects the killing of micro-organisms. The heat penetration into the food is affected by various factors as given below: Nature of material by which container of food is made, affects heat penetration eg. heat penetrates easily into a smaller and slim can than through larger and wider can. In the case of open heating of food, agitation or stirring of food facilitates heat penetration.

Pasteurization

Pasteurization is a selective or gentle heating technique originally developed by Lewis Pasteur for milk and milk products. Nowadays, pasteurization is also used for the preservation of other beverages like beer, fruit juice, etc. Pasteurization is a selective heating technique and it kills some but not all micro-organisms present in milk. If milk is contaminated by some disease-causing micro-organisms, milk cannot be heated at boiling temperature to kill these micro-organisms. It is because heating at high temperature brings several undesirable changes in milk eg.. it kills beneficial lactic acid bacteria present in milk. Similarly, casein protein precipitate, lactose caramelizes, flavor of milk changes. Such milk becomes unsuitable for the production of milk products like ice cream, cheese. Similarly, beverages like beer and fruit juice cannot be heated at higher temperatures because it brings several undesirable changes in flavor, taste etc. In pasteurization, milk and beverages are heated at lower temperatures that kills spoilage or pathogenic organisms without bringing undesirable changes in the product. There are three techniques of pasteurization of milk which differ in their time-temperature relationship". Time-temperature relations of pasteurization are selected on the basis of three bacteria that are transmitted through milk (*Mycobacterium bovis, Coxiella buretii* and *Brucella abortus*). Low temperature holding (LTH) or Vat pasteurization: In this method, milk is heated at 62.8°C for 30 minutes. In this method, milk is heated at 71.7°C for 15 seconds. Ultra-pasteurization: In this method, milk is heated at about 137.8°C for 2 seconds.

Canning

Canning is the process of preservation of food by keeping them in a sealed container and then heating the container. Canning techniques were originally developed by Nicoles Appert. There are two types of canned food i.e. heated and unheated canned food. During canning at first raw food is washed and then processed. Then processed food is placed into a suitable can and the can is then heated. Sealed cans can be heated either by immersing them into boiling water or by a spray of super-heated steam or heated dry air.

Blanching or Scalding

Blanching is the process of washing vegetables or fruits in warm water before storage.

Advantages of Blanching Include

- It inactivates enzymes of food that cause self-decomposition.
- It reduces the number of micro-organisms present on the surface of food.
- It brings wilting of leafy vegetables and helps in packaging.
- It enhances the fixing of the intense green color of fruits and vegetables.

Roasting

It is used for meat and meat products. During roasting internal temperature of meat reaches (80-85)°C.

Baking

It is used for bread, cake and other bakery products. During baking temperature maximum of 97°C is reached.

Frying

It is used for vegetables, meat, fish, etc. During frying outer surface of food becomes very hot but the internal temperature of food never exceeds 100°C.

Cooking

During cooking temperature around 100°C is reached.

Simmering

It is the process of gentle boiling of food with a temperature of about 100°C.

Preservation of Food by Low Temperature

Low temperature is a commonly used physical method of preservation of food. "Low temperature preserves food by: Slowing down growth and activity of spoilage organism in food. By slowing down the rate of a chemical reaction or other enzymatic activity that causes self-decomposition of food. Sometimes low temperature kills spoilage organism in food eg. in case of deep freezing.

Different Methods of Food Preservation by Low Temperature

Common or Cellar Storage

Common storage is the technique of preservation of certain foods like potato, onion, etc. in a home by spreading them in a cold room. In common storage, storage temperature is slightly lower than that of outside air. But this technique food is not stored for a long time but it is still used where cold storage and facilities are not available.

Chilling or Cold Storage

In chilling or cold storage, storage temperature is usually between (0-10) °C. At this temperature, the growth and activity of most bacteria except psychrophiles are significantly slowed down. Therefore, by chilling storage, food can be stored for a long time than common storage. However, if psychrophilic organisms are present in food, they grow in food and cause spoilage. Selection of appropriate temperature is very important because the quality of some food is damaged at a lower temperature storage temperature of some food are (10-12.8)°C for sweet potato, (13.3-16.7)°C for banana, etc. This method is most commonly used for short term storage of highly perishable foods like meat, milk, fish, etc. Fish is stored by placing it in cold ice.

Freezing or Frozen Storage

In freezing storage, storage temperature is between (-15 to -29)°C. At this temperature, the growth of even psychrophilic bacteria is completely inhibited. Therefore, food kept in frozen storage remains fresh for a long time without undergoing microbial decomposition. Furthermore, self-decomposition of food occurs very rarely or not at all at this temperature. Sometimes, freezing also kills spoilage organisms present in food. Therefore, frozen storage is a very effective method of food preservation among these three techniques. The killing mechanism of freezing is due to the formation of ice-crystal and dehydration of microbial cells. When food is frozen, surrounding water freezes to form ice-crystals. Then cytoplasmic water from microbial cells continuously diffuses out and is added to the surrounding ice-crystal. The remaining cytoplasm in microbial cells gradually becomes more and more concentrated. High salt concentration in the remaining cytoplasm kills the microbial cell by causing mechanical damage to the cell. There are two methods of freezing the food i.e. quick freezing and slow freezing. Slow freezing is more effective in killing microbial cells but quick freezing is considered more appropriate for the preservation of food. When food is frozen very slowly, there is sufficient time for cytoplasmic water to diffuse out. In this case microbial cell is dehydrated to a greater extent. Similarly, larger ice-crystal are formed due to the continuous addition of diffused water into ice. Larger ice crystal gives more mechanical damage to microbial cell". However, if larger ice-crystal are formed in food, it gives more mechanical damage (causes cracking of food) to the food. Therefore, quick freezing is commonly used for food preservation. Although frozen storage completely prevents microbial decomposition, it brings several undesirable changes in some food.

Some such examples include Protein of food e.g. meat may become dehydrated so that the food becomes very tough and rigid. Myoglobin present in meat may be oxidized into brown-colored meat-myoglobin. The fat present in food may be hydrolyzed or oxidized. Ice-crystals formed on surface of food evaporate and give mechanical damage to the surface of the food. This damage is called freeze-burn.

Methods of Freezing

Quick Freezing

In quick freezing, the temperature of the food lowered very quickly. Food is frozen quickly by immersing food into the solution of refrigerant for eg. fish frozen by immersing it into the frozen solution of NaCl. By blowing frigid air (at -17.8 to -34.4°C) over the food to be frozen.

Slow Freezing

In this process, the temperature of the food is lowered very slowly. Food is frozen slowly by placing it inside the mechanical refrigerator. This technique is commonly used to freeze food in homes. However, slow freezing forms larger ice-crystal in food and gives more mechanical damage. Therefore, quick freezing is preferred over slow freezing.

3. BENEFICIAL MICROORGANISMS

Introduction

Generally, agricultural land gets impoverished after long-term cultivation, if not supplemented properly with inputs. To supplement the soil nutrient content under the conventional farming system, we need to apply high doses of agrochemicals which in turn pollute the ecosystem. "Therefore, in order to make agriculture sustainable, it is necessary to implement a balanced and responsible use of organic agriculture. The principles of organic farming also outline similar concepts where the soil health and biodiversity are built up to sustain plant growth in the longer term. The name itself is self explanatory. Biofertilizer is a ready-to-use live formulation of such beneficial microorganisms which on application to seed, root or soil, mobilize the availability of nutrients by their biological activity. They help build up the soil micro-flora and thereby the soil health. As we know, organic farming excludes the use of any chemical. The use of bio-fertilizer is recommended for improving soil fertility in organic farming. A simple form of classification of biofertilizers is given below:

Nitrogen

Rhizobium for legume crops. Azotobacter, Azospirillum for non legume crops. Acetobacter for sugarcane only. Blue -Green Algae (BGA) and aAzolla for low land paddy.

Phosphorous

Phosphatika for all crops to be applied with rhizobium, Azotobacter, Acetobacter. VAM (Vesicu1ar-arbuscular mycorrhiza).

Types of Biofertilizer

Nitrogen-fixing Biofertilizers

The nitrogen-fixing bacteria work under two conditions, symbiotically and as free-living bacteria (non-symbiotic). The symbiotic bacteria make an association with crop plants by forming nodules in their roots. The free-living bacteria do not form any association but live freely and fix atmospheric nitrogen. Now let us examine the features of these microbes in detail.

Rhizobium

Rhizobium lives in the root hairs of the legumes by forming nodules. The name *Rhizobium* was established by Frank in 1889. This genus has seven distinct species based on "Cross Inoculation Group Concept". More than twenty cross-inoculations groups have been established so far. A new classification has been established for *Rhizobium*" That is 'slow growing rhizobia' known as *Bradyrhizobium* and the other group is 'fast growing rhizobia' called *Rhizobium.* Still this classification is discretely not distinguishable because the bacteria of one group may infect to another group. This is called "the principle of cross inoculation" which relies on the assumption that legumes within a particular infection group may be nodulated by another species of nodule forming bacteria.

Rhizobium Inoculants

The seed treatment has been found to be the suitable method of "*Rhizobium* inoculation. Some adhesive is used to make proper contact between seeds and inoculants (bacteria). About 900 gm soil base culture is sufficient to inoculate the seeds for one-hectare area in case of legumes. A 10 % jaggery *(gur)* solution is used as sticker for *Rhizobium* cells to seed. First the solution is spread over the seeds and mixed to build up a thin coat over the seeds. After ascertaining the proper coating of slurry over the seeds, the inoculant is sprinkled over the seeds and the content is again mixed thoroughly. Then the content is dried in the shade by spreading thinly on a polythene sheet at least overnight. This is a free-living or non -symbiotic bacteria (does not form nodules but makes association by living in the rhizosphere). Azospirillum species establish an association with many plants particularly with C, plants such as maize, sorghum, sugarcane, etc. It is the most common organism and can form associative symbiosis on a large variety of plants. Azospirillum is recognized as a dominant soil microbe

Blue-green Algae (BGA)

This is another important class of biofertilizer. The Blue-green algae are small organisms and can be seen under the microscope as a single cell or large accumulation of cells (colonies) or strings of cells (trichomes). They have a similar external appearance to that of algae and Azolla growing in a pond their requirements for light, nutrients and carbon dioxide are also similar.

Azotobactor

Azotobactor is a heterotrophic free-living nitrogenfixing bacteria present in alkaline and neutral soils. *Azotobactor chrococcum* is the most commonly occurring species in arable soils of India. Apart from its ability to fix atmospheric nitrogen in soils, it can also synthesize growth promoting substances *viz.,* auxins, and gibberellins and also to some extent the vitamins. Many strains of *Azotobactor* also exhibit fungicidal properties against certain species of fungus. Response of *Azotobactor* has been seen in rice, maize, cotton, sugarcane, pearl millet, vegetable and some plantation crops. Field experiments were carried out on *Azotobacter.*

Features of Azotobactor

Azotobacter contributes to the moderate benefits. Azotobacter is the heaviest breathing organism and requires a large amount of organic carbon for its growth. It is a poor competitor for nutrients in soil. It can benefit crops by nitrogen fixation, release of growth promoting substances, and fungicidal substances. Azotobacter is less effective in soils with poor organic matter content". It improves seed germination and plant growth.

Acetobactor

Acetobactor diazotrophicus is a newly discovered nitrogen-fixing bacteria associated with sugarcane crop. This bacterium belongs to the alpha group of *proteobacteria.* "It was isolated from leaf, root, bud and stem samples of sugarcane. Acetobacter is located in the apoplastic fluid of sugarcane stem and to some extent in xylem vessels. It is an acid and high salt tolerant and sucrose loving bacteria that can fix up to 200 kg nitrogen per hectare.

Vesicular Arbuscular Mycorrhiza (VAM)

The term mycorrhiza was taken from Greek language meaning 'fungus root'. This term was coined by Frank in 1885. As indicated above, mycorrhiza is a mutualistic association between fungal mycelia and plant roots. VAM is an endotrophic (live inside) mycorrhiza formed by aseptated phycomycetous fungi.VAM help in nutrient transfer mainly of phosphorus, zinc and sulfur. They also mobilize different nutrients like Cu (copper), K (potassium), Al (aluminum), Mn (manganese), Fe (iron) and Mg (magnesium) from the soil to the plant roots. They posse's vesicles (sac like structure) for storage of nutrients and arbuscular for funneling them into root system.

Mechanism of Action

The VAM forms an association with plant roots. It penetrates in the root cortex and spreads around the roots of the plant. As the name indicates, they posse's sac like structure called vesicules which stores phosphorus as phospholipids. The other structure called arbuscule helps bringing the distant nutrients to the vesicules and root.

Actions of Mycorrhiza

The feeding areas of the plant root are as the hyphae spread around the roots are enhanced. Mobilizes the nutrients from distance to root. Stores the nutrients (phosphorus). Removes the toxic chemicals (example: phenolics) which otherwise hinder nutrient availability. It provides protection against other fungi and nematodes. This method is mostly used for fruit crops, sugarcane, and other crops where localized application is needed. At the time of planting of fruit trees, 20 gm of biofertilizer mixed with compost is to be added in the ring of one sapling. Sometimes, the biofertilizers are also broadcasted in the soil but we may require four to ten times more biofertilizers. Before broadcasting, the inoculants should be incubated with the desired amount of well decomposed granulated FYM for 24 hours.

Advantages

There are many advantages of using biofertilizers. They form an important association with other soil microbes and help in nutrient supply. Fixes atmospheric nitrogen Increase availability or uptake of nutrients through solubilization or increased absorption. Stimulate plant growth through hormonal or antibiotic action or by decomposing organic waste. They are cheap, hence, reduced cost of cultivation". Improves soil properties and sustains soil fertility. Lead to soil enrichment. It is compatible with long-term sustainability. Build up soil fertility in the long term. They are eco-friendly and pose no damage to the environment.

Disadvantage

As such there is no harmful impact of biofertilizers if it is used properly some constraints: specific to the plants. "*Rhizobiurn spp.* culture doesn't work well in high nitrate tolerant strains of soybean. The acceptability of biofertilizers has been rather low chiefly because they do not produce quick and spectacular responses. It is requiring skill in production and application. It is difficult to store.

Microbial Agents for Control of Plant Diseases

Rhizosphere microorganisms are found in the area surrounding and influenced by the plant roots and rhizoplane microorganisms are found on the surface of plant roots.

Exudates: carbohydrates and proteins secreted by roots; Attracts bacteria, fungi, nematodes, protozoa. Bacteria and fungi are like little fertilizer bags. Nematodes and protozoa eat and excrete the fertilizer. Biological control agents are expensive and host-specific. Chemical pesticides are cost-effective, easy to apply and broad spectrum

Chemical Pesticides

Implicated in ecological, environmental, and human health problems. It is requiring yearly treatments and broad-spectrum. It is toxic to both beneficial and pathogenic species.

Biological Control Agents

Non-toxic to humans, not a water contaminant concerns and host-specific Only affect one or few species

Biological Control of Plant Pathogens

Antagonism

It is biological agents that reduce the numbers or disease producing activities of the pathogen through one or more of the following mechanisms. "Antibiosis inhibition of one organism by another as a result of diffusion of a metabolic product such as antibiotic, lytic agents, enzymes, volatile compounds and other toxic substances. Antibiotic production is common in soil bacteria and fungi, for example: zwittermicin antibiotic production by *Bacillus cereus* against *Phytophthora* root rot in alfalfa. Nutrient competition between microorganisms for limited sources such as carbon, nitrogen, O_2, iron, and other nutrients or spaces. The most common way organisms do this is limit the growth of other destructive mycoparasitism. It occurs when the antagonist invades the pathogen by secreting enzymes such as chitinase, cellulase, glucanase, and other lytic enzymes. The parasitism of one fungus by another direct contact, cell wall degrading enzymes, some produce antibiotics example *Trichoderma harzianum,* used as seed treatment against pathogenic fungus *Trichoderma spp.* are present in nearly all agricultural soils antifungal abilities have been known since the 1930s, Mycoparasitism, nutrient competition, agriculturally used as a biocontrol agent and as a plant growth promoter. Root colonization by *Trichoderma* strains frequently enhances root growth and development. The crop productivity (increase up to 30%), resistance to biotic stress, uptake and use of nutrients. It is enhancing the solubilization of soil nutrients and increases root hair formation.

Biodegradation

Biodegradation – a biological activity that results in the breakdown of a specific contaminant.

Biodegradation

Bioavailability: bacteria are located at the oil-water interface, not inside the oil droplets. Smaller droplets give more surface area to increase biodegradation. Explosions in bacterial growth would deplete available nutrients and ultimately slow degradation. Oxygen consumption is possible dead zones. Sedimentation is deposition in anaerobic zones in sediment. Hazen *et al.* (2010) found 2-fold higher cell densities within the plume. Enrichment in plume for *Oceanospirillales*. Some oxygen depletion in the plume, not enough to create anoxic "dead zones". It is enough oxygen loss to indicate aerobic activity. Increased degradation genes in plume". Valentine *et al.* (2010) found oxygen depletion in the plume to be driven by ethane and propane metabolism.

Biogas Production

Biogas is generated by the activity of anaerobic bacteria. "Biogas is comprised of about 60% of methane, 40% of carbon dioxide, and small amounts of hydrogen sulfide, nitrogen, and hydrogen. The heating value of biogas is about 60% of natural gas and about 25% of propane. Biogas has corrosive nature and storage of biogas is not practical. The technology of biodigester is widely used in developing countries such as China, Vietnam, India, and Central and South America as well as in developed countries. The anaerobic digester can be used in remote farm areas to produce biogas from manure and protect water resources. The idea is to make a small-scale, low-cost biodigester plant so that anyone in the world can make it and produce biogas. Feed the bag with the effluent. It is cut the garden hose to an adequate length and pull the bag through the garden hose. Fold the bag against the surface of the garden hose. Attach the adapter to the garden hose over the bag. Connect the hose and the gas collector using adequate adapters.

Operation

The generator requires little maintenance besides occasional stirring when the batch is done, disconnect the garden adapter and unload the effluent. For the next batch, apply about 10% of the previous batch to activate the new batch. After loading the feedstock, connect the adapter again and repeat the process".

Plant-microbe Interaction

Heavy metals are the most important inorganic pollutants, which are not degraded and progressively accumulate in the environment. The use of plants for rehabilitation of heavy-metal-contaminated soils is an emerging area of interest, because it provides an ecologically and environmentally sound and safe method for restoration and remediation. Although a number of plant species are capable of hyper-accumulation of heavy metals, however, this approach is not applicable for remediating sites with multiple contaminants. "Biogeochemical capacities of microorganisms seem almost limitless and they can adsorb and accumulate metals in their cells and are being used in microbial leaching and also as agents of cleaning the environment. To overcome the metal stress, numbers of mechanisms have been evolved by microorganisms of agronomic importance by which they tolerate and promote the uptake of heavy metal ions. The best approach would be to combine the advantages of plant-microbe interactions within the plant rhizosphere into an effective cleanup technology. The activities of plants and plant-microbial associations may offer viable means of accomplishing the *in situ* remediation of contaminated soils. This portion examines the potential role of plant-microbe interactions in heavy metal contaminated soils towards phyto-bioremediation. Heavy metal pollution affects the production and quality of crops, the quality of atmosphere and water bodies and thus threatens human and animal health (Narula et. 2011). The metal species commonly found in the soils as a result of human activities include copper (Cu), zinc (Zn), nickel (Ni), lead (Pb), cadmium (Cd), cobalt (Co), mercury (Hg), chromium (Cr) and arsenic (As), etc. Some of these acts as micronutrients at small concentrations for living organisms for their normal physiological activities, but their accumulation is toxic to most life forms (Khan et al. 2005). The most common human activities resulting in the entry of heavy metal into land are the disposal of industrial effluents, disposal of waste such as sewage sludge, atmospheric deposition from industrial activities, mining activities, domestic and industrial wastes, landfill operations and use of agrochemicals. The release of heavy metals from various industrial sources, agrochemicals and sewage sludge present a major threat to the soil environment. Generally, heavy metals are not degraded biologically and persist in the environment indefinitely (Walker et al. 2003).

The toxic heavy metals inversely affect the microbial compositions, including plant growth-promoting rhizobacteria (PGPR) in the rhizosphere, and their metabolic activities. In addition, the elevated concentration of metals in soils and their uptake by plants adversely affect the growth, symbiosis and consequently the yields of crops (Wani et al. 2008a) by disintegrating cell organelles, and disrupting the membranes (Sresty and Madhava Rao 1999), acting as genotoxic substance disrupting the physiological process, such as photosynthesis (Wani et al. 2007b) or by inactivating the respiration, protein synthesis and carbohydrate metabolism (Shakolnik 1984). The remediation of metal-contaminated soils thus becomes important, as these are rendered unsuitable for sustainable agriculture. This review provides a short overview of plant-microbe interactions towards phyto- and bio-remediation.

Phytoremediation approach involves the cultivation of metal accumulating higher plants to remove contaminants from metal-polluted soils (Brooks 1998). In this approach, plants capable of accumulating high levels of metals are grown in contaminated soils. At development metal enriched above ground biomass is harvested and soil metal contamination is detached. Successful plant-based decontamination of even moderately contaminated soils would have the need for crops able to concentrate metals in excess of 1-2%. Accumulation of such high levels of heavy metals is highly toxic and would certainly kill the common non-accumulator plant. However, in hyper-accumulator species, such concentrations are attainable. Nevertheless, the extent of metal removal is ultimately limited by the plants' ability to extract and tolerate only a finite amount of metals. On a dry weight basis, this threshold is around 3% for Zn and Ni and considerably less for more toxic metals, such as Cd and Pb. The other biological parameter which limits the potential for metal phytoextraction is biomass production. With highly productive species, the potential for biomass production is about 100 tons' fresh weight/hectare. The values of these parameters limit the annual removal potential to a maximum of 400 kg metal/ha/yr. It should be mentioned, however, that most metal hyper-accumulators are slow-growing and produce little biomass. These characteristics severely limit the use of hyper-accumulator plants for environmental cleanup. Practices have been residential to increase the potential of common non-accumulator plants for Pb phytoextraction. Particularly, the uptake-inducing properties of synthetic chelates open the possibility of using high biomass-producing crops for Pb phytoextraction. Under chelate-induced conditions, maize (Huang and Cunningham, 1996) and Indian mustard (Blaylock et al., 1997) have been successfully used to remove Pb from solution culture and contaminated soil, respectively. Physical characteristics of soil contamination are also important for the selection of remediating plants. For example, for the remediation of surface-contaminated soils, shallow-rooted species would be appropriate to use, whereas deep-rooted plants would be the choice for more deep contamination.

The identification of metal hyperaccumulators, plants capable of accumulating high metal levels, demonstrate that some plants have the genetic potential to clean up metal contaminated soils. In general, the concentration of metals in hyperaccumulators is about 10-100 fold higher than most other plants growing on metal contaminated soils. It has been possible through bioengineering to develop plants (Raskin 1996) capable of removing methyl mercury from the contaminated soil. To detoxify this compound, such bioengineered plants express modified bacterial genes *merB* and *merA* which convert methyl mercury to elemental mercury. About 400 plant species have been identified as hyper-accumulator accumulators. The Indian mustard plant (*Brassica juncea*) can extract both heavy metals and radionuclides from the soil. Panwar et al. (2002) reported that *B. juncea* has the potential to be a hyper-accumulator of Ni. The survey of literature reveals that the rate of metal removal depends upon the plant species, soil contaminating heavy metal(s) and biomass harvested and metal concentration in harvested biomass as summarized in Table 3.1.

Table 3.1: Plants capable of hyperaccumulating metals

Plant species	Metal accumulated	Accumulated concentration mg/kg dry matter
***Thlaspi caerulescens* (Brassicaceae)**	Zn	**10,000**
***Seberatia acuminate* (Sapotaceae)**	Ni	**10,000**
***Alyssum lesbiacum* (Brassicaceae)**	Ni	**20,000**
***Arbidopsis halleri* (Brassicaceae)**	Cd	**1,000**
Thlaspi rotundifolium	Pb	**8,200**
***Astralagus* sp.(leguminosae)**	Se	**1,000**
***Pteris vittata* (Fern)**	**As**	**22,630**

Rhizosphere

The rhizosphere is the zone of soil surrounding a plant root where in biology and chemistry of the soil is influenced by the roots, rather it is an area of intense activity (biological, chemical and physical) influenced by compounds exuded by roots and by micro-organisms feeding on these compounds (Kumar et al. 2007). Generally, soil contains bacterial numbers in the range of 10^7 to 10^{10} cells per gm dry soil. But microbiological activity in the rhizosphere is much greater (10^8) than in soil away from plant roots (10^5) and also microorganisms provide or make available nutrients for the plants (Walker et al. 2003). Many of these microbes live there as a part of a distinct community surrounding plant roots. Heterotrophic bacteria are able to use organic compounds excreted in root exudates, whereas their metabolites can be used by other microbes, which in the end creates a network of closely associated microorganisms. This phenomenon of highly active micro-organisms in root-associated soil is known as the "rhizosphere effect". Thus the microbial population is one of the essential parts of the rhizosphere that affect the rhizosphere soil by its various activities such as water and nutrient uptake, exudation, and all the biological transformations. Among the fast-growing and early colonizing bacteria attracted by the plant, exudates are members of genera *Bacillus* and *Pseudomonas* besides N fixing bacteria *Azospirillum* and *Rhizobium*. Cultivation-based methods show that the *Pseudomonas* spp are generally more abundant in the rhizosphere than in the bulk soil. But the number of related clones to *Pseudomonas* is higher in the rhizosphere of ryegrass and white clover. On the basis of 16s rRNA gene clones, plant roots have further been selective effect towards r-proteobacteria leading to the majority of *Pseudomonas* spp in the rhizosphere.as compared to bulk soil. Free-living and associative diazotrophs are more abundant in the rhizosphere than in bulk soils, indicating their dependence on organic compounds exuded by roots in the rhizosphere. Various nitrogen-fixing micro-organisms have been found to be present in the rhizosphere of agricultural plants, but the contribution of fixed nitrogen to plant nutrition is controversial (Lima et al. 2006). Diazotrophs found in the soil or associated with roots include *Azotobacter chroococcum*, *Azospirillum brasilense* and *Gluconacetobacter diazotrophicus* (formerly *Acetobacter diazotrophicus*) and the positive responses of plants to inoculation with these bacteria are attributed to nitrogen (N_2) fixation besides, several other factors like phytohormone and/or ammonium production" etc. (Okon 1985). Plant-associated rhizobacteria and mycorrhizae may significantly increase the bioavailability of various heavy metal ions for their uptake by plants. Also, they are known to catalyze redox transformations leading to changes in heavy-metal bioavailability (Yang et al. 2005).

Plant association with diazotrophs or any colonizer indicates a high degree of adaptation between the host plant and the most abundant diazotrophs. The microhabitat provided by the host plant seems to generate a selection pressure in favour of the micro-organisms, which in turn best benefit the host. Nutrients and metals are typically present in the soil solution at low concentrations and tend to form sparingly soluble minerals (except nitrogen, sulfur, and boron), or may be adsorbed to a solid phase through ion exchange, hydrogen bonding, or complexation (White 2003). The extent to which they are transferred from the soil to the biota (i.e., microbes or plants) is dependent on the biogeochemical interactions (N,

P, S) and/or processes among the soil, plant roots, and microorganisms in the rhizosphere (Abbot and Murphy, 2003). At this interface, the presence of root exudates may influence chemical reaction kinetics within the soil environment and subsequently affect biological activities. The exudates act as messengers that "stimulate biological and physical interactions between root and soil organisms. In addition to the adsorption and conduction, roots also produce hormones and other substances that help to regulate the plant's development and structure, help in modifying the biochemical and physical properties of the rhizosphere (Abbot and Murphy, 2003) and contribute to root growth and plant survival.

PGPR (Plant Growth-promoting Rhizobacteria)

Plant growth-promoting rhizobacteria (PGPR) are capable of promoting plant growth by colonizing and establishing around the plant root (Narula et al. 2006). "PGPRs have been found to play a potential role in developing sustainable systems in crop production (Shoebitz et al. 2009). Soil bacteria have been used as biofertilizers for ameliorating soil fertility and enhancing crop production for decades. The main functions of these bacteria are (1) to supply nutrients to crops; (2) to stimulate plant growth, e.g., through the production of plant hormones; (3) to control or inhibit the activity of plant pathogens; (4) to improve soil structure; and (5) to act as bio-accumulator in microbial leaching of inorganics (Ehrlich 1990). More recently, bacteria have also been used in soil for the mineralization of organic pollutants, i.e. bioremediation of polluted soils (Zaidi et al. 2008). Generally, PGPRs function in three different ways (Glick 1995, 2001): a) synthesizing particular compounds for the plants (Dobbelaere et al. 2003), b) facilitating the uptake of certain nutrients from the soil (Çakmakçi et al. 2006) and c) lessening or preventing the plants from diseases. Besides their role in protecting the plants from metal toxicity, the plant growth-promoting rhizobacteria are also known for their role in enhancing soil fertility and promoting crop productivity by providing essential nutrients (Zaidi and Khan. 2006) and plant growth regulators (Kumar et al. 2007).

Glick et al. (2002) reported that these PGPRs also promote the growth of plants by alleviating the stress induced by ethylene-mediated impact on plants by synthesizing 1-aminocyclopropane-1-carboxylate (ACC) deaminase (Belimov et al. 2005). According to their relationship with the plants, PGPRs can be divided into two groups (Khan 2005). First group "Symbiotic bacteria" with species of *Rhizobium* (*Rhizobium, Mesorhizobium, Bradyrhizobium, Azorhizobium, Allorhizobium* and *Sinorhizobium*) have been successfully used worldwide to permit an effective establishment of the nitrogen-fixing symbiosis with leguminous crop plants (Bottomley and Maggard 1990; Lynch 1990). The second group "Free-living rhizobacteria" with non-symbiotic nitrogen-fixing bacteria such as *Azotobacter, Azospirillum, Bacillus* and *Klebsiella* sp. are also used to inoculate a large area of arable land in the world with the aim of enhancing plant productivity (Lynch 1983). In addition, phosphate solubilizing bacteria such as species of *Bacillus* and *Paenibacillus* (formerly *Bacillus*) have been applied to soils to specifically enhance the phosphorus status of plants.

The mechanisms have been evolved by microorganisms of agronomic importance by which they tolerate and promote the uptake of heavy metal ions. Such mechanisms include (1) the pumping of metal ions exterior to the cell (2) accumulation and sequestration of the metal ions inside the cell (3) transformation of toxic metal to less toxic forms (Wani et al. 2008b) and adsorption/desorption of metals (Mamaril et al. 1997). Due to these properties, when plant growth-promoting rhizobacteria including nitrogen fixers, used as seed inoculants (Narula et al. 2005a, 2005b), were applied to soil, either treated/amended intentionally with metals or already contaminated, have a substantial reduction in the toxicity of metals and concomitantly improved the overall growth and yield of chickpea (*Cicer arietinum*) (Gupta et al. 2004), green gram (*Vigna radiata* L. wilczek) and pea (*Pisum sativum*) (Wani et al. 2007a, 2007c). An experiment on the effect of rhizobial inoculation on the metal accumulation potential of the chickpea plant was done by Gupta et al. (2004). The chickpea plants were harvested and accumulation of metal was studied in root and shoot parts of the plant. Accumulation of Cu, Zn, Cr and Cd was more in shoot whereas Fe was more in the root part of the plant. These studies are significant while using the plant for metal decontamination of fly ash landfills. The use of such microbes possessing multiple properties like plasmid-borne (Saitia et al. 1989) metal resistance/reduction (Saitia and Narula 1989, Narula et al. 2011) and their ability to promote plant growth through different mechanisms in metal-contaminated soils are chosen as a suitable choice for bioremediation studies.

In the era of sustainable crop production, the plant-microbe interactions in the rhizosphere play an important role in the transformation, mobilization, solubilization, etc. of nutrients from a limited nutrient pool, and subsequent uptake of essential nutrients by plants to realize their full genetic potential.

An understanding of the mechanisms, which is important for the initiation and establishment of the association between host and bacterium, can be reached from the analyses of influences exerted by each interaction partner on the other. In addition, there is a need to know how diazotrophs may benefit the plant. Investigations on the production of phytohormones and the action of siderophores produced by *Azotobacter* strains might help to understand this aspect of interactions (Narula et al. 2006; Narula et al. 2005a, 2005b). Plant-microbe interactions are important for both the partners i.e., macro as higher plants and micro partners as the plant-associated bacteria (Somers et al. 2004). Microbial partners can induce antagonistic (in the case of phytopathogens) or symbiotic interactions. Different types of interactions involving plants roots in the rhizosphere have been reviewed by Bais et al. (2004, 2006). These include root-root, root-insect, and root-microbe interactions. The rhizosphere represents a highly dynamic front to study the interaction between roots and pathogenic as well as beneficial soil microbes, invertebrates, and root systems of competitors (Bais et al. 2004). In recent years several plant scientists have recognized the importance of root exudates in mediating these biological interactions. However, because plant roots are always hidden below ground, many of the interesting phenomena, their attractions, love and hate relationship in which they are involved have remained largely unnoticed. Especially the role of chemical signals (Peters et al. 1986) in mediating below ground interactions is only beginning to be understood. Chemical signaling between plant roots and other soil organisms, including the roots of neighboring plants, is often based on chemicals exuded from the roots. The same chemical signals may elicit dissimilar responses from different recipients. Most importantly, chemical components of root exudates may be the ones to deter one organism while attracting another, or two very different organisms may be attracted with differing consequences to the plant.

Therefore, the mechanisms used by roots to communicate and interpret these signals which they receive from other roots and soil microbes in the rhizosphere are largely unknown. The rhizosphere has some positive or negative and neutral associations. Much has to be done still to determine and elucidate whether the chemical signature of a plant root exudates will be perceived as a negative or a positive signal. However, evidence over the years suggest that root exudates are the determining factor to identify interactions in the rhizosphere and, ultimately, plant and soil community dynamics (Narula et al. 2009).

Biotic and Chemical Interferences of Heavy Metals

In the last decades, the essential role of soil microbiota has been emphasized for many processes that are directly related to soil bioremediation and, especially, phytoremediation, in which plant-microbe interactions have been documented to be of vital importance (Khan 2005). Phytoremediation is a process that uses actively growing plant roots to stimulate a diverse population of soil microorganisms, some of which have the capability to metabolize contaminants. In the case of phyto-extraction, inoculation of plants with PGPR may be beneficial mainly due to plant growth stimulation and a corresponding increase in the above-ground biomass, rather than to an increase in metal concentrations in plant tissues, which may actually be weakly influenced (Wu et al. 2006). Such highly persistent and hardly avoidable pollutants as heavy metals, in contrast to organic contaminations (Karthikeyan and Kulakow 2003), cannot be biologically or chemically degraded, but can only be deactivated by detoxification and/or immobilization) within the soil or removed from the contaminated site (Ruml and Kotrba. 2003).

On the other hand, they can directly or indirectly affect plant-microbe interactions in the rhizosphere. Their direct influence involves possible toxic or other detrimental effects of particular chemical metal species on the biochemistry and physiology of plant-associated microorganisms (Renella et al. 2005) and on the growth and development of plants and these effects comprise the biotic interferences. The chemical reactivity of particular metal species or their catalytic effects, which result in chemical binding or degradation of biogenic organics involved in plant-microbe interactions (i.e. remote exchange of molecular signals, quorum sensing within bacterial communities, release and exchange of nutrients, etc. (Gunatilaka. 2006), may thus be classified as indirect effects comprising chemical interferences". Biotic processes, induced in the rhizosphere by bio-geo-chemical interactions or formation of chemically active metal species, results in chemical depletion, inactivation or degradation of any biomolecules directly involved in plant-microbe interactions *via* their binding and/or by redox transformation, invariably affect these biologically specific interactions as described by various researchers (Lasat. 2000; Kamnev. 2008).

A metal uptake into roots is slow transport from soil particles to root surfaces. There are two mechanisms (Barber. 1984) responsible for metal transport from the bulk soil to plant roots: 1) convection or mass flow, and 2) diffusion. Due to convection, soluble metal ions move from soil solids to the root surface. From the rhizosphere, water is absorbed by roots to replace water transpired by leaves. Water uptake from the rhizosphere creates a hydraulic gradient directed from the bulk soil to the root surface. Some ions are absorbed by roots faster than the rate of supply *via* mass flow. Thus, a depleted zone is created in the soil immediately adjacent to the root. This generates a concentration gradient directed from the bulk soil solution and soil particles holding the adsorbed elements to the solution in contact with the root surface. This concentration gradient drives the diffusion of ions toward the depleted layer surrounding the roots. Plants have evolved specialized mechanisms to increase the concentration of metal ions in soil solutions. For example, at low ion supply, plants may alter the chemical environment of the rhizosphere to stimulate the desorption of ions from soil solids into solution. Such a mechanism comprises rhizosphere acidification due to H^+ extrusion from roots (Crowley et al. 1991). Protons compete with and replace metal ions from binding sites, stimulating their desorption from soil solids into solution. In addition, some plants can regulate metal solubility in the rhizosphere by exuding a variety of organic compounds from roots. Root exudates form a complex with metal ions keeping them in solution available for uptake into roots (Romheld and Marschner. 1986).

In cooperation microorganisms and plants have developed diverse strategies helping them to overcome or to adapt to unfavorable environmental conditions. A number of PGPR and other microorganisms have been reported to be resistant to relatively high concentrations of heavy metals (Ruggiero et al. 2005) and remain active in moderately acidic soils (Yasmin et al. 2004) which are wide-spread, comprising over 30% of only arable territories (Von Uexkull and Mutert 1995). Such phytostimulating rhizobacteria could contribute to plant growth promotion in metal-contaminated sites both indirectly, by increasing the overall fertility of the contaminated soil, as well as providing nutrients, phytohormones and exhibiting other PGPR-related traits (Rosenblueth and Martinez-Romero 2006) but also catabolizing certain organics and/or partly oxidized intermediate biodegradation products.

As for PGPR of the genus *Azospirillum*, studies that these phytostimulating micro-aerophilic rhizobacteria are relatively tolerant to sub milimolar concentrations of heavy metals at which, in many cases, growth of the bacterial culture is not significantly suppressed (Kamnev et al. 2006). Moderate concentrations of heavy metals in soil can often be found as a result of contamination and/or bioleaching of minerals, and assessing their impact on the bacterial metabolism is of importance for a deeper insight into their biology and presents an obvious biotechnological and agricultural interest, considering the plant-growth-promoting abilities of Azospirilla (Somers et al. 2004). In particular, *A. brasilense* represents a relatively rare chance to compare two strains of the same species which essentially differ in their mode of plant-root colonization. For example, *A. brasilense* strain Sp245 is known to be a facultative endophyte, whereas non-endophytic strain Sp7 colonizes the root surface only (Kirchhof et al. 1997). The levels of the auxin phytohormone (indole-3-acetic acid, IAA) production for each of the two strains were analyzed in a standard medium (control) and in the presence of 0.2 mM $CuSO_4$ or $CdCl_2$ under micro-aerobic conditions that are favorable for these micro-aerophilic bacteria. In parallel, the bacterial growth rate (in terms of colony-forming units, CFU s) was also assessed. Using strain Sp7, both copper (II) and cadmium ions were found to result in a significant decrease in the level of IAA production, whereas the bacterial growth rate was virtually not affected. In Sp245, the overall level of IAA production in the control was approximately 3 times higher than for Sp7. Nevertheless, in this respect, strain Sp245 appeared to be more sensitive to heavy metals: a noticeable decrease in IAA production was observed under the effect of both the metals, especially with Cu (II); note also that for strain Sp245 copper (II) was somewhat more toxic decreasing also the bacterial counts.

Thus, the decrease in IAA production by the bacteria in presence of copper (II) and cadmium (II) may directly affect the efficiency of associative plant-bacterial symbioses in heavy metal-polluted soils. To compare heavy-metal resistance of some plant-associated soil bacteria, the values of minimum inhibitory concentrations (MIC) and minimal lethal concentrations (MLC) for several PGPR grown on the agar medium (aerobic conditions) have been reported by Tugarova et al. (2006). MLC of cadmium (II) was found for both the two *A. brasilense* strains (5 mM) were much higher than those reported for the other associative rhizobacteria (0.35 to 0.6 mM), i.e. by an order of magnitude, while for the Cd-sensitive rhizobacterium *Flavobacterium* sp. L30 the MLC was 0.01 mM (Belimov et al. 2005). By their cadmium resistance, the following soil bacteria isolated from the rhizoplane of Indian mustard (*Brassica juncea* L. Czern.) plants (Belimov et al. 2005) were closer to *A. brasilense*: *Variovorax paradoxus* (MLC = 0.6 to 3.5 mM for different strains), a *Rhodococcus* sp., a *Ralstonia* sp., a *Flavobacterium* sp., a *Pseudomonas* sp. and some other strains (MLC = 2.0 to 4.0 mM). Soil microorganisms can also directly influence metal solubility by altering their chemical properties. For example, a strain of *Pseudomonas maltophilia* was reduce the mobile and toxic Cr (VI) to nontoxic and immobile Cr (III), and also to minimize environmental mobility of other toxic ions such as Hg (II), Pb (II), and Cd (II) (Blake et al. 1993; Park et al. 1999). In addition, it has been estimated that microbial reduction of Hg generates a significant fraction of global atmospheric Hg 0 emissions (Keating et al. 1997).

Simon et al. (2008) reported that pre-treatment of Indian mustard with *Pseudomonas fluorescens* before cadmium and nickel application slightly enhanced the Cd or Ni accumulation in roots. This enhancement was not observed when pseudomonads were applied simultaneously with metals. More cadmium was detected in the roots of Indian mustard when plants were pre-treated with Cd-tolerant than Cd-sensitive *Pseudomonas cepacia*. The cell number of pseudomonads in the surface of Cd-treated Indian mustard roots was significantly lower when plants were pre-treated with Cd-sensitive *P. cepacia.* Only for *Bacillus* and *Pseudomonas* species present in Se contaminated soils can be supposed to stimulate the Se phytoextraction and/or phytovolatization in the rhizosphere of higher plants (Simon et al. 2007). Pal et al. (2010) reported the influence of Pb and Hg on *Azotobacter spp.* and *Pseudomonas spp.* multiplication using Indian Mustard as a test crop on an artificially spiked sandy loam soil. The results showed that the minimum bacterial counts were observed in Pb_{200} and Hg_{40} treatments at 54 days after sowing of the crop.

In turn, rhizospheric microorganisms may interact symbiotically with roots to enhance the potential for metal uptake. Root growth affects the properties of the rhizospheric soil and stimulates the growth of the microbial consortium. In addition, some microorganisms may excrete organic compounds (Narula et al. 2009) which increase bioavailability and facilitate root absorption of essential metals, such as Fe (Crowley et al. 1991) and Mn (Barber and Lee 1974), as well as nonessential metals, such as Cd (Salt et al. 1995). The rhizosphere effect is often expressed quantitatively as the ratio of a number of microorganisms in rhizosphere soil to the number of microorganisms in nonrhizosphere soils (R/S ratio). This type of increased microbial activity in the rhizosphere may be responsible for the increased metabolic degradation rate of certain xenobiotic compounds in the rhizosphere. The actual composition of the microbial community in the rhizosphere is dependent on root type; plant species, plant age and soil type as well as other factors. The bacteria clearly showed a preference for one exudate over the other. This might well be related to the substances found in each exudate. It might be argued that the energy yield from C_6-compounds is better, thus resulting in observed preferences (Kumar et al. 2007). Generally, the rhizosphere is colonized by gram-negative bacteria. The interaction between plants and microbial communities in the rhizosphere is complex and has evolved to the mutual benefit of both organisms. Plants sustain a large microbial population in the rhizosphere by secreting substances such as carbohydrates and amino acids through root cells and by sloughing root epidermal cells. The magnitude of rhizodeposition can be quite large. In addition, root cells secrete mucigel, gelatinous substances that are lubricants for root penetration through the soil during growth. This mucigel along with other cell secretions consists of root exudates. Beneficial effects of the root exudates are stimulated by adding organic acids to the soil. This enhances the microbial population and the evolution of greater amounts of carbon dioxide. Microbial counts are generally more in the soil from vegetation and plants, this explains the help given by root exudation. Microbial activity, biomass and degradation are greater than in the non-vegetated soil.

Chemical Interferences

Microorganisms have been role to alter metal bioavailability. For example, microbes have been documented to catalyze redox reactions leading to changes in metal mobility in soil. A strain of *Xanthomonas maltophyla* was catalyze the reduction and precipitation of highly mobile Cr (VI) to Cr (III), significantly less mobile and environmentally less hazardous component. Ionic Hg (II) which is susceptible to methylation and subsequent volatilization generally of toxic methyl mercury was reduced by the same bacteria to elemental Hg which possess less environmental risk. A heavy metal resistant *Bacillus circulans* have been isolated from soil which is capable of biosorption of 25, 22, 20, 13, 12 mg/l of Mn, Zn, Cu, Ni and Co, respectively.

Besides the biotic reduction of iron (III) mediated by specific soil bacteria dissimilatory reducers, a range of secondary metabolites produced both by soil microorganisms and by plant roots can abiotically reduce Fe^{III} (Kovacs et al. 2006). Another rapidly developing field within the "underground hemisphere" of plant-microbe interactions, that can be affected by chemical interferences in soil, includes biosynthesis and exudation of intercellular bacterial signaling (quorum-sensing molecules), as well as a variety of already known, undisclosed biomolecules by which plant root-microbe signaling and interactions are implemented (Sugawara et al. 2006) or by which plants mimic, quench or otherwise interfere with bacterial signals (Teplitski et al. 2004; Adonizio et al. 2006). In particular, noteworthy is the active role of PGPR cell-surface polysaccharide-containing macromolecular complexes (Somers et al. 2004) and cell-surface lectins, in bacterial interactions with host-plant roots involving also plant lectins. As an example, wheat lectin (wheat germ agglutinin, WGA), which is exposed on the root surface and can also dissolve therefrom in the aqueous medium, has been documented to induce a number of metabolic responses in Azospirilla at low concentrations and therefore is proposed to act as a molecular signal for these wheat-associated bacteria (Narula et al. 2011, in press).

Soil micro-flora is an important and active participant in biogeochemical processes, leading to diverse chemical transformations of soil minerals. In this respect, the rhizospheric microbial communities supported by host plants (along with a contribution from plant-root exudates) can demonstrate even higher activities (Mukerji et al. 2006). "Rhizosphere, as compared to the bulk soil, is highly populated by various microorganisms mainly comprising bacteria (predominating Gram-negative bacteria) and mycorrhizal fungi (Karthikeyan and Kulakow 2003; Lucy et al. 2004) showing higher metabolic activity even in polluted soils (Biró et al. 2006). Some of the diverse routes leading to microbially mediated bioleaching and biotransformation of heavy metals and radionuclide entrapped within soil minerals (Zachara et al. 2001; Kamnev et al. 2006) are: (i) microbial dissimilatory reduction and concomitant solubilization of metal ions (e.g., Fe, Mn and many other less abundant redox-active metals) *via* different mechanisms as reviewed by (Lovley and Coates 2000). (ii) siderophores, carboxylic acids, acidic exopolysaccharides and other organics production and excretion capable of binding and/or reducing metal cations (Faraldo-Gomez and Sansom 2003) as well as (iii) contact interactions of microbial cells with metal complexes and solid compounds *via* microbial cell-wall biomacromolecules, sometimes with further involvement of metal ions in cellular metabolism or their inactivation *via* special intracellular mechanisms, etc. (Langley and Beveridge 1999).

The bioavailability of iron, one of the few microelements vitally important for virtually all organisms (Sigel and Sigel 1998), which is commonly abundant in most soils, is yet well known to be extremely low within a wide range of environmental conditions. This is due to the very low solubility of ferric oxide or hydrated oxide species that are most stable under aerobic and microaerobic conditions. Thus, iron (III) largely represented by oxides or oxyhydroxides of variable crystallinity is an essential component of the majority of soil minerals. (iii) shuttle-assisted reduction involving specific compounds capable of redox cycling that serve as an electron shuttle between the bacterial cells and iron (III)-bearing mineral. All three mechanisms are clearly fundamentally different. Nevertheless, as for common features and distinctions, the first two mechanisms involving iron (III) reduction at the bacterial cell surface while the third one does not, which represents its main difference in the mode of electron transfer to iron(III). In addition to biotic (enzymatic) dissimilatory reduction by bacteria-reducers, humic substances, as well as their lower-molecular-weight precursors, are able to abiotically (chemically) reduce e.g., highly soluble Cr^{VI} It is interesting to note that some dissimilatory metal-reducing bacteria were found to be capable of sensing the depletion of soluble electron acceptors, with the subsequent synthesis of pili and flagella when growing on insoluble iron(III) or Mn(IV) minerals (but not on their soluble forms) (Lloyd 2003). Thereby the bacteria can adapt to the conditions by "switching on" their active chemotaxis mechanism to move towards the insoluble sources of electron acceptors (solid Fe^{III}- or Mn^{IV}-containing mineral particles), with subsequent attachment, thus passing from shuttle-assisted reduction (mechanism in view of shuttle deficiency, to contact reduction by mechanism).

The bacterial mobilization of heavy metals and radionucleotide are poorly understood processes Microorganisms may, in principle, be advantageous and biotechnologically beneficial when occurring in the rhizosphere of metal-tolerant plants (special hyperaccumulator plants), thereby facilitating phytoremediation processes. Interaction of metal ions with biological matter is essential as well as important for various biological processes for all organisms and in related fields (biogeochemistry, bioremediation and phytoremediation, biomining, biotechnology of metal extraction, sorption and recovery, etc.). It can hardly be over-emphasized that a lack of understanding of molecular mechanisms underlying the effects of soil microorganisms and plant-root exudates on the state of metal compounds in the rhizosphere is a serious impediment in the use of phyto-bioremediation technology" for cleaning soils contaminated with heavy metals. It can be overcome by widely using a variety of modern powerful physicochemical techniques in environmental and life sciences. On the other hand, knowing the mechanisms and routes of metal transformations may open ways for a variety of practical applications. It would be most imperative to genetically and/ or biotechnologically tailor plants for being hyper-accumulators, microbes to possess greater potential to convert the toxic form of heavy metals to less or non-toxic types and to harness the synergy of both partners to achieve better efficiencies for phyto-bioremediation of polluted soils through favorable plant-microbe interactions.

Subjective Questions

1. Write the contributions of Pasteur and Beijerink?
2. Explain briefly the structure and functions of the bacterial cell wall?
3. Briefly explain the concept of containment facility?
4. Write a commonly used method for isolation of pure culture of bacterium?
5. Distinguish between continuous and synchronous cultures?
6. Distinguish between bacteria and mycoplasmas?
7. Give a brief account of slime molds?
8. Briefly explain the morphological features of protozoa?
9. Discuss the kingdoms of organisms and phylogenetic trees?
10. Explain the importance of Beragy"s Mannual in bacterial taxonomy?
11. Write the special staining techniques for microscopic identification of microorganisms?
12. What is spontaneous generation theory?
13. Define the metabolism in bacteria?
14. What is chemoautotrophy?
15. What is photorespiration?
16. Differentiate between the lysogenic and lytic cycle?
17. What is viroids and prions?
18. What is nitrogen-fixation process?
19. What is composting?
20. Explain the microbial groups in soil?
21. Define the biofertilizer?
22. What is biodegradation?
23. Explain the biogas production?
24. Briefly explain the plant-microbe interactions?
25. What is the role of microbes in soil?
26. What is the role of microbes to food preservation?
27. Explain the fermentation process?
28. Define anabolism and catabolism?

Objective Questions

1. **Who demonstrated that open tubes of broth remained free of bacteria when the air was free of dust.**
   a. Abbc Spallanzani
   b. John Tyndall
   c. Francisco Redi
   d. Pasteur

Answer: B

2. **Reverse isolation would be appropriate for**
   a. a patient with tuberculosis
   b. a patient who has had minor surgery
   c. a patient with glaucoma
   d. a patient with leukemia

Answer: A

3. **The symptom "general feeling of illness and discomfort "is called**
   a. Cystitis
   b. Malaise
   c. Anaphylactic shock
   d. Arthritis

Answer: B

4. **On soybean which of the following forms symbiosis**
   a. *Azatobactor paspali*
   b. Rhizobium
   c. Nostoc
   d. Bradyrhizobium

Answer: D

5. **Who provided the evidence that bacteriophage nucleic acid but not protein enters the host cell during infection**
   a. Alfred D.Hershey & Leonard Tatum in 1951.
   b. Alfred D.Hershey & Zindar Lederberg in 1951.
   c. Alfred D.Hershey & Martha Chase in 1952.
   d. Alfred D.Hershey & Macleod in 1952.

Answer: C

6. Spirulina belongs to

a. Xanthophyceae

b. Cyanophyceae

c. Rhodophyceae

d. Phaeophyceae

Answer: B

7. The first antibody to contact invading microorganisms was

a. IgG

b. IgM

c. IgA

d. IgD

Answer: B

8. The light emitted by luminescent bacteria is mediated by the enzyme

a. Coenzyme Q

b. Luciferase

c. Lactose dehydrogenase

d. Carboxylase reductase

Answer: C

9. Pick out the vector used in the Human Genome project

a. Phagemid vector

b. Yeast artificial chromosomes

c. Cosmid vectors

d. Yeast episomal plasmids

Answer: B

10. Salt and sugar preserve foods because they

a. Make them acid

b. Produce a hypotonic environment

c. Deplete nutrients

d. Produce a hypertonic environment

Answer: D

11. In a fluorescent microscope the objective lens is made of

a. Glass

b. Quartz

c. Polythene

d. None of these

Answer: C

12. Fixation of atmospheric nitrogen is by means of

a. Biological process

b. Lightning

c. Ultraviolet light

d. All of the above

Answer: D

13. Which one of the following fungi is the most serious threat in a bone marrow transplant unit?

a. *Candida albicans*

b. Aspergillus

c. Blastomyces

d. Cryptococcus

Answer: B

14. Direct microscopic count can be done with the aid of

a. Neuberg chamber

b. Anaerobic chamber

c. Mineral oil

d. Olive oil

Answer: A

15. The image obtained in a compound microscope is

a. Real

b. Virtual

c. Real inverted

d. Virtual inverted

Answer: B

16. Enzymes responsible for alcoholic fermentation

a. Ketolase

b. Zymase

c. Peroxidase

d. Oxidase

Answer: B

17. Which type of spores is produced sexually?

a. Conidia

b. Sporangiospores

c. Ascospores

d. None of these

Answer: C

18. Bacterial transformation was discovered by

a. Ederberg and Tatum

b. Beadle and Tatum

c. Griffith

d. None of these

Answer: A

19. Father of microbiology is

a. Louis Pasteur

b. Lister

c. A.V. Leeuwenhock

d. Robert Koch

Answer: C

20. The antiseptic method was first demonstrated by

a. Lwanowski

b. Lord Lister

c. Edward Jenner

d. Beijerinck

Answer: B

21. Smallpox vaccine was first discovered by

a. Robert Koch

b. Louis Pasteur

c. Lister

d. Edward Jenner

Answer: D

22. The term mutation was coined by

a. Pasteur

b. Darwin

c. Hugo devries

d. Lamark

Answer: C

23. Compound microscope was discovered by

a. Antony von

b. Pasteur

c. Johnsen & Hans

d. None of these

Answer: C

24. Father of Medical Microbiology is

a. Pasteur

b. Jenner

c. Koch

d. A.L. Hock

Answer: C

25. Disease that affects many people at different countries is termed as

a. Sporadic

b. Pandemic

c. Epidemic

d. Endemic

Answer: B

26. Prophylaxis of cholera is

a. Protected water supply

b. Environmental sanitation

c. Immunization with killed vaccines

d. All of these

Answer: B

27. In an electron microscope, what material is used as an objective lens?

a. Magnetic coils

b. Superfine glass

c. Aluminum foils

d. Electrons

Answer: A

28. The main feature of a prokaryotic organism is

a. Absence of locomotion

b. Absence of nuclear envelope

c. Absence of nuclear material

d. Absence of protein synthesis

Answer: B

29. The stalked particles on the cristae of mitochondria are called

a. Glyoxysomes

b. Peroxisomes

c. Oxysomes

d. Spherosomes

Answer: B

30. Antiseptic methods were first introduced by

a. Lord Lister

b. Iwanowski

c. Beijernick

d. Edward Jenner

Answer: A

31. Kuru disease in Humans is caused by

a. Bacteria

b. Viroids

c. Prions

d. Mycoplasma

Answer: C

32. Phagocytic phenomenon was discovered by

a. Louis Pasteur

b. Alexander Fleming

c. Metchnikof

d. Robert Koch

Answer: C

33. Meosomes are also known as

a. Mitochondria

b. Endoplasmic reticulum

c. Plasmids

d. Chondroids

Answer: D

34. Hybridoma technique was first discovered by.

a. Kohler and Milstein

b. Robert Koch

c. 'D' Herelle

d. Land Steiner

Answer: A

35. The minimum number of bacteria required to produce clinical evidence of death in a susceptible animal under standard conditions is called

a. LD50

b. ID

c. MLD

d. All of these

Answer: C

36. In an Electron Microscope source of electrons is from

a. Mercury lamp

b. Tungsten metal

c. both a and b

d. None of these

Answer: B

37. Griffith (1928) reported the phenomenon of transformation first in

a. H. influenza

b. Bacillus species

c. Pneumococci

d. E. coli

Answer: C

38. Hanging drop method for motility study was first introduced by

a. Robert Koch

b. Louis Pasteur

c. Jenner

d. Leeuwenhock

Answer: D

39. Electron microscope gives magnification upto

a. 100 X

b. 2000 X

c. 50,000 X

d. 2,00,000 X

Answer: D

40. Term vaccine was coined by

a. Robert Koch

b. Pasteur

c. Needham

d. None of these

Answer: B

41. The inventor of Microscope is

a. Galileo

b. Antony von

c. Pasteur

d. Koch

Answer: B

42. First Pasteur conducted fermentation experiments in

a. Milk

b. Food material

c. Fruit juices

d. Both a and c

Answer: C

43. Modern concepts of chemotherapy were proposed by

a. Paul Ehrlich

b. Joseph Lister

c. Elie Metchnikoff

d. None of these

Answer: A

44. Viral infection of bacteria was discovered by

a. De'Herelle

b. F.W. Twort

c. Beijernick

d. Jwanoksy

Answer: B

45. Eye cannot resolve any image less than

a. 1ìm

b. 2ìm

c. 7ìm

d. 5ìm

Answer: D

46. Compound Microscope was discovered by

a. A.V. Lewenhoek

b. Pasteur

c. Janssen and Hans

d. None of these

Answer: C

47. Electron Microscope was discovered by

a. Prof. Fritz

b. Janssen and Hans

c. Knoll and Ruska

d. None of these

Answer: C

48. Magnification range of light microscope is

a. 1000x – 5000x

b. 1000x – 2000x

c. 500x – 1000x

d. None of these

Answer: B

49. Condensation of light in light Microscope is by

a. Objective

b. Condensor

c. Ocular

d. All of these

Answer: B

50. Who perfected a magnetic lens in 1927

a. Gabor

b. Broglie

c. Busch

d. None of these

Answer: A

51. The magnification power of electron microscope developed by Knell and Ruska is

a. 10,000x

b. 12,000x

c. 15,000x

d. 20,000x

Answer: B

52. In an electron microscope source of electrons is from

a. Mercury lamp

b. Tungsten metal

c. Both a and b

d. None of these

Answer: B

53. The electrons passed out from the specimen are called

a. Primary electrons

b. Secondary electrons

c. Tertiary electrons

d. None of these

Answer: B

54. Mycorrhiza was first observed by

a. Funk

b. Frank

c. Fisher

d. Crick

Answer: B

CHAPTER 8

CELL METABOLIC

1. INTRODUCTION TO METABOLISM: ANABOLISM AND CATABOLISM

Introduction

A metabolic pathway has many steps that begin with a specific molecule and end with a product that are each catalyzed by a specific enzyme. They are two metabolic pathways.

Catabolic Pathways

It breaks down complex molecules into simpler compounds eg. amylase breaks complex starches into simple sugars and it is the process of cellular respiration.

Anabolic Pathways

It consumes energy to build complicated molecules eg. Anabolic steroids to build muscle. The building of a protein from amino acids figure 1.1.

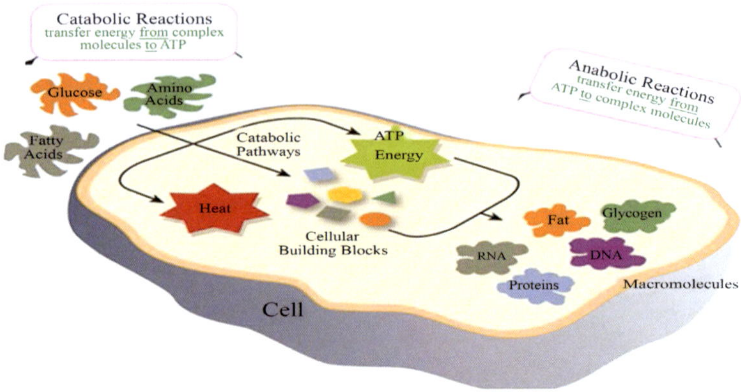

Figure 1.1: Metabolism.

Different Types of Energy (energy=the ability to do work or cause change).

Potential

The energy stored in an object. It is measured in joules and chemical energy is the potential energy of a chemical reaction.

Kinetic

The energy of an object in motion. It is measured in joules. Thermal energy is the kinetic energy of atoms or molecules.

Laws of Thermodynamics

1st Law: Energy can be transferred and transformed but it can't be created or destroyed.

2nd Law: Every energy transfer or transformation increases the entropy of the universe.

Thermodynamics is the metabolic reactions in organisms. It must occur at body temperature. The body temperature does not get substrates to their transition state. The active site of enzymes lowers the amount of energy needed to reach a transition state.

Mechanism of Transport

Nonmediated transport: It has through simple diffusion.

Mediated transport: It has required specific carriers.

Passive mediated transport: Specific molecules flow from high to low concentrations. The net flow will cease when the concentration gradient is no longer in existence.

Active transport: The movement is from low to high concentration and endergonic processes. It must be coupled to a sufficiently energetic (exergonic) process to make it possible.

Classification

Uniport is **the** movement of a single type of molecule at a time.

Symport is **the** simultaneous movement of two different molecules in the same direction.

Antiport is **the** simultaneous movement of two different molecules in opposite directions.

Nonmediated transport is through simple diffusion, driven by (electro)chemical potential gradient. The substance diffuses in the direction that eliminates its concentration gradient. The rate of diffusion is: (a) proportional to the magnitude of the (electro)chemical gradient [For many nonelectrolytes, their fluxes across erythrocyte membranes increase linearly with $(A)_{out}$ -$(A)_{in}$ (b) dependent on the solubility of the substance in the membrane's nonpolar core.

Mediated transport is through the action of a specific protein or other molecules referred to as carriers, permeases, channels, and transporters.

A. Passive – Mediated Transport

1. **Ionophores:** It has small organic molecules, many are antibiotics and vastly increase the permeability of membranes to particular ions. It can be carriers or channel formers. The efficiency of a carrier can be drastically reduced at lower T but not so for the channel formers. **Carriers:** The carrier-ionic complexes are soluble in non-polar solvents. It can move from one side to the other side of a lipid-bilayer membrane. Valinomycin contains 6 oxygen atoms complexing with a K^+. Channel Former: gramicidin two molecules together form a channel. Facilitates the passage of cations (eg. Na^+, K^+, H^+) but is blocked by Ca^{2+}.

2. **Bacterial Porins:** Channel forming protein. X-ray structure known (*E. coli* OmpF porin). Some are cation-selective, others are anion-selective.
3. **Ion Channels:** It has many ion-specific channels for the rapid passage of ions. K^+ passively diffuses from cytoplasm to the extracellular space through transmembrane proteins "K^+ Channels." (High selectivity for K+ over Na^+ are "Gated." The physiological functions of ion channels depend on their ion specificity, speed of transport, and the ability to gate. Mechanosensitive channels: it responds to physical stimuli (touching and sound, etc.), which cause local deformations in membranes. Ligand-gated channels: respond to extracellular chemical stimuli such as a neurotransmitter. Signal-gated channels: open upon intracellular binding of Ca^{2+} or other signal molecules. Voltage-gated channels: respond to changes in membrane potential, such as in the generation of nerve impulses.
4. **Aquaporins:** Water passes through biological membranes extremely fast and aquaporins were discovered in 1992 by Peter Agre.
5. **Transport Proteins:** They function in mediated passive and active transport.

2. CARBOHYDRATE METABOLISM

Introduction

The flow of intermediates through metabolic pathways is controlled by four mechanisms

- Substrate availability
- Allosteric activation or inhibition
- Covalent modification of enzymes.
- Induction/repression of enzyme synthesis

The types of transporters and enzymes that exist in a certain cell type.

Concentration Effects on Enzymatic Steps

High K_m of glucokinase determines the rate of glucose catabolism and insulin secretion in pancreatic beta cells, High K_m of glucokinase allows liver cells to rapidly phosphorylate glucose at high concentrations.

Allosteric Effector

Glucose stimulates the release of glucokinase from the "glucokinase regulatory protein (GKRP) and glucose inhibits liver glycogen phosphorylase. Glucose stimulates gene transcription through, glucose-responsive regions of gene promoters. High glucose concentration stimulates the expression of the pyruvate kinase in the liver. The control of flux through a pathway is distributed among all the enzymes. A key regulatory enzyme is characterized by a large change in flux when the enzymatic activity in the cell changes eg. phosphofructokinase in glycolysis. Once one enzymatic bottleneck for flux is removed, another enzyme may become limiting for the flux. Theoretically, if the enzyme activities of all the enzymes in the pathway were changed by the same degree, then the flux would change by the same degree. An efficient increase of flux through the pathway is brought about by increasing the enzyme activities or enzyme concentrations of all regulatory enzymes in a pathway. In the fed state with excess carbohydrates, the liver cells can synthesize fatty acids from glucose and fructose, and esterify them into triacylglycerols". Triacylglycerols are assembled together with cholesterol and protein into very-low density lipoprotein particles (VLDL) that are secreted into the circulation.

Conclusion

Glucose up-regulates genes encoding enzymes and proteins involved in the entire pathway of lipogenesis. ChREBP-Mlx is the principal mediator that transcriptionally regulates glucose-responsive genes.

3. CARBON REDUCTION IN C3 PLANTS AND CARBON REDUCTION IN C4 PLANTS

Introduction

The net overall equation for photosynthesis is. Photosynthesis occurs in 2 "stages": The light reactions (or Light-Dependent Reactions). The Calvin cycle (or Calvin-Benson cycle or dark reactions or light-independent reactions) figure 3.1.

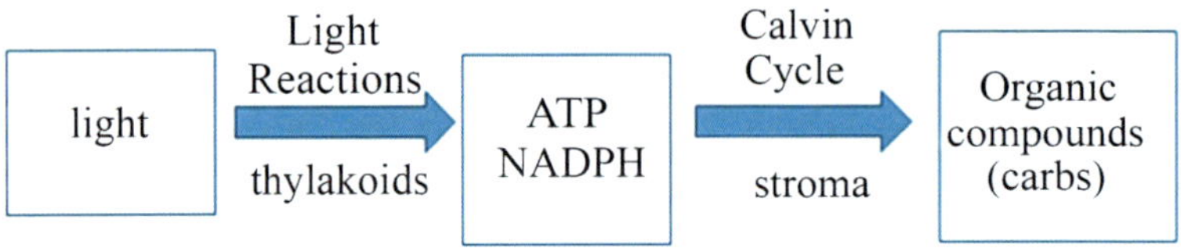

Figure 3.1: Photosynthesis.

In the Calvin cycle, chemical energy (from the light reactions) and CO_2 (from the atmosphere) are used to produce organic compounds (like glucose). The Calvin cycle occurs in the stroma of chloroplasts. The Calvin cycle involves the process of carbon fixation. This is the process of assimilating carbon from a non-organic compound (CO_2) and incorporating it into an organic compound.

Step 1: Carbon Fixation

There are three molecules of CO_2 (from the atmosphere) are joined to 3 molecules of RuBP (a 5-carbon sugar) by Rubisco (an enzyme also known as RuBP carboxylase).

Step 2: Reduction

The three 6-carbon molecules (very unstable) split in half, forming six 3-carbon molecules. These molecules are then reduced by gaining electrons from NADPH. ATP is required for this molecular rearranging There are now six 3-carbon molecules, which are known as G3P or PGAL. Since the Calvin cycle started with 15 carbons (three 5-carbon molecules) and there are now 18 carbons, we have a net gain of 3 carbons. One of these "extra" 3-carbon G3P/PGAL molecules will exit the cycle and be used to form ½ a glucose molecule. Once the Calvin cycle "turns" twice (well, actually 6 times), those 2 molecules of G3P (3-carbon carbohydrate) will combine to form 1 molecule of glucose (6-carbon carbohydrate molecule) or another organic compound **figure 3.2**.

Step 3: Regeneration of RuBP

The remaining 5 G3P molecules (3-carbons each) get rearranged (using ATP) to form 3 RuBP molecules (5-carbons each).

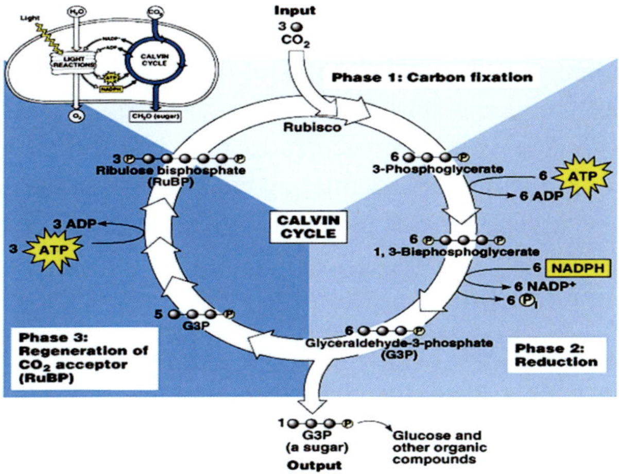

Figure 3.2: Calvin cycle.

In the Calvin cycle, energy and electrons from the light reactions (in the form of ATP and NADPH) and carbon dioxide from the atmosphere are used to produce organic compounds. The Calvin cycle occurs in the stroma inside the chloroplasts (inside the cells). The carbon dioxide, ATP, and NADPH are required (reactants). The organic compounds (G3P) are produced. So, as a broad overview of photosynthesis. The Light reactions (Phase 1) capture the energy in sunlight and convert it to chemical energy in the form of ATP and NADPH through the use of photosystems, electron transport chains, and chemiosmosis. The Calvin cycle (Phase 2) uses the energy transformed by the light reactions along with carbon dioxide to produce organic compounds.

The Rate (or speed) of Photosynthesis can vary, based on Environmental Conditions

- Light intensity
- Temperature
- Oxygen concentration

Light Intensity

As light intensity increases, so too does the rate of photosynthesis. This occurs due to the increased excitation of electrons in the photosystems. However, the photosystems will eventually become saturated. It is above this limiting level, no further increase in photosynthetic rate will occur.

Temperature

The effect of temperature on the rate of photosynthesis is linked to the action of enzymes. As the temperature increases up to a certain point, the rate of photosynthesis increases. The molecules are moving faster & colliding with enzymes more frequently and facilitating chemical reactions. However, at this temperature higher than this point, the rate of photosynthesis decreases.

Oxygen Concentration

As the concentration of oxygen increases, the rate of photosynthesis decreases. This occurs due to the phenomenon of photorespiration.

Photorespiration

Photorespiration occurs when Rubisco (RuBP carboxylase) joins oxygen to RuBP in the first step of the Calvin cycle rather than carbon dioxide. Whichever compound (O_2 or CO_2) is present in higher concentration will be joined by Rubisco to RuBP. Photorespiration prevents the synthesis of glucose AND utilizes the plant's ATP. Photorespiration is primarily a problem for plants under water stress. When plants are under water stress, their stomata close to prevent water loss through transpiration. However, this also limits gas exchange. O_2 is still being produced (through the light reactions). Thus, the concentration of O_2 is increasing. CO_2 is not entering the leaf since the stomata are closed. Thus, as the CO_2 is being used up (in the Calvin Cycle) and not replenished, the concentration of CO_2 is decreasing. As the concentration of O_2 increases and the concentration of CO_2 decreases (due to the closure of the stomata to prevent excessive water loss), photorespiration is favored over photosynthesis. Some plant species that live in hot, dry climates (where photorespiration is an especially big problem) have developed mechanisms through natural selection to prevent photorespiration C_4 plants and CAM plants.

C_3 Plants

C_3 plants, which are "normal" plants, perform the light reactions and the "Calvin cycle in the mesophyll cells of the leaves. The bundle sheath cells of C_3 plants do not contain chloroplasts. C_4 plants and CAM plants modify the process of C_3 photosynthesis to prevent photorespiration. C_4 plants perform the Calvin cycle in a different location within the leaf than C_3

plants. CAM plants obtain CO_2 at a different time than C_3 plants. Both C_4 and CAM plants separate the initial fixing of CO_2 (carbon fixation) from the use of CO_2 in the Calvin cycle. Plants that use C_4 photosynthesis include corn, sugar cane, and sorghum. In this process, CO_2 is transferred from the mesophyll cells into the bundle-sheath cells, which are impermeable to CO_2. This increases the concentration of CO_2. Thus, the Calvin cycle is favored over photorespiration. The bundle-sheath cells of C_4 plants do contain chloroplasts. C_4 plants use the Hatch-Slack pathway prior to the calvin cycle PEP carboxylase adds carbon dioxide to PEP, a 3-carbon compound, in the mesophyll cells. This produces a 4-carbon compound. This 4-carbon molecule then moves into the bundle-sheath cells via plasmodesmata. In the bundle sheath cells, the CO_2 is released and the Calvin cycle begins. Plants that use CAM photosynthesis include succulent plants (like cacti) and pineapples. In CAM (crassulacean acid metabolism) photosynthesis, plants open their stomata at night to obtain CO_2 and release O_2. This prevents them from drying out by keeping their stomata closed during the hottest & driest part of the day. When the stomata are opened at night, the CO_2 is converted to an organic acid (via the C_4 pathway) and stored overnight. During the day when light is present to drive the light reactions to power the Calvin cycle – carbon dioxide is released from the organic acid and used in the Calvin cycle to produce organic compounds. Even though the CO_2 is taken in at night, the Calvin cycle cannot occur because the light reactions can't occur in the dark. Both C_4 and CAM plants – which are primarily found in hot, dry climates have evolutionary adaptations which help prevent photorespiration. C_4 plants perform the Calvin cycle in the bundle sheath cells. CAM plants open their stomata at night and store the CO_2 until morning".

Introduction

Photosynthesis is the process governing energy capture by ecosystems, and thus the entry of organic carbon into the biosphere controls on GPP: across ecosystems – LAI, N, growing season length. within ecosystems (daily and seasonal) – light, temp, nutrients, photosynthesis. Light-harvesting converts light energy to chemical energy (ATP, NADPH) splits H_2O and produces O_2 as a byproduct. Carbon fixing uses ATP and NADPH from light reactions to fix CO_2 into sugar figure 3.3 and 3.4.

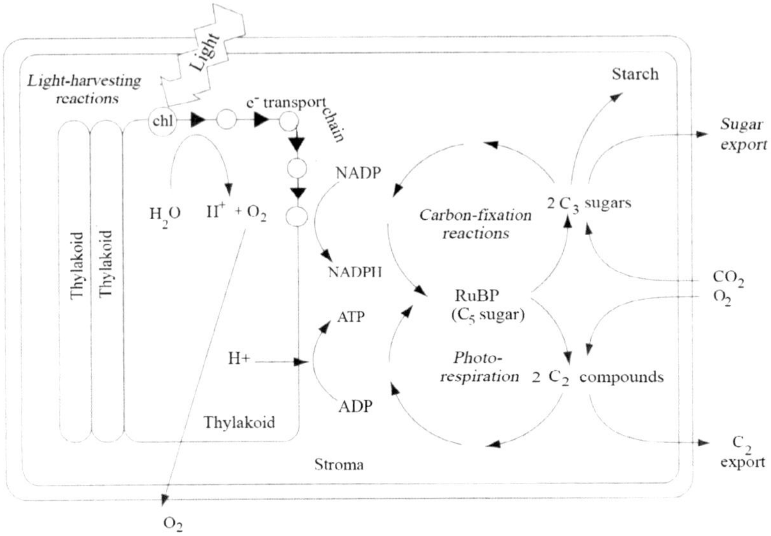

Figure 3.3: Photosynthesis reaction.

CO_2 Enters Leaf via Diffusion

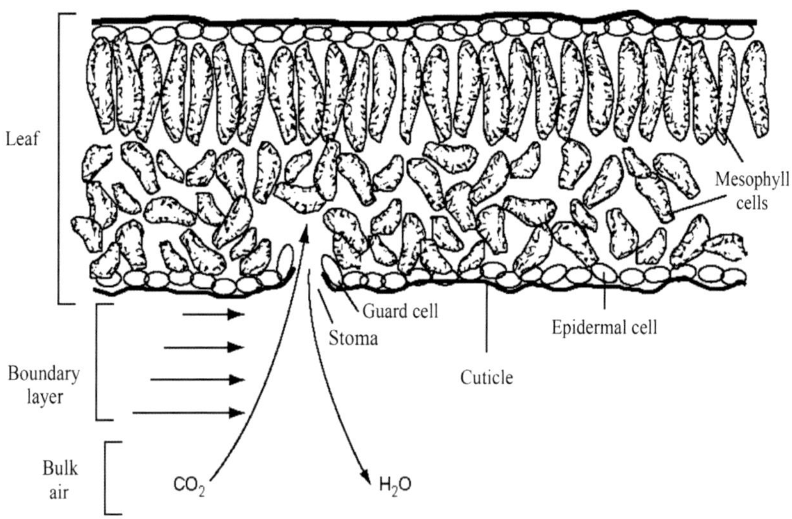

Figure 3.4: Photosynthetic pathway.

Stomata (plural), stoma (singular): biggest and most variable component of diffusion barrier.

Water evaporates on the cell surface, diffusing out through the stomata, leaf boundary layer, and into the bulk atmosphere is called transpiration.

Stomata: carbon gain, water loss inevitable trade-off. Three major photosynthetic pathways. C3 photosynthesis 85% of plants. C4 photosynthesis: spatial separation of 2 carbon fixation paths. CAM (Crassulacean Acid Metabolism): The temporal separation of 2 carbon fixation paths.

3 Photosynthetic Pathways

C4 and CAM reduce water loss and reduce photorespiration because PEP carboxylase has a higher affinity for CO_2 and no affinity for O_2. These adaptations are most important in hot and dry environments.

The C4 and CAM pathways evolved to reduce the impact of oxidation by RUBISCO (photorespiration). The C4 photosynthetic pathway concentrates CO_2 in the vascular bundle sheath cells of leaves to significantly reduce competition with O_2 for the active site on RUBISCO. The CAM photosynthetic pathway further separates absorption of CO_2 in time from its incorporation in the Calvin cycle. C4 & CAM are more efficient than C3 at yielding energy above about 85 degrees F–Rubisco favors oxygenation the higher the temp because the concentration of CO_2 in leaf tissue is less the higher the temperature figure 3.5.

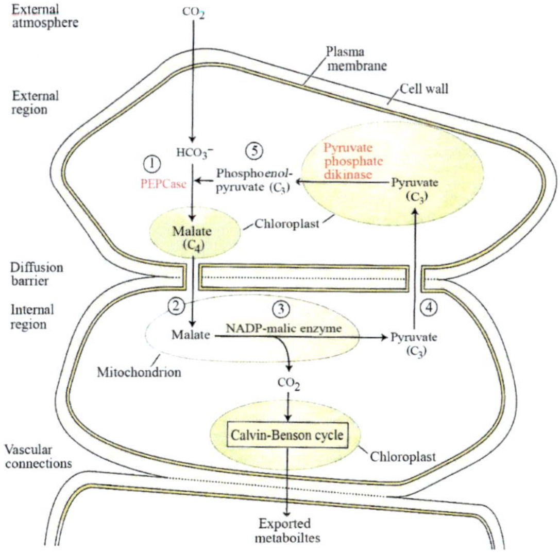

Figure 3.5: Calvin-Benson cycle.

At 86° F C4 becomes more efficient than C3 due to photorespiration. The starch rich chloroplasts of the Kranz anatomy lack grana, the site of the light reactions. They differ from the chloroplasts of the outer bundle sheath (dimorphic chloroplasts). In addition to being more efficient like C4 at higher temperatures, CAM conserves water by allowing stomata to be closed during the heat of the day. "Since every CO_2 molecule has to be fixed twice, first by 4-carbon organic acid and second by RuBisCO, the C4 pathway uses more energy than the C3 pathway. The C3 pathway requires 18 molecules of ATP for the synthesis of one molecule of glucose, whereas the C4 pathway requires 30 molecules of ATP. This energy debt is more than paid for by avoiding losing more than half of photosynthetic carbon in photorespiration as occurs in some tropical plants, making it an adaptive mechanism for minimizing the loss".

4. SUCROSE METABOLISM

Introduction

It is a non-reducing disaccharide. It is composed of glucose and fructose linked via their anomeric carbons. It is synthesized in the cytosol of plant cells. It is synthesized from UDP-glucose and fructose 6-phosphate. The sucrose is translocated from its site of synthesis in mature leaves to various metabolic tissues, where it is used to support the growth and synthesis of reserve materials such as starch.

Sucrose is synthesized from UDP-Glucose and Fructose-6-P in the cytosol by sucrose 6-phosphate synthase and sucrose 6-phosphate phosphatase. "Sucrose biosynthesis is beginning with dihydroxyacetone phosphate exported from the chloroplast by Pi-triose phosphate antiporter. Dihydroxyacetone phosphate is then converted to glyceraldehyde 3-phosphate by triose phosphate isomerase. Pi-triose phosphate antiporter, a transport system exports triose phosphates from the chloroplast and imports phosphate: P_i-triose phosphate antiporter simultaneously moves P_i into the chloroplast and moves triphosphate into the cytosol. Sucrose synthesis releases P_i. If this exchange was blocked, triose phosphate synthesis would quickly deplete the available P_i in the chloroplast. A second potential source of energy is the ATP and NADPH generated in chloroplast". Pi-triose phosphate antiporter system has the indirect effect of moving ATP equivalents and reducing equivalents.

Trans-aldolase Reaction

The condensation of dihydroxyacetone phosphate and glyceraldehyde -3-phosphate by transaldolases. Transaldolase reaction is identical to aldolase reaction in glycolysis/gluconeogenesis; other is unique to carbon assimilation. "Sucrose synthase (Km, 15 mM) has a much lower Km for sucrose compared with the neutral invertase (Km, 65 mM). Consequently, the sucrose synthase pathway may be relatively more important when sucrose availability is limiting. This pathway is also more energetically efficient, as the energy contained in the glycosidic linkage of the sucrose molecule is preserved. Thus, to metabolize one molecule of sucrose to the level of Triose-P requires the input of three ATP in the sucrose synthase pathway, compared with four ATP in the invertase pathway. Consequently, it may be beneficial to the cells to have the most efficient pathway operate when carbon supplies are limiting".

Starch Metabolism: Starch Biosynthesis, Regulation and Degradation

Introduction

It is a polymer of α-D-glucose. It occurs in two main forms: amylose, consisting of predominantly linear chains of glucose monomers linked by α1,4-glycosidic bonds, and amylopectin, in which the chains are branched by the addition of α1,6-glycosidic bonds. The fewer branches than glycogen. It is synthesized in the chloroplast (stroma). The precursor is activated ADP-glucose.

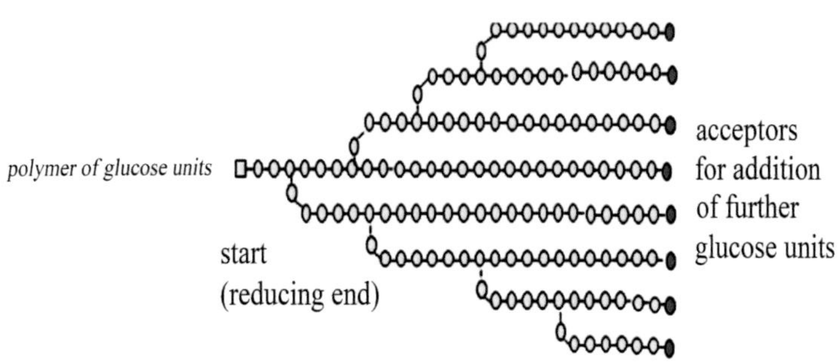

Starch granules are classified as transitory or reserve. The starch granules accumulate for only a short period of time before they are degraded eg. starch forms in leaf chloroplasts during the day. Hydrolyzed and transported to other plant parts at night as a simple sugar. The reserve starch, energy storage for germination, a major component of food and feed, and an industrial starting material, is formed in amyloplasts. Starch is made in photosynthetic and non-photosynthetic cells.

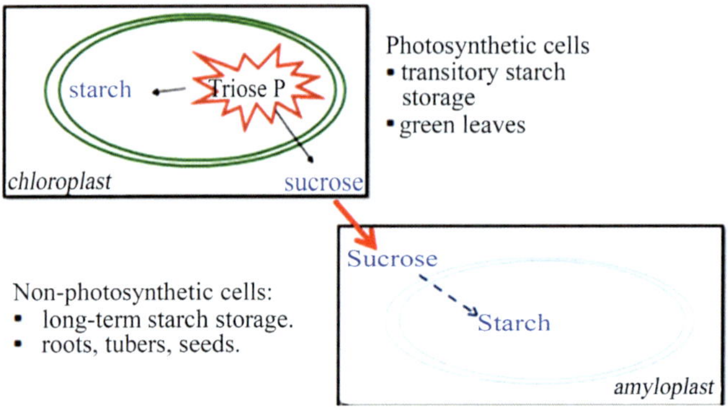

Amylose

A linear polymer of α-D-glucose with α1,4-linkage. It may have a low level of branching (~one branch per 1000 residues) with an α1,6-linkage. It comprises between 11 and 37% of the starch found in plants (depending upon the species and the site of storage). It is much lower in molecular weight than amylopectin.

Amylopectin

It is a highly branched polymer of α-D-glucose with α1,4 & α1,6 linkages. It consists of 10,000 - 100,000 glucose units. It is highly branched, 20 - 25 glucose residues/branch. It makes up ~65% of starch. It has a much higher molecular weight than amylose.

Starch Biosynthesis

ADP-glucose is used as the precursor. The starch synthase transfers the glucose unit to the non-reducing end of a preexisting primer. The branches in amylopectin are synthesized using branching enzymes. The synthesis of ADP-Glucose, catalyzed by ADP-glucose pyrophosphorylase, is rate-limiting. Starch Synthase (SS) catalyzes a 1,4- linkage between the non-reducing end of the glucan chain & Glc from ADP-Glc. SS can use both amylose and amylopectin as acceptors. The priming event not known: some evidence for protein primer, some evidence for de-novo synthesis.

Starch Degradation

The starch granule is attacked by the endoamylase α-amylase, which releases soluble linear and branched glucans. These are acted on by the debranching enzyme limit dextrinase and the exoamylase β-amylase to produce maltose. Maltose is then hydrolyzed to glucose by an α-glucosidase (maltase).

5. METABOLISM OF NITROGEN-AMINO ACIDS AND NUCLEOTIDES

Introduction

Nucleic acids are essential for information transfer in cells. The information encoded in a DNA molecule is transcribed via the synthesis of an RNA molecule. The sequence of the RNA molecule is "read" and is translated into the sequence of amino acids in a protein. The compound contained C, N, O, and a high amount of P. It was an acid compound found in nuclei therefore named nucleic acid. In 1944, Oswald, Avery, MacLeod and McCarty demonstrated that DNA is the molecule that carries genetic information. 1953 Watson and Crick proposed the double helix model for the structure of DNA nucleic acids are long polymers of nucleotides.

Nucleotides and Nucleic Acids

The amino acids sequence and nucleotide sequence in RNA is specified by a nucleotide sequence in the cell's DNA. **Gene:** a segment of a DNA molecule that contains the information required for the synthesis of a functional biological product whether a protein or RNA. The cell contains thousands of genes and DNA molecules. **Nucleic acids** are polymers of nucleotides. Deoxyribonucleic acid - DNA - storage of genetic information. Ribonucleic acid - RNA - carriers of genetic information and catalysis. Nucleotide structure: Nitrogenous base+ ribose phosphate figure 5.1 and 5.2.

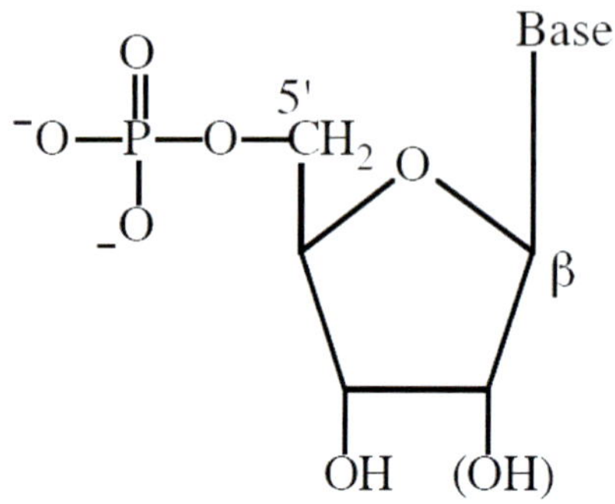

Figure 5.1: Nucleotide structure.

Nucleotides Structure

The nucleotide has three characteristic components. A nitrogenous base, Pentose sugar and Phosphate.

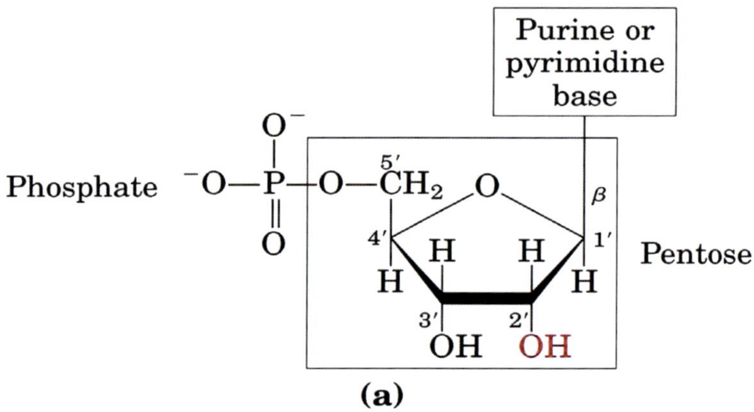

Figure 5.2: Nitrogenous base, Pentose sugar and Phosphate.

Nitrogenous Bases

Nitrogenous base: derivatives of purines and pyrimidines, DNA and RNA contain the same purine bases and the pyrimidine base cytosine but thymine is found only in DNA and uracil is found only in RNA figure 5.3.

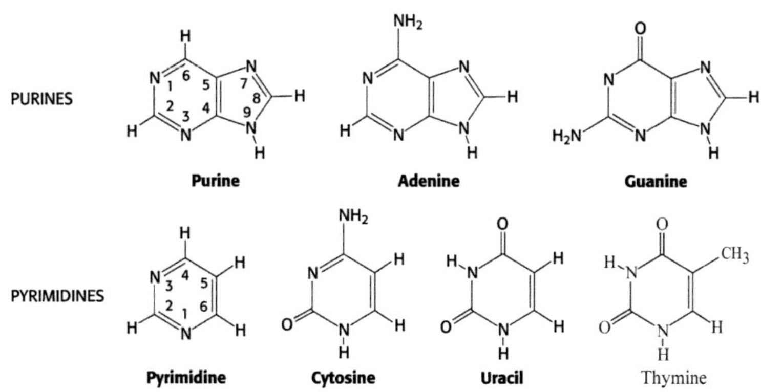

Figure 5.3: Purines and Pyrimidines.

Pentoses of Nucleotides

D-ribose (in RNA), 2-deoxy-D-ribose (in DNA), The difference - 2'-OH vs 2'-H, this difference affects secondary structure and stability. The bases are attached by b-N-glycosidic linkages to 1 carbon of pentose sugar – (Nucleoside). The base is linked via a b-N-glycosidic bond. The carbon of the glycosidic bond is anomeric. The named by adding -idine to the root name of a pyrimidine or -osine to the root name of a purine sugars make nucleosides more water-soluble than free bases figure 5.4.

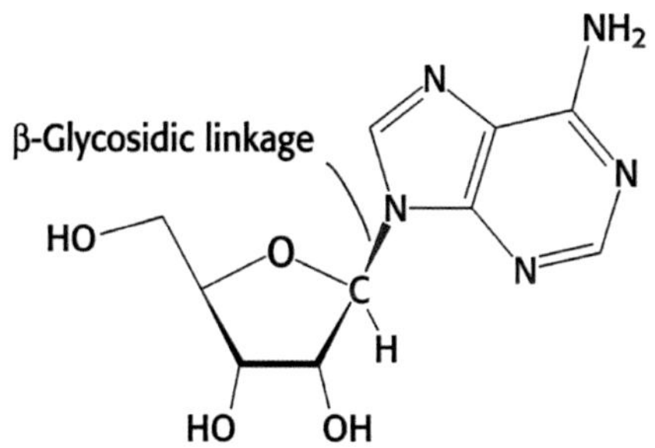

Figure 5.4: Glycosidic linkage.

Ribonucleotides

- Adenosine 5'-monophosphate, **A**denylate, AMP
- Guanosine 5'-monophosphate, **G**uanylate, GMP
- Cytidine 5'-monophosphate, **C**ytidylate, CMP
- Uridine 5'-monophosphate, **U**ridylate, UMP

Deoxyribonucleotides

- Deoxythymidine 5' monophosphate
- Deoxythymidine, dTMP.
- NMP = Nucleoside mono phosphate.
- Numbering of Ribose sugar is given 1', 2', …, 5'

Unusual Nucleotides

The modified nucleotides are found in some viral DNA and in transfer RNA. These modifications include methylation, hydroxymethylation, glycosylation, acetylation.

Functions of Nucleotides

Nucleoside 5'-triphosphates are carriers of energy, bases serve as recognition units, cyclic nucleotides are signal molecules and regulators of cellular metabolism and reproduction. The structural component of some coenzymes, e.g CoA, FAD, NADH, NADPH, ATP is central to energy metabolism, GTP drives protein synthesis, CTP drives lipid synthesis, UTP drives carbohydrate metabolism.

De Novo Synthesis of Purine Nucleotides

The atoms of the purine ring are contributed by a number of compounds including amino acids (aspartic acid, glycine and glutamine) CO2, and tetrahydrofolate. These compounds donate N and C to construct Ribose 5-phosphate.

Synthesis of 5-phosphoribosyl-1-pyrophosphate (PRPP)

Ribose 5-phosphate is synthesized from HMP (Hexose monophosphate pathway), ribonucleotides are first synthesized then may reduce to deoxyribonucleotides.

Synthesis of 5-phosphoribosyl-1-pyrophosphate (PRPP)

The amide group of the glutamine replaces the pyrophosphate group attached to PRPP, this reaction is mediated by Glutamine: phosphoribosyl pyrophosphate amidotransferase. This enzyme is inhibited by the end product of this pathway purine 5'-nucleotides AMP, GMP and IMP. This reaction is the committed step in purine nucleotide synthesis.

Synthesis Inosine Monophosphate (IMP)

IMP is the parent purine nucleotide. the synthesis of IMP requires 4 ATP molecules.

Inhibitors of Purine Synthesis

Specific inhibitors that inhibit the growth of rapidly growing microorganisms eg. Sulfonamides. Structural analogues for folic acid (methotroxate).

Conversion Nucleoside Monophosphate (NMP) to Nucleoside Diphosphate (NDP) and Triphosphate (NTP)

NDP and NTP are synthesized from the corresponding NMP by Nucleoside monophosphate kinases. These kinases don't discriminate between ribose or deoxyribose in the substrate. ATP is the source of the transferred phosphate examples:

AMP + ATP $\leftrightarrow$ 2 ADP $\rightarrow$ adenylate kinase (highly active in the liver)

GMP + ATP $\leftrightarrow$ GDP + ADP $\rightarrow$ Guanylate kinase

Nucleoside diphosphates and triphosphates are interconverted by nucleoside diphosphate kinase

GDP + ATP $\leftrightarrow$ GTP + ADP

CDP + ATP $\leftrightarrow$ CTP + ADP

Salvage Pathway for Purines

Salvage Pathway

Purines that result from the normal turnover of cellular nucleic acids or that are obtained from the diet and not degraded can be reconverted into nucleoside triphosphates and used by the body. The two enzymes are involved such as Adenine phosphoribosyl transferase (APRT) and hypoxanthine-guanine phosphoribosyl transferase (HGPRT). Both enzymes are utilizing PRPP as the source of the ribose-5-phosphate group.

Degradation of Purine Nucleotides

Purines are sequentially degraded into uric acids (in humans). Several steps will be involved in this catabolic pathway.

Degradation of Dietary Nucleic Acids in the Small Intestine

Ribonucleases and deoxyriboncleases secreted in the pancreatic juice can hydrolyze RNA and DNA into oligonucleotides. Oligonucleotides are further hydrolyzed by pancreatic phophodiesterases producing a mixture of 3'-and 5' mononucleotides. A family of nucleotidases removes the phosphate group releasing nucleosides that may be absorbed from GIT. Dietary nucleotides are not used to large extent in cells because they are converted into uric acid in the small intestine and are also used by the normal flora.

Pyrimidine Nucleotides Synthesis

Purine ring is synthesized on an existing ribose 5-phosphate pyrimidine ring is synthesized then attached to ribose 5-phosphate donated by PRPP the sources of carbon atoms in pyrimidine rings are glutamine, CO_2, and aspartic acid.

Synthesis of Carbamoyl Phosphate

The committed step of this pathway in mammalian cells is the synthesis of carbamoyl phosphate from Glutamine and CO_2

2ATP + CO2 + Glutamine ↔ Carbamoyl phosphate + 2ADP + Glutamate

This reaction is mediated by Carbamoyl Phosphate Synthetase II (CPS II). CPS II is inhibited by UTP and activated by ATP and PRPP. Carbamoyl phosphate is the precursor of Urea and the pyrimidine synthesis occurs in the cytosol while the urea production occurs in the mitochondria by Carbamoyl Phosphate Synthetase I (CPS I), CPS I uses ammonia as a source of nitrogen, CPS II uses the amide group of glutamine, glutamine is required in the synthesis of both purines and pyrimidine.

Degradation of Pyrimidines

Purines are not cleaved in human cells. Pyrimidines rings can be opened and degraded to highly soluble structure, such as β-alanine and β-aminoisobutyrate that can serve as precursors of acetyl CoA and succinyl CoA. Pyrimidine can be salvaged and converted into nucleotides by the enzyme pyrimidine phosphoribosyltransferase and it utilizes the PRPP.

Conversion of Ribonucleotides to Deoxyribonucleotides

2'-deoxyribonucleotides are synthesized from ribonucleoside diphosphatase. Ribonucleotide reductase is multi-subunit enzyme (2B1 and 2B2 subunits)

that catalyzes the reduction of NDP (ADP, GDP, CDP, UDP) into dNDP (dADP, dGDP, dCDP, dUDP). The immediate donors of hydrogen atoms needed for the reduction are two–SH groups of the enzyme itself. The reduced form of the enzyme should be regenerated. The reducing agent is a peptide coenzyme of ribonucleotide reductase called Thioredoxin. The thioredoxin contains two cysteine residues that can be oxidized to reduce the ribonucleotide reductase enzyme. The oxidized thioredoxin is reduced back by NADPH, this reaction is mediated by Thioredoxin reductase

Regulation of Deoxyribonucleotides Synthesis

The regulation of this enzyme is complex. The activity is regulated but also substrate specificity. The binding of dATP to an allosteric site called activity site inhibits the enzyme while the binding of ATP to this site activates the enzyme. The binding of NTP to an allosteric site called substrate specificity will increase the conversion of different NTP to their corresponding dNTP according to the need of the cell. When ATP or dATP are bound to this site for the reduction of UDP, and CDP is favored. When dTTP, or dGTP is bound for the reduction of GDP, ADP is stimulated. Synthesis of Thymidine monophosphate from dUMP. dUMP is converted into dTMP by thymidylate synthetase which utilizes $N^5 N^{10}$–methylene tetrahydrofolate as the source of methyl group and 2 hydrogens is oxidation into dihydrofolate. inhibitors of thymidylate synthetase (5-fluoro uracil) act as anti-tumor. DHF can be reduced into THF by DHF reductase, which can be inhibited by methotrexate to inhibit purine synthesis and decrease the supply of THF so prevents methylation of dUMP to dTMP.

Nucleotide Biosynthesis

Nucleotides can be synthesized de novo ("from the beginning") from amino acids, ribose-5-phosphate, CO_2, and NH_3. Nucleotides can be salvaged from RNA, DNA, and cofactor degradation. Many parasites (e.g., malaria) lack de novo biosynthesis pathways and rely exclusively on salvage. The compounds that inhibit salvage pathways are promising antiparasite drugs.

de-Novo Biosynthesis of Nucleotides

All organisms were studied about bases synthesized while attached to ribose approximately the same. Glu provides most amino groups. Gly is a precursor for purines. Asp is the precursor for pyrimidines. Nucleotide pools are kept low, so cells must continually synthesize them. This synthesis may actually limit rates of transcription and replication. Adenine and guanine are synthesized as AMP and GMP. Synthesis begins with the reaction of 5-phosphoribosyl 1-pyrophosphate (PRPP) with Glu. Purine ring builds up following the addition of three carbons from glycine. The first intermediate with a full purine ring is inosinate (IMP).

de-Novo synthesis of pyrimidine nucleotides (1, 2 and 3). Unlike purine synthesis, pyrimidine synthesis proceeds by first making the pyrimidine ring and then attaching it to ribose 5-phosphate. Aspartate and carbamoyl phosphate provide the atoms for the ring structure. After the addition of ribose-5-phosphate via PRPP, the resulting nucleotide (orotidylate) is decarboxylated to form uridylate (UMP), the first possible pyrimidine. UMP is phosphorylated to UTP. After the formation of UTP, amination can convert UTP to CTP.

Ribonucleotides are Precursors to Deoxyribonucleotides

2'C-OH bond is directly reduced to a 2'-H bond without activating the carbon. catalyzed by ribonucleotide reductase. Two H atoms are donated by NADPH and carried by proteins thioredoxin or glutaredoxin.

Reduction of Ribonucleotides by Ribonucleotide Reductase

NADPH serves as the electron donor. It is directed through glutathione (a) or thioredoxin pathways (b).

dTMP is made from dUTP

- dUTP is synthesized.
- dUTP → to dUMP by dUTPase
- dUMP → dTMP by thymidylate synthase
- - adds a methyl group from tetrahydrofolate

Thymidylate synthase is a target for some anticancer drugs.

Folic Acid Deficiency Leads to Reduced Thymidylate Synthesis

Folic acid deficiency is widespread, especially in nutritionally poor populations. It reduced thymidylate synthesis causes uracil to be incorporated into DNA. There are repair mechanisms that remove the uracil by creating strand breaks that affect the structure and function of DNA. It is associated with cancer, heart disease and neurological impairment.

Antibiotics also Target Nucleotide Biosynthesis

Allopurinol, and so on studied against African sleeping sickness (trypanosomiasis) because the trypanosomes lack enzymes for de novo nucleotide synthesis. Trimethoprim inhibits bacterial dihydrofolate reductase but binds human enzymes several orders of magnitude less strongly.

6. PROTEIN SYNTHESIS AND LIPID METABOLISM

Introduction

Proteins are important, 44% of the dry weight of the human body and 5% of human caloric intake goes for protein synthesis. It is catalyzing most of the reactions in living organisms. It is serve many roles (enzymatic, structural, transport, regulation) protein synthesis is tightly regulated by environmental stimuli as well as intrinsic processes (e.g., hormonal, developmental). The dysregulation can cause disease and many antibiotics act at the level of protein synthesis.

Central Dogma

The central dogma states that once "information" has passed into protein it cannot get out again. The transfer of information from nucleic acid to nucleic acid, or from nucleic acid to protein, may be possible, but transfer from protein to protein, or from protein to nucleic acid is impossible. Information means here the precise determination of sequence, either of bases in the nucleic acid or of amino acid residues in the protein.

Genetic Code

The first step in decoding genetic messages is "transcription, during which a nucleotide sequence is copied from DNA to RNA. The next step is to join amino acids together to form a protein. The order in which amino acids are joined together determines the shape, properties, and function of a protein. The four bases of RNA form a language with just four nucleotide bases: adenine (A), cytosine (C), guanine (G), and uracil (U). The genetic code is read in three-base words called codons. Each codon corresponds to a single amino acid (or signals the starting and stopping points of a sequence). In transcription, a DNA sequence is rewritten, or transcribed, into a similar RNA. In eukaryotes, the RNA molecule must undergo processing to become a mature messenger RNA (mRNA). In translation, the sequence of the mRNA is decoded to specify the amino acid sequence of a polypeptide". The name translation reflects that the nucleotide sequence of the mRNA sequence must be translated into the completely different "language" of amino acids.

Lipids

The substances in dairy foods, known as fats and oils, are hydrophobic and soluble in ethanol like alcohol, petroleum and chloroform, it is classified into two general types. Fats and waxes (can be hydrolyzed with ester linkages) cholesterol and steroids (cannot be hydrolyzed).

Type of Lipids

The saturated fat, animal oil like meat, milk, butter, vegetable oil like coconut and palm kernel oil, Polyunsaturated fat, plan sources like safflower, corn, cottonseed, sunflower oil and soybean oil Monounsaturated fat, plant and animal product like olive oil, canola oil, avocado and peanut oil.

Lipids Functions

- Excellent energy reserves
- Structure of cell membranes
- Organ padding
- Body thermal insulation
- Essential fatty acids (EFA)
- Hormone synthesis
- Fat-soluble vitamin absorption

Lipids Disorder

- Lipids deficiency (Shortage in Lipids intake)
- Lipids exceeding (Overtaking in Lipids intake)

Lipids Deficiency

- Fat should comprise 3% of total calories to prevent fatty acid deficiency.
- Fatty acid deficiency syndromes.

- Dry scaly skin, dermatitis (Linoleic acid deficiency).
- Hand tremors (Prostaglandin deficiency).
- Inability to control blood pressure.

Lipids Exceeding

Fat should comprise not more than 30% of total calories to prevent lipids exceeding, to prevent overtaking, we should consume fat breakdown (% total calories) <8% from saturated fat. It is 10% from polyunsaturated fat, 10-15% from monounsaturated fat. Fatty acids are taken up by cells. They may serve as: precursors in the synthesis of other compound fuels for energy production substrates for ketone body synthesis. Ketone bodies may be exported to other tissues are used for energy production. Some cells synthesize fatty acids for storage or export. Fats are an important source of calories. Typically, 30-40% of calories in the American diet are from fat. Fat is the major form of energy storage. Typical body fuel reserves are: fat: 100,000 kcal. protein: 25,000 kcal. carbohydrate: 650 kcal. It provides 60% of the energy needs for the body at rest. TAG reserves would enable someone to survive starvation for ~30 days. Digestion and absorption of lipids, the 98% of ingested lipids are triacylglycerols (TAGs), Digestion in the mouth: enzymes are aqueous -little effect on lipids. Digestion in the stomach causes a large physical change: churned into droplets: "Chyme". Gastric Lipase begins actual the lipid digestion. ~10% of TAGs are hydrolyzed in the stomach. Chyme stimulates cholecystokinin (CCK) to release bile from the gall bladder. Bile is an emulsifier. Fatty acid micelle hydrophobic fatty acids & monoacylglycerols are in the interior. Bile salts on the exterior. Micelles are small enough to penetrate the membrane of intestinal cells. Free fatty acids & monoacylglycerols are reformed into triacylglycerols. TAGs are combined with membrane & water-soluble proteins to form a chylomicron, a lipoprotein. Chylomicrons carry TAGs from intestinal cells into the bloodstream via the lymph system. Triacylglycerols reach the bloodstream & are hydrolyzed down to glycerol and fatty acids. These are absorbed by cells and processed further for energy by forming acetyl CoA or stored as lipids in fat cells (adipose tissue).

Triglyceride Storage & Mobilization

Storage of triacylglycerol is in adipocytes. Fatty acids are stored primarily as triacylglycerol. Triacylglycerol is hydrolyzed to release fatty acids when needed. Adipocytes are found mostly in the abdominal cavity and subcutaneous tissue. Store energy, insulation against heat loss, shock absorber for organs. Adipocytes are metabolically very active: triacylglycerol is constantly hydrolyzed & re-synthesized.

Hormonal control of lipolysis is the breakdown of triglycerides by lipases is under hormonal control. Hormones involved such as Epinephrine, glucagon, and insulin. Epinephrine & glucagon are promoting the breakdown of fat (lipolysis) Insulin inhibits lipolysis. Triacylglycerol mobilization is hydrolyzing lipid reserves in adipose tissue for energy triggered by hormones. ~10% TAGs are replaced in adipose tissue daily as they get used up for energy. Hydrolysis of stored triacylglycerols in adipose tissue is triggered by hormones that stimulate cAMP production within adipose cells. The third time is a charm. TAGs hydrolyzed a 3rd time to form fatty acids. Triacylglycerol lipase, Diacyclglycerol lipase, Monoacylglycerol lipase, only triacylglycerol lipase is activated by epinephrine.

7. CATABOLISM, GLYCOLYSIS, GLUCONEOGENESIS, CITRIC ACID CYCLE, ELECTRON TRANSPORT CHAIN AND OXIDATIVE PHOSPHORYLATION

Introduction

Metabolism is considered into metabolic pathways that break down the molecules that generate energy (catabolism) and the molecules that consume energy by structuring larger molecules (anabolism). Energy metabolism mentions additional exactly to the metabolic pathways that release or store energy **figure 7.1.**

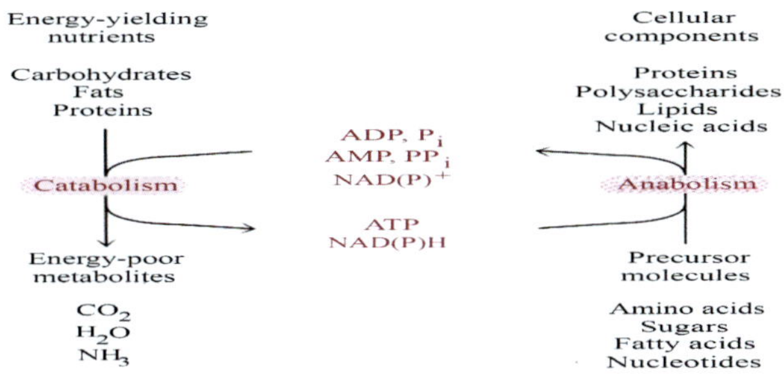

Figure 7.1: Metabolism.

The exergonic reactions of catabolism sustain ATP, NADH and NADPH formation. ATP and NADPH provide energy and reducing equivalents for the endergonic reactions of anabolism figure 7.2.

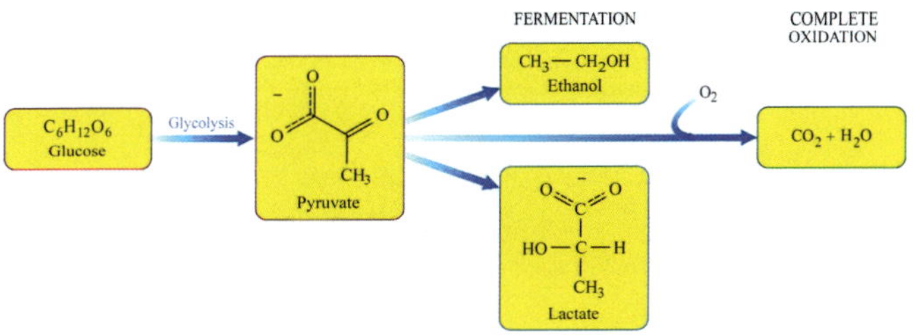

Figure 7.2: Fate of glucose.

The fate of glucose is differing from physiological conditions, tissues, and organisms.

Glycolysis

Stage 1: Investment of ATP. Glucose is phosphorylated. The negative charge concentrates glucose in the cell and glucose becomes less stable.

Stage 2: The 6 carbon sugar is split into two 3-carbon fragments.

Stage 3: Energy yielding phase. The oxidation of the 3-carbon fragments yields ATP **figure 7.3.**

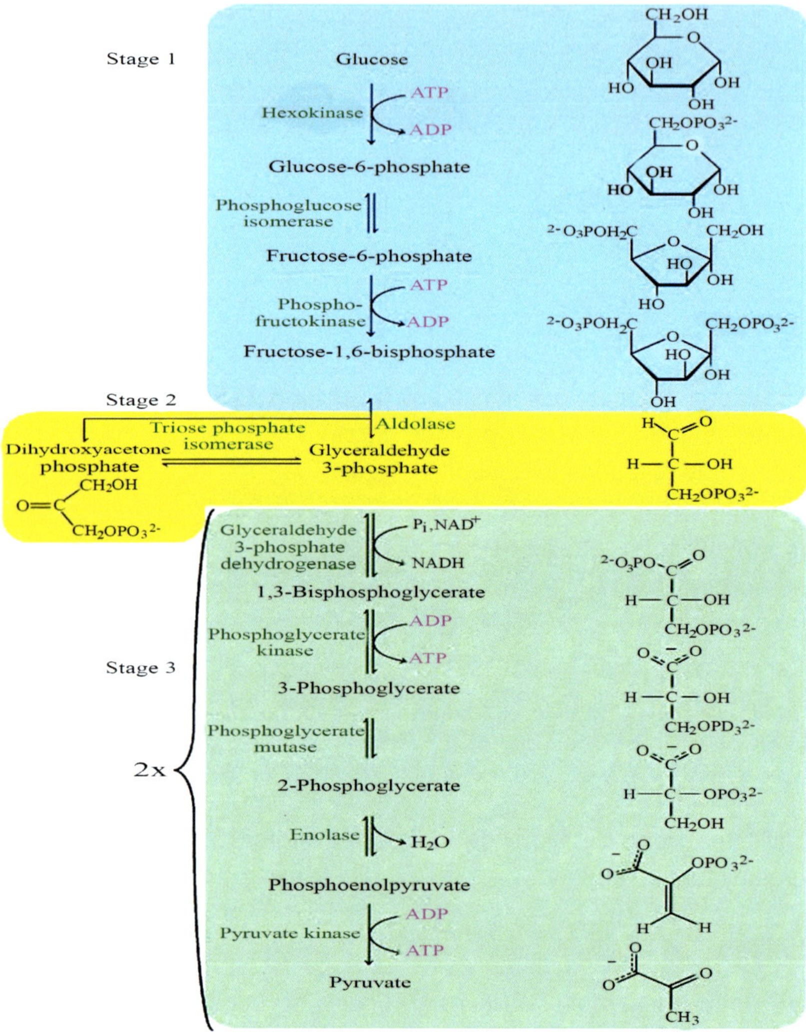

Figure 7.3: Glycolysis.

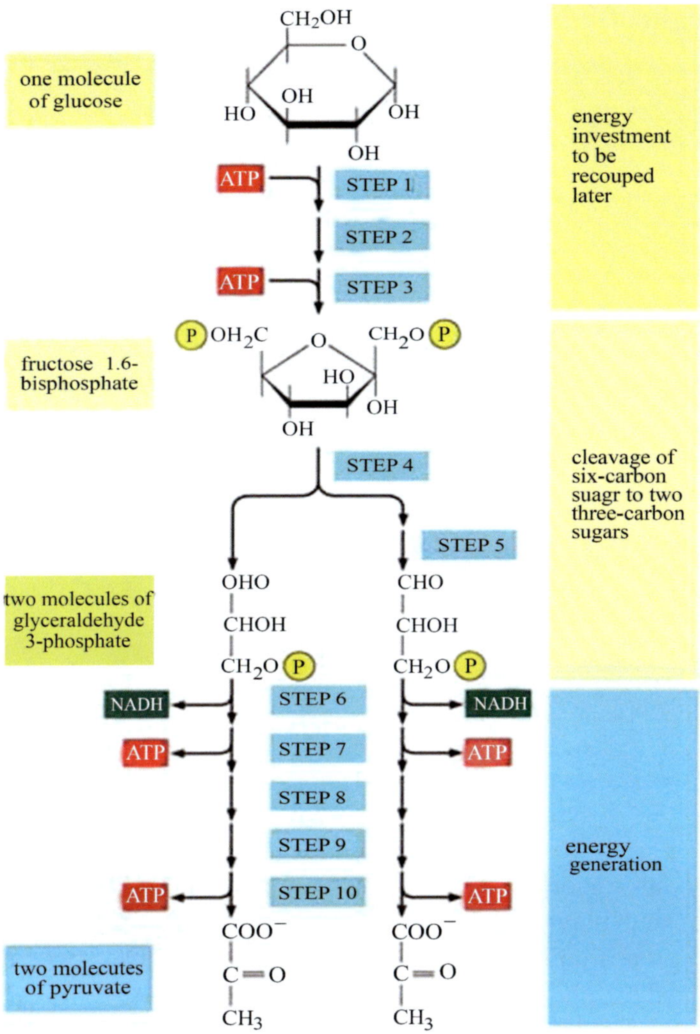

ΔG°'= -4.0 kcal mol-1

Phosphoryl Transfer Reaction: Kinases transfer phosphate from ATP to an acceptor. Hexokinase has a more general specificity in that it can transfer phosphate to other sugars such as mannose.

ΔG°'= .40 kcal mol-1

Phosphoglucose Isomerase

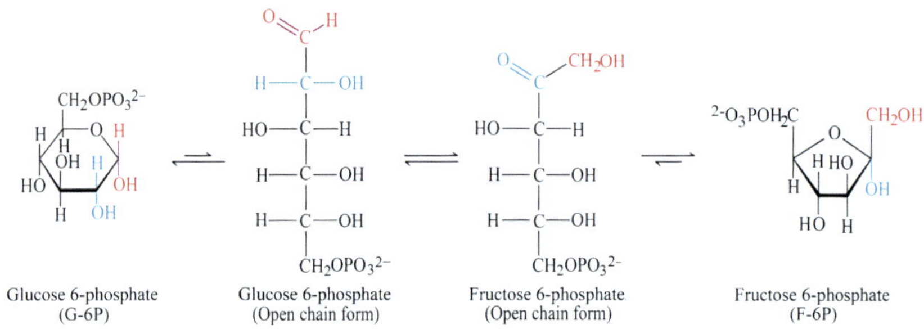

The enzyme opens the ring, catalyzes the isomerization, and promotes the closure of the five-member ring.

Phosphofructokinase-1

The 2nd investment of ATP in glycolysis.

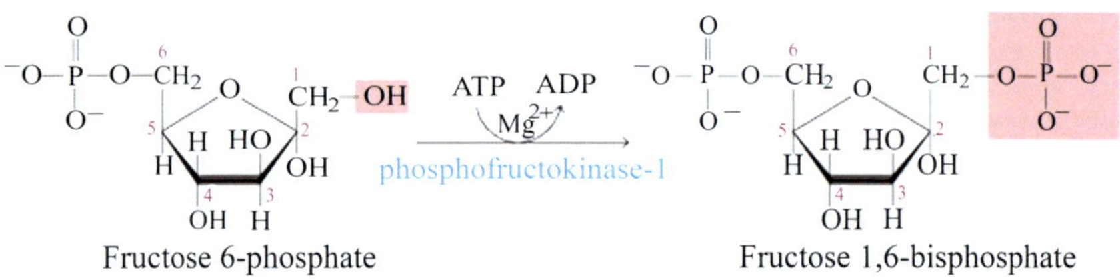

$\Delta G'^{\circ} = -14.2$ kJ/mol

PFK is an important allosteric enzyme regulating the rate of glucose catabolism and plays a role in integrating metabolism. **Bis** means two phosphate groups on two different carbon atoms. Di means two phosphate groups linked together on the same carbon atom.

Phosphofructokinase-1

The PFK reaction is the first unique and irreversible step in the glycolytic pathway. It is the "committed" in the pathway. In general, the enzyme catalyzing the committed step in a metabolic pathway is the most important control component in the pathway.

2nd stage: Aldolase

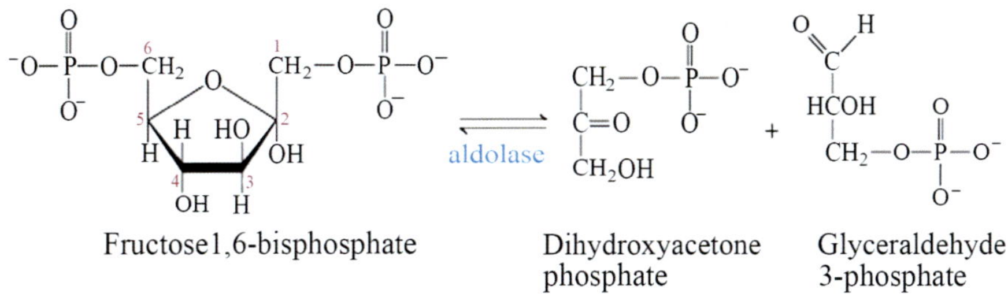

$\Delta G'^{\circ}$=23.8 kJ/mol

Reverse aldol condensation; converts a 6 carbon atom sugar to 2 molecules, each containing 3 carbon atoms.

Triose Phoshate Isomerase

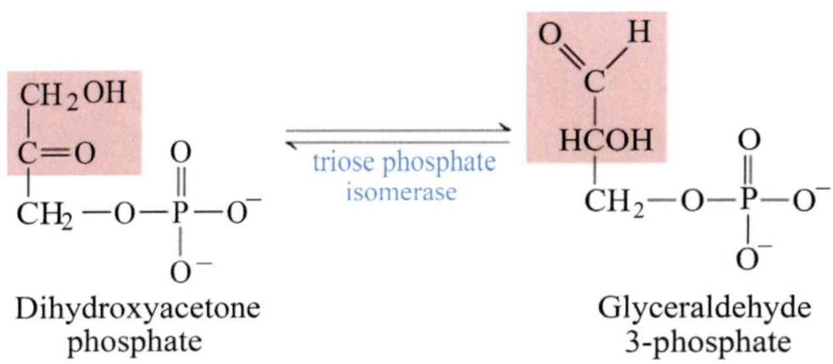

$\Delta G'^{\circ}$= 7.5 kJ/mol

$\Delta G^{\circ\prime}$ = 1.8 kcal mol^{-1}

All the DHAP is converted to glyceraldehyde 3-phosphate. Although, the reaction is reversible it is shifted to the right since glyceraldehyde 3-phosphate is a substrate for the next reactions of glycolysis. Thus, both 3-carbon fragments are subsequently oxidized. The fate of glyceraldehyde 3-phosphate Glyceraldehyde 3-phosphate DH.

Stage 3: The energy-yielding phase

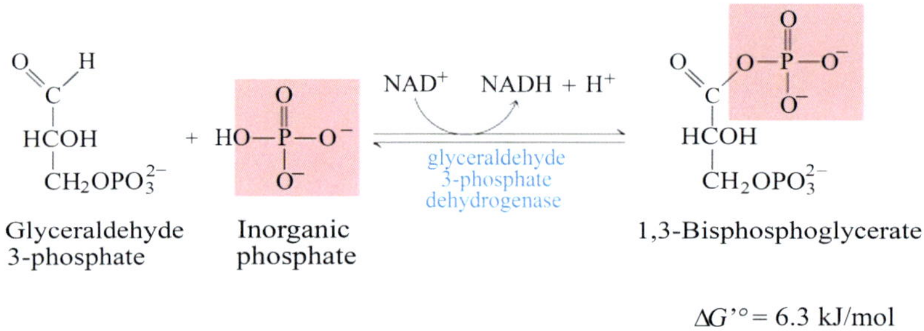

An aldehyde is oxidized to carboxylic acid and inorganic phosphate is transferred to form acyl-phosphate. NAD^+ is reduced to NADH.

$\Delta G°' = 1.5$ kcal mol^{-1}

1,3-BPG has a high phosphoryl-transfer potential. It is a mixed anhydride. **Notice,** under anaerobic conditions NAD^+ must be re-supplied.

Phosphoglycerate Kinase Substrate-level Phosphorylation

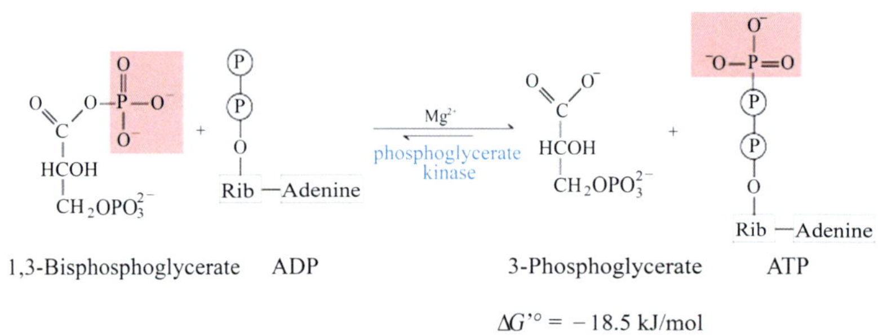

$\Delta G°' = -4.5$ kcal mol^{-1}

ATP is produced from P_i and ADP at the expense of carbon oxidation from the glyceraldehyde 3-phosphate DH reaction. The two molecules of ATP are produced per glucose. At this point, 2ATPs were invested and 2ATPs are produced.

Phosphoglycerate Mutase

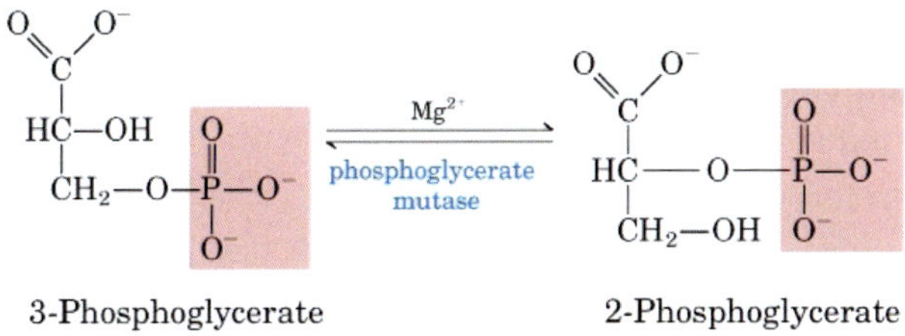

$\Delta G'^{\circ} = 4.4$ kJ/mol

$\Delta G^{\circ\prime} = 1.1$ kcal mol^{-1}

Enolase

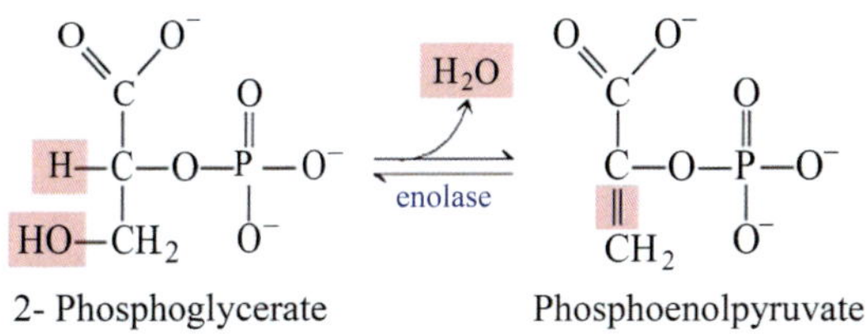

$\Delta G'^{\circ} = 7.5$ kJ/mol

$\Delta G^{\circ\prime} = .4$ kcal mol^{-1}

PEP

Pyruvate kinase 2nd example of substrate level phosphorylation. The net yield from glycolysis is 2 ATP.

unstable Enol form→ more stable keto form

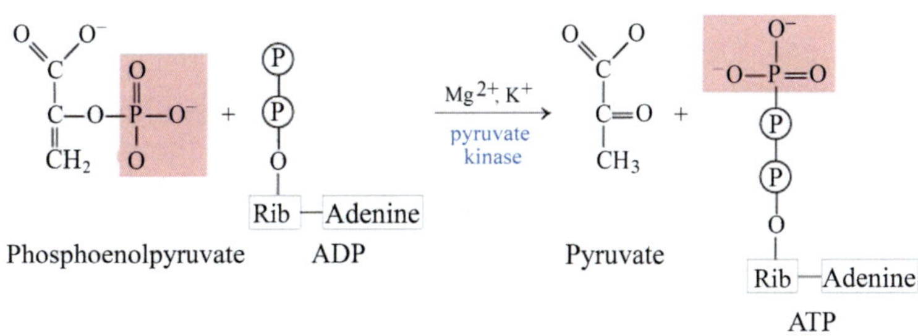

$\Delta G'^{\circ} = -31.4$ kJ/mol

Substrate level phosphorylation is the synthesis of ATP from ADP that is not linked to the electron transport system. The conversion of glucose to pyruvate

Glucose + 2 P_i + 2 ADP + 2 NAD^+ → 2 pyruvate + 2 ATP + 2 NADH +2 H^+

The energy released from the anaerobic conversion of glucose to pyruvate is -47kcal mol^{-1}. Under aerobic conditions, much more chemical bond energy can be extracted from pyruvate. Under anaerobic conditions, pyruvate is converted to lactate. Exercising muscle is an example. The NAD^+ that is consumed in the glyceraldehyde 3-phosphate reaction is produced in the lactate DH reaction. The redox balance is maintained. The activities of glyceraldehyde 3-phosphate DH and lactate DH are linked metabolically.

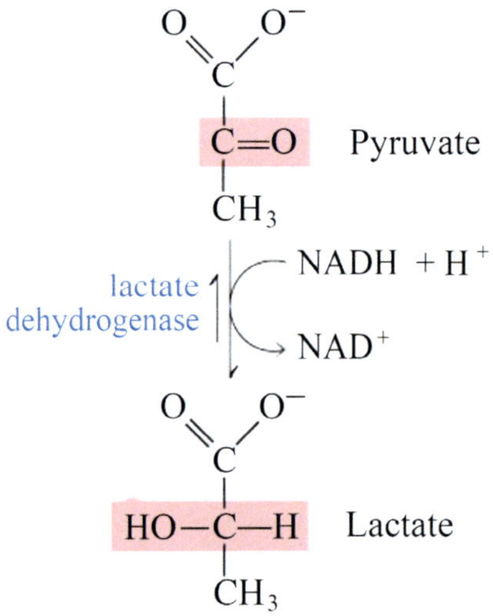

$$\Delta G'° = -25.1 \text{ kJ/mol}$$

The NAD^+ that is consumed in the glyceraldehyde 3-phosphate reaction is produced in the lactate DH reaction. The redox balance is maintained. The activities of glyceraldehyde 3-phosphate DH and Lactate DH are linked metabolically. The NAD^+ that is consumed in the glyceraldehyde 3-phosphate reaction is produced in the lactate DH reaction. Thus, redox balance is maintained. The NADH that is produced in the glyceraldehyde 3-phosphate reaction is consumed in the lactate DH reaction. Thus, redox balance is maintained.

Glucose + 2 P_i +2 ADP → 2 lactate + 2 ATP + 2 H_2O

Fructose and galactose feed into the glycolytic pathway. In the liver, when fructose enters glycolysis the PFK reaction is bypassed.

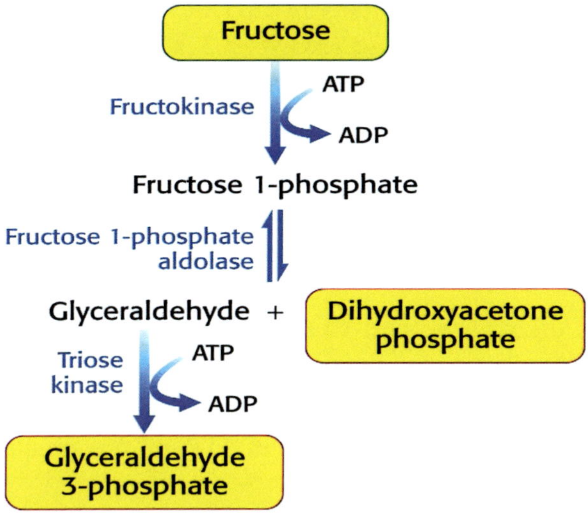

Galactose is converted to glucose-6P via a four-step reaction involving UDP-glucose

Gluconeogenesis

Gluconeogenesis is the synthesis of glucose from non-carbohydrate precursors. Glucose stores are depleted during periods of starvation or fasting beyond a day. Since the brain relies on glucose (120g/d) as a source of energy, glucose must be synthesized from molecules other than carbohydrates.

PYRUVATE → GLUCOSE

So any molecule that can be converted to pyruvate is considered glucogenic. Lactate and alanine are glucogenic. Glycerol is also glucogenic.

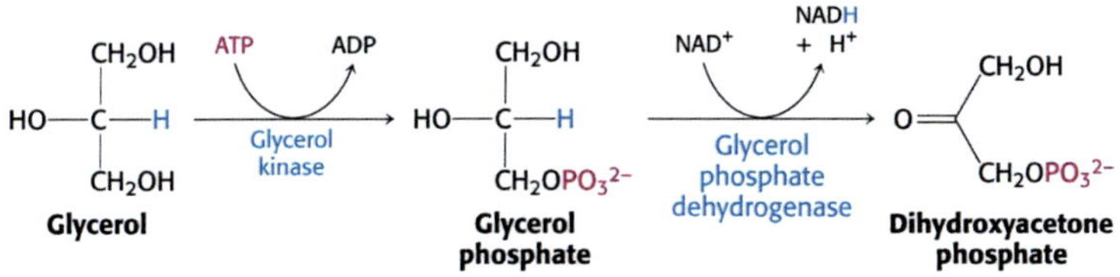

Gluconeogenesis: Pyruvate → Glucose

The enzymes in red belong to the gluconeogenic pathway. These reactions overcome the high negative free energy of the irreversible reactions of glycolysis. The enzymes in blue are held in common between the two pathways. The irreversible glycolytic enzymes are: hexokinase (ΔG =-8 kcal mol^{-1}) phosphofructokinase (ΔG = -5.3 kcal mol^{-1}) pyruvate kinase (ΔG = -4.0 kcal mol^{-1}). The enzymes of gluconeogenesis are: pyruvate carboxylase (ATP) phosphoenolpyruvate carboxykinase (GTP) fructose 1,6-bisphosphatase, glucose 6-phosphatase.

Pyruvate Carboxylase

Pyruvate + CO_2 + ATP + H_2O → oxaloacetate + ADP + P_i + 2 H^+.

Pyruvate Carboxylase fixes CO_2. Enzymes that fix CO_2. require the cofactor BIOTIN. Biotin is a vitamin and is always involved in CO_2 fixation. This reaction takes place in the mitochondrial matrix.

Notice there are no components of ATP in the structure of biotin. Phosphoenolpyruvate Carboxykinase

Oxaloacetate + GTP → phosphoenolpyruvate + GDP + CO_2

This reaction takes place in the cytosol. PEP is now synthesized and the sum of the two reactions is:

Pyruvate + ATP + GTP + H_2O →

PEP + ADP + GDP + P_i + H^+.

Pyruvate is carboxylated in the mitochondria. Pyruvate Carboxylase. Oxaloacetate can't pass out of the mitochondria. Oxaloacetate decarboxylated and phosphorylated in the cytosol. Phosphoenolpyruvate Carboxykinase.

Fructose 1,6-bisphosphatase

fructose 1,6-bisphosphate + H_2O → fructose 6-phosphate + P_i

Fructose 1,6-bisphosphatase is an allosteric enzyme and regulates gluconeogenesis. fructose 6-phosphate is easily converted to glucose 6-phosphate.

Glucose 6-phosphatase

glucose 6-phosphate + H_2O → glucose + P_i.

Liver can send glucose to blood to maintain homeostasis. Glucose 6-phosphate is also a precursor to glycogen. Glucose 6-P is valuable; a precursor for glycogen synthesis. Glucose 6-phosphatase is present only in tissues responsible for maintaining blood glucose levels, liver and kidney. In the liver, glucose 6-phosphatase is highly regulated.

In Liver

The lumen of the ER: It takes 5 proteins to convert glucose 6-phosphate to glucose.

Gluconeogenesis Stoichiometry

2pyruvate + 4ATP + 2GTP + 2NADH + $6H_2O$ → glucose + 4ADP + 2GDP $+6P_i$ $+2NAD^+ + 2H^+$; G°' = -9kcal mol^{-1}.

Just the reverse of glycolysis, G°' = 20kcal mol^{-1}. Note: it takes 6 nucleotide triphosphate molecules to synthesize glucose. Only 2 nucleotide triphosphate molecules are generated from glycolysis. So it takes four extra high phosphoryl-transfer potential molecules to drive the unfavorable gluconeogenesis pathway.

Glycolysis and Gluconeogenesis are Reciprocally Regulated

A high [AMP] indicates that the energy charge is low and signals the need for ATP. High [ATP] and [citrate] indicate the energy charge is high and intermediates are abundant.

Lactate from active muscle is converted to glucose in the liver. Carl and Gerty Cori received the Nobel Prize in Physiology and medicine 1947: "for their discovery of the course of the catalytic conversion of glycogen"

Lactate and Alanine are Glucogenic

In muscle, alanine is produced from pyruvate by transamination. In the liver, alanine is converted back to pyruvate. In active muscle lactate builds up, passes through the blood and is converted to pyruvate in the liver. Thus, part of the metabolic burden of active muscle is shifted to the liver.

pyruvate + glutamate → alanine + α-ketoglutarate

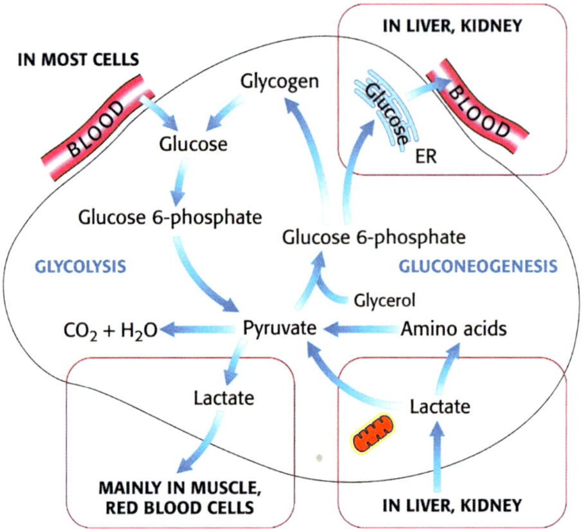

Gluconeogenesis

The source of pyruvate and oxaloacetate for gluconeogenesis during fasting or carbohydrate starvation is mainly amino acid catabolism. Some amino acids are catabolized to pyruvate, oxaloacetate, or precursors of these. Muscle proteins may break down to supply amino acids. These are transported to the liver where they are deaminated and converted to gluconeogenesis inputs. Glycerol, derived from hydrolysis of triacylglycerols in fat cells, is also a significant input to gluconeogenesis. Gluconeogenesis is the process whereby precursors such as lactate, pyruvate, glycerol, and amino acids are converted to glucose. Fasting requires all the glucose to be synthesized from these non-carbohydrate precursors. Most precursors must enter the Krebs cycle at some point to be converted to oxaloacetate. Oxaloacetate is the starting material for gluconeogenesis. Main sites of gluconeogenesis: Major site (Liver), Minor site: (Kidney) Very little: (Brain, Muscle (skeletal and heart). In the liver and kidney, it helps to maintain the glucose level in the blood so that brain and muscle can extract sufficient glucose from it to meet their metabolic demands.

Gluconeogenesis versus Glycolysis

- Steps are shared between glycolysis and gluconeogenesis. 3 essentially irreversible steps shift the equilibrium far on the side of glycolysis.

- Most of the decrease in free energy (consuming energy) in glycolysis takes place during these 3 steps.

Pyruvate is converted to Oxaloacetate before being changed to Phosphoenolpyruvate

1. Pyruvate carboxylase catalyzes the ATP-driven formation of oxaloacetate from pyruvate and CO_2
2. PEP carboxykinase (PEPCK) concerts oxaloacetate to PEP that uses GTP as a phosphorylating agent.

Gluconeogenesis is not just the reverse of Glycolysis

Several steps are different so that control of one pathway does not inactivate the other. However, many steps are the same. Three steps are different from glycolysis.

1 Pyruvate to PEP

2 Fructose 1,6- bisphosphate to Fructose-6-phosphate

3 Glucose-6-Phosphate to Glucose.

Biotin is an Essential Nutrient

There are hardly any deficiencies for biotin because it is abundant and bacteria in the large intestine also make it. However, deficiencies have been seen and are nearly always linked to the consumption of raw eggs. Raw eggs contain Avidin a protein that binds biotin with a $K_d = 10^{-15}$ (that is one tight binding reaction!) It is thought that Avidin protects eggs from bacterial invasion by binding biotin and killing bacteria.

Acetyl-CoA Regulates Pyruvate Carboxylase

Oxaloacetate concentrations increase then increase the activity of the Krebs cycle and Acetyl-CoA is an allosteric activator of the carboxylase. However, when ATP and NADH concentrations are high and the Krebs cycle is inhibited, oxaloacetate goes to glucose.

Transport between the Mitochondria and the Cytosol

Generation of oxaloacetate occurs in the mitochondria only, but, gluconeogenesis occurs in the cytosol. PEPCK is distributed between both compartments in humans, while in mice, it is only found in the cytosol. In rabbits, it is found in the mitochondria. PEP must be transported across the membranes or oxaloacetate has to be transported. PEP transport systems are seen in the mitochondria but oxaloacetate cannot be transported directly in or out of the mitochondria.

Hydrolytic Reactions bypass PFK and Hexokinase

The hydrolysis of fructose-1,6-phosphate and glucose-6- phosphate are separate enzymes from glycolysis. Glucose-6-phosphatase is only found in the liver and kidney. The liver is the primary organ for gluconeogenesis.

Glucose + $2NAD^+$ + 2ADP + 2Pi

2Pyruvate +2NADH + $4H^+$ + 2ATP + $2H_2O$

2Pyruvate +2NADH + $4H^+$ + 4ATP + 2GTP + $6H_2O$

glucose + $2NAD^+$ + 4ATP + 2GDP + 4Pi

2ATP + 2GTP + $4H_2O$ 2ADP + 2GTP + 4Pi

Low blood [glucose]

↓

Increased glucagon secretion

↓

Increased [cAMP]

↓

Increased enzyme phosphorylation

↓

Activation of FBPase-2 and inactivation of PFK-2

↓

Decreased [F2,6P]

↓

Inhibition of PFK and activation of FBPase

↓

Increased gluconeogenesis

The Glyoxylate Pathway

Only plants have the ability to convert Acetyl-CoA to Oxaloacetate directly without producing reducing equivalents of NADH. This is done in the glyoxyzome, separate from the mitochondria and allows replenishment of oxaloacetate. Isocitrate lyase - cleaves isocitrate into succinate and glyoxylate. The succinate goes to the mitochondria. Malate synthase makes malate from glyoxylate and Acetyl-CoA. Oxaloacetate can go directly to carbohydrate synthesis.

Glycogen Storage

Glycogen is a D-glucose polymer, a(1→4) linkages, a(1→6) linked branches every 8-14 residues

The Citric Acid Cycle

It is called the Krebs cycle or the tricarboxylic and is the "hub" of the metabolic system. It accounts for the majority of carbohydrate, fatty acid and amino acid oxidation. It also accounts for a majority of the generation of these compounds and others as well. Amphibolic - acts both catabolically and anabolically:

3NAD+ + FAD + GDP + Pi + acetyl-CoA 3NADH + FADH + GTP + CoA + $2CO_2$

By 1930 it was established that the addition of lactate, acetate succinate, malate, a-ketoglutaric acid (dicarboxylic acids) and citrate and isocitrate (tricarboxylic acids) when added to muscle mince that they stimulated oxygen consumption and release of CO_2. 1935 Albert Szent-Gyorgyi showed that

Succinate → Fumarate → Malate → Oxaloacetate

Carl Martius and Franz Knoop Showed

Citrate → cis-aconitate→ isocitrate → a ketoglutarate

succinate → fumarate → malate→ oxaloacetate

Martius and Knoop showed that pyruvate and oxaloacetate could form citrate non-enzymatically by the addition of peroxide under basic conditions. Krebs showed that succinate is formed from fumarate, malate or oxaloacetate. This is interesting since it was shown that the other way worked as well. Pyruvate can form citrate enzymatically. The interconversion rates of the intermediates were fast enough to support respiration rates.

Pyruvate + oxaloacetate = citrate + CO_2

The citric acid cycle enzymes are found in the matrix of the mitochondria. Substrates have to flow across the outer and inner parts of the mitochondria. CoA acts as a carrier of acetyl groups. Acetyl-CoA is a "high energy" compound: The DG°' for the hydrolysis of its thioester is -31.5 kJ• mol^{-1} making it greater than the hydrolysis of ATP. Pyruvate dehydrogenase converts pyruvate to Acetyl-CoA and CO_2.

Pyruvate Dehydrogenase

Multienzyme complexes are group of non-covalently associated enzymes that catalyze two or more sequential steps in a metabolic pathway.

1. Enzymatic reactions rates are limited by diffusion, with a shorter distance between subunits in an enzyme can almost direct the substrate from one subunit (catalytic site) to another.
2. Channeling metabolic intermediates between successive enzymes minimizes side reactions
3. The reactions of a multienzyme complex can be coordinately controlled.

Pyruvate Dehydrogenase

- Pyruvate dh decarboxylates pyruvate using a TPP cofactor forming hydroxyethyl-TPP.
- The hydroxyethyl group is transferred to the oxidized lipoamide on E2 to form Acetyl dihydrolipoamide-E2.
- E2 catalyzes the transfer of the acetyl groups to CoA yielding Acetyl-CoA and reduced dihydrolipoamide-E2.
- Dihydrolipoyl dh E3 reoxidizes dihydrolipoamide-E2 and itself becomes reduced as FADH2 is formed.
- Reduced E3 is reoxidized by NAD^+ to form FAD and NADH The enzymes SH groups are reoxidized by the FAD and the electrons are transferred to NADH.

Electron Transport Chain

In eukaryotes => Electron transport and oxidative phosphorylation => inner **mitochondrial** membrane. These processes => re-oxidize NADH and FADH2 <= from the citric acid cycle (mitochondrial matrix), glycolysis (cytoplasm) and fatty acid oxidation (mitochondrial matrix) and => trap the energy released as ATP.

Oxidative phosphorylation => major source of ATP in the cell.

In prokaryotes => electron transport and oxidative phosphorylation components => in the plasma membrane.

Redox Potential

Oxidation => loss of electrons.

Reduction => gain of electrons.

In chemical reaction: if one molecule is oxidized => another must be reduced i.e. oxidation-reduction reaction => transfer of electrons. When NADH => oxidized to NAD+ => it loses electrons. When molecular oxygen => reduced to water => it gains electrons. Oxidation-reduction potential, E, (redox potential). a measure of the affinity of a substance for electrons and is measured relative to hydrogen.

Positive redox potential substance => higher affinity => electrons than hydrogen so would accept electrons from hydrogen, e.g., Oxygen, a strong oxidizing agent.

Negative redox potential substance has a lower affinity for electrons than does hydrogen would donate electrons to H+, forming hydrogen e.g., NADH, a strong reducing agent.

For Biological Systems,

Total voltage changes of the reaction (change in electric potential, ΔE) => is the sum of voltage changes of individual oxidation-reduction steps. **Standard free energy change** of a reaction at pH 7 => **ΔG0'**=> calculated from the change in redox potential ΔE0' of substrates and products:

ΔG0' = -n F ΔE0'

Where n = number of electrons transferred,

ΔE0' in volts (V),

ΔG0' in kilocalories per mole (kcal mol-1) and

F-constant called Faraday (23.06 kcal V-1 mol-1).

A reaction with a positive ΔE0' has a negative ΔG0' (i.e., is **exergonic**). Thus for the reaction

Electron Transport from NADH

- NADH oxidation and ATP synthesis → not occur in a single step.
- Electrons → not transferred from NADH → oxygen directly.
- Electrons are transferred from NADH → oxygen → along a chain of electron carriers → called **electron transport chain** (**respiratory chain**).

Electron Transport Chain

It consists of 3 large protein complexes embedded in the inner mitochondrial membrane**.** NADH dehydrogenase complex (Complex I). Succinate Q reductase. The cytochrome bc1 complex (Complex II) cytochrome oxidase (Complex III). Electrons flow from NADH to oxygen through these three complexes. Each complex contains several electron carriers that work sequentially to carry electrons down the chain. There are 2 free electron carriers are also needed to link these large complexes.

Ubiquinone {coenzyme Q (CoQ)}

Cytochrome c

ATP Synthesis (Oxidative Phosphorylation)

NADH and FADH2 are oxidized by electron transport through the respiratory chain Synthesis of ATP. The energy liberated by electron transport is used to create a proton gradient across the mitochondrial inner membrane that is used to drive ATP synthesis (chemiosmotic hypothesis) in presence of ATP synthase. Thus the proton gradient couple's electron transport and ATP synthesis.

(not a chemical intermediate as in substrate-level phosphorylation.)

Summary

Electron transport down the respiratory chain from NADH oxidation causes H+ ions to be pumped out into the inter membrane space by three H+ pumps NADH dehydrogenase, cytochrome bc1 complex and cytochrome oxidase. Free energy change in transporting an electrically charged ion across a membrane leads to the formation of an electrochemical proton gradient. The pumping out of H+ ions generates a higher concentration of H+ ions in inter membrane space and an electrical potential on the side of the inner mitochondrial membrane facing the inter membrane space positive. Protons flow back into the mitochondrial matrix according to electrochemical gradient through ATP synthase drives ATP synthesis.

The ATP synthase is driven by proton-motive force which is the sum of pH gradient (the chemical gradient of H+ ions) and membrane potential (electrical charge potential across the inner mitochondrial membrane). FADH2 is reoxidized via ubiquinone its oxidation causes H+ ions to be pumped out only by the cytochrome bc1 complex and cytochrome oxidase so the amount of ATP made from FADH2 is less than from NADH. Measurements define that 2.5 ATP molecules are synthesized per NADH oxidized whereas 1.5 ATPs are synthesized per FADH2 oxidized.

Oxidative Phosphorylation

Carbohydrates, lipids, and amino acids are the main reduced fuels for the cell. Electrons from reduced fuels are transferred to reduced cofactors NADH or $FADH_2$. In oxidative phosphorylation, energy from NADH and $FADH_2$ is used to make ATP.

Chemiosmotic Theory

$ADP + P_i \rightarrow ATP$ is highly thermodynamically unfavorable.

Phosphorylation of ADP is not a result of a direct reaction between ADP and some high-energy phosphate carriers. Energy needed to phosphorylate ADP is provided by the flow of protons down the electrochemical gradient. The energy released by electron transport is used to transport protons against the electrochemical gradient.

Chemiosmotic Energy Coupling requires Membranes

The proton gradient needed for ATP synthesis can be stably established across a membrane that is impermeable to ions: plasma membrane in bacteria, inner membrane in mitochondria and thylakoid membrane in chloroplasts. The membrane must contain proteins that couple the "downhill" flow of electrons in the electron-transfer chain with the "uphill" flow of protons across the membrane. The membrane must contain a protein that couples the "downhill" flow of protons to the phosphorylation of ADP.

Structure of a Mitochondrion

Double membrane leads to four distinct compartments:

- **Outer membrane:** Relatively porous membrane; allows passage of metabolites.
- **Intermembrane space (IMS):** Similar environment to cytosol and Higher proton concentration (lower pH).

Inner membrane: Relatively impermeable, with proton gradient across it, Location of electron transport chain complexes **and** Convolutions called cristae serve to increase the surface area. **Matrix:** Location of the citric acid cycle and parts of lipid and amino acid metabolism and lower proton concentration (higher pH)

Electron-transport Chain Complexes contain a Series of Electron Carriers

Each complex contains multiple redox centers consisting of: Flavin mononucleotide (FMN) or flavin adenine dinucleotide (FAD), Cytochromes *a*, *b*, or *c* and Iron-sulfur cluster.

Complex I

Complex I catalyzes the transfer of a hydride ion from NADH to FMN, and from the FMN, two electrons pass through a series of FeScenters to the Fe-S Center N-2 in the matrix arm of the complex. Electron transfer from N-2 to ubiquinone on the membrane arm forms QH2, which diffuses into the lipid bilayer. This electron transfer also drives expulsion from the matrix of four protons per pair of electrons. Proton flux produces an electrochemical potential across the inner mitochondrial membrane (N side negative, P side positive). The detailed mechanism that couples electron and proton transfer in Complex I is not yet known, but three of the membrane subunits are structurally related to a known Naƨ-Hƨ anti-porter, and the path of proton movement may be similar in both cases. The fourth putative proton pathway is through an integral subunit closer to the Q-binding site. A long helix (not visible in this view) lying along the surface of the membrane arm may coordinate the action of all four proton pumps when Q is reduced.

Complex II

This complex (porcine) has two transmembrane subunits, C and D; subunits A and B extend into the matrix. Just behind the FAD in subunit A is the binding site for succinate. Subunit B has three Fe-S centers, ubiquinone is bound to subunit B, and heme b is sandwiched between subunits C and D. Two phosphatidylethanolamine molecules are so tightly bound to sub-unit D that they show up in the crystal structure. Electrons move (blue arrows) from succinate to FAD, then through the three Fe-S centers to ubiquinone. The heme b is not on the main path of electron transfer but protects against the formation of reactive oxygen species (ROS) by electrons that go astray.

Complex III

The complex (bovine) is a dimer of identical monomers, each with 11 different subunits. The functional core of each monomer consists of three subunits: cytochrome b (green), with its two hemes (bH and bL); the Rieske iron-sulfur protein (purple), with its 2Fe-2S centers; and cytochrome c1 (blue), with its heme. The cytochrome c1 and the Rieskeiron-sulfur protein project from the P surface and can interact with cytochrome c (not part of the functional complex) in the intermembrane space. The complex has two distinct binding sites for ubiquinone, QN and QP, which correspond to the sites of inhibition by two drugs that block oxidative phosphorylation. Antimycin A, which blocks electron flow from cytochrome b to cytochrome c1, specifically from heme bH to Q, binds at QN, close to heme bH on the N (matrix) side of the membrane. Myxothiazol, which prevents selectron flow from QH2 to the Rieske iron-sulfur protein, binds at QP, near the the2Fe-2S center and heme bL on the P side. The dimeric structure is essential to the function of Complex III. The interface between monomers forms two caverns, each containing a QP site from one monomer and a QN site from the other. The ubiquinone intermediates move within these sheltered caverns. Complex III crystallizes in two distinct conformations. In one, the Rieske Fe-S center is close to its electron acceptor, the heme of cytochrome c1, but relatively distant from cytochrome b and the QH2-binding site at which the Rieske Fe-S center receives electrons. On the other, the Fe-S center has moved away from cytochrome c1 and toward cytochrome b. The Rieske protein is thought to oscillate between these two conformations as it is first reduced, then oxidized.

Complex IV

This complex (bovine) has 13 subunits in each identical monomer of its dimeric structure. Subunit I (yellow) has two heme groups, a and a3, near a single cop-per ion, CuB (not visible here). Heme a3 and CuB form a binuclear Fe-Cu center. Subunit II (purple) contains two Cu ions complexed with the OSH groups of two Cys residues in a binuclear center, CuA, that resembles the 2Fe-2S center s of iron-sulfur proteins. This binuclear center and the cytochrome c–binding site are located in a domain of subunit II that protrudes from the P side of the inner membrane (into the inter membrane space). Subunit III (blue) is essential for rapid proton movement through subunit II. The roles of the other 10subunits in mammalian Complex IV (green) are not fully understood, although some function in assembly or stabilization of the complex. The binuclear center of CuA. The Cu ions (blue spheres) share electrons equally. When the center is reduced, the ions have the formal charges Cu11Cu11; when oxidized, Cu1.51Cu1.51. Six amino acid residues are ligands around the Cu ions: Glu, Met, two His, and two Cys.

Proton-motive Force

The proteins in the electron-transport chain created the electrochemical proton gradient by one of three means: actively transporting protons across the membrane Complex I and Complex IV, chemically removing protons from the matrix, reduction of CoQ and reduction of oxygen, releasing protons into the intermembrane space and oxidation of QH_2.

Chemiosmotic Model for ATP Synthesis

Electron transport sets up a proton-motive force. The energy of proton-motive force drives the synthesis of ATP.

Mitochondrial ATP Synthase Complex

It contains two functional units: F_1: soluble complex in the matrix and individually catalyzes the hydrolysis of ATP. F_0: integral membrane complex, transports protons from IMS to the matrix, dissipating the proton gradient, the energy transferred to F_1 to catalyze the phosphorylation of ADP.

Regulation of Oxidative Phosphorylation

Primarily regulated by substrate availability NADH and ADP/P_i. Inhibitor of F_1 (IF_1): prevents hydrolysis of ATP during low oxygen and only active at lower pH, encountered when electron transport it stalled (i.e., low oxygen). Inhibition of OxPhos leads to accumulation of NADH. It causes feedback inhibition to cascade up to PFK-1 in glycolysis.

Interrelations helps Metabolic Pathways

Learned about the catabolism of glucose, which provides energy to living cells. But living things consume more than glucose from food. This happens because all of the catabolic pathways for carbohydrates, proteins, and lipids eventually connect to glycolysis and the citric acid cycle pathways. Metabolic pathways should be thought of as porous; that is, substances enter from other pathways, and intermediates leave for other pathways. These pathways are not closed systems. Many of the substrates, intermediates, and products in a particular pathway are reactants in other pathways. Like sugars and amino acids, the catabolic pathways of lipids are also connected to the glucose catabolism pathways. Glycogen, a polymer of glucose, is an energy-storage molecule in animals. When there is adequate ATP present, excess glucose is shunted into glycogen for storage. Glycogen is made and stored in both the liver and muscles. The glycogen is hydrolyzed into the glucose monomer, glucose-1-phosphate (G-1-P), if blood sugar levels drop. The presence of glycogen as a source of glucose allows ATP to be produced for a longer period of time during exercise. Glycogen is broken down into G-1-P and converted into glucose-6-phosphate (G-6-P) in both muscle and liver cells; this product enters the glycolytic pathway.

Excess amino acids are converted into molecules that can enter the pathways of glucose catabolism. Metabolic pathways should be thought of as porous; that is, substances enter from other pathways and intermediates leave for other pathways. These pathways are not closed systems. Many of the substrates, intermediates, and products in a particular pathway are reactants in other pathways. Proteins are a good example of this phenomenon. They can be broken down into their constituent amino acids and used at various steps of the pathway of glucose catabolism. Proteins are hydrolyzed by a variety of enzymes in cells. Most of the time, the amino acids are recycled into the synthesis of new proteins or are used as precursors in the synthesis of other important biological molecules, such as hormones, nucleotides, or neurotransmitters. However, if there are excess amino acids, or if the body is in a state of starvation, some amino acids will be shunted into the pathways of glucose catabolism. Each amino acid must have its amino group removed (deamination) prior to the carbon chain's entry into these pathways. When the amino group is removed from an amino acid, it is converted into ammonia through the urea cycle. The remaining atoms of the amino acid result in a keto acid: a carbon chain with one ketone and one carboxylic acid group. In mammals, the liver synthesizes urea from two ammonia molecules and a carbon dioxide molecule. Thus, urea is the principal waste product in mammals produced from the nitrogen originating in amino acids; it leaves the body in urine. The keto acid can then enter the citric acid cycle. When deaminated, amino acids can enter the pathways of glucose metabolism as pyruvate, acetyl CoA, or several components of the citric acid cycle. For example, deaminated asparagine and aspartate are converted into oxaloacetate and enter glucose catabolism in the citric acid cycle.

Connecting Lipids to Glucose Metabolism

Lipids can be both made and broken down through parts of the glucose catabolism pathways. Like sugars and amino acids, the catabolic pathways of lipids are also connected to the glucose catabolism pathways. The lipids that are connected to the glucose pathways are cholesterol and triglycerides. Cholesterol contributes to cell membrane flexibility and is a precursor to steroid hormones. The synthesis of cholesterol starts with acetyl groups, which are transferred from acetyl CoA, and proceeds in only one direction; the process cannot be reversed. Thus, the synthesis of cholesterol requires an intermediate of glucose metabolism.

Triglycerides

Triglycerides, a form of long-term energy storage in animals, are made of glycerol and three fatty acids. Animals can make most of the fatty acids they need. Triglycerides can be both made and broken down through parts of the glucose catabolism pathways. Glycerol can be phosphorylated to glycerol-3-phosphate, which continues through glycolysis. Fatty acids are catabolized in a process called beta-oxidation that takes place in the matrix of the mitochondria and converts their fatty acid chains into two-carbon units of acetyl groups, while producing NADH and $FADH_2$. The acetyl groups are picked up by CoA to form acetyl CoA that proceeds into the citric acid cycle as it combines with oxaloacetate. The NADH and $FADH_2$ are then used by the electron transport chain.

Biochemistry of Seed Germination and Development

It provides protection against entry of parasites, it protects against mechanical injury, in some seeds, it buffers against unfavorably high or low temperatures. It stops germination until the right time. Inside of the seed coat, an embryo is protected. The embryo is an immature plant with all of the parts of an adult plant. A close look shows leaves and roots, albeit very small leaves and roots. The leaves of the embryo are called plumules. The leaves are sheathed by a cotyledon. The embryonic roots are called radicles. The embryonic stem is called the hypocotyl. The seed is filled with nourishment for the new plant; this source of nutrition is called the endosperm.

Germination

The first step of germination is the absorption of water a lot of water. The absorption of water activates enzymes in the seed that stimulate growth. These enzymes break down starches in the endosperm into sugars that can be used for energy. The deciding factor in whether or not a seed germinates is whether or not energy is available for growth and cell division.

Stages Leading to Cell Division

Seed Dormancy

If moisture is the key to starting seed germination, it should be clear what is necessary for seed dormancy, a waterproof seed coat, no oxygen and chemical inhibitors. For germination to occur, these must be negated, coat broken down; oxygen available and water absorbed and growth promoters activated, inhibitors are inhibited.

Germination

Stage	Events
Pregermination	(a) Rehydration – imbibition of water. (b) RNA & protein synthesis stimulated. (c) Increased metabolism – increased respiration. (d) Hydrolysis (digestion) of food reserves by enzymes. (e) Changes in cell ultrastructure. (f) Induction of cell division & cell growth.
Germination	(a) Rupture of the seed coat. (b) Emergence of seedling, usually radicle first.

Post germination	(a) Controlled growth of root and shoot axis. (b) Controlled transport of materials from food stores to the growing axis. (c) Senescence (aging) of food storage tissues.

Germination Factors

Water is clearly the most important factor in germination an adequate continuous supply of water is necessary for complete emergence. Water functions a triggering enzyme for starch conversion into sugar, turgor pressure for moving the radicle root down and the cotyledons up and for transporting nutrients and enzymes within the seed. Light is another key germination factor; light can either stimulate or inhibit seed germination. Some crops have a requirement for light to assist seed germination (eg. begonias, impatiens, lettuce). Others germinate best in the dark. This is determined by how the seed would naturally be sown. Small seeds must sprout on the surface of soil because they lack a suitable endosperm to supply the needed nutrients; these are typically aided by light exposure. Large seeds contain enough nutrition to grow underground when photosynthesis is not possible. These seeds are more likely to germinate in dark conditions. Oxygen is the third factor. Cellular respiration is necessary for plants to grow; oxygen is necessary to complete respiration. Oxygen removes metabolic waste from the cell. Waste is not removed without oxygen and the cellular metabolism is slowed. If oxygen supply is limited during germination, emergence may not occur due to inhibited growth. A favorable temperature is necessary to allow for plant growth. Temperature not only affects the germination percentage but also the rate of germination. For every species of seed, there is an optimal soil temperature for germination. At that temperature, the maximum number of seeds will germinate and in less time than at any other temperature. Many seeds germinate best around room temp. Germination is a complex process. A pre-formed plant (embryo) inside of the seed coat must turn the endosperm (starch) into sugar. This sugar powers cell division (mitosis); the addition of cells will cause the embryonic roots, leaves, and stems to grow, expand, and develop. During germination, the radicle (embryonic root) emerges due to mitosis fueled by the breakdown of starch into sugar. Under warm conditions, this process will take 4-5 days. Under cool conditions, this process takes longer. Initially, the radicle grows in what direction the kernel tip is pointing. Later, smaller roots will emerge from the radicle at varying angles. These roots will absorb the nutrients necessary for growth and development when the endosperm is completely consumed.

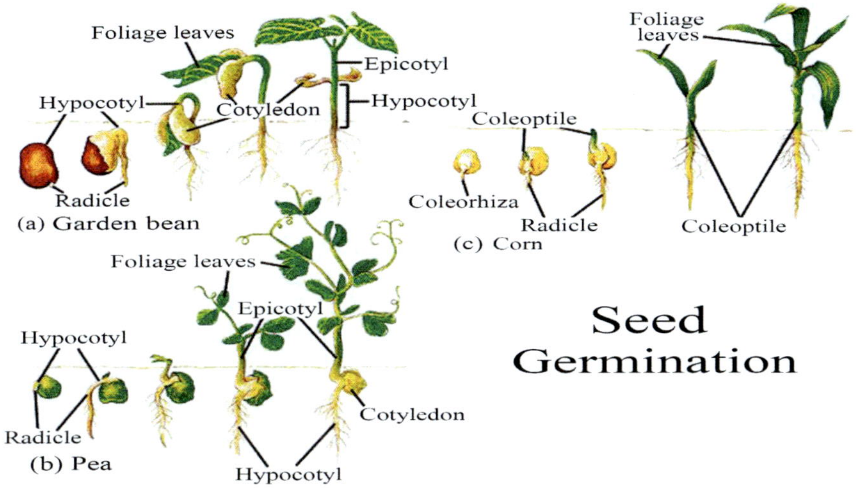

Randy Moore, Dennis Clark, and Darrell Vodopich, Botany Visual Resource Library, MacGraw-hill companies, Inc. (1998).

The coleoptile, a protective sheath covering the shoot, pushes through the soil until it reaches light. Upon reaching the light, the plumule emerges. If the kernel is planted too deep, emergence will not occur. The time between planting and emergence is determined mostly by temperature; warmer temps are reduced emergence time. This is only true to an extent; excessively high temps can also increase emergence time. Soil compaction can increase the time to emergence. Lack of moisture can also increase emergence time.

Sprouting Time

Many temperate-zoned species use chemical inhibitors to induce dormancy in seeds. In fall the seeds are exposed to abscisic acid (ABA) by the plant to prevent the seeds from sprouting before winter. Over winter, enzymes in the seed degrade the abscisic acid and by spring it is gone.

Vernalization

Seeds of some species can be caused to sprout earlier by exposure to cool or cold temperatures. Seeds of temperate species that were exposed to the cold will sprout earlier in a greenhouse than those planted directly. This process is called vernalization. Vernalization is the intentional exposing of seeds to cool conditions to increase the breakdown of chemical inhibitors and stimulate the production of growth or flowering enzymes. Many annuals are facultative and vernalization is not necessary for development but does speed it up.

Biochemistry of Fruit Ripening

Ripening is the process by which fruits attain their desirable flavor, quality, color, palatable nature and other textural properties. Ripening is associated with a change in composition i.e. conversion of starch to sugar. On the basis of ripening behavior, fruits are classified as climacteric and non-climacteric fruits. Climacteric: Climacteric fruits are defined as fruits that enter 'climacteric phase' after harvest i.e. they continue to ripen. During the ripening process, the fruits emit ethylene along with an increased rate of respiration. Ripe fruits are soft and delicate and generally cannot withstand rigors of transport and repeated handling. These fruits are harvested hard and green, but fully mature and are ripened near consumption areas. A small dose of ethylene is used to induce ripening under controlled conditions of temperature and humidity. These fruit in fully ripe state are too delicate to withstand transportation over long distances and should preferably be ripened near the consumption area. Non-climacteric: Non-climacteric fruits once harvested do not ripen further. Non-climacteric fruits produce a very small amount of ethylene and do not respond to ethylene treatment. There is no characteristic increased rate of respiration or production of carbon dioxide.

Technologies for ripening of fruits ack of easier and rapid methods for uniform ripening poses a major problem in the fruit industry. "Almost all methods of ripening, either conventional or the modern chemical methods, come with their own merits and demerits. There are several simple technologies and methods available today for farmers for proper ripening. Normally, the number of days taken for edible ripening varies for different fruits and prevailing climatic conditions. For instance, it takes about 5 to 6 days for mangoes and 6 to 7 days for sapotas to ripen. Under natural conditions, ethylene, a ripening hormone produced by the plant plays a major physiological role in the ripening process. A simple technology practiced in households to trigger ripening is to keep unripened and ripened fruits together inside an air tight container. Since the already ripened fruits release ethylene, ripening will be faster. Another method is to place the fruits intended for ripening inside an air tight room and induce ripening through smoking inside smoke chambers. Smoke emanates acetylene gas. Several fruit traders follow this technique to achieve uniform ripening especially in edible fruits like banana and mango. But the major drawback of this method is that the fruits do not attain uniform color and flavor. In addition, the persistence of smoke odor on the product impairs its quality. Spreading unripe fruits as layers over paddy husk or wheat straw for a week to ripen is another alternative.

Another practice is that some farmers dip unripe mature fruits in 0.1 percent ethrel solution (1 ml of ethrel solution in 1 liter of water) and wipe it dry. The fruits are then spread over a newspaper without touching each other and a thin cotton cloth is covered over this. In this method, the fruits will ripen within two days. In one of the simple and harmless techniques, 10 ml of ethrel and 2 gm of sodium hydroxide pellets are mixed in five liters of water taken in a wide-mouthed vessel. This vessel is placed inside the ripening chamber near the fruits and the room is sealed air tight. About a third of the room is filled with fruits leaving the remaining area for air circulation. Ripening of fruits takes place in about 12 to 24 hours. In order to reduce the cost of chemical, some ethylene releasing fruits such as papaya and banana can also have kept in the same room". Ethylene gas filled in pressurized cans promotes fruit ripening in 24-48 hours.

Subjective Questions

1. Define Carbohydrates. Classify them with suitable examples.
2. Describe the structure and functions of Mucopolysaccharides.
3. Define polysaccharides. Classify them and write their importance.
4. Explain why sucrose is called as invert sugar
5. Give a diagrammatic representation of proteoglycans aggregate.
6. Discuss:

a) Structure and properties of starch.
b) Osazone formation.
c) Clinical importance of Inulin and Dextran.
d) Structure and functions of three biochemically important disaccharides.
e) Sugar derivatives.
f) Explain the structure of starch and glycogen.

7. Define lipids and classify them with suitable examples. Classify fatty acids in various ways with suitable examples.

8. Define phospholipids. Classify them with suitable examples and state their functions.
9. Chemistry and functions of cholesterol 2. Lipoproteins 3. Prostaglandins 4. Essential fatty acids.
10. Explain in detail the structural organization of proteins.
11. Classify aminoacids in various ways with suitable examples.
12. Classify proteins in various ways with suitable examples.
13. Give the functional classification of proteins.
14. Explain the alpha-helical structure of the protein with examples.
15. Describe the beta-pleated structure of the protein with examples.
16. Write a short note on the functions of albumin.
17. Write a short note on the isoelectric pH of protein and its importance.
18. Write a short note on the biologically important peptides.
19. Write a short note on nonstandard aminoacids.
20. Write a short note on plasma proteins and their functions.
21. Define enzymes. Explain nomenclature and IUBMB classification with suitable examples.
22. Define enzyme inhibition. Explain in detail the different types of inhibitions with suitable examples.
23. Define isoenzymes and explain their structure, organ distribution and diagnostic importance.
24. Explain the different theories proposed for mechanism of enzyme substrate complex formation.
25. Explain factors affecting enzyme activity.
26. Explain the Watson and Crick model of DNA. Add a note on different forms of DNA (DNA polymorphism)?
27. Differentiate between anabolism and catabolism?
28. What is thermodynamics and their types?
29. Define carbohydrate metabolism?
30. What is photosynthesis? Explain the cyclic and non-cyclic process?
31. Define electron transport chain?
32. Explain the role of mitochondria?
33. What is sucrose metabolism, elaborate with explanation?
34. Explain the metabolism of nitrogen and amino acid?
35. Explain de novo-salvage pathway?
36. Explain the protein synthesis?
37. Define lipid metabolism?
38. Differentaite between saturated and unsaturated fatty acid?
39. Explain glycolysis?

40. Briefly explain citric acid cycle?
41. Differentiate between C3 and C4 cycles?
42. Describe oxidative phosphorylation?

Objective Questions

1.) Which among the Following Enzymes does not Participate in Galactose Metabolism?

a. Glucokinase
b. Galactokinase
c. Galactose-1-Phosphate Uridyl transferase
d. UDP-Galactose 4- epimerase

Answer: A

2.) Which of the Following Enzymes is Considered as Defective in Galactosemia- a Fatal Genetic Disorder in Infants?

a. Glucokinase
b. Galactokinase
c. Galactose-1-Phosphate Uridyl transferase
d. UDP-Galactose 4- epimerase

Answer: C

3.) Erythrocytes Undergo Glycolysis for Production of ATP. The Deficiency of Which Enzyme Leads to Hemolytic Anemia-

a. Glucokinase
b. Phosphofructokinase
c. Phosphoglucomutase
d. Pyruvate Kinase

Answer: D

4.) In the Liver, the Accumulation of which among the Following Metabolites Attenuates the Inhibitory of ATP on Phosphofructokinase?

a. Glucose-6-Phosphate
b. Citrate
c. Fructose-1,6-Bisphosphate
d. Fructose-2,6-Bisphosphate

Answer: D

5.) The Most Active Site of Protein Synthesis is the-

a. Nucleus

b. Ribosome

c. Mitochondrion

d. Cell sap

Answer: A

6.) How many Total Molecules of ATP are Synthesized from ADP via Glycolysis of a Single Molecule of Glucose?

a. 36

b. 38

c. 2

d. 4

Answer: A

7.) Mutation in which of the Following Enzymes Leads to a Glycogen Storage Disease Known as Tarui's Disease?

a. Glucokinase

b. Phosphofructokinase

c. Phosphoglucomutase

d. Pyruvate Kinase

Answer: B

8.) Cancer Cells have High Energy Demands for Replication and Division. Increased Flux of glucose into Glycolysis Replenishes the Energy Demand. Which of the Following Enzymes Plays an Important Role in Tumor Metabolism?

a. Glucokinase

b. Phosphofructokinase

c. Phosphoglucomutase

d. Pyruvate Kinase M2

Answer: D

9.) The Rate of Absorption of Sugars by the Small Intestine is Highest for -

a. Pentoses

b. Disaccharides

c. Polysaccharides

d. Hexoses

Answer: D

10.) Which of the Following is not a Polymer of Glucose?

a. Glycogen

b. Cellulose

c. Amylase

d. Insulin

Answer: D

11.) An Essential for the Conversion of Glucose to Glycogen in Liver is -

a. UTP

b. GTP

c. Pyruvate kinase

d. Guanosine

Answer: A

12.) Which of the Following Glucose Transporter (GLUT) are Important in Insulin-Dependent Glucose Uptake?

a. GLUT1

b. GLUT2

c. GLUT3

d. GLUT4

Answer: D

13.) Which of the Following Metabolites Negatively Regulates Pyruvate Kinase?

a. Fructose-1,6-Bisphosphate

b. Citrate

c. Acetyl CoA

d. Alanine

Answer: D

14.) Which of the Following Glycolytic Enzymes is Inhibited by Accumulation of Long-Chain Fatty Acid in the Liver?

a. Hexokinase

b. Glucokinase

c. Phosphofructokinase

d. Pyruvate kinase

Answer: B

15.) Glycogen Synthesis is Increased by -

a. Cortisone

b. Insulin

c. GH

d. Epinephrine

Answer: B

16.) Gluconeogenesis Occurs in the Liver and Which Organ -

a. Kidney

b. Muscle

c. Heart

d. Fat

Answer: C

17.) Adrenaline acts on which Enzyme in Glycogenolysis?

a. Glucokinase

b. Hexokinase

c. Phosphorylase

d. Glucose diphosphate

Answer: C

18.) Which among the Following Statements about Phosphofructokinase-2 (PFK-2) is False?

a. PFK-2 is a bifunctional enzyme having kinase domain, phosphatase domain and regulatory domain

b. Activated protein kinase A phosphorylates PFK-2 and activates the phosphatase domain

c. PFK-2 catalyzes the conversion of fructose-6-phosphate to fructose 2-6 bisphosphate

d. PFK-2 phosphatase activity is activated by the insulin signaling pathway

Answer: B

19.) Which among the Following Hormones Decreases Blood Glucose and Increases Uptake of Glucose in Various Tissues like Skeletal Muscle, Adipose Tissues?

a Glucagon

b. Epinephrine

c Cortisol

d Insulin

Answer: D

20.) The First Product of Glycogenolysis is -

a. Glucose-6-phosphate

b. Glucose 1,6 diphosphate

c. Glucose-1-phosphate

d. Fructose-1-phosphate

Answer: B

21.) Enzymes Concerned with the Citric Acid Cycle are Found in the -

a. Nucleus

b. Ribosomes

c. Mitochondria

d. Nonparticulate cytoplasm

Answer: C

22.) Which among the Following Statement is True?

a. Glycolysis occurs only in mammalian cells

b. Glycolysis occurs in mitochondria

c. Glycolysis occurs in the presence and absence of oxygen

d. Glycolysis occurs when ATP concentration is high

Answer: C

23.) What high Energy Phosphate Compound is Formed in the Citric Acid Cycle through Substrate level Phosphorylation?

a. ATP

b. TTP

c. ITP

d. GTP

Answer: D

24.) The Conversion of Glucose-6-Phosphate to Fructose-6-Phosphate is an Example of which of the Following Reactions?

a. Phosphate transfer

b. Isomerization

c. Dehydration

d. Aldol cleavage

Answer: B

25.) The Net Gain of ATP during Conversion of Glucose to Pyruvate is -

a. 1 ATP

b. 2 ATP

c. 1 ATP +1 GTP

d. 4 ATP

Answer: B

26.) During the Conversion of Glucose to Pyruvate, Two NADH Molecules are Generated. Which of the Following Steps Generates NADH?

a. Conversion of fructose-6-phosphate to fructose-1-6-bisphosphate

b. Conversion of glyceraldehydes-3-phosphate to 1-3-bisphosphoglycerate

c. Conversion of 3-phosphoglycerate to 2-phosphoglycerate

d. Conversion of phosphoenolpyruvate to pyruvate

Answer: B

27.) Which One of the Following is a Monosaccharide?

a. Maltose

b. Sucrose

c. Fructose

d. Starch

Answer: C

28.) Which among the Following Step is Inhibited during Arsenate Poisoning?

a. Glyceraldehyde-3-phosphate dehydrogenase

b. 3-phosphoglycerate mutase

c. Enolase

d. Pyruvate kinase

Answer: A

29.) The End Product of Glycolysis under Anaerobic Conditions is -

a. Lactic acid

b. Pyruvic acid

c. Acetoacetic acid

d. Oxaloacetic acid

Answer: A

30.) Gluconeogenesis is the Production of Glucose from Non-Carbohydrate Molecules. Which of the Following is not a Substrate for Gluconeogenesis?

a. Lactate

b. Alanine

c. Glycerol

d. Acetyl CoA

Answer: D

CHAPTER 9

BIOTECHNOLOGY

1. BIOTECHNOLOGY, BASIC CONCEPTS AND HISTORY OF BIOTECHNOLOGY AND DIFFERENT BRANCHES OF BIOTECHNOLOGY

Introduction

The biotechnology term was coined in 1917 by a Hungarian Engineer, Karl Ereky, to describe a process for the large-scale production of pigs. According to him all types of work are biotechnology by which products are produced from raw materials using living organisms. During the end of the 20th -century biotechnology emerged as a new discipline of biology integrating with technology, but the route of biotechnology from biology.

There was no sudden sprout of this discipline, but some of the methods for the production of products were developed centuries back. Therefore, biotechnology is concerned with the exploitation of biological components for the production of useful products.

- Biotechnology is defined by different organizations in different ways. It has been broadly defined as "the development and utilization of biological processes, forms, and systems for obtaining maximum benefits to man and other forms of life".
- Biotechnology is "the science of applied biological process" (Biotechnology A Dutch Perspective, 1981). Following are some

of the definitions given by other organizations.

- Biotechnology is the "controlled use of biological agents such as microorganisms or cellular components for beneficial use" **(U.S. National Science Federation).**
- The definition was given by OECD "scientific and engineering principles" refer to microbiology, biochemistry, and genetics, etc.
- Biological agents: we can call biological agents such as microorganisms, enzymes, plant, and animal cells. The meaning of these three definitions and others given by many organizations are more or less similar.
- The definition of genetic engineering has been given by **Smith (1996)** as "the formation of new combinations of heritable material by the insertion of nucleic acid molecules produced by whatever means outside the cell, into any virus or bacterial plasmid or other vector systems so as to allow their incorporation into host organisms in which they do not naturally occur but in which they are capable of continued propagation".

Importance

- Biotechnology is widely used in pharmacy to create more efficient and less expensive drugs. Recombinant DNA technology is used for the production of specific enzymes, which enhance the rate of production of a particular range of antibodies in the organism. The hormones such as somatostatin, insulin, and the human growth hormone can be synthesized easily and cheaply.
- The first human hormone to be synthesized by genetic engineering was somatostatin. Somatostatin is a brain hormone originating from the hypothalamus.
- It acts to inhibit the release of human growth hormone and insulin is related to the treatment of diabetes, pancreatitis, and a few other conditions. Genetech, a California-based company, has produced human growth hormone (hGH) from genetically engineered bacteria.
- Human insulin is the first genetically engineered pharmaceutical product, developed by Eli Lilly and company in 1982. Bovine Somatotropin (BST) is produced for a large quantity of milk production in cows.

- Antibiotics are chemical substances produced by several microorganisms.
- Recombinant DNA technology has helped in the increased production of antibiotics for example; the rate of penicillin produced at present is about 150,000 unit/ml against about 10 unit/ml in the 1950s.
- Antibiotics produced using such technology have very specific effects and causes fewer side effects. Currently, scientists are working on vaccines for fatal illnesses such as AIDS, hepatitis, malaria, flu, and even some forms of cancer.
- Interferon, an anti-viral protein, is prepared from mammalian cells by recombinant DNA technology. By cloning cDNA to genes for human interferon, it has been found that there are large numbers of interferons differing in amino acid sequences and properties.
- A large number of interferon is prepared in yeast cells by the fermentation process. Expects that in the near future vaccines will come in more convenient ways "some will come in the form of mouthwash; others will be swallowed in time-release capsules, avoiding the need for boosters".

Biotechnology Century

- Biotechnology is the technology applied to biology, molecular biology, genetics, and many other subfields of biology. Biotechnology utilizes cellular and biomolecular processes to create technologies and products that help improve our lives and nature.
- By making useful food, such as bread and cheese, and preserving dairy products, we have done these for many years by now.
- Recent biotechnology develops breakthrough products and technologies to fight diseases, reduce our environmental harm, feed the hungry, and useless and cleaner energy, and have safer, cleaner, and more efficient industrial manufacturing processes.
- More than 250 biotechnology health care products and vaccines have been made available to patients, many for previously untreatable diseases.
- More than 13.3 million farmers around the world use agricultural biotechnology to increase yields, prevent damage from insects

and pests and reduce the damage done to the environment due to farming and more than 50 biorefineries are being built across North America to test and refine technologies to produce biofuels and chemicals from renewable biomass, which can help reduce greenhouse gas emissions.

- Biotechnology is the third wave in biological science and represents such an interface of basic and applied sciences, where the gradual and subtle transformation of science into technology can be witnessed.
- Biotechnology is defined as the application of scientific and engineering principles to the processing of the material by biological agents to provide goods and services.
- Biotechnology comprises a number of technologies based upon an increasing understanding of biology at the cellular and molecular levels.
- The Bible already provides numerous examples of biotechnology. Namely, it deals with the conversion of grapes to wine, of dough to bread, and of milk to cheese.
- The oldest biotechnological processes are found in microbial fermentations, as born out by the Babylonian tablet dated circa 6000 B.C explaining the preparation of beer.
- The Sumerians were able to brew as many as twenty types of beer in the third millennium B.C. In about 4000 B.C., leavened bread was produced with the aid of yeast.
- During the Vedic period (5000-7000 B.C) Aryans had been performing daily *Agnihotra* or *Yajna*. In *Ayurved*, production of **'Asava' and 'Arista'** using different substrates and flowers of mahua (***Madhuca indica***) or dhataki (***Wodfordiafructicosa***) has been well characterized till today since the Vedic period.
- One of the materials used in *Yajna* is animal fat (i.e. ghee) which is a fermented product of milk.
- The term 'biotechnology' was described in a Bulletin of the Bureau of Biotechnology published in July 1920 from the office of the same name in Leeds in Yorkshire. The articles in this bulletin described the varied roles of microbes in the leather industry to pest control.

There are numerous Sub-fields of Biotechnology

- **Red biotechnology** is biotechnology applied to medical processes. Some examples are the designing of organisms to produce antibiotics, and the engineering of genetic cures to cure diseases through genomic manipulation.
- **White biotechnology**, also known as **grey biotechnology**, is biotechnology applied to industrial processes. An example is the designing of an organism to produce a useful chemical. White biotechnology tends to consume less resources than traditional processes when used to produce industrial goods.
- **Green biotechnology** is biotechnology applied to agricultural processes. An example is the designing of an organism to grow under specific environmental conditions or in the presence (or absence) of certain agricultural chemicals. One hope is that green biotechnology might produce more environmentally friendly solutions than traditional industrial agriculture. An example of this is the engineering of a plant to express a pesticide, thereby eliminating the need for external application of pesticides. Whether or not green biotechnology products such as this are ultimately more environmentally friendly is a topic of considerable debate.
- The term **blue biotechnology** has also been used to describe the marine and aquatic applications of biotechnology, but its use is relatively rare.

Broadly Biotechnology can be divided into Two Major Branches

Non-gene biotechnology– deals with whole cells, tissues or even individual organisms

Gene biotechnology– involves gene manipulation, cloning, etc.

- Non-gene biotechnology is a more popular practice, and plant tissue culture, hybrid seed production, microbial fermentation, production of hybridoma antibodies or immunochemicals are widespread biotechnology practices.
- For centuries humans have used microorganisms to produce foods and drinks without understanding the microbial processes underlying their production.

- In recent years the understanding of the biosynthetic pathways and regulatory control mechanisms used by microorganisms for the production of several metabolites has been increased by developing the knowledge of the biochemistry of industrially important organisms.
- Notable biotechnologies for food processing include fermentation technology, enzyme technology, and monoclonal antibody technology.
- Beneficial microbes participate in fermentation processes, producing many useful metabolites such as enzymes, organic acids, solvents, vitamins, amino acids, antibiotics, growth regulators, flavors, and nutritious foods.
- Some leading food bioprocessing technologies are dairy processing, alcohol, and beverage processing. Production of alcoholic beverages includes: wine, beer, whiskey, rum, shake, etc. utilizing microorganisms like *Clostridium acetobutylicum*, *Lecuonostoc mesenteroides*, *Aspergillus oryzae*, *Saccharomyces cerevisiae*, *Rizopus sp.* and *Mucor sp.*
- Biotechnologically produced organic acids like citric acid, acetic acid, gluconic acid, D-Lactic acid, fumaric acid also have a very high market value.
- The application of biotechnology can result in (a) new ways of producing existing products with the use of new inputs, and (b) new ways of producing new products.
- Examples of the former include the production of gasoline from ethanol which in turn is produced from sugar; the production of insulin using recombinant DNA technology; the production of hepatitis B vaccine using recombinant DNA technology and the extraction of copper using mineral leaching bacteria.
- The alternative inputs are oil for gasoline, porcine pancreases for insulin, human blood for hepatitis vaccine, and the conventional mining techniques for copper.
- Examples of the latter include possible medicinal substances that are produced in minute quantity in the human body and which cannot be synthesized such as insulin, interleukin, or Tissue Plasminogen Activator (TPA).

- A wide variety of microorganisms are now being employed as tools in biotechnology to produce useful products or services.
- Raw materials can be converted to useful finished products both by ordinary chemical processes and by biological means.
- Generally, the costs of chemical conversion are quite high as the reactions require high temperature or pressure.
- In contrast, biological alternatives, using microbes or cultured animal or plant cells, operate at physiologically normal conditions of temperature, pressure, pH, etc.
- During the next few decades, biotechnology would have overtaken chemical technology and many such chemicals that are today produced chemically would be made through biotechnology.
- Enzyme technology is an area of considerable current interest and development. Enzymes are biological catalysts and have been used for many years as isolated agents particularly in food e.g. rennin, papain, and invertase.
- These enzymes have increasingly replaced plants and animal enzymes thus amylase from *Bacillus* and *Aspergillus* has substituted those of malted wheat and barley in brewing, baking, and biscuit-making and also in the textile industry, etc.
- Today, enzyme technologies have four distinct areas of application in cosmetics, therapy, the food and feed industry, and for diagnostic purposes.
- One very important recent application is the production of foodstuffs from non-traditional raw materials, for instance, the development of the sweetener, high fructose corn syrup (HFCS), also called isoglucose.
- Another recent application is the use of phytase in animal feed.
- Nowadays, interest in the traditional fermentation technology for food processing has greatly increased because of emphasis placed upon plant materials as human foods.
- **Single-cell protein** (SCP) is a term generally accepted to mean the microbial cells (algae, bacteria, actinomycetes, and fungi) grown and harvested for animal or human food.
- During World War II, when there were shortages in proteins and vitamins in the diet, the Germans produced yeasts and a mold (*Geotrichum candidum*) in some quantity for food.

- Research on SCP has been stimulated by a concern over the eventual food crisis or food shortages that will occur if the world's population is not controlled.
- Many scientists believe that the use of microbial fermentations and the development of an industry to produce and supply SCP are possible solutions to meet a shortage of protein if and when the amount of protein produced or obtained by agriculture and fishing becomes insufficient.
- The roots of molecular biology were established only after the British biophysicist Francis Crick and the American Biochemist James Watson, in 1953, proposed the structure of DNA (deoxyribonucleic acid) molecule which is well known as the chemical bearer of genetic information of most of the organisms.
- We really began understanding and utilizing molecular biotechnology (or gene biotechnology) only after recombinant DNA technology was developed in the 1970s.
- Daniel Nathans (in 1971) of John Hopkins University utilized the restriction enzyme to split DNA of monkey tumor virus, *Simian Virus* (SV40).
- Recombinant DNA technology, often referred to as genetic engineering or gene manipulation, involves the extraction of a particular gene of interest from one organism and then the insertion of the gene into other organisms.
- Genetic manipulation may be defined as the extracellular (i.e. *in vitro*) creation of new forms of arrangements of DNA in such a way as to allow the incorporation or continued propagation of altered genetic conditions in nature.
- Among the first scientist to attempt genetic manipulation was Paul Berg of Stanford University who in 1971 along with his co-workers opened the DNA molecule of SV40 and spliced it into a bacterial chromosome and constructed the first recombinant DNA molecule.
- Genetic engineering techniques are useful tools for genetic research.
- They can help to gain in the structure, function and regulation of genes.

- They also help to prepare the physical maps of the viral genome.
- Maps of several viruses have been made available like SV40, Polyomavirus and adenovirus.
- Another goal in genetic engineering is to design a super bug that can degrade most of the major hydrocarbon components of petroleum.
- The different strains of ***Pseudomonas putida*** contain a plasmid that has genes coding for enzymes that digest a single family of hydrocarbons.
- By crossing the various strains of this bacterium, a superbug has been created.
- The multiplasmid bacterium is able to grow on a diet of crude oil.
- The superbug has the potential for clearing up oil spills.
- One of the best- known applications of genetic engineering is that of the creation of genetically modified organisms (GMOs).
- There are potentially momentous biotechnological applications of GM, for example, oral vaccines produced naturally in fruit, at very low cost.
- This represents, however, a spread of genetic modification to medical purposes and opens an ethical door to other uses of the technology to directly modify human genomes.
- A genetically modified food is a food product derived in whole or part from a genetically modified organism (GMO) such as a crop plant, animal or microbe such as yeast. Genetically modified foods have been available since the 1990s.
- The principal ingredients of GM foods currently available are derived from genetically modified soybean, maize, and canola. Between 1996 and 2001, the total surface area of land cultivated with GMOs had increased by a factor of 30, from 17,000 km^2 (4.2 million acres) to 520,000 km^2 (128 million acres).
- The value for 2002 was 145 million acres (587,000 km^2) and for 2003 was 167 million acres (676,000 km^2).
- Future applications of GMOs include bananas that produce human vaccines against infectious diseases such as Hepatitis B, fish that mature more quickly, fruit and nut trees that yield years earlier, and plants that produce new plastics with unique properties.

- Now scientists have transformed Tobacco Mosaic Virus (TMV) to infect host plants and produce immunizing proteins rather than debilitating leaf shrivel, turning greenhouse tobacco into a biofactory for plague vaccine.
- Genetic diseases could be treated through the use of genetic engineering. Defective genes in an organism cause genetic disorders.
- If a defective gene could be identified and located in a particular group of cells – it could be replaced with a functional one.
- The transgenic cells are then planted into the organism, resulting in a cure for the disorder. Cloning is a relatively new sector of biotechnology, but it promises answers to very important problems related to surgery.
- Tissues and organs could be cloned for surgical purposes. If scientists could isolate stem cells and then direct their development, they would be able to create any kind of tissue, organ or even a whole part of a body.
- Another revolutionizing tool of biotechnology is DNA fingerprinting. DNA fingerprints are useful in several applications of human health care research, as well as in the justice system. DNA fingerprinting is used to diagnose inherited disorders in both prenatal and newborn babies in hospitals around the world.
- These disorders may include cystic fibrosis, hemophilia, Huntington's disease, familial Alzheimer's, sickle cell anemia, thalassemia, and many others.
- Early detection of such disorders enables the medical staff to prepare themselves and the parents for proper treatment of the child.
- In some programs, genetic counselors use DNA fingerprint information to help prospective parents understand the risk of having an affected child.
- DNA fingerprint information can also help in developing cures for inherited disorders.
- DNA fingerprints help to link suspects to biological evidence blood or semen stains, hair, or items of clothing found at the scene of a crime and help in solving crime.

- Another important use of DNA fingerprints in the court system is to establish paternity in custody and child support litigation.
- The U.S. armed services have just begun a program to collect DNA fingerprints from all personnel for use later, in case they are needed to identify casualties or persons missing in action or for suspect verification.
- Due to the revolutionary development of biotechnology during the last couple of decades agriculture has drastically advanced.
- Sensational achievements were made in both plant cultivation and animal husbandry.

Plants have been improved in four Different Ways

- Enhanced potential for more vigorous growth and increasing yields
- Increased resistance to natural predators and pests, including insects and disease-causing microorganisms.
- Production of hybrids exhibiting a combination of superior traits derived from two different strains or even different species.
- Selection of genetic variants with desirable qualities such as increased protein value increased content of limiting amino acids, which are essential in the human diet, or smaller plant size, reducing vulnerability to adverse weather conditions.
- Another important area of biotechnology is the improvement of livestock. Improvement in disease control, the efficiency of reproduction, yields of livestock products i.e. meat, milk, wool, eggs, the composition of livestock products i.e. leaner meat, feed value of low-quality feeds i.e. straw; are some of the applications of biotechnology.
- One of the major scientific revolutions of the twentieth century was the breaking of the genetic code and the development of tools that enable scientists to probe the molecules of life with incredible precision.
- Now, in the twenty-first century, these developments in biology are being married with the use of ever-increasing computer power to help us face the challenges that the new century brings.

- Bioinformatics is the name given to the new discipline that has emerged at the interface of biology and computing.
- A huge amount of genetic data (DNA, RNA, amino acid, and protein sequences) of various organisms, from bacteria to humans, being generated worldwide is stored in a computer database. Specialized software programs are used to find, visualize, and analyze the information, and most importantly, communicate it to other people.
- Various computer tools are used to predict protein structure which is valuable information for the development of vaccines, diagnostic tools as well as more effective drugs.
- Bioinformatics can help in easy and early detection of various diseases like cancer, diabetes, and many more with the help of **microarray** chips (microarrays are miniature arrays of gene fragments attached to glass slides).
- Bioinformatics also helps scientists to construct a phylogenetic tree based on molecular biology and ultimately contribute to the study of evolution.
- Computer simulations model such things as population dynamics or calculate the cumulative genetic health of a breeding pool (in agriculture) or endangered population.
- One very exciting potential of this field is that entire DNA sequences or genomes of endangered species can be preserved.
- Biotechnology has a promising future. In the future, biotechnology will be accredited for some revolutionary technology.
- Recent advances in bioenergy, bioremediation, synthetic biology, DNA computers, virtual cell, genomics, proteomics, bioinformatics, and bio-nanotechnology have made biotechnology even more powerful.
- The recent discovery of the conduction of electricity by DNA and its behavior as a superconductor has opened a new realm in modern science. In the future biotechnology will have a profound impact on the world economy.
- Biotechnology is a golden tool to solve some of the key global problems like global epidemics, fatal diseases, global warming, rising petroleum fuel crisis, and above all poverty.

- For all the positive effects of biotechnology, there are some possible side effects. Nobody knows what ecological hazards could be caused by transgenic organisms.
- Some even speculate that some transgenic organisms could fall into wrong hands to develop bioweapons.
- The opposition of genetic engineering says that "the science is very young and needs a lot more research".
- The path from a test tube to the field is not a straight highway. Both intellectual and financial resources should be realized before new discoveries pave their way to industrial applications.
- In conclusion, biotechnology has also proved to be extremely productive and innovative and the 21st century should be the century of biotechnology.

History of Biotechnology

The origin of biotechnology is as old as human civilization. The development of biotechnology can be studied considering its growth that occurred in two phases

- Traditional Biotechnology (old)
- New Biotechnology (Modern)

Origin and Definition: Old Vs new Biotechnology

- The origin of biotechnology can be traced back to prehistoric times when microorganisms were already used for processes like fermentation.
- Although a molecular biologist may consider the cloning of DNA to be the most important event in the history of biotechnology, later has actually been rediscovered in the 1970s for the third time during the present century. In the 1920s *Clostridium acetobutylicum* was used by **Chaim Weizmann** for converting starch into butanol and acetone, the latter was an essential component of explosives during World War I.
- This raised hopes for the commercial production of useful chemicals through biological processes and may be considered as the first rediscovery of biotechnology in the present century.
- Similarly, during World War II (in the 1940s), the production of penicillin : an antibiotic discovered by **Alexandar Fleming** in

1929 on a large scale from cultures of *Pencillium notatum,* the second rediscovery of biotechnology.

- This is the beginning of an era of antibiotic research. The third rediscovery of biotechnology is its recent reincarnation in the form of **recombinant-DNA technology**, which leads to the development of a variety of gene technologies and is considered to be the greatest scientific revolution of the century.
- Thus, even though biotechnology had its origin in the past, it has assumed special significance only in the 1970s and the 1980s.
- Biotechnology is the product of an interaction between the science of biology and technology.
- This relationship between science and technology has been observed to be complex, so that not only science has influenced technology, but technology has also influenced science.
- Because of this complex relationship and its major impact on human welfare, it is believed that biotechnology in the future may become a major force for human existence.
- Already, the products of biotechnology (including, diagnosis, prevention, and cure of diseases; new and cheaper chemical products e.g. pharmaceutical drugs and the new food sources; devices for environment protection and energy conservation, etc are playing a very important role in employment, productivity, trade, economics and the quality of human life throughout the world. Biotechnology is being such an important area of study; attempts have been made to define it.
- It has been recognized that a complete definition of biotechnology is difficult due to such a wide range of its usage.
- Fermentation, by some microorganisms, the formation of yogurt (curd) and cheese from penicillin from certain fungi, and the processes of baking and brewing are often included in describing what is called **old biotechnology**.
- The other example of biological processes involving the techniques of recombinant DNA and polymerase chain reaction (PCR) and bioprocessing which become possible only through the researches in molecular biology has been described **as new biotechnology**.

- However, in recent literature, no effort as made to distinguish between old and new biotechnology. Following are some available definitions of biotechnology, and one of them may be really adequate.
  - Biotechnology is "the application of biological organism, system or processes to manufacturing and service industries" (British biotechnology).
  - Biotechnology is " the integrated use of biochemistry, microbiology, and engineering sciences in order to achieve the technological application of the capabilities of micro-organisms, cultured tissue cells and parts thereof"
  - Biotechnology is "a technology using biological phenomena for copying and manufacturing various kinds of useful substances".
  - Biotechnology is "the controlled use of biological agents, such as micro-organisms or cellular components, for beneficial use"

Traditional Biotechnology

- It is kitchen technology developed by our ancestors using fermenting bacteria. It is as old as human civilization.
- The Aryans are performing daily ***Agnihotra or Yojna*** during the ***Vedic*** period (*5000-7000 BC*). The material used in Yajna is animal fat (ghee) which is a fermented product of milk.
- Similarly, the divine "soma" had been offered to God (a fermented microbial product used as a beverage). Summarians and Babylonians (6000 BC) had the beer. Egyptians were baking leavened bread by 4000 BC.
- The preparation of curd, ghee, wine, beer and vinegar, etc. were the kitchen technology. In spite of all development, preparation of curd, ghee, vinegar and alcoholic beverages, idli, jalebi, dosa (Table 1.1) have become an art of the kitchen of all Indians.
- Traditional biotechnology refers to conventional biotechnology which has been used for many centuries.
- Beer, cheese, wine and other many foods have been produced using traditional biotechnology. Thus the traditional

biotechnology includes the processes that are based using traditional biotechnology.

- It includes the process that is based on the natural capabilities of microorganisms. It has established a huge and expanding world market. In monetary terms, it represents a major part of all biotechnology financial profits.
- There is one story related to the butter **"Makhan Chor"** (butter stealing) by Lord Sri Krishna during the Mahabharat period. The butter has been produced following the same kitchen method or art.

Table 9.1: Fermented foods prepared in different parts of India

Types of foods	Substrate	Regions	Quality and Uses
Wari	Black gram	North	The spongy cake used as a snack
Vadai (wada)	Black Gram	North	Deep-fried cake used as snack
Tharra	Mahua	North	Sweet alcoholic beverage obtained through distillation
Tari	Date Palm	East	Sweet alcoholic beverage
Sinki	Radish taproot	Himalaya	Sun-dried sour soup pickles
Srikhand	Milk	West	Concentrated sweetened preparation
Rabdi	Mixture of butter-milk/ wheat/ barley/ peralmillet	West	Cooked paste used as Staple food
Paneer	Milk	North	Soft cheese used as a fried curry
Papad	Black gram flour (besan)	North	Circular wafers used as a snack
Nan	Wheat Flour (maida)	North	Leavened flat baked bread used as a staple food
Mishti dahi	Milk	East	Thick sweet gel
Mesu	Bamboo shoot	Himalaya	Sour pickles
Khalpi	Cucumber	Himalaya	Sour pickles
Khaman	Bengal gram	West	Spongy cake, used in breakfast
Jalebi	Wheat Flour (maida)	North	Crispy, deep-fried used as a sweet confectionery
Idli	Rice and Black gram	South	Steamed Spongy cake used in Breakfast

Gundrum	Leafy Vegetables	Himalaya	Sun-dried, sour taste, used as-soup or pickles
Dhokla	Bengal-gram	West	Spongy cake used as a snack
Dahi	Milk	North	Sour, thick gel with Whey
Dosa	Rice and black gram	South	Spongy, shallow fried used as a staple food
Chhurpi	Milk	Himalaya	Cheese like mild sour, soft mass used as a curry
Ambali	Millet and rice	South	The steamed cake used in snack
Bhatura	Wheat flour (maida)	North	Flat fried, leavened bread used with chhola

- The Egyptians during the 2000 BC period used to prepare vinegar from crushed dates by keeping it for a longer time.
- But the crushed dates produce intoxicants at first. In Egypt, Mesopotamia and Palestine during 1500 BC the art of production of wine from crushed grapes, and beer from the germinated cereals (malt) using a bread leaven (a mass of yeast) were established.
- In Indian Ayurveda, the production of "**Asava**" using different substrates and flowers of **mahua** (***Madhuca indica***) or dhataki (***Wodfordia fructicosa***) has been well characterized till today since the Vedic period. In these methods, various substrates are transformed into a number of products.
- Hence the odor, color and taste of final products are changed. Moreover, the use of salts for preservations of different foods has been earlier developed even in Europe. The preservation methodology is still of the mummies of Egypt are noteworthy.
- The possible mummification has been done through dehydration of a dead body followed by the use of a mixture of salts largely sodium carbonate. Traditional technology is an art rather than a science.
- The cause of fermentation could be discovered after observing microorganisms using a microscope by Antony van Leeuwenhoek (1673-1723) at Delt (Holland). A significant contribution was done by chemists on the process and the products of fermentation during the 18th century. The milky precipitate was obtained when the gas evolved from fermentation was passed through the lime water in 1757. It was called the lime water test.

- A similar gas was coming out from the burning of charcoal. Henry Cavendish demonstrated that the gas evolved from brown sugar in water treated with yeast was absorbed by sodium hydroxide solution.
- **Schele and J. Priestley** in 1772-1774 confirmed the identity of oxygen gas present in air. Using an analytical technique for carbon estimation, Antoine Lavoisier gave the chemical basis of alcoholic fermentation.
- A French Man, **Nichola Appert** in 1810 was described as the method of food preservation. In the same year, Peter Durand also gave the use of the tin container for food preservation.
- It was done by putting an air-tight vessel containing food material in the boiling water.
- It increased the importance of the canning industry.
- Lack of oxygen in such a closed and heated vessel was reported by Gay Lussac. He concluded the oxygen required for initiation of alcoholic fermentation, but not for further progress of fermentation.
- **Charles B Astier** gave the concept that air is the carrier of all kinds of germs after 1830. Theodore Schwann after a series of experimentations demonstrated that the development of the fungus (sugar fungus) on fruit juice causes fermentation.
- He was the first to observe and describe the yeast in the growing process. Charles Cagiard-Latour (1838) observed yeast budding using a microscope allowing 300-400 power magnification.
- **Justus von Liebig** in 1839 a well-known chemist proclaimed that all the activity of yeast cells was the result of chemical and physical reactions going on in the medium.
- Microbes also called microorganisms are invisible creatures too small to be seen with the unaided eye and a microscope is needed to study them.
- They are diverse, present everywhere and exist in a range of environments from mountains and volcanoes to deep-sea vents and hot springs.
- ***Thermus aquaticus*** is an example of an unusually heat -stable bacterium isolated from hot springs from which heat resistant enzyme **Taq polymerase** was isolated.

- Microorganisms can be found everywhere like in the air, water, in the food we eat and even in our body.
- In fact, we come in contact with countless numbers of microorganisms every day.
- It plays a vital role in the ecology of the earth.
- They degraded dead bodies of plants and animals and recycle chemical elements to be used by living plants and animals.
- Some are used commercially in the production of industrially important chemicals such as alcohol, organic acids and antibiotics.
- It is also responsible for food fermentations thus enhancing their quality and shelf life.
- Using microbes to make products such as different foods and chemicals is called biotechnology. Using recombinant DNA, microbes can produce important substances such as proteins, vaccines and enzymes.
- Microbes include bacteria, archaea, fungi, algae, protozoa, viruses, viroids and prions.
- While bacteria and archaea are classed as prokaryotes and the term was introduced by Edouard Chatton for the cells in which the nuclear material is not surrounded by a haploid.
- The fungi, algae, and protozoa are eukaryotic whose genomic DNA is contained in a nucleus surrounded by a membrane, others don't have regular cell structures and are classified separately (Aneja, 2014).
- Microbes range in size from ultramicroscopic viruses 20 nm diameter to large protozoans 5 mm or more in diameter.
- In simple words, the largest microbes are as much as 2, 50,000 times the size of the smallest ones.
- The study of all organisms in the microscopic range is called microbiology.
- The naming of organisms called the binomial system of nomenclature was devised by Carolus Linnaeus in 1735.
- In this system each living organism is assigned two names- the genus and the specific epithet, both of which are underlined or italicized e.g ***Escherichia coli.***

- Microorganisms are relevant to all of us in different ways.
- Sometimes the influences of microorganisms in human life are beneficial apart from those other times it is detrimental or harmful to human beings.
- For example, microorganisms are required for the production of cheese, beer, bread, yogurt, alcohol, wine, antibiotics, vitamins and vaccines and many ore important products that are useful in human life.
- Microorganisms are essential components of our ecosystem.
- Fungi and bacteria play an important role in the recycling or organic and inorganic material through their role in the carbon, nitrogen and sulfur cycles.
- They also are a source of nutrients at the base of all ecological food chains and webs.
- A recent study discovered by Professor Jonathan Rhodes with his colleagues at the University of Liverpool, North West England, says that common edible mushrooms contain a protein lectin that can stop cancer cell multiplication.
- This discovery finally could lead to new targets for therapy.
- ***Taxomyces andreanae***, an endophytic fungus known to occur in the U.S.A is being used to produce taxol, an antitumor diterpinoid used in the treatment of some cancers.
- Taxol was originally obtained from the bark of Pacific yew (***Taxus brevifolia***).
- It was in the year 1347 when plaque or Black Death struck Europe and within 4 years killed 25 million people that is 1/3rd of the population.
- Over the years the disease struck, again and again, wiping out 75% of the European population by 1431. This disease is believed to have changed European culture.
- It was Robert Koch 1843-1910, a German bacteriologist, who in 1876, first of all, proved that anthrax is caused by a microbe ***Bacillus anthracis***.
- The past 30 years have witnessed an increase in new and emerging infectious agents such as the AIDS virus (HIV-human immunodeficiency virus) and ***Escherichia coli*** and several other older diseases regularly appear to be on the increase.

- Recently correlation define between gastric ulcers and the bacterium (***Helicobactor***) that invades the stomach, between type 1 diabetes and certain coxsackieviruses and between coronary artery disease and cytomegalovirus infection.
- Microbes play a role in the decomposition of waste material and help other animals digest grass such as cattle, sheep, and termites, so they can also balance nature.
- Microbes affect human health.
- The environment provides the development of biodegradable products and safe drinking water which use bacteria to clean up oil spills through the bioremediation process.
- Microbes participation in agriculture research has led to healthier livestock and crops free from disease.
- Microbes play a role in industries such as (foodstuffs, beer, wine, cheese, and bread), antibiotics, genetic engineering, and insulin.
- It is a specialized area of biology that deals with the study of microorganisms or microbes. Roughly speaking, organisms with a diameter of one mm or less are microorganisms and fit into the broad domain of microbiology (Aneja, 2014; Garg et al, 2013).

The branches of Microbiology are Categorized into Pure and applied Sciences such as:

- **Bacteriology:** It is the study of bacteria.
- **Mycology:** It is the study of fungi.
- **Protozoology:** It is the study of protozoans.
- **Virology:** It is the study of viruses.
- **Algology or Phycology:** It is the study of algae.
- **Parasitology:** It is the study of parasitism or parasites.
- **Microbial Ecology:** It is the study of the interrelationship between the environment and microbes.
- **Microbial Morphology:** It is the study of the detailed structure of microbes.
- **Microbial Taxonomy:** It is concerned with the classification, naming, and identification of different microbes.
- **Microbial Physiology:** It is the study of the metabolism of microbes at the cellular and molecular levels.

- **Microbial Genetics:** It is the study of the genetic material of microbes.
- **Molecular Microbiology:** It is the study of structure and function and the biochemical reactions of microbial cells involved in metabolism and growth.
- **Industrial Microbiology:** It is concerned with the industrial uses of microbes in the production of vitamins, alcoholic beverages, amino acids, enzymes, antibiotics, and other drugs.
- **Food Microbiology:** It is the study of the interaction between food and microbes in relation to food bioprocessing, food spoilage, and their prevention.
- **Agricultural Microbiology:** It is the study of the interrelationship of crops and microbes with an emphasis on control of plant diseases and improvement of yield.
- **Dairy Microbiology:** It deals with the production of and maintenance of quality control of dairy products.
- **Aquatic Microbiology:** It is the study of microorganisms and their activity concerning human and animal health in fresh, estuarine and marine waters.
- **Air Microbiology:** It deals with the role of aerospora in contamination and spoilage of food and discrimination of plant and animal disease by air.
- **Exomicrobiology:** It is the study of exploration for microbial life in outer space.
- **Medical Microbiology:** It deals with the fundamental principles and techniques involved in the study of pathogenic organisms as well as their application in the diagnosis of infectious diseases.
- **Immunology:** It deals with the immune system that protects against infections and attempts to understand the many phenomena that are responsible for both acquired and innate immunity.
- **Biotechnology:** It is the scientific manipulation of living organisms especially at the molecular and genetic level to produce useful products.
- **Public Health Microbiology:** It is concerned with monitoring, control and spread of diseases in communities.

- The first living things to appear on earth and the study of fossil remains indicate that microbial infections and epidemic diseases existed thousands of years ago.
- It was **Rogen Bacon (1220-125)** who in the 13th century postulated that disease is produced by invisible living creatures. This idea was made again in 1546 by a physician "**Girolamo Fracastoro (1478- 1553)** of north Itlay.
- Fracastoro wrote a treatise-De Contagion in which he said the disease was caused by a minute **"seed" or "germ"** and was spread from person to person. As early as 1658, A monk named **Athanasius Kircher (1601-1680)** referred to "Worms" invisible to the naked eye in decaying bodies, meat, milk and diarrheal secretions.
- He examined thin cork slices, the bark of an oak tree, and found that cork was made of tiny boxes that Hooke referred to as "Cells" The work of Hook was followed by **Matthias Schleiden and Theodore Schwann** who examined a variety of organisms and in 1838-1839 reported that all forms of life are composed of cells.
- An observation later became the foundation of **"Cell Theory".** The cell theory since been modified to all life is composed of cells that originate from other cells. The discovery of microbiology as a discipline could be traced along with the following historical eras.
- It is the period concerned with the discovery of the microbial world that has been dominated by **Antony van Leeuwenhoek**.
- He was the first person to observe and accurately describe microorganisms in 1676. He also examines blood and other human tissues including his own tooth scrapping, insects, minerals, and plant materials.
- He was the first person to describe bacteria and protozoa using a microscope he made himself, due to this extraordinary contribution to microbiology he was considered as the father of bacteriology and protozoology.
- Transition Period Later by **Leeuwenhoek** time, there are two notes worthy contributions such as (1) controversy over spontaneous generations that said that living organisms could develop from non-living matter called as abiogenesis and (2) disease transmission Francesco Redi (1626-1697) In 1665, **Redi**

put the theory of spontaneous generation to rest by conducting an experiment in which he placed meat in three jars. One jar was covered with fine gauze, the second was covered with paper and the third was not covered.

- Flies entered the jars that were open to air i.e left uncovered and landed on the meat where they laid their eggs that later developed into maggots. The other two pieces of meat did not produce maggots spontaneously.
- However, flies were attached to the gauze-covered jars and laid their eggs on the gauze and maggots developed without access to the meat and indicating that maggots were the offspring of the flies and did not arise from some vital source in the meat as previously believed.
- **John Needham** (1713-1781) was a greater supporter of the theory of spontaneous generation. In 1749, He proposed that tiny organisms the animalcules arose spontaneously on his mutton gravy. He had covered the flasks with cork as done by Redi and even heated some flasks.
- Still, the microbes appeared on the mutton broth. **Lazzaro Spallanzani** (1729-1799). In 1775, he attempted to refute **Needham's** work by performing experiments.
- He boiled the beef broth for a longer period and removed the air from the flask and then sealed the container.
- After the incubation time which he was followed, no growth was observed by him in these flasks. When he was accused of destroying the vegetative force of the nutrients by over heating, showed that the heated nutrients could still grow animalcules when exposed to air by simply making a small crack in the neck so he was disproved the spontaneous generation.
- **Georg Schroeder and Theodor von Dusch** in 1854 were conducting experiments to disprove the theory of spontaneous generation by simply passing air through cotton into flasks containing heated broth.
- No growth was observed on the infusions due to the filtering out of microscopic organisms by cotton. He was the first to introduce the idea of using cotton plugs for plugging microbial culture tubes in 1854.

- He finally resolved the controversy of spontaneous generation versus biogenesis and proved that microorganisms are not generated spontaneously; it is from inanimate matter but arise from other microorganisms.
- **John Tyndall** (1820-1893) was the final blow to spontaneous generation in 1877. He has conducted experiments in an aseptically designed box to prove that dust indeed carried the germs. He demonstrated that if no dust was present, sterile broth remained free of microbial growth for an indefinite period even if it was directly exposed to air.
- The diversity of our life is present on our planet with new species of all types being constantly discovered.
- The formal system of organizing classification and naming living things is called taxonomy. The naming of organisms called the binomial system of nomenclature was devised by **Carolus Linnaeus** in 1735.
- He gave the binomial system of nomenclature i.e naming of an organism by two names such as genus and species. Therefore, the Linnaeus system provides each organism's names and arranges them into groups called taxa with other similar organisms to reflect their phylogeny and evolutionary relatedness.
- The main taxa or groups in a classification scheme are organized in several descending ranks beginning with the kingdom the largest and followed by "Phylum or Division, Class, Order, Family, Genus and Species".
- The classification of microorganisms began in 1674 with the invention of the light microscope and today is discipline-based on increasingly complex criteria.
- The first phylogenetic trees of life were constructed on the basis of just two kingdoms such as Plantae and Animalia. In 1886, a German scientist **Ernest Haekel** proposed that the bacteria, algae, fungi and protozoa that lacked tissue differentiation be removed from the plant and animal kingdoms and be separated in a third kingdom Protista.
- In 1940, in some microorganisms e.g. typical bacteria the genetic material was not enclosed by a nuclear membrane. This remarkable discovery, the absence of nuclear membrane-bound internal structure in one group of Protista (bacteria) and the

presence of bound structure in the majority of others such as algae, fungi and protozoa was discovered significantly.

- This results in the division of these organisms into two groups Prokaryotes and Eukaryotes. Therefore, several systems of classification of living organisms have been developed between 1940 and 1969 but the one that is most widely accepted was developed in 1960 by Robert Whittaker of Cornell University.
- This system places all living things except viruses into five kingdoms based on cellular organization and nutritional patterns such as (1) prokaryotic or Monera; (2) Protista; (3) Myceteae or Fungi; (4) Plantae and; (5) Animalia.
- The three kingdom domain which was discovered by scientists **Carl Woese and George Fox** have proposed a new system of classification that assign all organisms to one of the three super kingdoms (**Figure 1.1**) called domain such as (1) Archae; (2) Bacteria; and (3) Eukaryotic. It is believed that these three kingdoms arose from an ancestor most similar to the archaebacteria.
- There are several classifications that have been proposed for bacteria of which Bergey's Manual of Systematic Bacteriology has wide acceptance.
- The **Bergey's Manual of Determinative Bacteriology** published by Williams and Wilkins Co. Baltimore, USA (1974), is the international standard for bacterial taxonomy. It deals with the classification of bacteria and contains a description of all the species known at the time of its publication.
- In this bacteria are assigned 19 parts based primarily on the following five features such as (1) Morphology and Gram Stain Reaction; (2) Gaseous requirements; (3) Energy and carbon sources; (4) Mode of locomotion; and (5) ability to produce endospores. It serves as a practical guide for the identification of bacteria.
- After some time the manual was revised between 1984 -1989 and appeared in volumes with a new name Bergey's Manual of Systematic Bacteriology.
- In these four volumes, bacteria have been classified in 33 sections based on the following characteristics such as general shape and morphology, gram staining properties, presence of endospores,

oxygen relationship, motility and mode of energy production.

- Therefore, in the four volumes, the bacterial groups are divided as follows (1) gram-negative bacteria of general, medical and industrial importance; (ii) Gram-positive bacteria other than actinomycetes; (iii) Gram-negative bacteria with distinctive properties, cyanobacteria and archaeobacteria; and (iv) Actinomycetes (gram-positive filamentous bacteria).
- The Bergey's Manual of Systematic Bacteriology places all bacteria in the kingdom of prokaryotic which in turn is divided into four divisions as follows:
- The **Gracilicutes (thin skin):** Prokaryotes with a complex cell wall structure characteristics of gram-negative bacteria e.g Nonphotosynthetic bacteria.
- **Firmicutes (Strong skin):** Prokaryotes with a cell wall structure characteristics of gram-positive bacteria (Rods and cocci, Actinomycetes and related organisms).
- **Tenericutes (soft or tender skin):** Prokaryotes that lack a cell wall (e.g mycoplasms).
- **Mendosicutes (Skin with faults):** Prokaryotes with unusual cell walls e.g Archaeobacteria. There are groups that do not reflect the phylogenetic relationship between the organisms; each division is separated into classes for a total of 7 classes of bacteria.
- The system used in Bergey's manual recognizes 7 classes under the 4 divisions.
- Groups of Bacteria as per Bergey's Manual: Division I Gracilicutes Gram-negative bacteria Class I Scotobacteria Gram-negative non-photosynthetic bacteria Class II Anoxyphotobacteria Gram-negative photosynthetic bacteria that do not produce oxygen (green or purple bacteria) Class III Oxyphotobacteria Gram-negative photosynthetic bacteria that evolve oxygen (cyanobacteria) Division II Firmicutes Gram-positive bacteria Class 1 Firmibacteria Gram-positive rods or cocci Class II Thallobacteria Gram-positive branching cells (actinomycetes) Division III Tenericutes Class I Mollicutes Bacteria lacking a cell wall (mycoplasmas) Division IV Mendosicutes Class I Archaeobacteria Bacteria with typical compounds in the cell wall and membranes.

- A Bacteria (singular bacterium), the microscope, single-celled (unicellular), prokaryotic organisms, reproduce asexually, are haploid and do not have their DNA localized within a nucleus surrounded by a unit membrane (i.e., without a nucleus). There are two types' of bacteria gram-positive and gram-negative.
- They lack the membrane-enclosed intracellular structures found in most other cellular organisms. The shapes of bacteria are spherical or ovoid (coccus), rod-like (bacillus), corkscrew or curved (spiral), comma, helical and filaments. Individual bacteria may form pairs, chains, clusters or other groupings.
- Such arrangements are usually characteristics of a particular genus or species of bacteria. All bacteria have a cell wall composed of a carbohydrate and proteins complex called peptidoglycan. Bacteria reproduce asexually by a process called binary fission. i.e., by dividing into two equal cells.
- They possess flagella moving appendages for their motility. For nutrition, most bacteria absorb nutrients from their environment from the dead or living organisms, some can manufacture their own food by photosynthesis, and some can derive nutrition from inorganic substances. The study of bacteria is called bacteriology.

Archaea Bacteria

- It is like bacteria, consists of unicellular prokaryotic cells, but do not have peptidoglycan in their cell walls.
- They often live in extreme environments and carry out unusual metabolic processes. These are found in hot springs, geothermally heated marine sediments and submarine hydrothermal vents. They have no nucleus and any other organelles.
- It is classified as bacteria and called archaebacteria. They are further divided into phyla. It has the same structure size as bacteria such as flat and square-shaped cells.
- They use more energy sources than eukaryotic organisms such as sugar, metal ions and ammonia or even hydrogen gas. They reproduce asexually by budding, fragmentation or binary fission. It ranges from 0.1 μm to 15 μm in diameter and various shapes such as rods, spheres and spirals or plates.

Fungi

- Fungi are eukaryotic with a different nucleus containing the genetic material (DNA) surrounded by an envelope, the nuclear membrane.
- True fungi have cell walls composed of chitin, a polysaccharide also found in the exoskeleton of arthropods such as ticks and spiders.
- It reproduces both methods asexually and sexually and also separately.
- Fungi have generally a mycelium, a loosely organized mass consisting of thread-like hyphae.
- It consists of three groups of organisms- yeasts, molds and mushrooms. The study of fungi is called mycology.

Protozoa

- Protozoa are unicellular eukaryotic organisms that have no cell wall, no photosynthesis, no fruiting bodies, reproduce sexually and asexually have various ways of locomotion such as pseudopodia, flagella or cilia.
- Protozoa are classified into four groups on the basis of their motility.
- Protozoa are found in soil and water and as normal microbiota of animals.

Algae

- Algae are photosynthetic eukaryotes containing chloroplast an organelle carrying out oxygenic photosynthesis.
- The body of algae is called a thallus.
- Most algae are single-celled, simple microscopic organisms.
- But some are multicellular which can attain an enormous size.
- The cell walls of most algae are composed of cellulose.
- They reproduce both sexually and asexually.
- It is abundant in freshwater and saltwater, in soil and in symbiotic association with fungi called lichens. The study of algae is called algology (phycology).

Viruses

- Viruses are acellular (non-cellular) entities, ultramicroscopic, ranging in size from 30 nm to 300 nm.
- Hence they can be seen only with an electron microscope.
- They lack the characteristics of life because of which viruses are considered to be non-loving entities.
- A virus particle (called virion) has either DNA or RNA, but never both.
- The DNA and RNA found in viruses may be single-stranded or double-stranded.
- Indeed, viruses can be crystalized and stored in a container on a shelf for years, but they retain the ability to invade cells.
- A virus can reproduce only by using the cellular machinery of their host.
- Viruses are obligate intracellular parasites of bacteria (bacteriophages), protozoa, fungi, algae, plants and animals and responsible for diseases of humans (AIDS, polio, smallpox, swine flu, H1N1 influenza, bird flu) and plants (tobacco mosaic, cauliflower mosaic).
- The study of viruses is called virology.
- Viroids are small circular pieces of naked single-stranded RNA, only 300 to 400 nucleotides long, with no protein coat.
- They are the smallest known infective agent causing diseases in plants only.
- They retard the growth of the plant on which they multiply interfering with the mRNA processing. One of the worst diseases caused by viroids is potato spindle tuber viroid causing losses of millions of dollars from crop damage.

Prions: Prions also called virions are proteinaceous infections particles having no nucleic acid. They cause infectious diseases the deadliest disease in cattle is the **"mad-cow disease"** also called bovine spongiform encephalopathy (BSE) and Creutzfeldt Jacob disease (CJD) of humans which involves the degeneration of brain tissue. Prion diseases are the result of altered proteins; the cause can be a mutation in the normal gene for **PrPc** (normal host glycoprotein) or contact with an altered protein (PrPsc).

- The study of microbiology was started with the first report of **Louis Pasteur** (1857) on lactic acid fermentation from sugar.
- He isolated the microorganisms (lactic yeast) that were associated with lactic acid and formed curd.
- The cells of lactic yeast were smaller than that of beer yeast.

Lactic acid production got increased when he added chalk powder to the fermentation medium.

Pasteur showed the presence of lactic acid in curd by using a polarimeter.

In 1860, Pasteur provided a detailed report on the use of a synthetic medium for microbiological studies. He concluded that there are many steps in a fermentation process such as

- Fermentation is carried out anaerobically by the living cells.
- During the alcoholic fermentation in a synthetic medium, the yeast increased the weight with an increase in carbon and nitrogen contents of the overall batch.
- The increase in yeast protein in the synthetic medium was accompanied by a related decrease in the ammonium nitrogen in the medium.
- Fermentation sugar was required for the multiplication of yeast cells.
- A similar phenomenon occurred in the fermentation of lactic acid, tartaric acid and butyric acid, etc.
- When a pure culture is not used; some of the fermentation processes were stopped.
- Growth and physiology of yeast differ when they are grown under anaerobic growth conditions. A large amount of sugar was converted into alcohol under anaerobic growth conditions, while in aerobic conditions, a large amount of sugar was converted into yeast cell mass.
- Pasteur suggested that the high percentage of microbial population is killed by heating the juice at 62.8° C which is now called Pasteurization.
- Robert Koch in 1881 gave the method for establishing the relationship of a pathogen with a disease which is known as "***Koch's***" postulates or "pathogenicity test".

- He proved that anthrax disease is caused by "***Bacillus anthracis***".
- The first demonstrates enzymatically- mediated fermentation reaction was by Edward Buchner in 1897. He showed that the cell-free yeast juice mixed with concentrated sugar solution evolved upon incubating carbon dioxide and produced ethyl alcohol.
- A similar product was also produced in an aqueous solution of the other sugars such as glucose, sucrose, fructose and maltose. Buchner called the dissolved substance in juice responsible for sugar fermentation "zymase".
- This work showed improved techniques of fermentation. Fermentation could also be demonstrated in test tubes without a living organism.
- The work of chemistry had more roles in understanding the fermentation phenomenon than microbiology.
- **Hans Buchner and Martin** developed a more effective method of getting cell- free extracts after disrupting the microbial cells.
- They ground the microbial cells with quartz and adding Kieselguhr. Thus sucrose fermentation using cell-free extracts from yeast cells obtained by the above method was demonstrated.
- The discovery of viruses and their role in disease was possible when **Charles Chamberland** in 1884 constructed a porcelain bacterial filter.
- Edward Jenner in 1798 used vaccination by taking out liquid material from cowpox lesions and introducing into people having smallpox.
- But Pasteur and Chamberland developed toxins produced by *Corynebacterium diphtheria* (causes diphtheria in humans) its antitoxin was developed by Emil von Behring in 1890 and Shibasaburo Kitasato.
- He injected the inactivated toxin into a rabbit that induced to produce antitoxin in the blood.
- The antitoxin inactivated the toxin and protected the disease. Similarly, tetanus antitoxin was developed by Behring.
- Traditional biotechnology is really the kitchen technology developed by our ancestors using fermenting bacteria. Kitchen technology is as old as human civilization.

- During the Vedic period (5000-7000 BC), Aryans had been performing daily Agnihotra or Yagya One of the materials used in Yagya a is animal fat (ghee) which is a fermented product of milk Similarly, the divine soma, a fermented microbial products used as a beverage, had been offered to God.
- Summarians and Babylonians (6000 BC) were drinking beer. Egyptians were baking leavened bread by 4000 BC.
- Preparation of curd, ghee, Wine, beer, vinegar, etc, was the kitchen technology.
- In spite of all development, preparation of curd, ghee, vinegar alcoholic beverages jalebi, idli, dosa, have become an art of the kitchen of all Indians.
- Traditional biotechnology refers to conventional technology which has been used for many centuries.
- Beer, wine, cheese and many foods have been produced using traditional biotechnology.
- Thus, traditional biotechnology includes the process that is based on the natural capabilities of microorganisms.
- Traditional biotechnology has established a huge and expanding world market. In monetary terms, it represents a major part of all biotechnology financial profits.
- Who can forget the story of *Makhan Chori* (butter stealing) by Lord Sri Krishna during the Mahabharat period? Butter would have been produced following the same kitchen art.
- Besides breeding strong productive animals, the selection of desirable seeds for enhanced crop production has been a part of human activity since time immemorial.
- The Egyptians (about 2000 BC) used to prepare vinegar from crushed dates by keeping it for a longer time.
- But these crushed dates produce intoxicants at first.

Modern Biotechnology

- There are two major features that differentiate modern biotechnology from classical biotechnology such as the capability of science to change the genetic material for getting new products for specific requirements through recombinant DNA technology.
- The ownership of technology and its socio-political impact.

- Now the conventional technology industries, pharmaceutical industries, agro-industries, etc are focusing their attention to produce biotechnology-based products.

Emergence of Modern Biotechnology

- The new or modern biotechnology embraces all methods of genetic modification by recombinant DNA and cell fusion technologies.
- It also includes the modern developments of traditional biotechnology processes.
- The new aspects of biotechnology founded in recent advancements of modern biology, genetic engineering and fermentation process technology are now increasingly finding a wide industrial application. But the rate of application will depend on:

1. Adequate investment by the industries
2. Improved system of biological patenting
3. Marketing skill
4. Economics of the new methods
5. Public perception about the biotechnology products.

- The traditional biotechnology associated with fermentation was gradually industrialized at the end of the 19th century. This resulted in the gradual growth of industries producing beer, whisky, wine, canned food and rum, etc.
- The first time in 1920 the Leeds City Council (UK) established the Institute of Biotechnology. In the late 1960s, OECD was set up to promote policies for the sound economic growth of the member countries. In 1978, The European Federation of Biotechnology was set up.
- Biotechnology brought industries and agriculture together in the 20th century. The fermentation processes were developed which produced acetone from starch and paint solvent from automobile industries during World War I.
- The antibiotic penicillin was discovered during World War II. Manufactured of penicillin shifted the biotechnological focus towards pharmaceuticals.
- Linking fermentation with biochemistry, bioprocess, chemical engineering and instrument designing helped substantially in the progress of industries.

- The work on Microorganism dominated the preparation of biological welfare, antibiotics and fermentation process.
- Suspected preparation of biological and chemical warfare led to the US attack on Iraq in 2003 during the **Gulf War** (1991).
- After the discovery of double helix DNA by **Watson and Crick** (1953), Werner Arber in 1971 discovered a special enzyme in bacteria which he called the restriction enzymes.
- These enzymes can cut the DNA strand and generate fragments.
- The cut ends of two fragments are single-stranded sticky ends because these ends having identical base pairs can rejoin S. Chen and H. Boyer in 1973 removed a specific gene from a bacterium and inserted into another bacterium using restriction enzymes.
- This discovery marked the start of recombinant DNA technology or genetic engineering.
- **Baltimore** successfully transferred the human growth hormone gene into a rabbit in 1976.
- The European Federation of Biotechnology was created in 1978.
- In 1978, a U.S. company "Genetech" used genetic engineering techniques to produce human insulin in *E. coli*. In 1980, a trial of new hormones was conducted in the U.S.A., France, Japan, and the United Kingdom.
- The US Food and Drug Administration gave marketing approval to "Humulin" *i.e* human insulin made by Eli Lilly (U.S.A) by the end of 1982.
- Another hormone somatotropin was produced on an industrial scale.
- The first genetically engineered tomatoes in 1993, FlavrSavr, were sold in the market.
- In 1996, the first clone lamb "Dolly" was borne successfully by the efforts of scientists of Scotland. Thereafter several cloned animals were produced.
- The sequence of the human genome was published in Nature and Science in 2001. The human genome project was completed by March 2003 and in 2002, a designer baby was born to cure the genetic disease of her elder sister.

- A claim was made for the birth of a clone baby "Eve" on December 27, 2002, by the scientist of "Human Cloning Society", the Clonaid of France.
- In February 2005 in the meeting of the United Nations many countries supposed gene cloning in humans, while a few countries supported with a request to allow for the sake of research only.
- In May 2005, scientists in South Korea have used a method called therapeutic cloning to produce stem cell lines.
- These are genetic matches for patients. Such stem cell lines could be used for disease research. The U.S.A. condemned this approach.
- In this method, human embryos were produced through cloning and stem cells were obtained from a blastocyst.
- The excised stem cells could be grown *in vitro* and used further. It is therefore interesting that the scientists are engaged in doing the most challenging task mass production of growth hormones, insulin, vaccines, immunogenic proteins and polypeptides, gene therapy, biofertilizers, biopesticides, producing disease and stress -resistant plants, biomass, enzymes, antibiotics, acids, fuels, etc.
- Manu biochemical companies such as National Pituitary Agency (U.S. A), E. Lilly (U.S.A), Kabi Vitram A B (Sweden), Genetech Co. (U.S.A), Biogen (Switzerland), etc. are producing/trying to produce some of the above products by using genetic engineering techniques. Many Nobel Laureates including **Dr. Har Govind Khorana** are associated with these companies.
- Biotechnology is not a sudden discovery, rather a coming of age of a technology that was initiated several decades ago.
- By the middle of the 20th century, there had been tremendous growth in the area of chemistry, physics and biology.
- Further knowledge of each branch has been advanced substantially. Multidisciplinary strategies were made for the solution of various problems. A novel spectrum of investigations occurred through the true interdisciplinary synthesis.
- This led to the evolution of biotechnology which is an outcome of integrated effort of biology with technology the root of which lies in biological science. The key difference between biology and biotechnology is their scale of operation.

- Usually, the biologist works in the range of nanograms to milligrams. Biotechnologists working on the production of a vaccine may be satisfied with milligram yields, but many other projects aim at kilograms or tonnes. Thus the main objective of biotechnologists consists of scaling up the biological processes (Smith, 1996).
- The main discipline of biology is microbiology, biochemistry, genetics, molecular biology, immunology, cell and tissue culture **Figure 9.1**. However, on the engineering such as large scale cultivation of microbes and cells, their upstream and downstream process, etc. These processes can be separated into five major operations:
- Strain selection and improvements
- Mass culture
- Optimization of cell responses
- Process operation
- Downstream processing

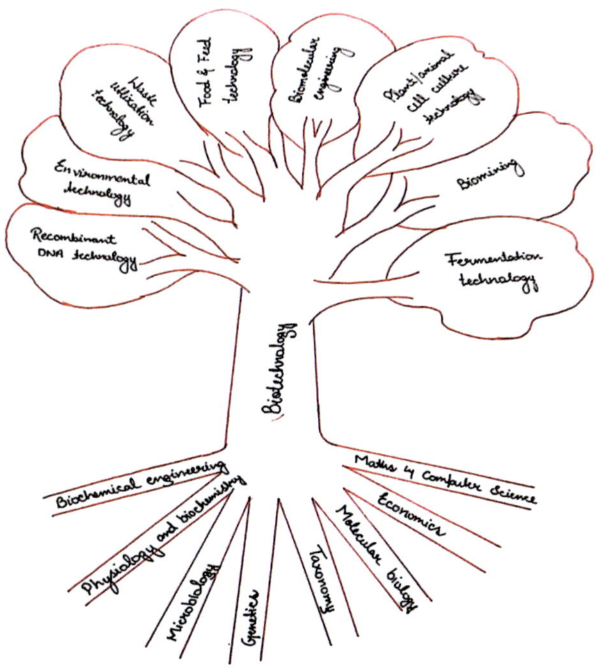

Figure 9.1: A tree of Biotechnology.

- Many areas of biotechnology have arisen through the interaction between various parts of biology and engineering, biochemistry, biophysics, cell biology, colloid chemistry, embryology, ecology, genetics, immunology, molecular biology, medical chemistry, pharmacology, polymer chemistry, thermochemistry and virology.
- Modern biotechnology has developed several technologies extracting the basic knowledge from biology.
- Many areas of biotechnology have arisen through the interaction between various parts of biology and engineering, biochemistry, biophysics, cell biology, colloid chemistry, embryology, ecology, genetics.
- Immunology, molecular biology, medical chemistry, pharmacology, polymer chemistry, thermo-chemistry and virology.
- The modem biotechnology has developed several technologies extracting the basic knowledge from biology.

Global Impact and Current Excitement of Biotechnology

- Each and every organism performs its function within its optimum limits.
- The excitement about modern biotechnology is that scientific methods (such as genetic engineering) have enhanced the natural capabilities of the natural production of organisms.
- What a miracle is that a mouse the size of a rabbit can be produced. Bacteria like *E.coli* are producing mammalian, hormones such as insulin, somatostatin, somatotropin, etc.
- Yeast cells have been genetically manipulated to produce a vaccine against the hepatitis B virus (hepatitis disease).
- Myeloma cells (cancerous cells of bone-marrow) and B-cells of immunized mice were hybridized to produce hybrid cells that consist of the characteristics of both the cells which were cell division and antibody production.
- Now the hybrid cells (hybridomas) are being used for the production of monoclonal antibodies. In 1982, interferon (C, B and Y) were produced by genetically engineered *E. coli* cells.
- Techniques have been developed to produce rare and medicinally

valuable molecules to change hereditary traits of plants and animals, to diagnose diseases, to produce useful chemicals and, to clean up and restore the environment. In this way, biotechnology has a great impact in the fields of health, food/agriculture and environmental protection.

- Due to rapid development, the present situation is that there is no difference between pharmaceutical firms and the biotechnology industry.
- However, approved products in the pipeline and renewed public confidence made it one of the most promising areas of economic growth in the future.
- India offers a huge market for the products as well as a cheap manufacturing base for export (Padh, 1996). Following are some of the areas where biotechnology has done the best.

Biotechnology Impact on Health Care

- The maximum benefits of biotechnology have been utilized by health care. Biotechnology-derived proteins and polypeptides form the new class of potential drugs.

For example, insulin was primarily extracted from slaughter animals.

- Since 1982, human insulin (Humulin) has been produced by microorganisms in a fermenter. Similarly, hepatitis B vaccine. Recombivax HB" (from Merk), Guni (from Shantha Biotechnics Ld, Hyderabad), Shanvac (Biological E. Laboratory), etc, are the genetically engineered vaccines produced biotechnologically. Since 1987, the number of biotechnology-derived new protein drugs has surpassed the number of new chemical drugs (Padh, 1996).

Some of the Important Products Produced through Genetically Modified Organisms.

- With the advancement of gene manipulation in organisms, science has led to a new revolution in biology which is called the gene revolution.
- Obviously, it is a third revolution in science after the industrial revolution and computer revolution. Thus the roots of today's biotechnology lie in chemistry, physics and biology.

- Similarly, about 600 biotechnology diagnostics practices. About 130 gene therapy protocols have been approved by the US authorities imports of many immunodiagnostic are worldwide available in clinical areas.
- The major landmark in human history is the human genome sequence.
- The HGP is an international research program.
- Almost the whole human genome has been sequenced and a chromosome map has been developed in various laboratories worldwide through coordinated efforts.
- Human chromosome mapping was completed by March 2003.
- There are about 3,000 functional genes in humans. More than 97% of genes are non-functional.
- They do not encode for any polypeptide chain.
- Objectives of the human genome project are to: construct the detailed genetic and physical map of the human genome, (i) determine the complete nucleotide sequence of human DNA, (ii) tore information in databases, (iii) locate the estimated 50,000-10,000 genes within the human genome. (iv) Address the ethical, legal and social issues (LELSI) that may arise from the project, and (v) perform similar analysis on the genomes of several other organisms.

Biotechnology Impact on the Environment

- The natural biodegradability of pollutants present in the environment has increased with the use of biotechnology.
- Bioremediation technologies have been found successful to combat pollution problems. Bioremediation is the use of microorganisms to detoxify pollutants, present in the environment usually as soil or water sediments.
- These pollutants cause several health problems. Microorganisms that show potential to degradation of oil, pesticides and fertilizers belong to the genera of bacteria ***Pseudomonas, Micrococcus, Bacillus, and fungi Candida, Cladosporium, Torulopsis, Trichoderma***, etc.
- Computer-based study and designing of the genome is called genomics.

- Genomics deals with the sequencing of the complete genome of a particular organism. Similarly, the study of proteins presents on the genome using a computer is called proteomics.
- Proteomics can be defined as the study of all the proteins present in the genome of an organism.
- With the help of proteomics and genomics, the new molecules that can interact with the other partners could be identified.
- This gives us deep insight into the biological pathway. Now the whole genome is available to biologists for scrutiny.
- There may be new small molecules as potential drug candidates.
- Therefore, the interaction of some molecules can be studied in greater detail.
- The 33,000 genes of human beings are on a microchip.
- It has helped to design specific drugs for genetic diseases, for which there is no cure so far, for example, a specific gene expresses in breast cancer patients.
- A designed drug (Herceptia) is good for the treatment of breast cancer.
- Thus, the field of genomics has helped the growth of pharmacologic, toxicologic and protein studies on animals, therefore, and the new areas are called pharmacogenomics, toxicogenomics and proteogenomics, respectively.
- **Bioinformatics:** It is a new field of biotechnology linked with information technology. Bioinformatics may be defined as the application of information sciences mathematics, statistics, and computer sciences) to increase the understanding of biology, biochemistry and biological data.
- The most remarkable success of bioinformatics to date has been its use in the 'shotgun sequencing' (breaking of a large piece of DNA into smaller fragments) of the human genome.

Biotechnology impact on Agriculture

- Biotechnology is making new ground in the food/agriculture area.
- Current public debate about BSTC bovine somatotropin (a hormone administered to cows to increase milk production) example of biotechnology product testing public acceptance.

- Similarly, the FlavrSavr tomato (produced by transgenic plants engineered antisense technology to preserve flavor. texture and quality) is a new breed of value- added foods.
- Food biotechnology offers a valuable and alternative to food problems and a solution to nutritionally influenced diseases such as diabetes, hypertension, cancer, heart diseases, arthritis, etc.
- Transgenic golden rice has been produced by introducing three genes for the production of **vitamin A in Taipei rice**.
- Several insect resistant transgenic Bt plants have been produced insecticidal toxin -producing Bt gene of ***Bacillus thuringiensis*** into the desired plants.
- A transgenic cotton named•Ingaurd' was released in Australia which contained the Bt gene and provided resistance against insects.
- We can recall the Bt cotton prepared for Andhra Pradesh but sown in
- Gujarat in 2001 raised debate throughout the country.
- Biopesticides are coming to the market and their sales are increasing.
- Molecular Pharming is a new concept where therapeutic drugs are produced in farm animals, for example, therapeutic proteins secreted in goat milk.
- There are about a dozen companies that produce lactoferrin, tPA, hemoglobin, melanin and interleukins in cows, goats and pigs (Padh, 1996).
- However, it is not surprising vegetables produce vaccines, insulin.
- Interferon and growth hormones would be available in the market in the 21st century, besides, human clones and several other miracles.

Biotechnology in India

- In most developing countries, recombinant DNA technology has become the major thrusts.
- In 1982, the Government of India set up an official agency.
- The National Biotechnology Board (NBTB) started functioning under the Department of Science and Technology (DST).

- In 1986, NBTB was replaced with a full--ledged department, the Department of Biotechnology (DBT), under the Ministry of Science and Technology for planning, promotion and coordination of various biotechnological programs.
- The DBT is making effort in promoting post-graduate education and research. Special M.Sc courses in Biotechnology in DBT.
- The selection of students is based on a National Test, In addition, it also provides trained manpower for the rapidly growing biotech industry.
- It has also raised the level of biology education n certain areas of biotechnology in the country, Moreover, a considerable amount of basic biochemical and molecular biology is imported in these courses.
- Today, India has the DBT, DST, CSIR, ICAR, ICMR and IARI, and others working under the government.
- These agencies and the other National and International Industries are manufacturing Biotech products and marketing them after clinical trials.
- A Technology Development Board (IDB) has been set up by the government.
- The TDB works with universities, industries and the National Institutes.
- The Technology Information Forecasting and Assessment Council (TIFAC) have prepared a vision 2020 document which consists of biotechnology also.
- There are over 30 companies in India that are producing modern biotech products such a Wipro, Reliance, Life Science, Pennetia Biotech Ltd.
- Since the 1980s, India has supported the biotechnology industry and its products.
- Teaching and research of biotechnology have been included in University's syllabi both at Under Graduate and Post- graduate levels DBT- supported departments are running in several Universities and institutions.
- It is expected that India will play a key role in the future as one of the largest producer of biotech products.

- The United Nations Industrial Development Organization (UNIDO) recognized the potential of genetic engineering and biotechnology for promoting the economic progress of developing countries.
- The initiation taken by UNIDO has led to the foundation of ICGEB.
- In 1981, in a meeting convened by UNIDO it was proposed to establish an international center of excellence to foster biotechnology in the developing world.
- In 1982, this concept was approved by a high- level conference of developed and developing nations in Belgrade.
- The statutes of the center were signed by 26 countries with the entry into force of statutes on February 3rd, 1994.
- The ICGEB has become a fully autonomous international organization composed of at present 33 member states.
- The ICGEB has two centers, one located in Trieste (Italy), and the other in Jawaharlal Nehru University, New Delhi (India).
- The Trieste component is currently occupying about 5,700 m^2 area, whereas the New Delhi component is occupying about 10,000 m^2 area.
- This center is functioning in a proper way since 1982.
- Many public and private institutions working under the Government departments and organizations have advised the DBT to formulate the biotechnology programs under the following areas: (i) Plant molecular biology and agricultural biotechnology, (ii) Biochemical engineering, process optimization and bioconversion, (iii) Aquaculture and marine biotechnology, (iv) Fuel fodder, biomass and green cover, (v) Medical biotechnology, (vi) Microbial and industrial biotechnology, (vii) Large scale use of biotechnology, (viii) Integrated systems in biotechnology, (ix) Veterinary biotechnology and (x) Infrastructural facilities.
- The pharmaceutical industry in India is very strong and vibrant with expertise in chemical drugs. It has little experience in biotech diagnostics and no experience in biotech therapeutics.
- Moreover, the pharma industry is located between Mumbai and Ahmedabad (90% of drug production in India is in Gujarat and Maharashtra).

- There is no government institution or university with expertise in this area to help the pharma industry.
- However, for a variety of reasons, the Indian pharmaceutical industry will sooner or later enter in manufacturing of biotechnology- based diagnostics and therapeutics (Padh, 1996).

Biotechnology in the World

- Biotechnology products are increasing in the world market day-by-day.
- The high value -added biotechnology product to be used Countries which have boosted U.K, Japan, France, Australia, Russia, Poland, Germany, as well as India (among the Developing Countries).
- The most effective international networks developed for applied microbiology/biotechnology are the Regional Microbiology Network for South-East Microbiological Resource Centers medical field are now in domination for the last few years.
- The biotechnology R & D during the past two decades are of promotes enterprises through policy development and support international cooperation is through networks (MICREN).
- To faster biotechnology inventions, the U.S.A. promotes enterprises. Funding for basic scientific research at the National Institute of Health (NIH) has been supported.
- The US administration has boosted up the process for improving new medicines so that these may be quickly and safely small business development has also been encouraged through the incentives.
- The U.S. administration has geared up to intellectual property rights protection.
- The international markets have been opened for biotechnology research and biotech products. One of the World's best examples of partnership has been established by developing public databases.
- It enables scientists to tie up with an enterprise.
- These have helped in developing a partnership among University research-Government and private industries.
- Science education has been improved. Guidelines have been prepared so that science- based regulatory programs (i) can

promote public biosafety, (ii) earn public confidence, and (iii) can guarantee a fair and open international market.

- For the promotion of technology products and bio-business, the other countries (such as U K. Japan, Germany, France, Australia, etc) have also prepared similar guidelines.

Future Development Achievements of Biotechnology

- A few developing countries like India have scientists' technologists related to biotechnology where national strategies of development in biotechnology could be implemented.
- The scientific and technical manpower has to be properly shifted towards new biotechnology with the aim to produce expertise in biotechnology.
- In a keynote address, Bachhawat and Banerjee (1985) have described the impact of biotechnology on third-world countries.
- They emphasized "Indian bioscientists must be trained to utilize their knowledge and expertise for application and orientation, for example, a microbiologist must he trained in microbial genetics to be really useful in fermentation technology trained in cell culture, protoplast fusion or DNA recombination for practical utility and similarly people from traditional disciplines in life science may be trained to reorient their knowledge towards application and process of training read readjusted according to need.
- It is, therefore, necessary to encourage the biotechnological programs' industrial and educational levels.
- In higher education, the teaching of biotechnology should be compulsory for undergraduate students to expand their understanding and knowledge of the scientific and engineering principles underlying biotechnology.
- Countries like USA France, Germany and Japan have taken this issue seriously or a botanist must be in India, most of the universities have started teaching biotechnology at under-graduate level.
- However, at postgraduate level teaching and research have been initiated only by a few universities/ institutes on all India entrance test basis.

- The government of India has selected many thrust areas of national and international relevance as described earlier.
- Undoubtedly, modern biotechnology is expected to solve many problems arising at the global level.
- In the 21st century growth and economy of a country will certainly depend on the operation of biotechnology.
- A revolution may occur in some of the areas like medical and health care, agriculture industry and environment.
- DNA fingerprinting has successfully helped forensic science in the search of criminals making the identity of individuals, solving parentage disputes, etc.
- Human diagnostics and therapeutic drugs have been discovered and commercialized. Gene therapy is hoped to solve the problems of genetic diseases; Biotechnology-based vaccines are the best the insulin.
- They have no side effects and pose no risk for the presence of a live form of viruses in vaccines. Many transgenic plants and animals have been genetically improved. Now they are capable of producing new or improved products.
- The questions may be raised on complex, ethical, spiritual and philosophical issues Hundreds of transgenic plants have been produced and many of them are being sown in the field and products are available in the market.
- Plant biotechnology will reduce the dependence of farmers on pesticides and will help to utilize the new technologies.
- In the near future papers and chemicals with less energy and less pollution may be produced through biotechnology.
- Abatement of pollution using potential microorganisms (i.e. bioremediation) has been used in many countries.
- Thus the potential microorganisms are looked for better health, better products and better environment for the future. Instead of chemical pesticides, biopesticides have been produced and commercialized by some industries in India and abroad.
- Similarly, biofertilizers (formulation of nitrogen-fixing bacteria and blue-green algae i.e. cyanobacteria and phosphate-solubilizing bacteria and fungi) have also been made available to farmers in India.

- On the other hand, biotechnology has generated new jobs for the youth and stimulated the growth of small businesses.
- It has also encouraged innovations in the industrial and agricultural sectors. In the U.S.A. alone more than 15 lakh youth have been employed in industries

Misuse of Biotechnology

- The prevention of the misuse of biotechnology is stressed in many countries because it may cause a lot of mischief when given free play in the hands of transnational companies engaged in ruthless pursuit of profits.
- One of these concerns is the rapid pace of genetic erosion.
- This will lead to a situation where the base genetic material is available to only a few multinational companies in their gene banks.
- However, this genetic erosion has to be checked by saving the genetic diversity in its own environment, but not gene bank/ germplasm bank, by the involvement of people proceeding from 'green revolution' to 'gene revolution.
- Many new and transgenic plants and animals are being produced, and seeds, embryos, sperms are preserved.
- A day may come when farmers would have to depend on genetically engineered seeds.
- Still, there is doubt whether such seeds will suit sustainable agricultural practices free from chemical poison.
- Moreover, genetically modified organisms (GMOs) should be carefully researched.
- In addition, concerted monitoring to ensure that the hazards to users and the environment will not action should be undertaken to ensure that necessary consideration is given to the ethical and social effects of such studies.
- People should be known about the impacts of GMOs and genetically engineered products.
- Such efforts are already being made in some of the countries.
- The German Green party has called for a 5-year moratorium on the commercial release of GMOs.

- The UK genetics forum is complaining about a particularly mentally irresponsible application of biotechnology.
- In India, there is a gene campaign against the patenting of life forms and the misuse of biotechnology.

Subjective Questions

1. What is biotechnology and discuss in brief the different areas of interest in biotechnology?
2. What is the achievement of biotechnology?
3. What is the importance of biotechnology in different fields?
4. Write an essay on the global impact of biotechnology?
5. Give a brief note on the potential of modern biotechnology?
6. Write short notes on the following:
a. ICGEB
b. OECD
c. NBTB
d. DBT
e. Lal Bahadur Shashtri center for Biotechnology
f. Biotechnology in India
g. Biodiversity
h. Traditional Biotechnology
i. Modern Biotechnology
j. Ban on Genetic food
7. Give a brief account of the emergence of modern biotechnology?
8. Why can microbiology, genetics and biochemistry, without combining with technology, not be called biotechnology?
9. Write an essay on the history of biotechnology?
10. How did traditional biotechnology help in the evolution of biotechnology?

2. TOOLS OF GENETIC ENGINEERING

Chimeric DNA

- In this chapter, we will describe different types of restriction enzymes used regularly as a routine in all DNA cloning and recombinant DNA enzymes.

- The term gene cloning can be defined as the isolation and amplification of an individual gene sequence by insertion of that sequence into a bacterium where it can be replicated.
- Cloning a fragment of DNA allows indefinite amounts to be produced from even a single original molecule.
- Cloning technology involves the construction of novel DNA molecules by joining DNA sequences from different sources. The product is often described as recombinant DNA.

The Basic Events in Gene Cloning have been described in the following Steps

- Isolation gene of interest.
- A fragment of DNA (gene of interest) to be cloned incorporated into a small replicating (usually circular) DNA molecule called a vector. This can be *E.coli* plasmid, a virus, a cosmid , etc. The vector with an incorporated gene is called a recombinant vector or molecule.
- The recombinant vector is introduced into a host cell by transformation.
- Cells that have acquired recombinant DNA molecules are selected.
- Recombinant DNA molecule is multiplied within the host to produce a number of identical copies of the cloned gene.

Restriction Enzymes for Cloning

- For cloning of DNA, often we need to cut DNA at specific sites, recognized and cleaved by specific enzymes.
- Also by locating the positions of the cleavage site of a number of restriction enzymes, restriction maps can be prepared.
- These restriction enzymes recognize short sequences of double-stranded DNA as targets for cleavage.
- Different enzymes recognize different but specific sequences, each ranging in length from 4 to 8 base pairs.
- The enzymes are named by three letters that identify their origin. Roman numerals are added to distinguish several enzymes with the same origin.

- For instance, EcoRI is derived from *E.coli* and *HpaI* is derived from *Haemophilus parainfluenzae.*
- Besides cleavage, modification in the form of methylation is also brought about by some enzymes called modification enzymes.
- This methylation distinguishes genes in different states of functioning.
- There are also some enzymes, which perform the functions of both restriction and modification.
- Based on the attributes, restriction enzymes have been grouped into two classes: (i) Type II restriction enzymes systems which have separate enzymes for modification and restriction; and (ii) type I and type III enzymes systems, in which the same enzyme possesses both activities, although the restriction and modification sites differ in position.
- The above two classes of restriction enzymes, type II enzymes are most important for cloning purposes.
- The target sites for these enzymes are 4-8 bp with symmetry.
- Most type II restriction enzymes cleave DNA at the unmethylated target site.
- Some enzymes introduce staggered cuts; others generate blunt ends.
- Enzymes with 4 bp target sites are used when frequent cuts are desired and those with 8 bp are used when rare cuts are desired to get long DNA segments.
- Otherwise, the majority of enzymes used, have 6 bp target sites.
- Some of them can cleave both methylated as well as unmethylated targets, but the majority of them cleave only unmethylated targets.

The Technique of Restriction Mapping

- The recombination frequencies, which are the function of distances between genes, are not completely independent of (i) the nature of mutants used, (ii) the position of these mutants on chromosomes, (iii) the genetic background, (iv) the environmental conditions and a variety of other factors.
- Consequently, the distances between genes on a genetic map may not correspond to the distances between them on the DNA

molecule of which they are a part at the molecular level.

- Gaps may also be present on a genetic map due to the non-availability of mutants in that region. At the molecular level, the fine structure of a gene can be studied through the determination of the nucleotide sequence of the concerned DNA segment.
- Sometimes this is done through isolation of DNA segment corresponding to a gene followed by sequencing of the DNA segment, which needs time and energy.
- Instead, we can prepare a map of the DNA by its cleavage at specific sites with the help of restriction endonucleases, which recognize very short specific DNA sequences and cleave the DNA at these specific sites.
- These sites of cleavage can be identified and mapped to give rise to a restriction map. In a restriction map, the nucleotide sequence of DNA between sites close to one another (300 bp or less) can be determined.
- These regions can then be connected into a sequence of the entire gene, whose nucleotide sequence, then can be compared with the amino acid sequence of the protein coded by it.
- This can be further extended to adjoining genes to saturate the physical map, an objective which is difficult but is being attempted for the human genome.
- On a restriction map, is found a linear sequence of sites, each for a specific enzyme and the distances between them are measured in terms of the number of base pairs of DNA.
- This technique can be used both in prokaryotes and eukaryotes, although, all regions of chromosomes can not be easily mapped in eukaryotes.
- Different steps involved in restriction mapping include the following:

Restriction Cleavage and Gel Electrophoresis

- If we take a particular DNA molecule or DNA sample, digest it with a specific restriction enzyme and then subject to it gel electrophoresis, we will notice a series of bands on the gel slab or cylinder.

- The bands can be observed under ultraviolet light after staining with ethidium bromide. The position of different bands will depend on DNA fragment size, such that the smaller the fragment, the more rapidly it will move, and the longer the segment, the more slowly it will move.
- It will mean that the fragment farthest away from the loading site will be smaller and that close to the loading site will be longer DNA fragments.
- The gel can be calibrated by using a mixture of DNA fragments of known length so that the position of the bands on this standard gel can be compared with the bands in the experimental DNA digest and the fragment length in each band of DNA digest can thus be determined sometimes using a computer device.
- The results of digestion of 5000 bp long hypothetical DNA molecule digested separately by two enzymes A and B.
- A cleaves the DNA into four fragments of lengths 2100, 1400, 1000 and 500 bp, while the enzyme B cleaves it into three fragments of length 2500, 1300, 1200 bp.
- These data with some additional experiments can be used to generate a restriction map.

Construction of a Restriction Map

- DNA digestion by more than one endonucleases as we discussed in the above lines can be utilized to arrange the sites of cleavage.
- This can be done by several methods.
- This method involves successive digests with two individual enzymes, where we digest with either enzyme A or enzyme B and then cleave it with the other enzyme.
- The original DNA sample is also digested by a mixture of both the enzymes to confirm the results of individual successive digests.
- When fragment A-2100 is obtained and digested with enzyme B, it is cut into two fragments of 1900 bp and 200 bp.
- Fragment B-2500 (obtained from individual digest with enzyme B) is similarly cut by enzyme A into two fragments of 1900 bp and 600 bp, suggesting an overlap of A-2100 and B-2500 in the region of the fragment of 1900 bp which is obtained by one cut with enzyme A and the other by enzyme B.

- The DNA fragment of 1900 bp is also available when DNA is digested with a mixture of both enzymes A and B (double digest).
- Using this information, we identify the overlapping regions in A and B digests and find out the sites of cleavage by A and B.
- This will then allow us to prepare the restriction map.

Use partial Digests, End-labeling Hybridization in Restriction Mapping

- The above technique of individual and double digests can be supplemented with other techniques for the actual construction of maps.
- For instance, by permitting incomplete digestion, fragments longer than those obtained by complete digestion may be obtained.
- These will be called partial digests. In the example used in the last section, in partial digests with enzyme A, instead of getting A-1000, A-2100, A-1400, A-500, we may get A-3100 A-1400 and A-500, which will suggest that A-1000 and A-2100 lie in the adjacent regions.
- In another partial digest if we get A-1000, and A-500, this will suggest that A-2100 and A-1400 are adjoining, which will mean that A-1000 and A-1400 are found on the two sides of A-2100s. This technique will thus allow the arrangement of fragments in a linear order.
- Another useful technique involves labeling of the ends (end labeling) of DNA ends that can be identified due to labeling, even after digestion.
- In the example, using end labeling, if A-1000 and A-500 are found to be radioactively labeled, these will be present at the two ends of the restriction map.
- Some of the features of a restriction map can also be confirmed by nucleic acid hybridization.
- A molecule before digestion, so that the fragments containing these 2100 and B-2500 have an overlapping region of 1900bp, they should hybridize with each other, which will confirm overlapping.
- The above technique will help in the accurate completion of a restriction map but will require that we have a complete set of

restriction fragments that make the entire DNA region being mapped.

- Once the restriction map (which is a physical map) is ready, it can be compared with the genetic map.
- Although large changes, already located on the genetic map can be easily located on the restriction map, point mutations cannot be always and easily located on the restriction map since the restriction sites often do not change due to mutation.
- In such a situation, one may like to determine the nucleotide sequences on individual fragments and compare them in normal and mutant individuals.
- Since sequencing of nucleotides is more laborious and cannot be easily undertaken for the whole genome, another technique of molecular genetic markers in the form of restriction fragment length polymorphisms (RFLPs) has been utilized during the 1980s to generate genetic linkage maps in human, mouse, rat, pig, sheep, fruitfly, (*Drosophila*), nematode (*Caenorhabditis elegans),* yeast, maize, wheat, rice, barley, tomato, potato, lettuce, rice, pepper, pea, etc.

Construction of Chimeric DNA

- DNA molecules are produced by inserting a foreign DNA segment into the DNA molecule of a vector.
- These chimeric DNA molecules are the means of cloning genes or other small DNA sequences. The construction of chimeric DNA, as we know it now, involves the joining of the broken ends of the vector with the two ends of the sequence to be cloned.
- There are at least three methods available for the construction of this chimeric DNA by palindromes and staggered cleavage.
- This is the most common palindromic sequence, a restriction enzyme causes staggered cuts producing short complementary single-stranded sticky ends (a palindromic sequence is one, which has complementary sequences at the two ends of a single strand, e.g. AT.GAT). This can be illustrated using the example of the enzyme EcoRI, which cuts the sequences at specific positions.
- When another DNA molecule is similarly cut by the same order enzyme, similar sticky ends having the same sequences in the

single-stranded ends will be produced, so that when it is mixed with the previous molecule similarly treated, the two will anneal producing a chimeric DNA or chi plasmid if plasmid DNA is involved.

- Enzyme DNA ligase helps in joining the bonds at the cut ends of two molecules. The above technique has the advantage of regenerating two restriction sites (eg; *EcoRI* site) in the chimeric DNA so that the foreign DNA segment can be retrieved rather easily from the cloned copies of chimeric DNA by cleavage again with the same enzyme.

There are also the following disadvantages with this technique.

(i) The two cleaved ends of a vector or of a foreign DNA may join end to end before getting inserted. Therefore, while isolating chimeric DNA; one will have to select chimeric molecules each having only a single insert. This can be achieved by separating molecules of different sizes by gel electrophoresis.

(ii) The recognition site, particularly in the sequence to be cloned may not lie at a convenient position so that sometimes only a part of the desired segment will be inserted.

Adding poly dA at the 3' ends of the vector and poly dT at the 3' ends of the DNA clone

- In this method, DNA can be cut at the desired position both in the vector and in the clone, without a staggered cleavage.
- Using the precursor dATP, poly dA is added at both the 3' cut ends of the vector with the help of enzyme **terminal transferase**.
- Similarly, using precursor dTTP, poly dT is added at both the 3' cut ends of the DNA sequence to be cloned.
- The vector and clone can then be joined by annealing the poly dA with poly dT tails and then ligating them using a DNA ligase enzyme.
- In this technique, there is no chance of reannealing between the two cut ends of the same DNA molecule or of the two similar DNA molecules so that one of the disadvantages of the first method has been overcome in this case.

- However, in this technique, it will not be easy to retrieve the cloned DNA, because the recognition site of the enzyme due to insertion of poly dA and poly dT has been lost in this case. But if poly dG and poly dC are used instead (which serve the same purpose), the cloned sequence can be retrieved, since poly dG: poly dC regenerate the recognition for the enzyme *Pstl.*

Blunt End Ligation by T4 DNA Ligase

- In this technique, a restriction enzyme is used to cut a duplex DNA at the same place in both the DNA strands.
- The broken ends are then used to join with the two ends of another DNA molecule irrespective of the sequences present at the broken ends of the two DNA molecules.
- The T4 DNA ligase is used for this joining reaction.
- The disadvantage of this technique is that any two broken ends may join including those belonging to the same DNA molecule.
- This leads to the production of a variety of products and one will have to select the desired product from a mixture of products.
- This blunt end ligation is used for developing a method in which the DNA to be cloned can be easily retrieved whenever required.
- This method makes use of short DNA duplexes (linkers), that contain the EcoRI palindrome or some equivalent palindrome, which is small in size, can be synthesized chemically.
- These linkers can be linked to the blunt ends of vector DNA of an insert by blunt end ligation. This will allow the creation of an *EcoRI* site in the linker region of the vector so that when a DNA segment is now cloned, the inserted DNA segment can be retrieved by with *EcoRI.*
- However, in this method, there is no restriction on the original choice of sites to generate the cut ends.
- Therefore, with this method, it is now possible to insert a foreign DNA segment at a particular site in the linker region of the vector and then retrieve this foreign segment whenever necessary.

DNA Cloning in Bacteria and Eukaryotes

Cloning in Bacteria

- Bacteria are often used as hosts for the multiplication of chimeric vectors or recombinant DNA molecules.
- Therefore, a bacterial strain should be available, which should be the host for the vector and should be modified or selected for high transformation efficiency. Bacteria can be classified into.

(i) gram -negative bacteria e.g. *Escherichia coli. Pseudomonas, Rhizobium* and

(ii) gram- positive bacteria e.g. *Bacillus subtilis* and *Streptomycetes* sp. Ordinarily, the same plasmid can not maintain itself in a stable form in both classes of bacteria and, therefore, plasmid vectors for two classes usually differ. While gram- negative bacteria can be used for plasmids, phages and cosmid vectors, gram -positive bacteria are used as hosts for only plasmids. Chimeric vector is inserted bacterial cells and transformed bacterial colonies are selected and for multiplication in suspension cultures. The chimeric DNA is retrieved whenever required.

Cloning in eukaryotes

- Since chromosomes found in the nucleus of eukaryotes are separated from the rest of the cell through the nuclear membrane, and since many of the genes are split genes with exons and introns, genetic engineering with eukaryotes requires new methods and tools.
- When eukaryotic genes are cloned prokaryotes, the split genes cannot be correctly expressed, because prokaryotes do not have the equipment necessary for splicing out the RNA transcribed from the introns of a gene.
- In view of this, eukaryotic cells may sometimes be needed for cloning and particularly for the expression of cloned eukaryotic genes.
- Among eukaryotes, DNA cloning has been done in yeast, mouse and to some extent even in some higher plant species.
- In yeast, a plasmid called 2u DNA (63 bp) is found, which is an appropriate cloning vehicle.

- An efficient transformation method is also available, which involves protoplast production followed by PEG-directed introduction of DNA into protoplast.
- In animal cells like mouse cells, special animal viruses were used as cloning vehicles. Simian virus 40 (SV40) is one such virus, in which the globin gene could be integrated.
- This gene integrated into SV40 could be transcribed and translated in mouse kidney cells.
- A variety of human cell lines in culture is also used for cloning foreign DNA, using some animal viruses as vectors.

Molecular Probes

- Molecular probes are small DNA or RNA segments that complementary sequences in DNA or RNA molecules and thus allow identification and isolation of these specific DNA sequences from an organism.
- Antibodies are also occasionally used as probes to recognize specific protein sequences. Although, initially, these probes were developed and used for genetic engineering research but are now frequently used for a variety of purposes including diagnosis of infectious diseases, identification of food contaminants, variety of microbiological tests, forensics (eg. fingerprinting of murderers or rapists), etc.
- Probes can also be used to identify strains of an organism eg. varieties of a crop species (a plant breeder likes to have a quick test to identify his variety to maintain his patent or breeder's right, so that a competitor may not use it in another name leading to infringement of rights).
- For basic studies in molecular biology laboratories, molecular probes are frequently used for the identification and isolation of genes or related sequences.
- In theory, any nucleic acid (or rarely protein, e.g. antibody) can be used as a probe, provided it can be labeled to permit identification and quantitation of the hybrid molecules formed between the probe and the sequence to be identified.
- In practice, double and single- stranded DNAs, mRNA and other RNAs synthesized *in vitro* are all used as probes. DNA/RNA

probe assays arc faster and sensitive, so that many conventional diagnostic tests for viruses and bacteria involving culturing of the organism, are being fast replaced by antibody and DNA probe assays.

- While culture tests can take days or even months, molecular probe assays can be performed within a few hours or minutes.
- Therefore, the use of DNA probes has become today's most sophisticated and sensitive technology for a variety of uses involving biological systems both in basic and applied studies including their commercial use.

Preparation of probes

Genomic DNA probes

- Simple, reproducible methods have been devised for the preparation of molecular probes. A random or specific DNA sequence (probe) can be isolated from the genomic DNA (DNA extracted from cell nuclei), using the following steps extract DNA from plant or animal tissue.

(i) Digest extracted DNA with a restriction enzyme (e.g. EcoRI or HindIII), which cuts DNA on specific sites or positions where a specific sequence recognized by the enzyme is found;

(ii) Run the digested DNA on an agarose or polyacrylamide gel for electrophoresis to separate fragments of different sizes;

(iii) Isolates DNA of specific fragment size from a particular band identified through Southern blots by hybridization with specifically labeled mRNA or cDNA molecules, if available. If random probes are needed, no hybridization of Southern blots is required (Southern blots are DNA bands transferred from gel to a nitrocellulose membrane with the help of a buffer rising through capillary action; to isolate specific DNA).

(iv) Clone this DNA in a vector to allow the chimeric vector to infect bacteria for multiplication, where it can make billions of copies.

The steps (ii) to (vi) above are also used for preparing cDNA probes. From the transformed bacteria, the chimeric vector can be obtained and used in one of the following ways:

(i) It may be directly used as a probe (the presence of vector DNA will not interfere so that it is not really necessary to remove it).

(ii) The segment may be separated (or retrieved), by using the same enzyme which was used for cloning. In the latter case, cleaved chimeric vector DNA will be again electrophoresed for separating the inserted segment on the gel. This inserted segment thus retrieved can now be used as a probe.

(iii) The chimeric DNA may be used for PCR, using flanking sequences as primers; the PCR probe) the product can be separated and used as a probe

cDNA probes

- A DNA sequence corresponding to a part of a gene(s) (an oligonucleotide) can be obtained by reverse transcription of mRNA. cDNA thus obtained can be cloned and used as a probe.

Synthetic Oligonucleotides as Probes

- DNA probes with nucleotide sequences can also be synthesized chemically using automated DNA synthesizers now available commercially.
- These synthetic probes will be efficient only when they are not more than 20 to 40 nucleotides in length.

RNA Probes OR Riboprobes

- High specific activity RNA probes or expression vectors.
- This is achieved through RNA synthesized *in vitro* (on DNA inserted in vector) and labeled simultaneously employing specific RNA polymerase and labeled nucleotides which are supplied by commercial firms in a kit for riboprobe preparation.
- RNA probes offer several advantages over DNA probes since these are single- stranded, and provide an improved signal or hybridization blots.
- However, the widespread presence of ribonucleases creates some problems in their preparation and riboprobes may also be synthesized from DNA templates cloned because these ribonucleases will cause degradation of RNA which is being synthesized.

- Therefore, extreme care is needed in the preparation of a riboprobe, keeping all glassware ribonuclease-free (no glassware should be touched with bare hands, because ribonucleases may be available on or skin).

Labeling of Probes

- The detection of homologous sequences after hybridization with the probe is like finding a needle in the haystack.
- Therefore, for the success of DNA probe assay, it is necessary to develop simple, safe and sensitive techniques for their use.
- As probes transmit no signal of their own, they have to (b) either labeled with radioactive isotopes or (i) methods are used to couple nonradioactive signal molecules to the probes.
- Without impairing the hybridization ability of these probes.
- These signal molecules may include fluorescent antibodies and enzymes that produce color changes and chemiluminescent catalysts.

Radiolabelled Probes

- Traditionally radioactively labeled probes are used for a variety of experiments. p is the most commonly used radioisotope, although other labels such as H, ^{35}S and $^{125}1$ have also been used.
- Conventional labeling replaces a proportion of the nucleotides in a nucleic acid sequence, with ^{32}P derivatives (e.g. ^{32}PUCTP) or adds ^{32}P as the end of the molecule.
- At least three methods are available for labeling of the probe (i) **nick translation** involves the creation of nicks in the probe DNA followed by extension of the broken ends using a labeled deoxyribonucleotide with help of DNA polymerase I (or Klenow fragment of this enzyme) (ii) **oligonucleotide labeling** involves the use of short random oligonucleotides, which are used as primers for copying the probe in presence of labeled deoxyribonucleotides ; (ii**) riboprobe preparation** this involves the synthesis of labeled RNA, using DNA probe as a template, in presence of a labeled ribonucleotide.

- After hybridization with the labeled probe, hybrids are detected by autoradiography. P^{32} has the advantage over other radioisotopes since it has high specific activity.
- However, in general, radioisotopes have some disadvantages:
- They are difficult to handle and expensive to dispose of.
- Detection by autoradiography, while sensitive, may take a long time if there are few counts in the hybrid.
- Furthermore, radioisotopes have a short half- life (half-life of 14.3 days) and therefore experiments should be completed preferably within one half-life

Non-radioactive Probes

(a) Biotin labeled probes

- Recent advances in nucleic acid technology now offer alternatives to radioactively labeled probes.
- One procedure that is becoming increasingly popular is the biotin labeling of nucleic acids. This system exploits the affinity, which the glycoprotein '**avidin**' has for biotin (vitamin H).
- Avidin is commonly found in egg white. Biotinylated probes are prepared through nick translation derivatives.
- Alter hybridization and washing, detection of hybrids is done by a series of cytochemical reactions which finally give a blue color whose intensity is proportional to the amount of biotin in the hybrid.
- There are several advantages of using biotinylated probes. For example, these assays employ in advance in bulk and stored at –20 °C for repeated uses.
- Detection technology is that very small probes non-toxic materials; whose half-life is longer. These probes can be prepared by hybrids is much faster than by radioactive probes.
- A limitation of this technology is that very small probes contain only a small number of biotinylated sites, limiting the intensity of the signal obtained.

- It has been solved by adding long tails of biotinylated nucleotides to the probes through enzymatic methods.
- Sometimes the probe does not need to be labeled with biotin, but only coupled with a tail.
- Another disadvantage of biotin-labeled probes is that the cytochemical visualization reactions lead to precipitation of insoluble material which cannot be removed and therefore, the filter cannot be reused, while in radiolabelled probes, the filters can be used repeatedly for hybridization with a number of probe one at a time.

(b) Digaxigeain labelled probes

- Digoxigenin is another chemical derived from plants and used for non-radioactive labeling of probes.
- An antibody associated with an enzyme (anti-digoxigenin-alkaline phosphatase conjugate) is used for the detection of the presence of digoxigenin.
- The probe may be labeled with digoxigenin 11 dUTP supplied with a digoxigenin kit (these kits are available from any commercial firm, e.g. Boehringer Mannheim).
- The labeled and denatured probe may be used for hybridization with denatured DSA on Southern blots or for in-situ hybridization.
- After hybridization, the membrane or the slide may be washed and the membrane or the slide is transferred into a detection buffer containing 20 mg/ml of anti-digoxigenin fluorescein and 5 % (w/v) BSA (bovine serum albumin).
- The system is incubated for 1h at 37°C, and then the membrane/ slide is washed in detection buffer three times (8 min each at 37° C).
- Alkaline phosphatase activity can be detected using 0-17 mgml BCIP (5-bromo-4 loro-3 indolyl phosphate) and 0.33 mg/ml NBL (nitroblue tetrazolium) as a dye substrate.

(c) Alternatives to biotin and digoxigenin labeling

- The techniques of non-radioisotopic labeling have been further expanded and new methods have been devised for attaching other ligands (e.g hapten determinant 2, 4-dinitrophenol, arsenate derivatives, etc.) to nucleotides without hampering their ability to be incorporated into DNA.

- These alternatives require binding of attached ligands to specific proteins that can be to monitor many probes simultaneously by using several different ligands since each ligand would yield a different signal.
- A chemiluminescent (light-emitting) probe system has also been developed in which two different probes complementary to a continuous segment of DNA hybridize adjacent segments of a gene.
- The first label is a chemiluminescent complex that emits light at a specific wavelength; this emission excites the label Molecule on the second probe to emit light at a different wavelength which can be detected using a photomultiplier device.
- This process called non-radioactive energy transfer can occur only if the two probes hybridize correctly and the two labels are close to each other.
- This system has great fidelity and provides basic technology for a homogeneous assay. In addition, DNA does not need to be immobilized and no washing steps are necessary which may be an additional advantage for large- scale testing.

Amplification of DNA Probe Signals

- In recent years, DNA probe assays and immunoassays (using antibodies) have been competing with each other for a variety of purposes.
- While DNA probes have an advantage of detecting the presence of gene rather than its product, the immunoassay has the advantage of getting amplified target (the gene product) firstly due to transcription (several mRNA molecules can be synthesized on a single copy of the gene) and secondly due to translation (several protein molecules can be produced from same mRNA molecule).
- The amplified target gives an amplified signal. Therefore, if DNA probes have to compete with monoclonal antibodies for immunoassay, DNA probe signals need to be amplified.

Following devices have been used for the Signal Amplification

- Probes have been developed for tRNA since rRNA genes are present in thousands thus amplifying the target (as done by companies like Gene-Probe and Gene-Trak systems in the USA).
- Polymerase chain reaction (PCR) technology (which utilizes a thermostable polymerase) is used to generate millions of copies of target DNA sequences, which can then be identified by standard methods.
- The signal to generate a capacity of the probe can also be increased by anyone of the following methods: (i) concentrating more labels at the site of the target molecule by attaching multiple enzyme molecules to each of its DNA probes or (i) by using multiple probes, or (ii) by attaching (details consult next Chapter).
- Amplification probe technology uses multiple secondary probes (each having one or more enzyme molecules) that hybridize to multiple target specific primary probes. The secondary probe is independent of the target but helps in amplifying the signal. These approaches have been multiple enzymes to each of the multiple probes. Described as Christmas Tree or Christmas Forest approaches. The above amplification systems will play a significant role in the use of DNA probes in the future for a variety of purposes.

Techniques used in Molecular Probing

Separation of DNA fragments using agarose or polyacrylamide gel electrophoresis.

- When genomic DNA, extracted from any tissue of a plant or animal species, is digested with a restriction enzyme, it is cleaved into segments.
- The segments of different sizes can be separated through **gel electrophoresis** before a molecular probe is used to detect the segments which have sequences similar to those in the probe.
- Gel electrophoresis involves the movement of fragments or molecules under a high voltage electric current.
- The mixture of DNA fragments is loaded in a well created on one edge of the gel. The gel may be a cylinder or a slab (usually a slab

for cloning expt.), about 10 cm long and 0-5 cm thick.

- The rate of movement of fragments is inversely correlated with the size of fragments molecules so that heavier fragments will remain closer to the site of loading and the lighter fragments will move away.
- Fragments of different sizes will appear as bands on the gel and can be examined or isolated for further study.
- More often agarose gels are used, but for the separation of fragments differing by a few base pairs, polyacrylamide gels are used.
- Polyacrylamide gels are more commonly used for DNA sequencing experiments.

Separation of large DNA molecule molecules using (PFGE)

- The technique of gel electrophoresis is used for the separation of DNA molecules of different sizes. However, DNA molecules of large size are not handled in this technique.
- In recent years, a new technique called the separation of large DNA molecules (whole chromosomes) using PFGE.
- The technique of gel electrophoresis is used for the separation of DNA pulsed- field gel electrophoresis (PFGE) has been used, for separating large -sized DNA molecules, sometimes representing whole chromosomes.
- Using this technique, separation of DNA molecules belonging to each of individual chromosomes of yeast has become a possible technique, short pulses of electricity are used in two different directions and DNA is embedded and used in the form of agarose plugs to avoid fragmentation of large DNA molecules.
- Using the technique: PFGE and a more refined technique **CHEFE (contour clamped homogeneous electric field electrophoresis)**, the genome of several fungi could be resolved into chromosomal bands and used for mapping of DNA and sequences on specific chromosomes.

Southern, Northern and Western blotting

- A mixture of DNA, RNA or protein fragments can be separated by gel electrophoresis and the separated bands can be stained and visualized directly in the gel.
- "However, to confirm the identity of these bands or to find similarity of one or more of these bands with a known and available molecular probe, it is possible to hybridize these bands with a labeled probe.
- To facilitate this hybridization, the bands are often transferred to a nitrocellulose technique described as blotting.
- When called Southern blotting (after the name of E.M. Southern); when RNA bands are thus transferred it is described by the jargon term Northern blotting and similarly when protein bands are transferred, the technique is described as Western blotting".

(a) Southern blotting

- For Southern blotting, the DNA sample is first digested with a restriction enzyme and the digested sample is gel electrophoresed.
- The DNA bands in the gel are denatured into single strands with the help of an alkali solution.
- "Subsequently, the gel is laid on top of a buffer saturated filter paper, placed on a solid support (e.g. glass plate), with its two edges immersed in the buffer.
- A sheet of and a stack of many A weight of about 0.5 nitrocellulose membrane is placed on top of the gel papers (paper towels) on top of this membrane is placed on top of paper towels.
- The buffer solution is drawn up by filter paper wick, and passes through the gel to the nitrocellulose membrane and finally to the paper towels.
- While passing through the gel, the buffer carries with it single-stranded DNA, which binds to the nitrocellulose when the buffer passes through it to the paper towels.
- After leaving this arrangement for a few hours or overnight, paper towels are and discarded.
- The nitrocellulose membrane with single-stranded bands blotted onto it is baked at 80°C for 2.3 hours to fix the DNA; when RNA

is used, the term Northern is used.

- This membrane now has DNA bands from agarose gel and can be used for hybridization with radioactively labeled DNA or RNA probes.
- The membrane may then be washed to remove any unbound DNA and X-ray film is exposed to the hybridized membrane to get autoradiographs".

(b) Northern blotting

- Initially, the technique of Southern blotting used for DNA transfer from the gel to the membrane could not be used for blot-transfer of RNA.
- Instead, mRNA bands from the gel were blot-transferred into a chemically reactive paper, prepared by diazotization of amino benzyloxymethyl paper.
- The technique being related to Southern blotting was called Northern blotting (not after the name of any scientist as in Southern blotting).
- Later, it was suggested that mRNA bands can be blotted firstly onto the nitrocellulose membrane, a technique that has been widely adopted.
- The mRNA bands blótted onto nitrocellulose membrane can be hybridized with a labeled DNA or RNA probe.
- The single-stranded regions of the probe are removed by nuclease (mungbean nuclease or $-1 nuclease), so that quantitative estimation of hybridized mRNA can also be made.

(c) Western blotting

- This technique is used to detect proteins of a particular specificity. Particularly when a transferred gene expresses in transformed cells, the translated product in the form of protein can be identified by this technique.
- The extracted proteins are subjected to polyacrylamide gel electrophoresis (PGE) and are then transferred onto nitrocellulose, to which they bind.

- Nitrocellulose membrane is then used for probing with a specific labeled antibody (antibody will not hybridize with the protein, but bind with it).
- The antibody may be labeled with and the signal is detected again with autoradiography. If a radioactive label is not used, a bound antibody may be detected by a second antibody tagged with an enzyme.

Dots and Slot Blots

- Another variety of Southern or Northern is dot blots or slot blots.
- In dot blots, cloned or pure extracted DNAs to be tested are spotted adjacent to each other on a nitrocellulose membrane.
- DNA blots thus produced arc immobilized and denatured so that they are bound on the membrane as single-stranded DNA blots.
- "The membrane is then hybridized with a radioactively labeled probe (DNA, or RNA). The dot representing sequence related to the probe will light up autoradiography.
- The intensity of the dot will indicate the relative concentration of sequences related to the probe in the DNA sample used for a particular dot.
- No steps involving digestion with an enzyme concentration of a sequence related to the probe in the DNA electrophoresis or transfer from the gel to the membrane are needed in this technique".
- Therefore, it is considered much more convenient, when o restriction sites are needed to be studied.
- Sometimes, an apparatus is used for placing spots on the membrane through slots made in this equipment.
- The spots made thus are in form of oblong slots rather than round blots.
- These slots are used just like dot blots and are described as slot blots.

Applications of Molecular Probes

- Molecular probes can be used for both basic and applied studies in the field of molecular biology and biotechnology. These uses will be discussed in detail in the different subsequent chapters of this book at appropriate but will be described briefly in this section.

Use of molecular probes in restriction fragment length polymorphisms RFLPs) and related analyses.

(a) What are RFLPs

- When genomic DNAs from each of several individuals belonging to one or more species are digested separately with restriction enzymes, electrophoresed, blotted a membrane, and probed with a labeled DNA clone, polymorphism the hybridization pattern is sometimes revealed and attributed to sequence differences between the individuals. Such variation has been termed as "restriction fragment length polymorphism" (RFLP). Variation m one DNA fragment obtained with a specific enzyme is treated as RFLP.

(b) Detection of RFLPs

An RFLP can be demonstrated using the following steps:

(i) Extract and purify DNA from several individuals which may differ in some respects among themselves.

(ii) Digest with a restriction endonuclease.

(iii) Subject each "DNA digest to gel electrophoresis separately but on the same gel slab.

(iv) The DNA fragments of different lengths resulting from digestion can be separated by gel electrophoresis, but the differences in the distribution patterns of fragments in several DNA samples due to digestion by several endonucleases cannot be detected directly. This is because the number of fragments is large and the range in size is rather continuous, such that they form a continuous smear on the gel. Electrophoresis is, therefore, followed by the following steps.

(v) Fragments are transferred from the gel to nitrocellulose filters using the technique of Southern blotting.

(vi) The filters with small fragments, fully or partially homologous to the probe, can be detected by autoradiography after hybridization.

- Since repetitive DNA may hybridize with many fragments, they will give a smear on autoradiography. Therefore, unique DNA sequences are generally used as probes for detecting RFLPs. The DNA fragments hybridizing with the probe can be visualized by autoradiography and will be called RFLPs" at the phenotypic level. where RFLPs are detected by using four plants (which differ by point mutation as well as by insertions) and three enzymes individually and in combination.

(c) RFLPs for evolutionary studies

- The restriction fragment length polymorphisms (RFLPs) can be studied in a set of related species, using a random or a specific DNA probe. The similarities and differences can be used to infer phylogenetic relationships. This has actually been done in a number of cases, both in plants and in animals.

d) RFLP maps and linkage of RFLPs with specific genes

- RFLPs have been used to prepare chromosome maps in humans, mice, fruitfly and in Mendelian markers for genetic mapping is sometimes limited due to the non-availability of mutants.
- The list of markers can be extended by using molecular markers, which are examined in the form of RFLPs plants including maize, tomato, lettuce and rice.
- The use of mendelian markers for genetic mapping is sometimes limited due to the non-availability of mutants.
- The list of markers can be extended by using molecular markers, which may be examined in the form of RFLPs.
- "Once a large number of RFLPs are available in a species, the parents, F, and F2's can be used to study their inheritance and linkage relationship and genetic linkage maps can be prepared.
- They have been assigned to specific chromosomes using monosomies in maize and efforts are being made to relate them with morphological and economic traits so that they can be used for practical plant breeding.
- A difference in restriction maps between two individuals or an RFLP can be used as a genetic marker in much the same way as any other phenotypic marker. The polymorphism at the molecular

level may be due to the presence or absence of a restriction site.

- It shows how the presence of an additional site will produce two fragments and its absence will produce only one fragment.
- Although, there may be a linkage between a molecular marker and a phenotypic trait, the change in the molecular map (restriction map) may not affect the phenotype.
- In order to study linkage relation between a molecular marker and present in the DNA parents that differ morphologically as well as for molecular markers by we may examine the restriction fragment patterns and the phenotypes of making crosses between two such parents, we can estimate the recombination frequencies between molecular markers and phenotypic markers using" a test cross.
- We used a hypothetical example showing the linkage of red and white eyes with different molecular markers in Drosophila.
- "It can be seen that as a result of recombination, in test cross, 15 % progeny shows the association of the red eye with a molecular marker associated the white eye in the parent and vice versa.
- Therefore, it is concluded that the restriction marker is 15 map units away from the marker eye color although it has no causal relation with eye color.
- Such a linkage between the molecular marker and phenotypic marker allows the identification of genetic loci at the molecular level.
- A tight linkage of such a molecular marker with disease resistance may also be recommended as a tool for the identification of disease- resistant cultivars in many of our crop plants and will be increasingly utilized in the future. Molecular markers are also being used for the selection of plants at the seedling stage for plant breeding purposes".

Use of Molecular Probes in Molecular Cytogenetics

(a) Isolation of gene or a related sequence

This can be facilitated through the use of a probe carrying a known sequence.

For this purpose, sequences may be hybridized following any one of the following techniques:

- Southern blots can be prepared (from genomic DNA or cDNA) to isolate DNA corresponding to a band that hybridizes with the probe.
- This DNA can then be used for cloning experiments to get a partial genomic library or cDNA library (see later in this chapter for details).
- The genomic or cDNA library can be screened through colony hybridization or plaque hybridization to identify and isolate the clone carrying the desired sequence.
- Sometimes the step (ii) is not used for complete genomic/cDNA libraries can be prepared and used in this step.
- The identified clone should carry the gene or sequence of interest, that can be multiplied and the sequence or gene retrieved according to a standard procedure.

In Situ Hybridization (ISH) for the Location of Sequences Chromosomes

- A molecular probe can also be used for in situ hybridization "where chromosomes in miotic or meiotic preparation are used for hybridization with the probe (without extracting DNA).
- In a number of cases, rye chromosomes in the wheat background have been identified using this technique.
- Sat- chromosomes with NORs (nucleolar organizing regions) are recognized using probes for ribosomal DNA as done in many cases. in wheat and can be identified using a D genome-specific probe.
- Telomeres have been identified in human eukaryotes using a telomeric sequence as a molecular probe, showing all chromosomes carrying the same sequence.
- In D genome telomere- related species, chromosomes of a genome -specific probe (pASI separated from Aegilops be cited where the use of the molecular probe in hybridization served a very useful purpose.

Use of Molecular Probes in DNA Fingerprinting

- With the help of probes prepared from human minisatellite DNA dispersed throughout the human genome, it has become possible to identify individual humans, in the possibility of any two humans (except identical twins) showing the same pattern of fingerprinting is remote". This aspect is already playing an important role in identifying criminals like murderers and rapists. This subject will also receive detailed treatment.

Construction and Screening of Genomic and cDNA Libraries

- In order to isolate one or more related genes from a genome, we prepare a mixture of clones each carrying DNA derived either from the genomic DNA or from cDNA (derived from the mRNA isolated from a specific metabolically active tissue of an organism).
- This mixture may contain thousands of clones, which when derived directly from the genomic DNA are collectively called a genomic library.
- Similarly, when these clones are derived from cDNA, they are collectively called a cDNA library construction and the use of these libraries will be discussed in this section.

Genomic Library by Shotgun Experiment

- Cloning an entire genome in the form of a library of random genomic clones (without identifying them) is often called a shotgun experiment.
- "In this experiment, genomic DNA is extracted, broken into fragments reasonable size by a restriction endonuclease and then inserted into a cloning vector to generate a population of chimeric vector molecules.
- A set of fragments cloned in this manner is called a genome library.
- Once such a library is available, then clones can be perpetuated indefinitely in a plasmid vector and retrieved whenever needed for a variety of purposes, including identification and isolation of a gene, when a specific probe is available.
- Genomic libraries can be prepared by using a number of restriction endonucleases, one at a time, so that fragments of varying sizes having different places of the genome will be available.

- However, this may lead to cuts at inconvenient places, including sites within a gene, so that is having complete genes will be difficult to obtain.
- In order to overcome this difficulty, we use the following strategy in the shotgun experiment:

(i) We use restriction endonucleases, which have short (4 bp) recognition sequences, so that such a sequence may be frequently distributed.

(ii) Conditions are used which give only partial digests, so that a particular restriction site is only occasionally cleaved, and long fragments without having any breaks on recognition sites available within a gene can be easily obtained.

- This technique of shotgun experiment leads to the construction of a random genomic library, in which all fragments have the same fragment ends thus helping retrieval of a fragment from the vector with the help cuts of the same enzyme".
- The number of fragments representing every sequence of the genome increases with genome size.
- For instance, for a probability level of 99 that all sequences are present in our library of a species, we may need 1,500 cloned fragments for *E. coli*, 4,600 for yeast, 48,000 for *Drosophila melanogaster* and 8,00,000 for a mammal like the human being, Libraries reaching these desired limits have been prepared in all these cases.

cDNA Library from mRNAs

- Complementary DNA (cDNA) libraries can also be prepared by isolating. "As from tissues which are actively synthesizing proteins, like roots in plants, ovaries or reticulocytes in mammals, etc.
- The mRNA for copying it into cDNA through the use of reverse transcriptase. The mRNA is used for copying it into cDNA through the use of cDNA molecules, it can be made double -stranded, and cloned.
- However, cDNA clones will differ from genomic clones in lacking the introns present in split genes, but the advantage of being capable to be expressed bacteria, which do not have the

machinery to process HnRNA (obtained cDNA clones from split genes) mRNA".

Colony (plaque) Hybridization for the Screening of Libraries

- Once a genomic library or cDNA library is available, we may like to it for the isolation of a gene sequence.
- This can be achieved by the colony hybridization technique. In this technique, bacterial transformed chimeric vectors are grown into colonies, which are lysed on nitrocellulose filters.
- Their DNA is denatured in situ and fixed on the filter, which is hybridized with a radioactively labeled probe carrying sequence related to the gene to be isolated (usually a cloned cDNA for the screening of a genomic library).
- "Colonies carrying this sequence will be identified by dark spots after autoradiography so that the original chimeric vector carrying the desired gene sequence can be recovered from one or more colonics in the original master plate and used for further experiments. This technique is described as colony hybridization.
- It is possible that the probe may identify more than one clone or that a gene is fragmented in the library.
- In such a case, one needs to reconstruct the desired sequence using several overlapping sequences available in the library.
- This is a very routine exercise whenever we like to isolate specific DNA sequences from the genome of a species, or from cDNA derived from mRNA of a specific tissue of species.
- Sometimes the library may be available not in the form of bacteria is formed with chimeric DNA molecules, but in the form of chimeric particles carrying the cloned segments.
- In such a situation, a bacterial is infected with a mixture of chimeric phage particles (i.e the library) and a large number of plaques develop overnight".
- These plaques can be treated just like the colonies in colony hybridization to identify and isolate the chimeric phage particle carrying the gene of interest. This technique is then described as plaque hybridization.

Chromosome Walking and Characterization of Chromosome Segments

- When a probe is used for the identification of a gene sequence in a genomic library, the probe may hybridize with a number of clones, each carrying a part of a large gene fragmented during the preparation of a genomic library.
- "If we obtained partial digests (by digesting the DNA only partially) from the genome, different genomes (from a large number of cells) may give fragments which have overlapping sequences, because sites cleaved different genomes of the same organism, will differ being random.
- Since one of these fragments may have the entire sequence represented in the probe, overlapping, sequences may be used to construct the original genome sequence. Identification of fragments with overlapping sequences may be a key to the reconstruction or characterization of large chromosome regions.
- This is achieved by the technique popularly described as chromosome walking.
- The technique of chromosome walking involves the following
- (i) from the genomic library select a clone of interest (identified by a probe) and subclone a small fragment from one end of the clone (there is a technique available to subclone a fragment from the end).
- (ii) the subcloned fragment of the selected clone may be hybridized with other clones in the library and a second clone hybridizing with the subclone of the first clone is identified due to the presence of the overlapping region.
- (iii) the end of the second clone is then subcloned and used for hybridization with other clones to identify, a third clone having an overlapping region with the subcloned end of the second clone.
- (iv) the third clone identified as above is also subcloned and hybridized with clones in the same manner and the procedure may be continued.
- (v) Restriction map of each selected clone may be prepared and compared to know the region of overlapping so that identification of few overlapping restriction sites will amount to walking along the chromosome, or along a long in chromosome segment".

- Regions of chromosomes approaching 1000 kb have been mapped following the above technique.
- Restriction maps of entire chromosomes can be prepared in this manner following the technique of chromosome walking.

Reverse Genetics and Chromosome Jumping (or Hopping Libraries)

- In most cases, we identify the genes by their protein products, which help ln the isolation and cloning of a specific gene.
- However, for many human disorders which are each controlled by a single gene (e.g. Huntington's disease, cystic fibrosis, Duchenne muscular dystrophy, etc.), the gene product is not known despite intensive investigations.
- Such genes are cloned processes often called identified primarily by their map position.
- Once the gene is cloned, it can be utilized for getting information about the encoded protein, the gene mutation and the metabolic defect that may be associated with the disease.
- Since the information about the is coming in the reverse order, this is popularly described his reverse genetics approach involves the following two steps: (i) locate the gene to a particular chromosome through RFLP linkage analysis; (ii) map the position of the gene with respect to molecular marker, both through recombinational analysis (in terms of recombination cM=centiMorgan units) and by physical mapping (in terms of kilobase pairs) with pulsed-field gel electrophoresis (PFGE) and preparation of restriction map.
- Even if the distance is 1 cM, which is the limit of resolution, this will be equivalent to 1,000 kb.
- Therefore, if the segment carrying the gene is to be cloned using the linked molecular marker such a segment will be too large to be cloned.
- In order to bring the molecular marker close to the gene of interest, 'chromosome jumping' approach has been utilized.
- The technique of 'chromosome jumping' is based on the following steps depending upon the distance between the gene and the marker, decide about the distance of jumps' or 'hopsize'

(eg 100 kb or 200 kb): (i) genomic DNA molecules in the range of selected size (say 80 kb-130 kb in case of 100 kb 'hop size are 16)-240 kb for 'hopsize' of 200 kb) are selected through pulsed-field gel elcctrophoresis (ii) for circularization of DNA segments, ligation been two ends of each long linear DNA molecule was allowed using T4 ligase in the presence of supF^{+} (iv) DNA circles obtained in (ii) above are digested with EcoRI;(v) the vector iCh3A A lac, an amber mutated phage vector (supF"is also cut with *EcoRI* and used for cloning mal DNA fragments representing the junctions of the circularized genomic molecules and carrying sup F (vi) the cloned DNA fragments and in(v) above represent the jumping library, which can be plated on the host and screened through the technique of plaque hybridization described earlier.

- The above technique of chromosome jumping will help to narrow the gene and available molecular markers.
- After several cycles, the jumping is followed by cloning the regions that are very close to the desired gene . It is thus obvious that in the field of 'reverse genetics' using the technique of chromosome jumping, it will characterize genes, whose gene products are unknown.
- The gene for cystic fibrosis has recently been isolated using this technique. It could also be 'cystic fibrosis' disease is caused by a single base substitution in this gene. This technique of 'reverse genetics' will become a very active area of research, particularly in human genetics in the coming years.
- The term **'reverse genetics'** has however been redefined recently. Classical genetics, sometimes described as 'forward genetics' involves a study that starts from the phenotype, and follows through the identification of the gene and finally concludes with the isolation and sequencing of the gene.
- In view of this, if 'reverse genetics' has to use steps in the reverse order for a genetic study, then it should start with DNA molecules or clones with unknown effects and should conclude with the determination of the phenotype which it controls.
- This can be achieved by preparing random genomic DNA clones, which can be subsequently utilized for a genetic study in the reverse order. This will then be described as 'reverse genetics'.
- This definition of 'reverse genetics' differs from the earlier

conventional meaning of this term, where reverse genetics' was used to describe cases of genes with unknown products, which are isolated (with the help of phenotype and its linkage with molecular markers) and then studied in some detail.

Isolation, Sequencing and Synthesis of Genes

- Molecular manipulation of specific genes is of research in biotechnology.
- These genes, isolated or artificially synthesized before they are manipulated and used or transformation heading to the production of transgenic animals and plants that are discussed respectively, genes are also sometimes sequenced for a better understanding of its structure that needs to be manipulated.
- We, therefore, discuss the gene as one of the very popular areas that should be the isolation, sequencing, and synthesis of specific genes in this chapter.

Isolation of Genes

- During the last two decades, significant progress has been made in the techniques for isolation of a variety of genres, including those for (ribosomal RNA; (ii) specific protein products; (iii) phenotypic traits with an unknown product and those for (iv) regulatory functions e.g. promoter genes, etc.
- Different techniques have been used for the isolation of different types of genes and will be briefly described in this section.
- Early attempts for isolation of ribosomal RNA and proteins. As we know, ribosomes consist of ribosomal RNA (rRNA) and proteins. Ribosomal RNA (rRNA) can be of four sizes namely, 28S, 16S, 58S and 5S (S stands for Svedberg units, a measure of speed with which a molecule sediments on centrifugation).
- This ribosomal RNA makes 80 percent of the cellular RNA and is synthesized on ribosomal genes, which could be isolated.
- Isolation of ribosomal genes was considered convenient due to their following three characteristic features (i) availability of homogenous rRNA; (ii) differences between ribosomal RNA genes and other genes due to relatively high G + C content in rRNA (rRNA has 45-60 % G + C while the remaining RNA has

only 40 % G + C) ;(iii) ribosomal genes are resent in multiple copies, their number within several thousand. In view of the above, ribosomal RNA genes were isolated for the first time (in 1965) in an amphibian named *Xenopus*, by Hugh Wallace and Max L. Birnstiel, working at the University of Edinburgh (UK).

- Following steps were involved in this isolation : (i) RNA was isolated from ribosomes of *Xenopus* and made radioactively labeled, due to its replication in a medium containing tritiated uridine; (ii) ribosomal DNA was isolated by density gradient centrifugation followed by its denaturation (G +C content of rDNA differs from that of bulk DNA and helps in separation by centrifugation);(ii) single-stranded DNA was fixed on filter paper; (iv) labeled rRNA was added on filter paper carrying single-stranded DNA;(v) DNA-RNA hybridization was allowed to take place (vi) excess labeled RNA was washed (vi) radioactivity was measured and duplex hybrids isolated, which on denaturation, gave single-stranded DNA, which could be made double-stranded.
- Isolation of ribosomal DNA allowed the first characterization of ribosomal genes, which was followed by the isolation of 5S genes by Donald Brown.
- For a detailed account of isolation of ribosomal genes in *Xenopus* as originally done, readers are referred to an articlè Isolation of Genes in Scientific American (August 1973).

Isolation of Genes Coding for Specific Proteins

- Isolation of genes for specific proteins became possible only after the discovery of the reverse transcriptase enzyme in 1970.
- This enzyme can be easily used for the synthesis of copy DNA or complementary DNA (cDNA) from mRNA.
- This complementary DNA can then be used for the isolation of the corresponding gene from genomic DNA.
- It is, therefore, obvious that for the isolation of a specific gene, techniques should first be available for the isolation of specific mRNA.
- For this purpose, antibodies are produced against a specific protein for which the gene is to be isolated. Therefore, isolation of a gene coding for a specific protein involves the following basic steps.

- The purification of the protein product of gene;(ii) production of antibodies against this protein product by immunizing animals like rabbit or mouse; (ii) precipitation of polysomes engaged in synthesizing specific protein with the help of the antibody raised in step (i): (iv) isolation and purification of mRNA from the polysome fraction; (v) using mRNA for synthesizing cDNA with the help reverse transcriptase; (vi) cloning of cDNA thus synthesized for preparation of a cDNA library, (vi) immunological and electrophoretic analysis translation products of cDNA clones to identify specific cDNA meant for the specific protein in mind; (vii) use of the specific DNA probes selected in step (vii) the identification and isolation of the gene from genomic DNA through screening a complete or partial genomic library.
- In step (vi) above for the preparation of the cDNA library, cDNA may be cloned in expression vectors (vectors in which gene can be expressed due to the presence of promoter sequences for RNA polymerase enzyme) like ågtll, which can accept gene insertions into β- galactosidase gene.
- The chimeric vector will then produce the hybrid protein if correctly expressed. The hybrid protein can be identified using an antibody as above. Plaques belonging to the chimeric vector carrying the desired gene can also be identified by the reaction of the antibody attached to a radioactive protein.
- This procedure was used to clone a number of genes including a gene regulating zein synthesis in maize.
- The cDNA clones may also be used directly for gene manipulation I transformation experiments.
- However, when a cDNA clone is used, it can be described as a synthesis of the gene rather than as isolation of a gene, cDNA has been artificially synthesized, and not isolated from the organism.
- The above technique for the isolation of genes has now been successfully utilized both in plants and animals.
- However, the first genes isolated were those which existed in multiple copies and were expressed at high levels in specific tissues, e.g ovalbumin gene in chicks, globin and immunoglobulin genes in mouse, genes for storage proteins in cereals and legumes, amylase genes in barley, actin genes in some legumes, etc.

Isolation of Genes which are Tissue- Specific in Expression

- It is much easier to isolate genes that are expressed in specific tissues. For instance, genes for storage proteins are expressed only in developing seeds, ovalbumin gene is expressed in the oviduct or the globin gene is expressed.
- Such genes can be easily isolated because mRNA extracted from these specific tissues will either exclusively belong to the gene of interest or it will be rich in this species of mRNA.
- Other mRNA molecules in minor quantities can be eliminated since these can be identified through their isolation from tissues where this gene is silent.
- This strategy which was actually followed by isolation and cloning carrot genes expressed during the development of the somatic embryos, and also for the isolation of several genes for storage proteins in a crop is outlined.

Isolation of Genes using DNA or RNA Probes

- Specific molecular probes (whether DNA or RNA probes), if available can be used for the isolation of specific genes.
- These probes may be available either from another species for the same gene or maybe artificially synthesized using a part of the amino acid sequence of the protein product of the gene.

Use of Heterologous Probes

- The probes obtained from one species are heterologous probes. These have been found to be effective in identifying blots.
- For instance, the gene for chalcone synthase or on Southern clones during colony hybridization or plaque hybridization or on Southern blots synthase has been isolated from *Anuirhinum majus* and *Petunia hybrida* heterologous cDNA probe from parsley.
- Similarly, heterologous probes from maize were used for the isolation of barley genes in a cDNA library and not with the genomic library since the vector (e.g pGEM-4 blue) can also be used for getting the waxy gene) and Al (aleurone gene).
- Heterologous probes should be used in the latter case, unrelated genes or pseudogenes (which do not ay bo isolated and cloned.

- These heterologous clones if available have been found to be more sensitive and efficient.

Use of cDNA or Synthetic Probes

- We have already discussed the cDNA probes above. However, if a protein, purified using the technique of two-dimensional gel electrophoresis, is used for 5-15 consecutive amino acids, this information can be used for the synthesis of oligonucleotides (using automated DNA synthesizers).
- These oligonucleotides may then be directly utilized for screening of cDNA genomic libraries for isolation of specific genes.
- These can also be used for designing primers for the amplification of a gene in DNA extracted from a particular tissue. A 32 kilodalton (KD) glycoprotein (isolated from style) associated with the S2 incompatibility allele of *Nicotiana alaia*, was used for the synthesis of an oligonucleotide leading to the isolation of a related gene.

Isolation of Genes Coding for unknown Products

- In some cases, we are interested in the isolation of a gene whose phenotypic effect is known, but the gene product has not been identified or cannot be isolated.
- Such genes include those for morphological traits like dormancy photoperiodicity, disease resistance, etc.
- This area of research, in which genetics is studied by isolating the gene first without knowing the gene product is often described as reverse genetics.
- The method used for the isolation of these genes is different from those used for genes coding for known proteins.

Use of Transposable Elements (Transposons Tagging)

- The transposable elements (TEs), in some cases, have been effectively utilized for isolation, when the gene product is unknown.
- In this case, a transposon works as a mutagen and therefore as a gene tag. Following steps are involved in this procedure:
- (i) Clone a known and scorable phenotypic effect.

- (ii) TE is transposed to this gene to get an unstable allele.
- (iii) This unstable allele is cloned and TE is isolated with the unstable allele (this is to select a TE which can produce unstable as a gene with allele effect.
- (iv) This TE is transposed to a gene of interest known to produce an unstable allele.
- (v) The DNA is extracted.
- (vi) TE sequence is used as a probe to isolate and clone the mutant (Carrying inserted TE) so that we can then isolate the gene of interest.
- In maize, TE like Ac Ds, En/Spm and Mu1 have been isolated using Adh1. Similarly, TEs like Tam3 and Tam7 have been isolated from snapdragon (*Antirhinum mojius*). These TEs have been used in experiments leading to the isolation of genes.
- In maize, several P, Al, CI and C2 genes have been isolated successfully using the gene tagging method.
- For transposon tagging, often transposable elements endogenous to like maize and snapdragon have been used However, rarely transposons available from one plant species can be moved into the genome of another plant species, whose gene is to be isolated.
- For instance, the Ac element of maize has been transferred to tobacco, where it can integrate into any locus permitting transposon tagging and gene isolation.

Mutant Complementation

- In this technique, DNA clones from the selected which should be able to complement the wild type strain are mutants that are thus transformed into wild type.
- Once these DNA clones are available, the protoplasts derived from the mutant plant may be transformed using wild-type clones and the transgenic plants are produced.
- Once this is done, the gene of interest can be isolated from DNA extracted from the transformed plants using the wild-type complementary clone as a probe.

Use of RFLP Maps or Chromosome walking for Gene Isolation

- In recent years, RFLP (restriction fragment length polymorphism) maps have been prepared in maize, tomato, rice, etc.
- Linkages of RFLP loci with genes of interest are also being worked out as an aid in a plant. If RFLPs are known that are very close to the target gene (as may be done through RFLP maps or chromosome walking) then by locating the RFLPs on either side of the target gene, a long intervening DNA segment may be isolated.
- This fragment may be used for subcloning leading to the isolation of the desired gene.
- If RFLPs are examined in two near isogenic lines (NTLs) differing for a single gene, the difference in RFLP maps will help in locating the position of a gene on the molecular map which can then be utilized for isolation of the gene.

Use of Chromosome Jumping for Gene Isolation.

- We described the technique of chromosome jumping for walking long chromosome distances thus bringing the available known sequence (related to the available probe) closer to the gene of interest.
- Once this has been achieved, the techniques described in the previous section may be conveniently used for isolating this gene for further study.

Isolation of Novel Genes by Asis Datta at JNU, New Delhi

- Recently (in 1992), Prof Asis Datta of JNU, New Delhi was selected for the Birla Award for Science and Technology for cloning and characterization of two novel genes, relevant to human health.

Gene for Oxalate Decarboxylase

- In 1991, Prof **Asis Datta** reported for the first time, the cloning and characterization of the oxalate decarboxylase gene, which degrades and will thus reduce the content of oxalate or oxalic acid, which is harmful in many ways including the following.

- Several green leafy vegetables (e.g. Amaranthus, spinach, etc.) being rich in oxalate, if ingested in large quantities are toxic, since oxalate chelates calcium and also destroys renal tissues in the kidney.
- The attack and spread of a fungus (*Whetzelinidsclerot iorium*), which causes damage to crops like sunflower involve the accumulation of oxalic acid in the infected tissue.
- Oxalic acid is an essential substrate for the synthesis of the neurotoxin called 3-N-oxalyl 1-La, B diaminopropionic acid (ODAP) found in different parts of the plants of Lathyrus *sativus* (khesari dal). Consumption of this legume causes **neurolathyrism**, a well- known disease causing damage to leg muscles, etc.
- ODAP interferes with the metabolism of glutamic acid, which is involved in the transmission of nerve impulses in the brain so that despite its rich protein content, *L. sativus* cannot be used as a food source.
- In all the above cases (leafy vegetables, sunflower & *L. sativus*), the transfer of a single gene for oxalate decarboxylase to produce transgenic plants will improve these plant species by reducing the oxalate content.
- This new gene will also allow the development of a diagnostic kit to measure oxalate in blood and urine since prevalent methods for clinical tests are quite expensive.

Gene for a Seed -specific Protein with a Nutritionally Balanced Composition of Amino Acids

- Prof Datta's laboratory recently also reported for the first time (in 1992) the isolation of a gene from Amaranthus that encodes a seed -specific protein rich in all essential amino acids including lysine and S-amino acids.
- Interestingly, the amino acid composition of this protein corresponds to that of WHO -recommended protein standard for optimum human nutrition.
- It is known that seed proteins in cereals as well as in legumes (peas, etc.) are deficient, the cereal proteins being deficient in lysine and pulses being deficient in sulfur -containing amino acids.

- In view of this, the gene isolated from Amaranthus will be utilized in the future for compensating the amino acid deficiencies proteins, once it is genetically engineered into the target crop plants.

Sequencing of Gene or a DNA Segment

- Once a gene or a DNA fragment has been cloned, its further study involves DNA sequencing. (The technique of DNA sequencing was very laborious till 1975 when a breakthrough was made in DNA sequencing methods. Two different methods are now routinely used for the determination of DNA sequences.

Maxam and Gilbert's Chemical Degradation Method

- In this method, which is illustrated, the following steps are involved
- (i) Label the 3'ends of DNA with ^{32}P.
- (ii) Separate two strands, both labeled at the 3'ends.
- (iii) Divide the mixture into four samples, each treated with a different reagent having the property of destroying either only G, or only C, or 'A and. G' or T and C.
- The concentration of reagent is so adjusted - that 50 % of the target base is destroyed so that fragments of different sizes having ^{32}P are produced.
- (iv)Electrophoresis of each of the four samples in four different lanes of the gel.
- (v) Autoradiograph the gel and determine the sequence from positions of bands in four lanes.

Sanger's Dideoxynucleotide Synthetic Method

- **Fred Sanger** (who won the Nobel Prize twice) had initially developed a method for DNA sequencing, which utilized DNA polymerase to extend DNA chain length.
- This was termed the **plus-minus method**. Subsequently, he developed a more powerful method, utilizing single-stranded DNA as a template for DNA synthesis, in which 2',3' dideoxynucleotides were incorporated leading to termination of DNA synthesis.

- These dideoxynucleotides are used as triphosphates (dd TP) and can be incorporated in a growing chain, but they terminate synthesis since they cannot form a phosphodiester bond with the next incoming deoxynucleotide triphosphate (dNTP).
- Following steps are involved in Sanger's dideoxy method for DNA containing single- stranded DNA sample (cloned in M13 phage) to be sequenced, all four dNTPs (radioactively labeled) and an enzyme for sequencing.
- (i) Four reaction tubes are set up, each containing the single-stranded DNA sample to be sequenced, all four dNTPs and an enzyme for DNA synthesis.
- Each tube also contains a small amount (much smaller amount relative to four dNTPs) of one of the four dNTP, so that four tubes have each a different ddNTP, bringing about the nation at a specific base- adenine (A), cytosine (C), guanine (G) and thymine (T).
- (ii) The fragments, generated by random incorporation of a ddNTP leading to termination of the reaction, are then separated by electrophoresis on a high -resolution polyacrylamide gel. This is done for all the four reaction mixtures on adjoining lanes in the gel.
- (iii) The gel is used for autoradiography so that the position of different bands in each lane can be visualized.
- (iv) The bands on the autoradiogram can be used for getting the DNA sequence.
- A variant of the above dideoxy method was later developed, which has allowed the production of automatic sequencers.
- In this new approach, a different fluorescent dye is tagged to the oligonucleotide primer in each of the four reaction tubes (blue for A, red for C, etc.).
- The four reaction mixtures are pooled and electrophoresed together in a single polyacrylamide gel tube.
- A high sensitivity fluorescence detector, placed near the bottom of the tube, measures the amount of each fluorophore as a function of time.
- The sequence is determined from the temporal order of peaks corresponding to four different dyes.

Direct DNA Sequencing using PCR (also called Ligation-mediated PCR LMPCR)

- Polymerase chain reaction (PCR) has also been used for sequencing the amplified DNA product. This method of DNA sequencing is faster and more reliable and can utilize either the whole genomic DNA or cloned fragments for sequencing a particular DNA segment.
- The DNA sequencing using PCR involves two steps: (i) generation of sequencing template (double- stranded or single-stranded) using PCR and (ii) with the PCR products either with the thermolabile DNA polymerase or with the method or thermostable Taq DNA polymerase.
- Thus the DNA sequencing using PCR eliminates the need of cloning the DNA in single -stranded DNA phage vectors i.e. M13.
- Although double- stranded DNA product can be utilized for sequencing, this may cause difficulty due to reassociation after denaturation thus preventing the sequencing from annealing to its complementary sequence to allow the extension.
- To reduce this problem, either a variant of the standard method for sequencing double -stranded DNA is employed or single -stranded DNA templates are produced following asymmetric PCR earlier.
- A number of thermolabile DNA polymerases have been used for sequencing of *in vitro* amplified DNA Alternatively, **Taq** polymerase (thermostable enzyme) used for PCR, can also be used for the sequencing reaction.
- In either case, Sanger's synthetic method involving incorporation of dideoxynucleotide (ddNTP) for chain termination, is used for sequencing.
- However, the Maxam and Gilbert's chemical degradation method can also be used.
- In Sanger's method, as usual, four mixtures are prepared, each using one of the four ddNTPs.
- The sequencing primer is labeled with ^{32}P and the mixtures with amplified DNA, Taq polymerase and the appropriate buffer are incubated at 70°C for 5 min.

- The reaction is stopped by the addition of formamide stop solution in each tube and mixtures are run on polyacrylamide sequencing gel to obtain the ladders, which can be read by a computer or manually.
- This method using PCR helped in the automation of DNA sequencing.

Synthesis of Genes

- There are two approaches available for the synthesis of genes:
- (i) When the detailed structure of a gene is available, this gene can be synthesized by a purely chemical method as done by HG Khorana for the synthesis of the gene for a tRNA (reported for the first time in 1970).
- (ii) If the detailed nucleotide sequence of the gene is not available, one may utilize the 'RNA directed DNA polymerase' enzyme for the synthesis of the gene in question in the form of complementary DNA (cDNA) from the mRNA of the gene isolated in its pure form.
- As discussed above, cDNA can also be used for the isolation of a gene either
- (i) From DNA extracted from living cells utilizing the technique of Southern blot hybridization, or from
- (ii) A genomic library.

Chemical Synthesis of tRNA Genes.

- As outlined above, before one can start the chemical synthesis of a gene, the structure of the gene should be known.
- The structure of a gene earlier could not be worked out by direct chemical analysis, since there were no means for isolating a gene (techniques for isolation of genes were developed later and were described earlier in this chapter).
- The structure of the gene could, therefore, be inferred only from its product.
- For instance, if a gene is responsible for giving rise to a polypeptide chain and the structure of this chain is known, then from the genetic code dictionary, the structure of the gene could be easily inferred.

- Such genes were initially considered to be too long to be synthesized because an average gene contains about 1,500 base pairs.
- On the other hand, since tRNA molecules are fairly small in size (about 80 nucleotides), a gene responsible for giving rise to a tRNA molecule was within the reach for synthesis.

Synthesis of the Gene for yeast tRNA.

- As readers may know, the structures of a large number of tRNA molecules are now known. The first RNA whose structure could be known was yeast alanyl -tRNA. R.W. Holley who died in early 1993 and his coworkers (in 1965) gave the detailed structure of yeast alanyl tRNA.
- This information was immediately used by Khorana and his coworkers to deduce the structure of the gene for yeast alanyl tRNA.
- The structure of this gene would obviously be such that one of the two strands of DNA (gene) would be complementary to the base sequence of yeast alanyl tRNA.
- The other strand would then automatically have the same sequence as in tRNA except that in place of cytosine of tRNA there would be thymine in DNA.
- Synthesis of such a long chain (77 base pairs) of double -stranded DNA was rather difficult in 1965 (not in the 1990s) but Khorana and coworkers had extensive experience of synthesizing DNA of known sequences.
- It was decided that such a long chain could not be synthesized by adding a single base each time.
- Therefore, it was decided that small oligodeoxyribonucleotides ranging in length from 5 to 20 nucleotides should first be synthesized.
- These segments would be single- stranded and would cover the whole length of both strands of DNA.
- These would then be joined to form double -stranded DNA, 77 nucleotide pairs long. In the actual synthesis of the yeast alanyl tRNA gene, the following steps were involved:

(a) Synthesis of Oligonucleotides.

- Fifteen oligonucleotides ranging from pentanucleotide (5 bases) to an icosanucleotide (20 bases) were synthesized.
- Synthesis was conducted through condensation between the hydroxyl group at the 3' position of one nucleotide and the phosphate group at the 5' position of the second nucleotide.
- In order to bring about condensation, all other functional groups, not taking part in condensation were protected using specific protective groups.
- After protecting the groups, the reaction between a nucleotide with a protected 5' end and another nucleotide with a protected end proceeded.
- Subsequently, condensation was also brought about between groups of two, three or four nucleotides.

(b) Synthesis of three duplex fragments

- With the help of 15 nucleotides (single- stranded), three duplex fragments, each with a stranded end, were prepared.
- These three fragments consisted of the following:
  - The first 20 nucleotides, with the nucleotides (17-20) single-stranded, this fragment was called 'A'
  - The nucleotide residue is 17-50 with single-stranded regions 17- 20 and 46-50 segments called B.
  - The nucleotide residues 45-77 with single-stranded 17-20 as region 46-50; this segment was called 'C.'

(c) Synthesis of the gene from three duplex fragments

- The three fragments 'A',' B', and C" were joined to give complete gene in each of the two following ways; (i) In one scheme, 'A' was joined to 'B' taking advantage of the overlap in residues 17-20; C was then added, with the overlap in the region 46-50 residues.
- The complete double-stranded DNA with 77 base pairs representing the gene was thus ready. (ii) In the other scheme, B' was first added to 'C and to this A' was added in the end to obtain the complete gene.

- This gene, which was synthesized in 1970, was hopefully used for subsequent work since this gene could replicate and make its own copies.

(d) Synthesis of the gene for a true precursor tRNA.

- Before Khorana could complete the synthesis of the gene for yeast alanyl tRNA in 1970, as outlined above, it became obvious that tRNA was not the direct product of transcription. Instead, a precursor molecule is first synthesized which subsequently, after losing segments of RNA by cleavage, gives rise to tRNA.
- Obviously, therefore, the actual gene for yeast alanyl tRNA will be longer than the DNA duplex synthesized by Khorana.
- In view of this Khorana subsequently initiated the synthesis of a gene for *E coli* tyrosine suppressor tRNA precursor.
- DNA duplex which will give rise to this tRNA precursor was synthesized in the form of 26 small oligonucleotide segments.
- These were then arranged into six DNA duplex fragments having single-stranded ends. These six fragments gave rise to a presumed gene for *E. coli* tyrosine suppressor tRNA precursor.
- This gene, however, is the promoter region and other sequences essential for processing.
- Later, in 1979, Khorana reported the completion of the total synthesis of a biologically functional tyrosine suppressor transfer RNA gene to carry all regulatory sequences.
- This gene was 207 base pairs long and include the following: (i) a 51 base pairs long DNA promoter region;(ii) a 126 base pairs long DNA corresponding to the precursor tRNA and (ii) a 25 base pairs long duplex DNA corresponding to 16 base pairs adjoining CCA end of tRNA and the remainder, a modified sequence including EcoRI endonuclease specific sequence.
- The complete synthetic gene was cloned in the vector bacteriophage (lambda virus) by the gene cloning method discussed.
- On the transformation of *E. coli*, the phage could multiply with the cloned gene.

Gene Synthesis Machines

- Methods for the synthesis of DNA molecules of known base sequences have improved dramatically during the 1980s and 1990s. Therefore, genes can now be synthesized rapidly and in high yields.
- For instance, a gene for an IRNA took 20 man-years effort during 1965- 70 (described earlier in this chapter) but the same gene can now be synthesized within a matter of few days with the help of Gene Synthesis Machines developed in the 1980s.
- The key innovations which made the availability of gene synthesis machines possible are the following:
- (i) Development of silica-based supports, which provide support for the solid-phase synthesis of DNA chains and
- (ii) Development of stable protonated deoxyribonucleoside phoramidites as synthons, which are remarkably stable towards hydration and hydrolysis and thus are ideal for DNA synthesis.
- These two procedures can be used to prepare DNA molecules more than 100 nucleotides long within a few days.
- Different steps are involved in the solid-phase synthesis of a DNA chain.
- Several versions of gene machines are now available. These machines, under the control of a microprocessor, synthesize specific short sequences of single-stranded DNA automatically.
- The desired sequence is entered on a keyboard and the microprocessor automatically opens the valves of the containers of successive nucleotides, reagents and solvents needed at each step, into a synthesizer column, which is packed with tiny silica beads. These beads provide support on which DNA molecules are assembled.

Synthesis of Genes from mRNA

- H. Temin and D. Baltimore, in 1970 discovered the presence of "RNA directed DNA polymerase enzyme" which has the ability to synthesize DNA on an RNA template.
- This enabled molecular biologists to synthesize complementary DNA (cDNA) using mRNA as a template.

- If mRNA transcribed from a specific gene is made available in purified form, the elementary DNA (cDNA) synthesized with its help will represent the synthesized gene.
- To what extent this synthesized gene is a faithful copy of the native gene will depend on the fidelity of copying mRNA and also on the stability of DNA thus synthesized.
- Moreover, since the mRNA of a gene does not have the complete transcript of the gene *in vivo*, the synthesized gene will be smaller than the gene *in vivo*.
- By copying eukaryotic purified mRNA, several genes have been artificially synthesized. The most important of these genes are the genes for sea urchin histone proteins, ovalbumin gene in chicken and globin genes in mammals.
- The gene synthesized as cDNA from a B globin mRNA has been inserted into a plasmid in order to study its behavior. As pointed out earlier, this gene will lack any additional regulatory sequences absent in mRNA but present in a globin gene *in vivo*.
- This type of gene will also lack intron sequences found in eukaryotic split genes. Another factor causing a difference between the synthesized gene and the gene *in vivo* will depend on the sincerity with which the enzyme will synthesize cDNA because in many cases cDNA was found to be smaller than the mRNA itself.
- These genes (cDNA) have already become a very important tool in molecular biology experiments.

Gel Permeation

- It is a technique in which a polymeric organic compound is used to prepare a porous medium.
- The pore size is determined by the degree of cross-linking of polymeric chains.
- Solutes present in the mixture are separated on the basis of their size and shape when they pass through a column consisting of packed gel particles (Figure 2.1).

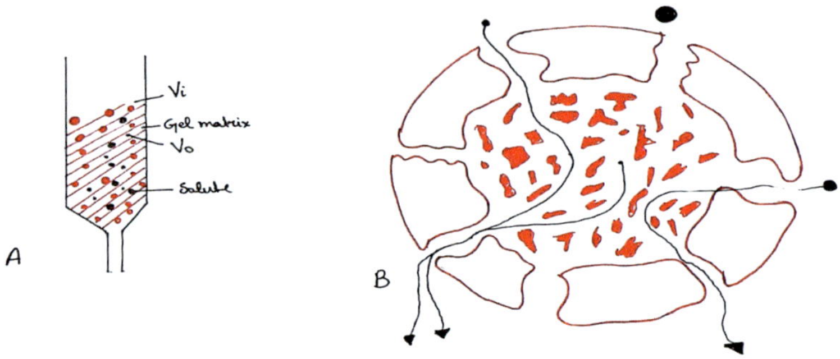

Figure 2.1: Gel permeation chromatography (a) spatial accessibility of molecules during gel filtration (b) movement of molecules of different sizes through the microporous gel-forming stationary phase.

Chromatography

- It is a technique of separation in which small-sized particles or liquid substances are separated in a buffer solution.

Ion Exchanged Chromatography

- When molecules are dissolved in a solvent, they dissociate into charged ions. Therefore, they develop polarity.
- On the basis of polarity i.e, the presence of charges they can be separated.
- Ion exchange chromatography is a powerful technique used for the separation of two proteins which are very similar to each other; see Figure 2.2.

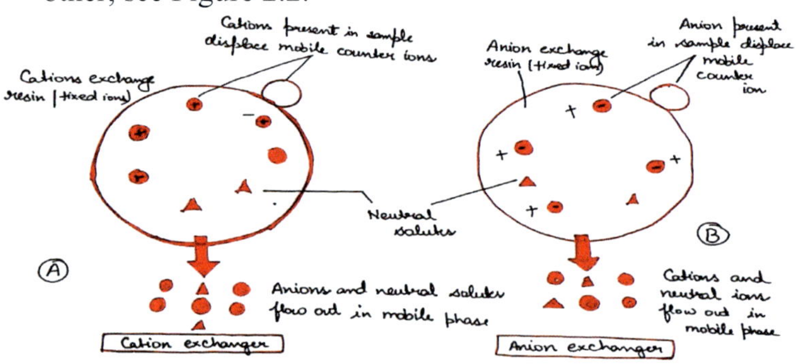

Figure 2.2: Ion exchange chromatography (a) cation exchanger (b) anion exchanger.

Electrophoresis

- It is a method of separation of charged molecules applying an electric field.
- When the charged molecules are placed in an electric field they migrate depending on their net charge, size, shape and applied current.

Agarose Gel Electrophoresis

- Agarose is a linear polymer of D- galactose and 3,6 –anhydro –L-galactose which is extracted from seaweed.
- At neutral pH, the negatively charged DNA migrates towards the anode after the application of an electric field across the gel.
- It is a standard technique of nucleic acids.

Pulsed-field Gel Electrophoresis (PFEG)

- It is a process to separate the several megabases long DNA molecules.
- PFEG varies a little from agarose gel electrophoresis and can resolve DNA molecules of 100-1000kb.
- It changes the direction of the electric field in such a way that larger DNA fragments are aligned more slowly with the direction of the new field than do the smaller DNA molecules.
- The pulsed-field electric fields are applied to force the DNA molecule to reorient before continuing to move like a snake through the gels; see Figure 2.3.

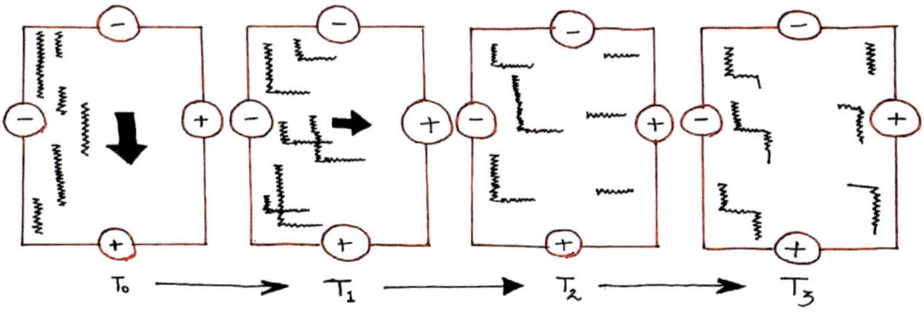

Figure 2.3: Pulse-field gel electrophoresis.

- Polyacrylamide Gel Electrophoresis (PAUL) is the most widely used electrophoresis for the separation and characterization of proteins and nucleic acids.
- Polyacrylamide gel offers several advantages such as inertness to chemicals, superior resolution stability over a wide range of pH, temperature and ionic strength.
- Polyacrylamide gel is prepared by using the following
- (i) monomers (e.g. acrylamide, N, N-methylene bisacrylamide,
- (ii) initiator NN.N. N-tetramethylethylene diamine (TEMED)
- (iii) propagators (ammonium persulphate riboflavin,
- (iv) terminator or inhibitor (oxygen/air).
- The pore size of polyacrylamide gel can be changed by varying the concentration of acrylamide and bi-acrylamide monomers in a fixed volume of gelatin solution. Free radicals formed by ammonium sulfate activate the acrylamide gel.
- A long polymer chain is thus produced after reacting, the activated acrylamide with successive acrylamide molecules. Further, bisacrylamide brings about gelation and cross-linking through polymerization. This results in the formation of a complex network of acrylamide.

Sodium Dodecyl Sulphate Polyacrylamide Gel Electrophoresis (SDS PAGE):

- Separation of some of the proteins does not occur due to similar charges: such proteins are treated first with an ionic detergent called sodium dodecyl sulfate (SDS) before the start and during the course of electrophoresis (PAGE).
- Therefore, such electrophoresis is called SDS-PAGE electrophoresis. Identical proteins are denatured by SDS resulting in their sub-units.
- The polypeptide chains get opened and extended. On the basis of their mass but not the charges, the molecules are separated. SDS-PAGE is a high- resolution method used universally for analyzing the mixture of proteins according to their respective size.
- SDS solubilized in soluble proteins makes possible the analysis of the other insoluble mixtures. Separation of the proteins does not occur due to a similar charge: mass ratio (z/m).

Two-dimensional Electrophoresis

- A mixture of protein is separated using two- dimensional electrophoresis.
- The first dimension uses iso-electric focusing, and the second dimension is sodium dodecyl sulfate-polyacrylamide gel electrophoresis (SDS-PAGE) Iso-electric Focusing (IEF).
- The biomolecules such as proteins, DNA and RNA have electric charges which depend on the molecule and the conditions of the medium (pH of buffer in which dissolved).
- Charged molecules can be separated by electrophoresis in gels.
- Due to the differences in amino acid composition proteins have a unique mass and charge.
- Hence the proteins have a net negative charge and net positive charge or iso-electric point (no charge) at a given pH of a buffer.
- The amphoteric substances such as proteins which differ in their isoelectric points can be separated by IEF.
- The isoelectric point is a pH value at which the net charges on molecules are zero Ampholytes (i.e. complex mixture of synthetic polyamino-polycarboxylic acids) are introduced in **Figure 2.4**.

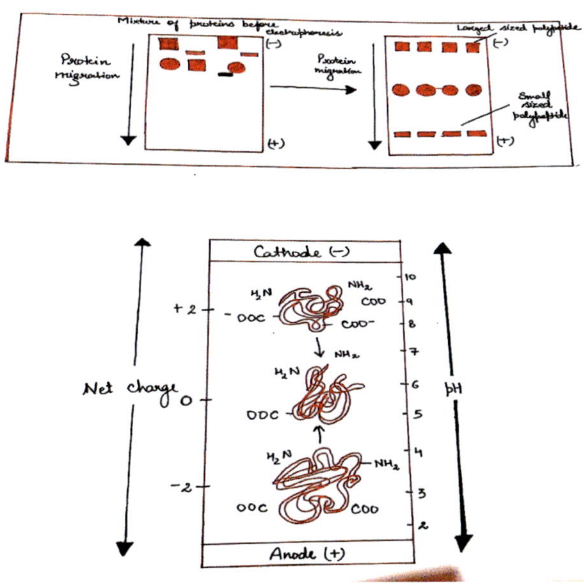

Figure 2.4: Isoelectric focusing.

Spectrometry

- In 1900, for the first time, J.I. Thomson introduced a mass spectrometer that employed fixed magnetic and electric fields to separate ions of different mass and energy.
- He recognized that charged particles differing in momentum behaved differently in an electric field, and used this property for the separation of ions with different masses.
- Its extensive use for research in various fields of biological science was started in the 1980s. During 1990-2000, mass spectrometry became an important technique for genomics and proteomics research leading to the 2002 Nobel Prize in Chemistry to J.B. Fenn and K. Tanaka. Two-dimensional electrophoresis is more powerful when coupled with MS.
- The unknown protein spot is cut from the gel and cleaved by trypsin digestion into a mass spectrometer and the mass of fragment is fragmented which are then analyzed by fingerprinting can be used to estimate plotted.
- This mass of the probable amino acid composition of each fragment and tentatively identifies the protein.
- The proteome and its changes can be studied very effectively by employing the two techniques together.
- The MS can also provide valuable information about covalent modifications of proteins which can affect their activity.
- Mass spectrometry is a very useful technique. It is used in the identification of unknown compounds, quantification of known compounds and determination of structural and chemical ties of compounds when present in small the production of ions of the materials in a sample, (ii) their separation on the basis of their mass: charge (m: e) ratio, and (iii) determination of the relative abundance of each ion.
- Therefore, the spectrophotometer consists of three components: the source. It does not directly measure the molecular mass but detects the m: e ratio.
- Mass is measured in terms of Dalton (Da). One Dalton = 1/12th of the mass of a single atom of isotonic carbon (3C) unit (106-10 g). This technique involves a mass of the ion, an analyzer, and a detector.

Matrix-assisted Layer Desorption-ionization (MALDI)

- MALDI was first developed by Karas and Hillenkamp in 1988. It co-precipitates the large excess of a matrix material with the analyte molecule.
- In this method, a sub-microlitre of the mixture (matrix +analyte) is pipetted onto a metal substrate and allowed to dry.
- The dried mixture is irradiated by a laser pulse (337 nm) which specifically absorbs the selected solid of the matrix.
- The irradiation causes energy transfer and desorption resulting.
- With the help of matrix ions, the non-absorbing analyte molecules also get desorbed into the gas phase and ionized.
- The charged molecular ions of the analyte are detected by MALDI-TOF mass spectrometer. Generally, MALDI is used for the study of molecules having mass above 500 Daltons (i.e. oligonucleotides and peptides).
- Proteome analysis will be crucial for understanding different biological processes in the post-genomic era.
- To achieve this objective mass spectrometry is extensively used for: peptide sequencing, identification of protein, protein expression in different tissues and conditions, identification and post-transcriptional modification (e.g. phosphorylation, glycosylation, etc.) of proteins in response to different stimuli, and characterization of protein interactions that include protein-ligand, protein-protein and protein-DNA interaction in the gas phase- matrix ions t: improvement in MS has significantly improved its application in the study of protein structure and function.
- For the study of macromolecules, the Indian Society for Mass Spectrometry (ISMAS) was established in 1978 with its headquarter at BARC (Bhabha Atomic Research center), Mumbai. This society conducts a workshop each year on mass spectrometry.
- In India, facilities for MALDI- TOF MS have been created at several centers including the center for Cellular and Molecular Biology (CCMB), Hyderabad, Indian Institute of Sciences, Bangalore, BARC, Rajiv Gandhi center for Biotechnology (RGCB), Thiruvananthapuram, etc.

Surface Enhanced Laser Desoption-Ionization (SELDI)

- It is a unique combination of affinity chromatography and time of flight (TOF) mass spectrometry on a single platform, a combination designated ionization time of flight mass spectrometry.
- This system enables protein capture, purification, ionization and analysis of complex biological mixtures directly on protein chips array surface and detection of the purified proteins by laser desorption-ionization time of flight (LDI-TOF), MS analysis. SELDI system consists of three parts: protein chip arrays (eight spots of 2 mm diameter). (ii) Protein chip reader (LDI-TOF MS) (iii) specialized software.

Electrospray Ionization (ESI)

- In the late 1980s, J.B. Fenn developed ESI mass spectrometry and awarded Nobel Prize in 2002 in chemistry.
- In this method, the sample is dissolved in liquid with mobile phase (i.e. water: acetonitrile: methanol) and pumped through a hypodermic needle at a high voltage (about 4,000 V which causes a strong electric field at the top of the nozzle of the metal needle).
- This disperses electrostatically electrosprays the highly charged small droplets of about one micrometer in size; see Figure 2.5.

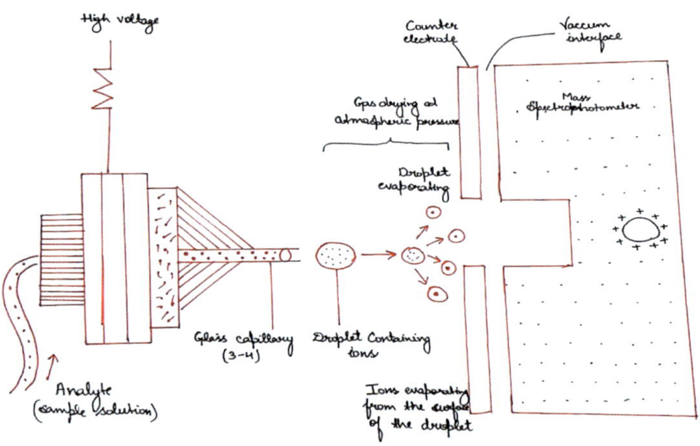

Figure 2.5: Electrospray ionization source.

- These droplets are electrostatically attracted to the inlet of MS. Dry gas or heat is applied. The solvent around these droplets soon evaporates and impart charge increases with a decrease in their size. Hence equal charges repel each other.
- This ionization is very gentle which causes no breakdown of analyte ions in the gas phase or onto the analyte molecules.

Fluorescence Spectroscopy

- Fluorescence refers to the emission of light from a molecule after absorption of radiant energy when a molecule absorbs radiant energy, its electrons are elevated to an excited state.
- If there is no chemical reaction, dissociation, rearrangement or energy transfer by collision, the molecule return to the ground state and excessive energy is lost as heat or light i.e. photons of energy (v) Emission of light is called fluorescence.
- A compound may absorb energy in the UV region and fluoresce in the visible region. In fluorescence, the absorbed excited light is of a smaller wavelength (more energy) and the wavelength of the emitted light is longer (less energy).
- More increase in wavelength of emitted light than the exciting light is called stoke shift.
- The fluorescence by the natural compounds is known as intrinsic fluorescence. There are certain chemical compounds that do not fluoresce. However, fluorescence can be detected after coupling them with a fluorescence probe or fluor.
- This phenomenon is called extrinsic fluorescence.

Polymerase Chain Reaction

- Polymerase chain reaction (PCR) provides a simple and ingenious method for exponentially amplification of specific DNA sequences by *in vitro* DNA synthesis.
- This technique was developed by Kary Mullis at Cetus Corporation in Emery Ville, California in 1985. Kary Mullis shared the Nobel prize for chemistry in 1993.
- This technique has made it possible to synthesize large quantities of a DNA fragment without its cloning. It is ideally suited where the quantity of biological specimen available is very low such as

a single fragment of hair or a tiny bloodstain left at the site of a crime.

- The details of PCR techniques and their mechanism are described by Erlich (1989) in his edited book 'PCR Technology'.
- The PCR technique has now been automated and is carried out by a specially designed machine.

The PCR requires the following

- **DNA Template:** Any source that contains one or more target DNA molecules to be amplified can be taken as a template.
- **Primers:** A pair of oligonucleotides of about 180-30 nucleotides with similar G+C contents act as primers. They direct DNA synthesis towards one another. The primers are designed to anneal on opposite strands of the target sequence so that they will be extended towards each other by the addition of nucleotides to their 3' ends.
- **Enzyme:** The most common enzyme used in PCR is a thermostable enzyme called Taq polymerase. It is isolated from a thermostable bacterium called *Thermus aquaticus*. It survives at 95°C for 1-2 minutes and has a half-life for more than 2 hours at this temperature. The other thermostable polymerases can also be used in PCR

DNA Amplification Fingerprinting (DAF)

- The DAF has been described by Caetano-Annoles and Gresshoff (1994) where a single arbitrary primer (about 5 bases long) is used to amplify DNA by using PCR. In this technique, the parameters are carefully optimized.
- But for early and easy determination of amplified products, automation and fluorescent tagging of primers are extremely amenable.
- DAF differs from RAPD as (i) it requires a higher concentration of shorter primers (5-8 nucleotides), (ii) two temperature cycles in contrast to three cycles in RAPD, (ii) it produces a very complex banding pattern, (iv) DAF products are used in Sed on acrylamide gels and detected by using silver staining.

Application of PCR

- After the discovery of PCR, modern biology has been revolutionized in each and every aspect. "Some of the areas of application of PCR have briefly been discussed herewith.

(a) Diagnosis of Pathogens:

- There are several pathogens that grow slowly. Therefore the cells are found less in number in the infected cells/tissues.
- It is difficult to culture them on an artificial medium. Hence, for their diagnosis, PCR-based assays have been developed.
- These detect the presence of certain specific sequences of the pathogens present in the infected cells/tissues.
- Besides, it is useful in the detection of viral infections before they cause symptoms or serious diseases.

(b) Diagnosis of Specific Mutation: in humans, there are thousands of genetic diseases. Mutations are also related to genetic diseases.

- The presence of a faulty DNA sequence can be detected before the establishment of the disease.
- By using PCR sickle cell anemia, phenylketonuria and muscular dystrophy can also be detected. The other diseases can also be diagnosed by using PCR. For example, PCR-based diagnostic for AIDS. Chlamydia, tuberculosis, hepatitis, human papillomavirus, and other infectious agents and diseases are being developed.
- The tests are rapid, sensitive and specific

(c) In Prenatal Diagnosis: It is useful in the prenatal diagnosis of several genetic diseases. If the genetic diseases are not curable, it is recommended to go for an abortion.

(d) DNA Fingerprinting:

- In recent years DNA fingerprinting is more successfully used in forensic science to search out criminals, rapists solving disputed parentage and unite the lost children to their parents or relatives by confirming their identity.
- This is done through making lines between the DNA recovered from samples of blood, semen, hairs, etc. at the spot of crime and the DNA of suspected individuals or between a child and his/her parents/relatives.

(e) In Research: In addition, DNA fingerprinting of new microorganisms isolated from various extreme environments (soil. water, sediments, air, extreme habitats, etc.) is also carried out to confirm their identity by comparing with the DNA sequences of known microorganisms.

- Their DNA and RNA can be amplified. Besides, it is also useful to determine the orientation and location or restriction fragments relative to one another (In Molecular Archaeology (Paleontology): PCR has been used to clone the DNA fragments from the mummified remains of humans and extinct animals such as the woolly mammoth and dinosaurs from the remains of ancient animals as recently epitomized in Michel Crighton's Jurassic Park.
- DNA from buried humans has been amplified and used to trace the human migration that occurred in ancient times.

g) Diagnosis of Plant Pathogens: It is also applied in the diagnosis of plant diseases. A large number of plant pathogens in various hosts or environmental samples are detected by using PCR, for example, viroids (associated with flower mosaic virus, bean yellow mosaic-virus, plum pox virus, potyviruses), mycoplasmas bacteria (*Agrobacterium tumifaciens, Pseudomonas solanacearum, Rhizobium leguminosarum, Xanthomon compestris*, etc), fungi (e.g., *Collectotrichum gloeosporioides, Glomus spp., Laccaria sp Phytophthora spp. Verticillium spp*)", and nematodes (e.g. *Meloidogyne incoginta, M. javanic*, etc), viruses (such as TMV, (Henson and French, 1993; Chawla, 1998).

3. ISOLATION, SEQUENCING AND SYNTHESIS OF GENE

Gene DNA Sequencing

- The structure of DNA (e.g. gene insert, a recombinant plasmid or entire genome) can be analyzed by determining the nucleotide sequences.
- In molecular cloning, the information of nucleotide sequences is essential.

- In 1965, Robert Holley and his research group at Cornell University completely sequenced nucleotides of tRNA (tRNA for yeast alanine).
- In 1977, the following two methods were ONA developed. Allan Maxam and Walter Gilbert developed a chemical method of sequencing.
- In this method, end-labeled DNA is subjected to base-specific cleavage reactions before gel separation. In routine sequencing of DNA, this method is not commonly followed.
- In the same year (1977) Frederick Sanger and co-workers developed. It is also called the dideoxynucleotide chain termination method because dideoxynucleotides are used in an enzymatic method of DNA sequencing of molecules.

Maxam and Gilbert's Chemical Degradation Method

- This method is not very popular because it is time-consuming and labor-intensive.
- In this method, the DNA molecule can be radiolabelled at either 5 'end by using polynucleotide kinase, or 3' end by terminal transferase.
- One end of radiolabelled double-stranded DNA is removed by using endonuclease.
- A base is modified chemically followed by cleavage of sugar-phosphate backbone of DNA. No specific reaction for the four bases is carried out, except specific reaction to G only and purine specific reaction, which removes A or G.
- A difference in these reactions indicates the presence of A.

Cleavage of Purines

- The mixture is separated into four sets, each treated with a different reagent which degrade only G or C or A and G or C and T.
- In one set, DNA is treated with acid followed by dimethyl sulfate.
- This causes methylation of A, (at 3 'position and G (at 7' position).
- Subsequently, the addition of alkali (-OH) and pyrimidine results in DNA cleavage and the removal of purines (A or G)

Cleavage of Pyrimidines

- Similar to the cleavage of purine, a pyrimidine (C or T) is cleaved in the presence of 1-2 M NaCl solution.
- It works only with C. Differences between these two indicate the presence of T in the DNA sequence. Cleavage of pyrimidines (C or T) through hydrazine hydrolysis.
- Partial chemical cleavage of DNA fragments as done above generates the populations of radiolabelled molecules extending from radiolabelled terminus to the site of chemical cleavage. These fragments of different sizes represent unique pairs of 5 and 3' cleavage products in the random collection.
- A complete set is formed by these products the length of each number is short by nucleotide. These can be separated by gel electrophoresis.
- The fragments containing labeled terminus can be observed by autoradiography of the gel. Following the order of fragments obtained from different digestions, the sequence of nucleotides is deduced and interpreted.

Sanger's Method (Dideoxynucleotide Chain Termination Method)

- Twice Nobel Prize winner Frederick Sanger et al. (1977) developed DNA sequencing that utilizes single-stranded DNA as a template.
- This method is also called a dideoxynucleotide chain termination method.
- The requirements are a free 3' -OH ends to primer start DNA synthesis, DNA polymerase and dNTPs.
- Shows the presence of a free 3'-OH group at 3' end in DATP and no 2',3' -OH in 2',3'ddATP In 2',3', ddATP, a hydrogen atom is attached at 2' and 3 carbons instead of -OH hydroxyl group. If any of the four ddNTPs binds, the chain elongation is terminated.
- Because ddNTPs do not have a free 3'-OH end which is required for chain elongation.
- Therefore, no phosphodiester bond will be formed.

Automatic DNA Sequencers

- Automatic DNA sequencing machines are based on Sanger's method.
- In this method, a different fluorescent dye is tagged to the ddNTPs. Using this technique, a DNA sequence containing thousands of nucleotides can be determined in a few hours.
- Each dideoxynucleotide is linked with a fluorescent dye that imparts different colors to all the fragments terminating in that nucleotide.
- All four labeled ddNTPs are added to a single capillary tube. It is a refinement of gel electrophoresis which separates fastly. DNA fragments of different colors are separated by their respective size in a single electrophoretic gel.
- A current is applied to the gel.
- The negatively charged DNA strands migrate through the pores of gel towards the positive end.
- DNA fragments migrate faster and vice versa. All fragments of a given length migrate in a single peak.
- The DNA fragments are illuminated with a laser beam.
- Then the fluorescent dyes are excited and emit light of specific wavelengths which is recorded by sequences that are read by determining the sequence of the colors emitted from specific peaks as they pass the detector.
- This information is fed directly to a computer that determines the sequence.
- A tracing electrogram of emitted light of the four dyes is generated by the computer. The color of each dye represents the different nucleotides.

Polymerase Chain Reaction

- Polymerase chain reaction (PCR) provides a simple and ingenious method for exponentially amplification of specific DNA sequences by *in vitro* DNA synthesis.
- This technique was developed by Kary Mullis at Cetus Corporation in Emery Ville, California in 1985. Kary Mullis shared the Nobel prize for chemistry in 1993.

- This technique has made it possible to synthesize large quantities of a DNA fragment without its cloning. It is ideally suited where the quantity of biological specimen available is very low such as a single fragment of hair or a tiny bloodstain left at the site of a crime.
- The details of PCR techniques and their mechanism are described by Erlich (1989) in his edited book 'PCR Technology'.
- The PCR technique has now been automated and is carried out by a specially designed machine.

The PCR requires the following

- **DNA Template:** Any source that contains one or more target DNA molecules to be amplified can be taken as a template.
- **Primers:** A pair of oligonucleotides of about 180-30 nucleotides with similar G+C contents act as primers. They direct DNA synthesis towards one another. The primers are designed to anneal on opposite strands of the target sequence so that they will be extended towards each other by the addition of nucleotides to their 3' ends.
- **Enzyme:** The most common enzyme used in PCR is a thermostable enzyme called Taq polymerase. It is isolated from a thermostable bacterium called *Thermus aquaticus*. It survives at 95°C for 1-2 minutes and has a half-life for more than 2 hours at this temperature. The other thermostable polymerases can also be used in PCR

DNA Amplification Fingerprinting (DAF)

- The DAF has been described by Caetano-Annoles and Gresshoff (1994) where a single arbitrary primer (about 5 bases long) is used to amplify DNA by using PCR. In this technique, the parameters are carefully optimized.
- But for early and easy determination of amplified products, automation and fluorescent tagging of primers are extremely amenable.
- DAF differs from RAPD as (i) it requires a higher concentration of shorter primers (5-8 nucleotides), (ii) two temperature cycles in contrast to three cycles in RAPD, (ii) it produces a very complex

banding pattern, (iv) DAF products are used in Sed on acrylamide gels and detected by using silver staining.

Application of PCR

- After the discovery of PCR, modern biology has been revolutionized in each and every aspect. "Some of the areas of application of PCR have briefly been discussed herewith.

(a) Diagnosis of Pathogens:

- There are several pathogens that grow slowly. Therefore, the cells are found less in number in the infected cells/tissues.
- It is difficult to culture them on an artificial medium. Hence, for their diagnosis, PCR-based assays have been developed.
- These detect the presence of certain specific sequences of the pathogens present in the infected cells/tissues.
- Besides, it is useful in the detection of viral infections before they cause symptoms or serious diseases.

(b) Diagnosis of Specific Mutation: in humans, there are thousands of genetic diseases. Mutations are also related to genetic diseases.

- The presence of a faulty DNA sequence can be detected before the establishment of the disease.
- By using PCR sickle cell anemia, phenylketonuria and muscular dystrophy can also be detected. The other diseases can also be diagnosed by using PCR. For example, PCR-based diagnostic for AIDS. Chlamydia, tuberculosis, hepatitis, human papilloma virus, and other infectious agents and diseases are being developed.
- The tests are rapid, sensitive and specific

(c) In Prenatal Diagnosis: It is useful in the prenatal diagnosis of several genetic diseases. If the genetic diseases are not curable, it is recommended to go for an abortion.

(d) DNA Fingerprinting:

- In recent years DNA fingerprinting is more successfully used in forensic science to search out criminals, rapists solving disputed parentage and unite the lost children to their parents or relatives by confirming their identity.
- This is done through making lines between the DNA recovered from samples of blood, semen, hairs, etc. at the spot of crime and

the DNA of suspected individuals or between a child and his/her parents/relatives.

(e) In Research: In addition, DNA fingerprinting of new microorganisms isolated from various extreme environments (soil. water, sediments, air, extreme habitats, etc.) is also carried out to confirm their identity by comparing with the DNA sequences of known microorganisms.

- Their DNA and RNA can be amplified. Besides, it is also useful to determine the orientation and location or restriction fragments relative to one another (In Molecular Archaeology (Paleontology): PCR has been used to clone the DNA fragments from the mummified remains of humans and extinct animals such as the woolly mammoth and dinosaurs from the remains of ancient animals as recently epitomized in Michel Crighton's Jurassic Park.
- DNA from buried humans has been amplified and used to trace the human migration that occurred in ancient times.

g) Diagnosis of Plant Pathogens: It is also applied in the diagnosis of plant diseases. A large number of plant pathogens in various hosts or environmental samples are detected by using PCR, for example, viroids (associated with flower mosaic virus, bean yellow mosaic-virus, plum pox virus, potyviruses), mycoplasmas bacteria (*Agrobacterium tumifaciens, Pseudomonas solanacearum, Rhizobium leguminosarum, Xanthomon compestris*, etc), fungi (e.g., *Collectotrichum gloeosporioides, Glomus spp., Laccaria sp Phytophthora spp.. Verticillium spp*)", and nematodes (e.g. *Meloidogyne incoginta, M. javac* , etc), viruses (such as TMV, (Henson and French, 1993; Chawla, 1998).

Site-directed Mutagenesis

- Now, techniques have been developed for the specific portion of the genome with a view to getting some novel products of enormous value.
- Mutations can be done by directing insertions or deletions to a site on the DNA but these minor changes may have some single amino acid is altered to improve its properties.
- For such modifications point- mutations i.e. changing of a single nucleotide is done on a specific portion in the gene.

- Thus, the most suitable method that could bring about point-mutation in a gene to mutate specific is unlikely to be of use. Importance. If a protein in order to oligonucleotide-directed mutagenesis.
- This technique has the potential for protein engineering, the engineered enzymes would be better than the wild-type ones. Change in a single nucleotide base pair is called 'point mutation'.
- Some useful properties in proteins (like stability in subtilisin) that result in the substitution of selected amino acids.
- Using this technique, the specific point of a gene can be mutated.
- Therefore, this method has been used to understand the function of many genes.
- Moreover, this technique can only be used when the nucleotide sequence of the gene is known.

Methods of Mutagenesis

- When it has been decided which base is to be Site-directed mutagenesis using short synthetic DNA molecules changed chemically (A).
- The oligonucleotide sequence has desired mutation at a specific site.
- The oligonucleotide corresponds to the mutated nucleotide and its neighboring regions, mainly between 15 and 20 nucleotides in length.
- A single-stranded clone of the wild-type gene is produced by using an M13 phage-based vector. The oligonucleotide is allowed to hybridize with the single -clone.
- The hybridization is performed under the conditions.
- The oligonucleotide will base pair with wild-type complementary sequence (B) and act as a primer to produce double-stranded DNA by using DNA polymerase (C).

Insertion Mutagenesis

It is a type of mutation involving the addition of genetic material. An insertion mutation can be small, involving a single extra DNA base pair, or large, involving a piece of a chromosome.

Deletion Mutagenesis

Deletion is a type of mutation involving the loss of genetic material. It is a type of mutation involving the loss of genetic material, genetic mutation, base pair, DNA chromosome.

Subjective Questions

1. What do you mean by a gene?
2. Discuss in detail the different units of a gene?
3. Give an account of the overlapping gene.
4. Discuss in detail different approaches made for the artificial synthesis of genes with emphasis on the works of H.G. Khorana?
5. What is chimeric DNA?
6. Write an essay on the organization of DNA in eukaryotes?
7. Write an essay on gene regulation by the lac operon model?
8. Write a short note on Chargaff's rule of equivalence?
9. What is electrophoresis? Describe the principle and working of the method?
10. Give a detailed account of working mass spectrometry?
11. What is pulsed electrophoresis?
12. What are the polymerase chain reaction and its application?
13. What is a restriction and modification system?
14. What are the DNA enzymes?
15. Give a detailed account of DNA polymerase and terminal transferase?
16. Write short notes on alkaline phosphatase, SI nuclease, Terminal transferases, sticky ends, palindromes, Linkers and adaptors, DNA ligases and Type III restriction enzyme?
17. What are plasmids?
18. What is cloning and cloning vector?
19. What is phagemid? Give a brief account of phagemids used as vectors?
20. Discuss in detail the yeast plasmid vectors?
21. What is a cloning vector?
22. How is the copy number of the plasmid vector and yield of the recombinant protein related to each other?
23. Can exonuclease be used while producing a recombinant DNA molecule?
24. What are the features of a plasmid being used as a cloning vector?

25. What are competent cells? What does the word "competent" refer to?
26. What do "Eco", "R" and "I" refer to in the enzyme EcoRI?
27. Why are proteases added while isolating the DNA?
28. If the "denaturation" step is missed during PCR, what would be its effect on the entire process?
29. Name a recombinant vaccine.
30. How is the Ti plasmid of *Agrobacterium tumefaciens* modified to convert it into a cloning vector?
31. Do biomolecules such as DNA, proteins exhibit biological activity in anhydrous conditions?
32. What is Biotechnology?
33. What is Genetic engineering?
34. What are the Principles of Biotechnology?
35. How is Biotechnology useful in developing food crops and in the agriculture process? What are the different types of biotechnology?
36. What do you understand by gene cloning?

Objective Questions

1. Restriction enzymes were discovered by

a. Smith and Nathans

b. Alexander Fleming

c. Berg

d. None

Answer: A

2. Bacteria protect themselves from viruses by fragmenting viral DNA with

a. Ligase

b. Endonuclease

c. Exonuclease

d. Gyrase

Answer: B

3. Klenow fragment is derived from

a. DNA Ligase

b. DNA Pol-I

c. DNA Pol-II

d. Reverse Transcriptase

Answer: B

4. **Southern blotting is**

a. Attachment of probes to DNA fragments

b. Transfer of DNA fragments from electrophoretic gel to a nitrocellulose sheet

c. Comparison of DNA fragments to two sources

d. Transfer of DNA fragments to electrophoretic gel from cellulose membrane

Answer: B

5. **ELISA is**

a. Using radiolabelled second antibody

b. Usage of RBCs

c. Using complement-mediated cell lysis

d. Addition of substrate that is converted into a coloured end product

Answer: D

6. **The Golden Rice variety is rich in**

a. Vitamin C

b. B-carotene and ferritin

c. Biotin

d. Lysine

Answer: B

7. **The DNA fragments have sticky ends due to**

a. Endonuclease

b. Unpaired bases

c. Calcium ions

d. Free methylation

Answer: B

8. **Plasmids are used as cloning vectors for which of the following reasons?**

a. Can be multiplied in culture

b. Self-replication in bacterial cells

c. Can be multiplied in laboratories with the help of enzymes

d. Replicate freely outside bacterial cells

Answer: B

9. The human genome project was launched in the year

a. 1980

b. 1973

c. 1990

d. 1989

Answer: C

10. The vaccines prepared through recombinant DNA technology are

a. Third generation vaccines

b. First-generation vaccines

c. Second-generation vaccines

d. None

Answer: A

11. Which is a genetically modified crop?

a. Bt-cotton

b. Bt-brinjal

c. Golden rice

d All

Answer: D

12. PCR technique was invented by

a. Karry Mullis

b. Boyer

c. Sanger

d. Cohn

Answer: A

13. The first transgenic plant to be produced is

a. Brinjal

b. Tobacco

c. Rice

d. Cotton

Answer: B

14. RNA interference helps in

a. Cell proliferation
b. Micropropagation
c. Cell defense
d. Cell differentiation

Answer: C

15. Which of the following is the quality of improved transgenic basmati rice?

a. Gives a high yield but no characteristic aroma
b. Gives a high yield and is rich in vitamin A
c. Does not require chemical fertilizers and growth hormones
d. Resistant to insects and diseases

Answer: B

16. The first clinical application of gene therapy over a 4-year-old girl was for

a. Adenosine deaminase deficiency
b. Adenosine deficiency
c. Growth deficiency
d. Adenine deficiency

Answer: A

17. Excision and insertion of a gene is called

a. Biotechnology
b. Genetic engineering
c. Cytogenetics
d. Gene therapy

Answer: B

18. The expression of a transgene in the target tissue is identified by a

a. Transgene
b. Promoter
c Enhancer
d. Reporter

Answer: D

19. ——— is used as a vector for cloning into higher organisms

a. Retrovirus

b. Baculovirus

c. Salmonella typhimurium

d. Rhizopus nigricans

Answer: A

20. Which bacterium is used in the production of insulin by genetic engineering?

a. Saccharomyces

b. Rhizobium

c. Escherichia

d. Mycobacterium

Answer: C

CHAPTER 10

PLANT BIOTECHNOLOGY

1. HISTORICAL PERSPECTIVE OF PLANT CELL/ TISSUE CULTURE, SCOPE AND IMPORTANCE IN CROP IMPROVEMENT

Introduction

- The beginning of plant biotechnology or plant tissue culture was made as early as 1898, when a German botanist, **G. Haberlandt** cultured successfully differentiated individual plant cells, which were isolated from different plant species.
- Further progress in cell culture research was made although embryo culture, roots and other tissues were successfully achieved in this period.
- The discovery of the importance of auxins and B-vitamins, the foundation of plant tissue culture was laid down by three scientists (**Nobecourt, White and Gautheret**) even though small pieces of tissue and not individual's differentiated cells could be grown in cultures during 1934-1939.
- To boost up these areas ICGEB in a workshop was held in New Delhi from September 18 to 20 1985, recommended the need for more research in developing countries on plant cell culture, differentiation, regeneration and transformation in tropical grain legumes, woody legumes and cereals.

- This was led to improve growth under stress conditions, pest and disease resistance, improved nutritional quality, nitrogen fixation and the control of partitioning within plants.
- Totipotency: The living cells of multicellular organisms are capable of independent development, when provided with suitable conditions, this statement given by White, 1963.
- According to Morgan, 1901 coined the term totipotency to denote this capacity of cells to develop into an organism by regeneration.
- However, the concept of totipotency is important for tissue culture.

History

- The experiments on wound healing in plants through spontaneous callus formation on the decorticated region of elm plants were observed by historically **Henri-Louis Duhamel du Monceau** in 1756.
- But the science is advanced related to cell and tissue culture after propounding the cell theory by **Schleiden and Schwann** (1839). Trecul in 1853 observed callus formation in a number of plants. **Hamberlandt** is regarded as the father of tissue culture. The organ culture was made attempts from 1902-1930.
- The isolated embryos of some crucifers and successfully grew on mineral salts and sugar solutions by **Hanning** in 1904. They successfully regenerated a bulk callus, buds, roots from a poplar tree on the surface of medium containing IAA which proliferated cell division by **Simon** 1908.
- The cultured plant material maintains its morphological identity, more or less with the same anatomy and physiology as *in vitro* of the parent plants by **Doods and Roberts,** 1985.
- The cultured tobacco tumor tissue from the hybrid Nicotiana Glauca and N. Langsdorffi by **White** 1939.
- **R.P White (USA), Gautheret (France) and Nobercourt (France)** independently cultured tissues excised from several plants on the defined nutrient media for a long period in 1930.
- The cultured cambium tissue of carrot on Knop's solution was supplemented with other chemicals in trace amount in 1939 by **Gautheret**.

- **Van Overbeek** in 1941 and coworkers used coconut milk for embryo development and callus formation in Datura.
- **Miller in 1955** was eventually a potent cell division factor from degraded DNA preparations was isolated and identified named as **Kinetin**.
- **Miller and Skoog in 1957** were advanced the hypothesis of organogenesis in cultured callus by varying the ratio of auxin and cytokinin in the growth medium. The shoot was formed by keeping the ratio of kinetin higher and the root developed when the ratio was lower.
- A cell can be propagated by subculturing by **Muir** in 1953 developed a successful technique for the culture of single isolated cells which is commonly known as the paper-raft nurse technique.
- Pfizer Inc. New York USA in 1952 got the US patent and started producing industrially the secondary metabolites of plants. The first natural production of the natural product shikonin by cell suspension culture was obtained.
- According to **Guha and Maheshwari** in 1966 developed techniques for the production of the vast numbers of embryos from the culture of pollens and sporogenous tissues of anthers. The methods to double the chromosome number in microspores of **Nicotiana and Datura** and collected seeds from the homozygous diploid plants within 5 months by **Nitsch** in 1974.
- Developing the following techniques of cell culture and somatic genetics of mainly grasses and cereals has emphasized by **Vasil** in 1982.
- The tissue culture was started in India during the mid-1950s at the Department of Botany (University of Delhi) by **Panchanan Maheshwari**. He is regarded as the father of embryology in India.
- **Sipra Guha Mukherjee and S.C Maheshwari** in 1964-1967 for the first time developed the haploid through another and pollen culture at the University of Delhi. The discovery of haploid production was a landmark in the development of plant tissue culture.
- The recombinant DNA technology made it possible to transform artificially cultured plant cells by introducing foreign genes.

Requirements for *in-vitro* Cultures

Tissue Culture Laboratory

- The plant cell culture and tissues *in vitro* is not an easy task. It requires all the nutrients and physiochemical factors in maintained in a laboratory.
- For the plant tissue culture, a good laboratory requires nutrient medium preparation, sterilization, cleaning and storage of supplies, aseptic condition for working the living materials, controlled environmental conditions for growth and development of culture, observation and evaluation of culture as hoped, recording the observation made during the experiment by **White** in 1963.

The most basic facilities that an individual's needs for tissue culture such as

- Washing and storage facilities
- Media preparation room
- Transfer Area

Washing and Storage facIlities

- A separate area is required which should have a large sink area with provision for hot and cold running water, distillation apparatus, washing machine, pipette washer, drier and cleaning brushes, keeping and weighing the chemicals, and putting glassware.

Media Preparation Room

- An area is required for the preparation of media. In such space, there should be provision for bench space for chemicals, labware, culture vessels, closures and miscellaneous equipment required for media preparation and dispensing.

Transfer Area

- The transfer of plant tissue is done in an open laboratory bench under clean and dry atmospheric conditions. Later on, a closed plastic box was constructed which consisted of a UV tube. UV light from the tube and the use of 95% ethanol helped to maintain sterile conditions inside the space of the plastic box.

Nutrient Media

- White in 1934 observed the unlimited growth of isolated root tissues when provided with nutrient medium containing inorganic salts, sucrose, vitamins, growth hormone and a few amino acids.

Inorganic Chemicals

- It includes macronutrients such as nitrogen, phosphorus, potassium, calcium, magnesium and sulfur in the form of salts in large amounts and microelements such as boron, molybdenum, copper, zinc, manganese, iron and chloride.

Growth Hormones

- There are several hormones are known to stimulate biological activity in cultured materials.
- Cytokines promote cell division and regulate growth and development similar to kinetin.
- Auxin resembles indole acetic acid (IAA) and stimulates shoot elongation.
- Gibberellins are used in apical meristem culture.
- The most widely used cytokines are adenine, kinetin, zeatin, benzyl adenine and auxins are IAA, NAA, 2,4-D.
- For the induction of callus, the amount of kinetin should be 0.1 mg/l. The structural formulae of cytokines and auxins are given in Figure 1.1.

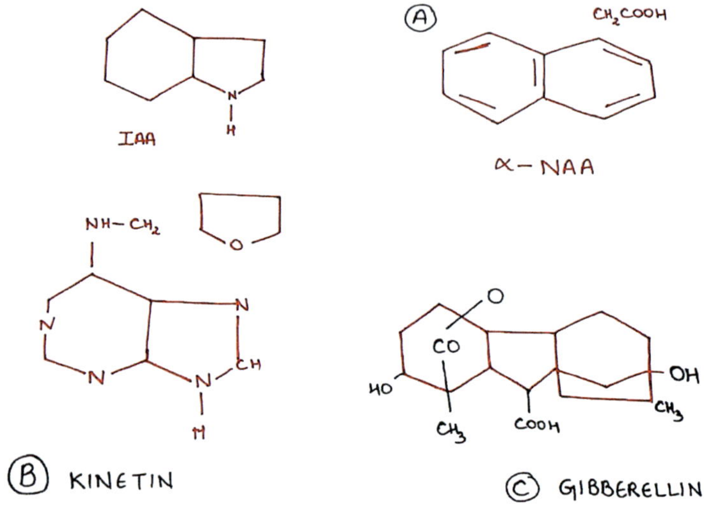

Figure 1.1: Structure of Growth Hormones (a) auxins (b) cytokinin (c) Gibberellin A_3.

Organic Constituents

- The organic compounds serve as a source of organic carbon and energy such as Sucrose, D-glucose, peptone, yeast extract, coconut water, malt extract, tomato juice, etc.

(1) Vitamins

- There are a number of vitamins that are required for culture such as Vitamin B_1, Vitamin B_2, Vitamin B_6, Vitamin C, Vitamin H, Vitamin B_{12}.

(2) Amino acids

- The most widely amino acids are used such as L-aspartic acid, L-asparagine, L-glutamic acid, L-glutamine and L-arginine.

Solidifying Agents

- Most commonly, agar- a polysaccharide obtained from a seaweed i.e a red alga, *Gelidium amansii* is used as a solidifying or gelling agent.

pH

- The pH affects the uptake ions optimum pH between 5.0 to 6.0 is required for the growth and development of cultured tissues. The pH should be maintained before the sterilization process or medium.

Maintenance of Aseptic Environment

- The contaminants produce toxic metabolites that inhibit the growth of cultured plant tissues. Therefore, each stem must be handled aseptically and with great care.

Sterilization of Glassware

- Glassware should be sterilized in a hot air oven at 160-180° C for 2-4 hours

Sterilization of Instruments

- The metallic instruments should be sterilized such as scalpels, forceps, needles, spatulas, etc dipping them in 75% ethanol followed by flaming and cooling. It is called incineration.

Sterilization of Culture Rooms

- The floor and room are washed first with detergent then 2% sodium hypochlorite or 95% ethanol. A large surface area is sterilized by exposure to UV light.

Sterilization of Nutrient Media

- The nutrient media are properly dispensed in a glass container plugged with cotton or sealed with plastic closures and sterilized by autoclaving at 15 psi for 30 minutes that gives 121°C for 30 minutes.
- During autoclaving plant extract, vitamins, amino acids and hormones are denatured so use Millipore filter paper which has 0.2 μm pore diameter for sterilization.

Sterilization of Plant Materials

- The plant materials have microbial contaminates so wash with sodium hypochlorite, hydrogen peroxide, mercuric chloride or ethanol should be used to make plant material sterilized.

Methods of Plant Cells, Tissue Organ Culture Basic Steps

- There are the basic steps as follows:
- Preparation of suitable nutrient medium.
- Sterilization of Explants
- Inoculation
- Incubations
- Regeneration
- Hardening
- Plantlet Transfer

Types of Culture

- There are different types of culture that are produced from plant material such as Explant culture, organ culture, cell or suspension culture, callus culture, protoplast culture, shown in **Figure 1.2.**

Explant Culture

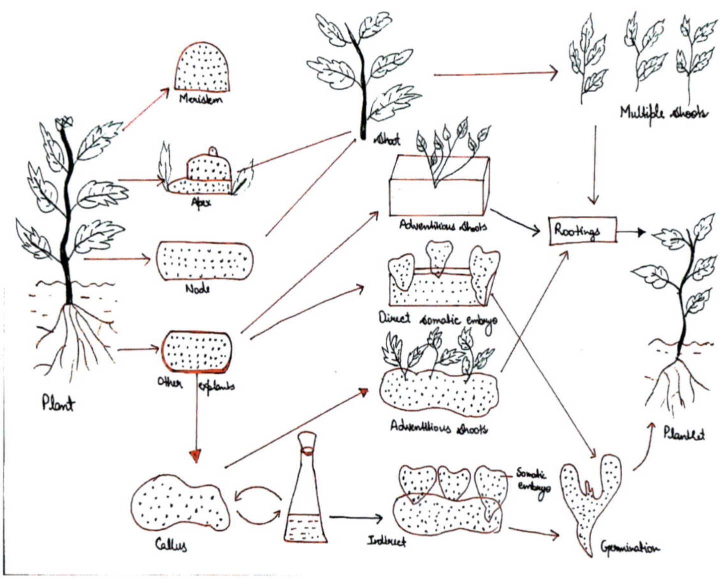

Figure 1.2: Regeneration of a plant through plant tissue culture.

- There are different varieties of seeds plants such as trees, herbs, grasses which exhibit the basic morphological units such as root, stem and leaves.
- There are various stages and differences in cells, tissues and their topography.
- Parenchyma is the most versatile of all types of tissue.
- They are capable of division and growth. However, the development of any tissue is characterized by three types of cell growth: cell division, cell elongation and cell differentiation.
- The healthy part of the plant should be used for culture purposes.
- Explant cultures are the cultures of plant material. Any part of a plant may be explant such as young and healthy pieces of stem, leaf, stem hypocotyl, cotyledons, etc.
- Explant cultures are generally used for induction of callus or regeneration of the plant.

Callus Formation

- The callus develops an infection of microorganisms from wounds due to stimulation by endogenous growth hormones, the auxins and cytokinins.
- However, it has been artificially developed by adopting tissue culture techniques.
- It is governed by the source of the explant, nutritional composition of medium and environmental factors.
- Explants of meristematic tissues developed cells more rapidly than thin three-walled and lignified cells of tissues.
- The callus is formed through three developmental stages: induction, cell division and differentiation.

Organogenesis

- Roots, shoots and leaves are the organs that are induced in plant tissue culture.
- Since the embryo is an independent structure and does not have a vascular supply, it is not supposed to be the plant organ.

- It starts with stimulation caused by the chemicals of the medium, substances carried over from the original explants and endogenous compounds produced by the culture by **Tomas and Davey, 1975**.
- Skoog and co-workers gave the concept of regulation of organogenesis by a balance between cytokinin and auxin.
- A high ratio of auxins and cytokinin stimulated the formation of the root in tobacco callus, but a low ratio of the same induced shoot formation.

Root Culture

- The root is cultured in a liquid medium. It has several advantages over solid media.
- The techniques of root culture give certain important information such as nutritional requirements, infection by Rhizobium and nodulation, physiological activities for eg, production of alkaloids, nicotine, etc by given Dodds and Roberts, 1985.
- **White in 1934** reported the first time successful organ culture e.g potentially unlimited growth of excised tomato roots.
- Subsequently, the roots of several species of gymnosperms and angiosperms have been successfully cultured.

Shoot Culture

- In the shoot culture, the apical meristem is cultured (the region of the shoot apex laying to leaf primordium). It is clearly different from shoot apex by having shoot apex and a few leaf primordial. The culture of relatively large tips (5-10 mm long sections), this technique is also known as meristem culture, Meristemming and mericlones (Murashige, 1974).

Micropropagation

According to Murashige in 1974 has described the procedure of development of micropropagation into three different development stages such as

- Establishment of explant aseptically
- Multiplication of propagules by repeated subcultures on a specific nutrient medium.
- Rooting and hardening of plantlets and planting into soil.

Cell-suspension Culture

- It is prepared by transferring a fragment of callus (500 mg) to the liquid medium (500 ml) and agitating them aseptically to make them cells free.
- It is difficult to have a suspension of single cells. However, the suspension includes single cell, cell aggregates residual inoculum and dead cells (Dodds and Roberts, 1985).
- **King (1980)** has described that a good suspension consists of a high proportion of single cells than a small cluster of cells.
- It is more difficult to have a good suspension than to find out optimum environmental factors for cell separation.

Benefits of Cell Culture

- Benefits for cell culture such as:
- Suspension can be pipetted.
- They are less heterogeneous and cell differentiation is less pronounced.
- They can be cultured in volumes up to 1.500 liters.
- They can be subjected to more stringent environmental controls.
- They can be manipulated for the production of natural products by feeding precursors by Kurz and Constable, 1979.

Somatic Embryogenesis

- It can be initiated in two ways
- Inducing embryogenic cells within the preformed callus
- Directly from the pre-embryonic determined cell (without callus)
- Which are ready to differentiate into embryoids by Sharp et al., 1980.
- In the first case, embryoids are initiated in callus from superficial cell aggregates where cells contain a large vacuole, dense cytoplasm, large starch granules and nucleus by Mc Willian et al., 1974.
- Two nutritional media of different compositions are required to obtain embryoids.

- The first medium contains auxins to initiate embryogenic cells.
- The second medium lacks auxin or a reduced level of auxin is needed for subsequent development of the embryonic cells into embryoids and plantlets.
- In both cases, a reduced amount of nitrogen is required (Ammirato, 1983).
- The embryogenic cells pass through 3 different stages such as globular heart-shaped and torpedo-shaped to form embryoids.
- These embryoids can be separated and isolated mechanically by using glass beads. When embryoids reach the torpedo stage they are transferred to a filter paper bridge and sucrose on which Whatman no.1.
- Filter paper is placed to make a bridge, described by Dodds and Roberts, 1985.

Somaclonal Variation

- The genetic heterogeneity of cells in a population represents a continuity of genotypes, whereas phenotypically the population is represented as a discrete sum of subclones.
- After cloning a single cell, from the population of the strain of *Dioscorea deltoidea*, there developed a subpopulation that defined their growth rate and sapogenin content.
- The mass occurrence of somaclonal variants, increase in the resistance, productivity and vital force of the plant has been explained by Carlson, 1983.
- This would be due to the dominance of nonlethal mutations that lead to heterozygosis with a wild -type allele which is thus phenotypically expressed as hybrid heterosis.
- There are some somaclonal variations in some plants such as *Brassica spp., Nicotiana tabacum, Oryza sativa, Saccharaum officinarum* and *Zea mays*.

Protoplast Culture

- Protoplast is the biologically active and most significant material of cells.
- When the cell wall is mechanically or enzymatically removed the isolated protoplast is known as a naked plant cell.
- The plant cell walls act as a physical barrier and protect the cytoplasm from microbial invasion and environmental stress.
- It consists of a complex mixture of cellulose, hemicellulose, pectin, lignin, lipids, protein, etc.
- For the dissolution of different components of the cell wall, it is essential to have the respective enzymes. There are a number of stages like:

Isolation of Protoplasts

- Surface sterilization of samples (Leaf)
- Rinsing in suitable osmoticum
- Plasmolysis of cells
- Peeling of the lower epidermis
- Isolation and purification of protoplasts

Regeneration

The protoplast regenerates a cell wall, undergoes cell division and forms a callus.

The callus can also be subcultured.

Embryogenesis begins from the callus when it is placed on a nutrient medium lacking mannitol and auxin.

The embryo developed into seedlings and finally, mature plants, shown in Figure 1.3.

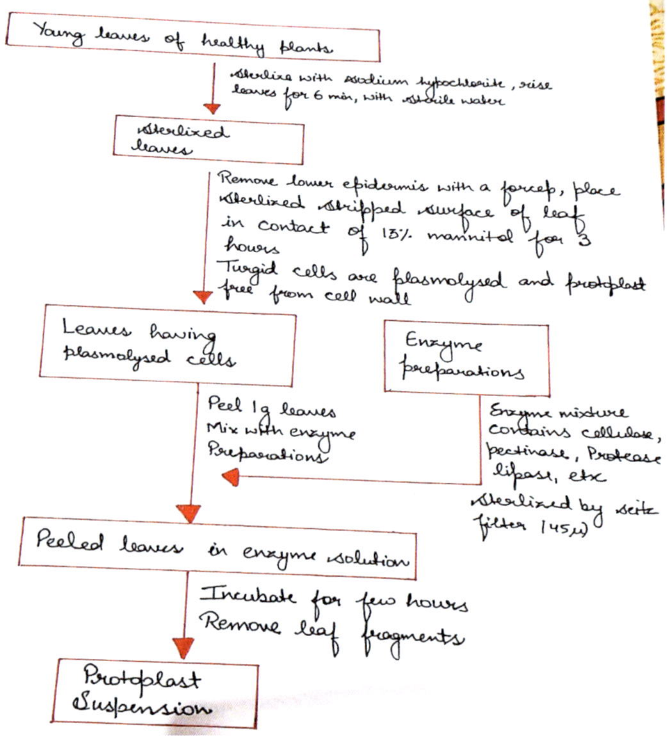

Figure 1.3: Isolation of Protoplasts.

Protoplast Fusion and Somatic Hybridization

Schieder in 1982 discussed four major aspects of protoplast fusion as: Production of fertile amphidiploid somatic hybrids of sexually incompatible species is achieved. The induced fusion of protoplasts from two genetically different lines of species must result in a variety of homo as well as heterokaryons fusion products. The selection of a few two somatic hybrid colonies from the mixed population of regeneration protoplast is a key step in a successful somatic hybridization technique. Production of heterozygous lines within one plant species which normally will be propagated only vegetatively e.g. potato. Transfer of only a part of genetic information from one species to another using the phenomenon of chromosome elimination. The transfer of cytoplasmic genetic information from one to a second line or species. It has been possible to transfer useful genes e.g. nif genes, disease resistance genes, rapid growth genes from one species to another, thereby widening the genetic base for plant breeding, shown in Figure 1.4.

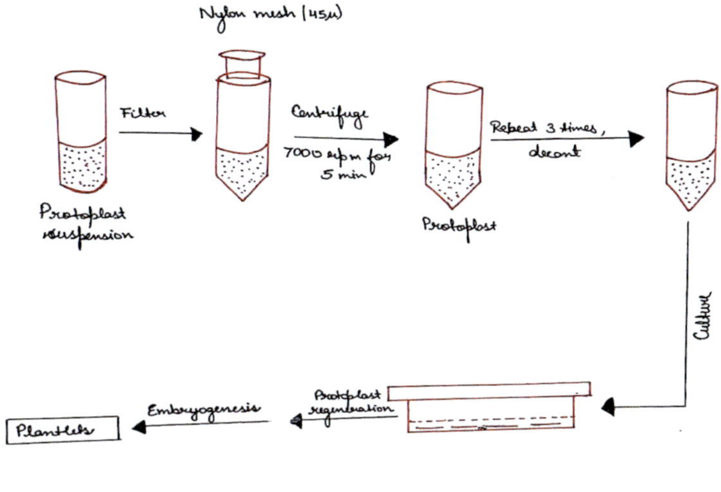

Figure 1.4: Purification, culture and regeneration of protoplasts.

Fusion Product

The fusion of the cytoplasms of two protoplasts results in coalescence of cytoplasms. The nuclei of two protoplasts may or may not use together even after the fusion of cytoplasms. The binucleate cells are known as heterokaryon or heterocyte when nuclei are fused the cells are known as hybrid or synkaryocyte and when only cytoplasm fuse and genetic information from one of the two nuclei is lost is known as cybrid i.e cytoplasmic hybrid or heteroplasts (Doods and Roberts, 1985), shown in Figure 1.5.

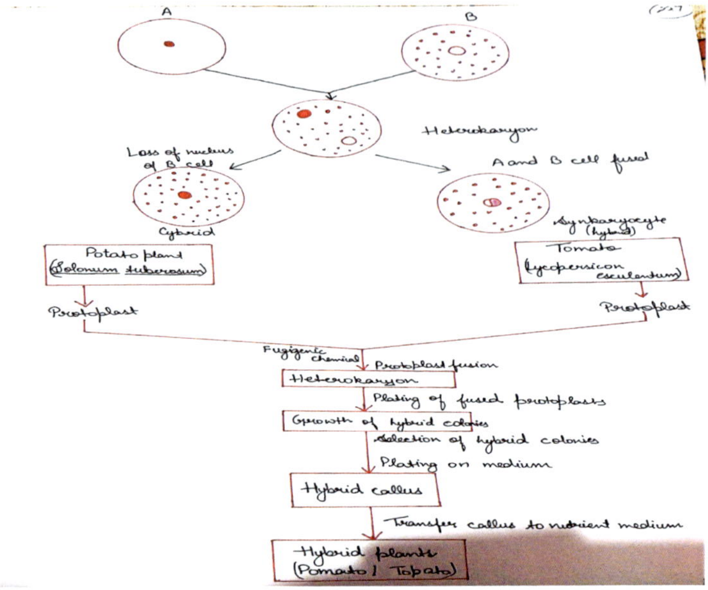

Figure 1.5: Cybrid/Hybrid productions.

Methods of Somatic Hybridization

- Isolation of protoplast from suitable plants.
- Mixing of protoplast in a centrifuge tube containing fugigenic chemicals i.e chemicals promoting protoplast fusion such as polyethylene glycol (PEG) 20% W/V, sodium nitrate ($NaNO_3$),; maintenance of high pH of 10.5 and temperature at 37° C.
- Wall regeneration by heterokaryotic cells.
- Fusion of nuclei of heterokaryon to produce hybrid cells.
- Plating and production of colonies of hybrid cells.
- Selection of hybrid, subculture and induction of organogenesis in the hybrid colonies.
- Transfer of mature plants from the regenerated callus.

Selection of Somatic Hybrids and Cybrids

- After the fusion, protoplasts of media regenerate cell walls and undergo mitosis.
- This results in a mixed population of parent cells, homokaryotic fusion products and hybrids.

- Hybrid cells should be differentiated from other cells.
- There are various selection methods used for the selection of fusion products. Selection methods are dependent on:
- Physical properties of fused cells.
- Biological properties of fused cells.
- Biological properties of colonies from fused cells.
- The somatic hybrids cannot be identified.
- For the first time, P.S Carlson and co-workers produced the first interspecies somatic hybrids between ***Nicotiana glauca* and *N. langsdorfii*** in 1972.
- The most somatic hybridization experiment, selection procedure includes full of chlorophyll-deficient (non-green) chloroplasts of one parent with the protoplasts of the other parents.
- This help in the visual selection of heterokaryon under a microscope.

Anther and Pollen Culture

- A male reproductive organ is haploid in chromosome numbers.
- As a result of microsporogenesis, tetrads of microspores are formed from a single spore mother cell.
- They are known as pollen grains after release from tetrads by Bhojwani and Bhatnagar, 1974. Haploid plants are very useful in direct screening of recessive mutation because in the diploid or polyploid screening of recessive mutation is not possible, development of homozygous diploid plants following chromosome doubling of haploid plant cells, shown in Figure 1.6.
- In China, the most widely grown wheat is a doubled haploid produced through homozygous diploid lines.
- Anther cultures of rice are also successfully grown.
- Haploid plants have been produced in tobacco, wheat and rice through pollen culture.

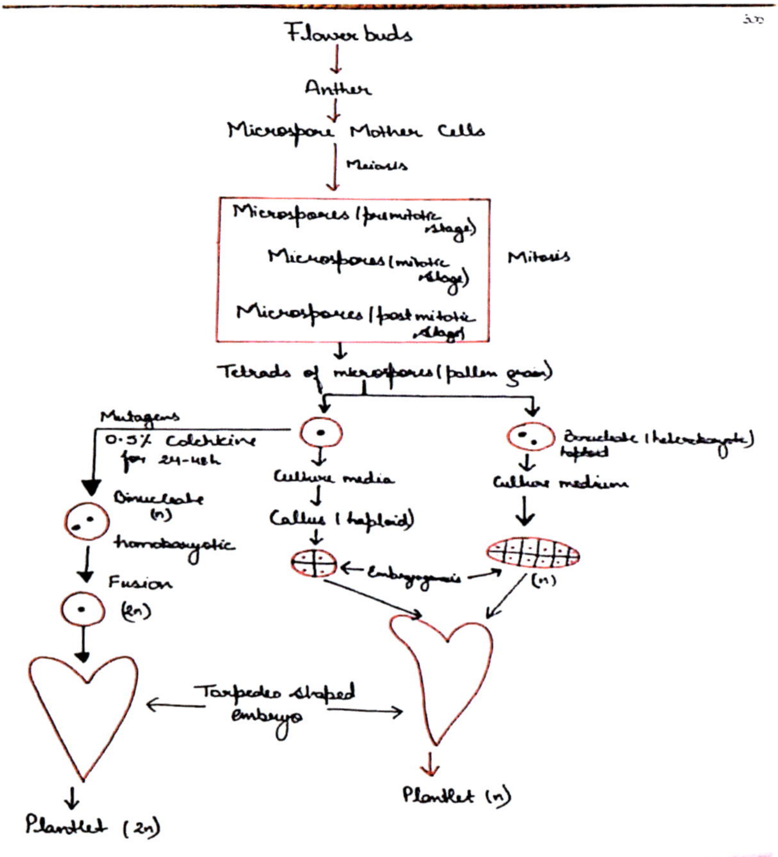

Figure 1.6: Anther Culture.

- More than 247 plant species and hybrids belonging to 38 genera and 34 families of dicots and monocots have been regenerated using anther culture technique.
- They include economically important crops and trees such as rice, wheat, coconut, rubber trees, etc. by Maheswari et al., 1983.

Culturing Techniques

Anthers are superficially sterilized and washed with double distilled sterile water. They are excised from the flower buds and their proper developmental stages are determined under a microscope. On confirmation of stages such as (a) anther are directly transferred on nutrient agar or liquid medium where induction of embryogenesis occurs (b) the pollen grains are aseptically removed from the anthers and cultured on liquid medium (Reinert and Bajaj,

1977). Methods of anther culture.

In vitro Androgenesis

- *In vitro* androgenesis is the formation of sporophytes from the male gametophyte on an artificial medium.
- It is most commonly found in families ***Solanaceae and Poaceae*** (Graminae).
- The success of *in vitro* androgenesis is based on adjustment of the development stage of pollen, minerals of culture medium and growth regulators as well as thermal shock or other treatments.
- Pollen grains are isolated from excised anther and extracted in a beaker.
- Well collected pollens are washed properly and centrifuged and decanted. They are inoculated in a liquid medium, then subcultured in a solid MS medium.
- In culture, pollen grains can be induced to produce callus or embryos from which whole plants are regenerated in one month.
- This technique is most successful in Brassica, Datura, petunia, etc. Mainly there are two ways of *in vitro* morphogenesis of pollen grains, direct and indirect.

Direct Androgenesis

- It is also called pollen derived embryogenesis. Here pollen directly acts as a zygote and therefore passes through various embryogenic stages similar to zygotic embryogenesis.
- Pollen grains reach the globular stage of the embryo, then the plants.
- Direct androgenesis is very common in many plants of the family Solanaceae and Brassicaseae, discussed by Prakash and Giles, 1987.

Indirect Androgenesis

- In indirect androgenesis, the pollen grains, instead of normal embryogenesis, divide erratically to develop a callus. Indirect androgenesis has been found in barley, wheat, Vitis, coffee, etc.

- Possibility of pollen morphology from the dividing pollen varies.
- The haploid callus, embryo or plantlets may originate from the continued division of vegetative cells of the pollen, multiple divisions of generative cells, division products of both generative and vegetative cells described by Sangwan, 1981.

Mentor Pollen Technology

- Pollination is a natural phenomenon that occurs in many plant species.
- However, there are certain plants where pollen grains remain viable but unable to set seed on their own pistil i.e, they are self -incompatible.
- Similarly, in several other plants, besides being viable they even fail to germinate on stigmas of another species i.e they are interspecific incompatible.
- After the birth of recent mentor pollen technology, the incompatibility problem of pollen grains can be overcome by altering the fertilization ability.
- The pollen which has been purposefully treated is called mentor pollen.
- Hence the mentor pollen is defined as the compatible pollen which has been treated in many ways to inhibit its fertilization ability and retain its power to stimulate incompatibility pollen to accomplish fertilization.
- The mentor pollen itself may or may not germinate on stigmas.
- As a result of this treatment, hybridization between closely related species or genera and pollination in cross -pollinating plants could be achieved.

Embryo Culture

- Embryo culture has also been done for the production of haploid plants. It is used for the recovery of plants from distinct crosses.
- It is useful when embryos fail to develop due to the generation of embryonic tissues.
- It is being used extensively in the extraction of haploid barley *Hordeum vulgare* from the crosses *H. vulgare* × *H. bulbosum*.

Das and Barman (1992) developed the method of regeneration of tea shoots from embryo callus.

- It produced somatic embryoids within 8 weeks of culture in the second medium which differentiated into buds after 2 weeks. Several shoots with 4-6 leaves developed after 16 weeks of culture.

Embryo Rescue

- Viable hybrids are produced as a result of sexual crosses between two varieties of the species. However, if sexual crosses are done between the species of the same genus or between two different genera, production of hybrid is rather difficult because of several barriers arising either during pollination, fertilization or embryogenesis.
- It has been observed that in some cases in spite of successful pollination and fertilization, the embryo does not develop.
- This is due to inherent deficiencies or incompatibility between the developing embryo or endosperms.
- In such cases, immature embryos are dissected out from the fruit (seed) and grown artificially on a medium that differentiates into shoot, root and plantlets.
- This technique of growing immature embryos is termed as **"embryo rescue".**
- The embryo rescue technique is very useful in wide hybridization, complete growth of the embryo in a plant, breaking dormancy of certain seeds where the dormancy period is very long. By using the embryo rescue technique wild varieties can be crossed with cultivars.
- As compared with cultivars, the wild species have greater resistance to pests and pathogens and produce grains of better quality.

Triploid Production

- The triploid primary endosperm (fusion of the second male gamete with two polar nuclei).

- Some examples of triploid plants raised from endosperm cultures such as Rice (*Oryza sativa*), Maize (*Zea mays*) and Barley (*Hordeni vulgare*).
- The triploids of poplar (*Populus tremuloides*) have better quality pulpwood. Therefore, it is important to the forest industry.

Protoplast Fusion in Fungi

- The protoplast fusion in fungi is the improvement of strains to be used for commercial purposes.

For example, two strains of *Cephalosporium acremonium* were crossed and *Cephalosporin C* was improved.

- Attempts to improve the citric acid production in *A. niger* were unsuccessful (Peberdy, 1989).

The Protoplast Techniques are applied in the following four Major Areas to Get the Crosses:

- Between different species and genera.
- Between apparently incompatible strains of the same species.
- Between different strains or isolates of a species primarily for breeding.
- Between isogenic strains may provide an opportunity for genetic mapping.

Intraspecific Protoplast Fusion

- It is the cross between the same species in an individual which involves the isogenic strains or the non-isogenic ones eg *Absidia glauca*, *Pleurotus ostreatus, Fusarium graminearum* and *Treesi* etc.

Interspecific Protoplast Fusion

- It is the crosses between two different species It is important in the area where new products are to be produced eg., *S. cerevisiae* × *S. fermentali*, *Pleurotus ostreatus* × *P. florida,* etc.

Application in Agriculture

- There are a number of applications in agriculture which has solved the problem of incompatibility of plants and scope of production of new verities within a short period of time.
- Improvement of Hybrids
- Production of Encapsulated Seeds
- Production of Diseases Resistant Plants
- Transfer of the *nif* gene to Eukaryotes

Bioethics in Plant Genetic Engineering

There are new varieties of production of a plant due to mutation. Cultivation of genetically modified crops by farmers is increasing fast thought out the world. It will support healthcare and industry and provide food, feed and fiber security on a global basis. The major concern about a genetically modified plant or crops foods is given below such as:

- Genetically modified crops may bring about changes in evolutionary patterns.
- Effect of genetically modified crops on biodiversity and the environment.
- Transfer of transgene from genetically modified food to pathogenic microbes.
- Risk of change in fundamental vegetable nature of plants.
- Effect of genetically modified on non-target and beneficial insects and microbes.
- Pollen transfer from genetically modified plants.
- The risk of transfer of allergies.
- Some Important contributions have been described below.

Year	Author
1902	Plant Tissue culture (Haberlandt)
1904	Embryo culture of selected crucifers (Hannig)
1922	Asymbiotic germination of orchid seeds invitro (Knudson)
1922	*In vitro* culture of root tips (Robbins)
1925	Use of embryo culture technique in interspecific crosses of Linum (Laibach).
1934	Successful culture of tomato roots (White).

1939	Successful establishment of continuously growing callus cultures (Gautheret, Nobecourt and White).
1941	Use of coconut milk containing a cell division factor for the first time in Datura (Van Overbeek).
1944	*In vitro* adventitious shoot formation in tobacco (Skoog).
1946	Rising of whole plants of *Lupinus and Tropaeolum* by shoot tip culture (Ball).
1952	Use of meristem culture to obtain virus- free dahlias (Morel and Martin).
1952	First application of micrografting (Morel and Martin).
1952	Got the US Patent on their claim that plant cells could be grown in a liquid nutrient medium and had the potential to produce useful compounds (Routien and Nickel of Pfizer and Co. Inc., New York).
1953	Production of haploid callus of the gymnosperm *Ginkgo biloba* from pollen (Tulecke).
1954	First plant from a single cell (Muir et al.,).
1955	Discovery of kinetin a cell division hormone (Miller et al.,).
1957	Discovery of the regulation of organ formation by changing the ratio of auxin: cytokinin (Skoog and Miller).
1958	Regeneration of somatic embryos *in vitro* from the nucellus of citrus ovules (Maheshwari and Rangaswamy).
1959	Regeneration of embryos from callus clumps and cell suspensions of Daucus carota (Reinert, Steward).
1959	Publication of the first handbook on plant tissue culture (Gautheret).
1960	First successful test tube fertilization in *Papaver rhoeas* (Kanta).
1960	Use of the microculture method for growing single cells in hanging drops in a conditioned medium (Jones et al.).
1960	Enzymatic degradation of cell walls to obtain a large number of protoplasts (Cocking).
1960	Filtration of cell suspensions and isolation of single cells by plating (Bergmann).
1962	Development of Murashige and Skoog nutrition medium (Murashige and Skoog).
1964	Production of first haploid embryos from pollen grains of *Datura innoxia* (Guha and Maheshwari).
1967	Androgenic haploid plants of *Nicotiana* were produced (Bourgin and Nitsch).
1968	Restriction endonuclease term coined to a class of enzymes involved in cleaving DNA (Meselson and Yuan).
1970	Selection of biochemical mutants *in vitro* by the use of tissue culture derived variation (Carlson).
1970	The first achievement of protoplast fusion (Power et al.).

1970	Discovery of first restriction endonuclease from *Haemophillus influenzae* It was later purified and named HindII (Kelly and Smith).
1970	In certain cancer -causing animal virus genetic information flow in reverse form. Discovery of reverse transcriptase (Temin).
1971	The first restriction map using HindII enzyme to cut circular DNA of SV 40 into 11 specific fragments was prepared (Nathans).
1971	Regeneration of first plants from protoplasts (Takebe et al.)
1972	Interspecific hybridization through protoplast fusion in two species of *Nicotiana* (Carlson et al.).
1972	First recombinant DNA molecule produced using restriction enzymes (Jackson et al.).
1972	Joining of two restriction fragments regardless of their origin produced by the same restriction enzyme by the action DNA ligase (Mertz and Davis).
1973	Use of Lobban and Kaiser technique to develop hybrid plasmid-insertion of EcoRI fragment of DNA molecule into circular plasmid DNA of bacteria using DNA ligase. Gene from African clawed toad inserted into plasmid DNA of bacteria (Cohen et al.).
1974	Ti plasmid is the tumor- inducing principle of Agrobacterium (Zaenen et al., Larebeke et al.,).
1975	Positive *in vitro* selection of maize callus cultures resistant to T toxin of *Helminthosporium maydis* (Gengenbach and Green).
1975	High resolution 2 Dimensional gel electrophoresis procedure developed that led to the development of proteomics (O'Farrel).
1976	Shoot initiation from cryopreserved shoot apices of carnation (Seibert).
1977	Successful integration of the Ti plasmid DNA from *A. tumefaciens* in plants (Chilton et al.,).
1977	Gene sequencing is based on the degradation of the DNA chain (Maxam and Gilbert).
1977	Discovery of split genes (Sharp and Roberts)
1978	Somatic hybridization of tomato and potato resulting in potato (Melchers et al.).
1979	A co-cultivation procedure was developed for the transformation of plant protoplasts with *Agrobacterium* (Marton et al.,)
1980	Commercial production of human insulin through genetic engineering in bacterial cells (Eli Lilly and Co).
1981	Somaclonal variation (Larkin and Scowcroft).
1982	Incorporation of naked DNA by protoplasts resulting in the transformation with isolated DNA (Krens et al.)
1984	Transformation of tobacco with *Agrobacterium* transgenic plants developed (De Block et al., Horsch et al.).
1986	A genetic fingerprinting technique was developed for identifying individuals by analyzing polymorphism at the DNA Sequence level (Jeffreys).

1990	Random amplified polymorphic DNA technique developed (Williams et al., Welsh and McClelland).
1991	DNA microarray system using light directed chemical synthesis system developed (Fodor).
1995	DNA fingerprinting by amplified fragment length polymorphism (AFLP) technique developed (Vos et al.,).
1997	Sequencing of *E.coli* genome
2001	Sequencing of the human genome successfully completed (Consortium and Venter et al.,)
2002	The draft sequence of the rice genome (*Oryza sativa L.ssp. japonica*)
2003	The draft sequence of the rice genome (*Oryza sativa L.ssp. indica*) (Yu et al., 2002).
2005	Map- based sequence of the rice genome (international Rice genome sequencing project).

Subjective Questions

1. What is Plant Biotechnology and its applications?
2. Write a brief history of the development of *in vitro* techniques of plant cell culture?
3. What are the requirements for establishing a tissue culture laboratory?
4. Give a brief account of sterilization and its methods?
5. How will you prepare the culture medium?
6. Write a note on the formulation and nutritional components of the culture medium?
7. What is an explant? How you will induce callus from it?
8. Write an essay on organogenesis and its application?
9. Give a brief account of benefits from root culture?
10. How will you prepare cell suspension? Discuss the benefits of using an aqueous medium over the solid one?
11. What do you know about micropropagation?
12. Write an essay on somatic embryogenesis?
13. How will you isolate protoplasts from the plant cells?
14. What are the benefits of pollen?
15. What is anther culture?
16. Write an essay on explant, callus, organogenesis, totipotency, cell culture, somaclonal variants, hybrid and cybrid, fugigenic chemicals, anther and pollen culture, micropropagation, protoplast fusion in fungi, *In vitro* androgenesis, Mentor pollen technology, embryo rescue and embryo culture?

Objective Questions

1. **Plant biotechnology involves**
   a. Production of valuable products in plants
   b. Rapid clonal multiplication of desired genotypes
   c. Production of virus free plants
   d. All of these

Answer: D

2. **The most common solidifying agent used in micropropagation is**
   a. Agar
   b. Dextran
   c. Mannan
   d. All of these

Answer: A

3. **The culturing of cells in the liquid agitated medium is called**
   a. Liquid culture
   b. Micropropagation
   c. Agar culture
   d. Suspension culture

Answer: D

4. **Which of the following is best-suited method for the production of virus-free plants**
   a. Embryo culture
   b. Meristem culture
   c. Ovule culture
   d. Anther culture

Answer: B

5. **Batch cultures are a type of suspension culture where**
   a. Medium is continuously replaced
   b. Medium is loaded only at the beginning
   c. No depletion of medium occurs
   d. Cellular wastes are continuously removed and replaced

Answer: B

6. **Immobilized cell bioreactors are based on**
   a. Cells cultures in solid medium
   b. Cells cultured in liquid medium
   c. Cells entrapped in gels
   d. All of these

Answer: C

7. **All are plant derived alkaloids except**
   a. Menthol
   b. Nicotine
   c. Quinine
   d. Codeine

Answer: A

8. **Elicitors are molecules that**
   a. Induce cell division
   b. Stimulate production of secondary metabolites
   c. Stimulate hairy root formation that accumulates secondary metabolites
   d. None of these

Answer: B

9. **All are plant derived elicitors except**
   a. Chitin
   b. Pectin
   c. Cellulose
   d. Pectic acid

Answer: A

10. **The modification of exogenous compounds by plant cells is called**
   a. Biotransformation
   b. Bioconversion
   c. Both a and b
   d. Biophytomodification

Answer: C

11. Artificial seeds are

a. Seeds produced in laboratory condition

b. Seeds encapsulated in a gel

c. Somatic embryos encapsulated in a gel

d. Zygotic embryos encapsulated in a gel

Answer: C

12. Hairy root cultures for secondary metabolite production are induced by transforming plant cells withs

a. Virus

b. Agrobacterium tumefaciens

c. Bacillus thuringiensis

d. Agrobacterium rhizogenes

Answer: D

13. The variation in *in vitro* culture is called as

a. *Invitro* variation

b. Mutation

c. Somaclonal variation

d. All of these

Answer: C

14. Haploid plants are produced in large numbers by

a. Anther culture

b. Ovary culture

c. Both a and b

d. Embryo culture

Answer: C

15. Cybrids are

a. Nuclear hybrids

b. Hybrid plants derived from cross-pollination

c. Cytoplasmic hybrids

d. Cytological hybrids

Answer: C

16. In plant tissue culture, what does the term ORGANOGENESIS mean?

a. Formation of callus culture

b. Formation of root and shoot from callus culture

c. Genesis of plants

d. None of the above

Answer: B

17. A recombinant DNA molecule is produced by joining together

a. One mRNA with a DNA segment

b. One mRNA with a tRNA segment

c. Two mRNA molecules

d. Two DNA segments

Answer: D

18. A gene produced for recombinant DNA technology contains a gene from one organism joined to the regulatory sequence of another gene. Such a gene is called

a. Oncogene

b. Junk gene

c. Chimeric gene

d. None

Answer: C

19. A group of genetically similar organisms obtained by asexual reproduction is called

a. Clone

b. Population

c. Assembly

d. None

Answer: A

20. To be useful in the preparation of recombinant DNA, a plasmid must have

a. No origin of replication

b. An origin of replication

c. The ability to alternate between the linear and circular forms

d. Restriction endonuclease activity.

Answer: C

CHAPTER 11

ANIMAL BIOTECHNOLOGY

1. HISTORY, TYPES OF TISSUE CULTURE, ADVANTAGES AND LIMITATIONS

Introduction

The most important areas of today's research that have potential economic value and prospects of commercialization are the cell and tissue culture based on the production of vaccines, monoclonal antibodies, etc.

History of Animal Cell Culture

Alex carrel used tissue and embryo extract as culture media. The fibrin clot of plasma served as an anchor for cell attachment and the extract provided growth factors and nutrients. A major innovation to get cell suspension was the use of trypsin for cell disaggregation from tissue explants. This allowed single-cell culture. Thus the technique of cell cultures differs from tissue culture. The use of biological fluids and extracts was a contaminated major problem because of getting contaminated. In 1907, Ross Harrison made the first attempt to culture animal cells and cultivated embryonic nerve cells of a frog by using the hanging drop method. Thereafter, this method was extended and a wide range of mammalian cells was cultured *in vitro*. However, after supplementing with the embryo of chick and plasma, cells proliferated well. These provide nutrients and proliferation factors. Moreover, the cell culture method was extracted quite improved after the discovery of antibiotics during the 1940s. Such studies provided an impetus for the culture of animal cells on a large scale. During this period many human carcinoma cell lines (e.g. HeLa cell line) were isolated and grown in culture. Alexis Carrel, the famous physiologist kept the chick embryo heart alive and its beating continued *in vitro* for about three months. Animal cell culture studies resulted in the total stop of using monkeys for the multiplication of animal viruses.

Requirements for Animal Cell and Tissue Culture

Animal Cell Growth in Culture

Unlike plant and microbial cells, animal cells can grow only to limited generations even in the best nutritive media. This growth also depends on the sources of tissue isolated. The special features of different cell cultures are briefly discussed (i) Neuronal cells constitute the nervous system, In culture, the neuronal cells cannot divide and grow.

The cells that form connective tissue (skin) is called a fibroblast. Fibroblast can divide and grow in culture to some generations. After completing several generations, they die. It means that all normal animal cells are mortal. (iii)In culture, the animal cells divide and grow. Consequently, they fill the surface of the container in which they are growing. Thereafter, they stop further growth. This phenomenon is termed contact inhibition i.e. inhibition of further cell growth after reaching the wall of the container. The environment of cell growth in culture differs from that of *in vitro*, for example:

- Absence of cell-cell interaction and cell-matrix interaction,
- Lack of three-dimensional architectural appearance, and
- Changed hormone and nutritional environment. The way of adherence to glass or plastic containers in which they grow, cell proliferation and shape of cell results in alterations.
- In culture, the cancer cells apparently differ from the normal cells. Due to uncontrolled growth and more rounded shape, they lose contact inhibition. Therefore, they are piled on each other. These features are exploited by cancer specialists i.e. the cancerologists to identify cancer cells from the normal cells.

Requirements for animal cell and tissue culture are the same as described for plant cell, tissue and organ culture Desirable requirements are: (i) air conditioning of a room, (i) hot room with temperature recorder

(iii) microscope room for carrying be installed. (iv) darkroom glassware and culture media, and should out microscope work, (v) plastics-shelving (vi) sealed container and drawers, (vii) specialized chemicals.

Substrates for Cell Growth

- There are many types of vertebrate cells that require support for their growth I anchorage-dependent cells e (e.g plastic, glass, palladium types may be used).

Plastic as a Substrate

- Disposable plastics are cheaper substrates as they are commonly used *in vitro* conditions.

Glass as a Substrate

- Glass apparatus etc. These are sterilized by chemicals, radiations, dry heat fin oven) and heat (in autoclaved test tubes, slides, coverslips flasks, rods, moist palladium deposited on agarose was used as For the first time and glia. the efficiency of cultured treated onto the surface

Culture Media

Culture for animal cells and tissue are prepared by certain chemical constituents from inorganic more difficult than that of microorganisms and substances. However, chemical media used for the culture of animal cells and the culture media provide the optimum growth factors (e.g pH, osmotic pressure) (a) Natural Media: Natural media are the natural sources of nutrient and proliferation of animal cells and tissue. It is used for a long time but is now available in the market in the form of liquid microbes).

Glassware, Equipment and Culture Media

The thoroughly washed glassware and equipment and carefully prepared culture media are millipore filter paper. The sterilized materials are removed sterilized by heat, steam or when the temperature cools down and used according to the procedure adopted.

Equipment required for Animal Cell Culture

The equipment required for animal cell culture are given below:

Laminar Air-flow (LAF)

LAF hood acts as animal cells, Culture manipulation in aseptic conditions protects from contamination by any microorganisms such as bacteria and fungi. The contaminated bacterial/fungal cells grow more rapidly than the cultured animal cells. Therefore, the growth of animal cells fails to occur in the presence of contaminants. The working area of the LAF hood is first made sterile by using 70% ethanol. Manipulation of any cell is done by keeping the LAF in ON position aseptic working table for inoculation of Blower The LAF hood performs two functions: t provides a sterile environment for cell manipulation (ie. protects tissue culture from operator) and (protects the operator from the potential infection risk from the culture. There are different types of LAF hoods, airflow hoods (cabinets) are commercially available in different size and shapes. In this apparatus, sterile air flows inside the space of the cabinet which maintains the sterile conditions required for all transfer work. Different types of laminar airflow hoods are available on the ba- dishes Flask sis of nature of the cells and kinds of organisms and tissue culture as below.

CO_2 Incubators

The CO, incubators provide suitable environmental conditions to the growing animal cells. A silicon gasket is used on the inner door which makes the chamber of the incubator airtight. It separates the internal environment from the external environment. The external environment possesses microbial contaminants, while the internal environment remains always contamination-free. The filtered air i.e. a high-efficiency particulate air (HEPA) is injected inside the chamber which maintains the internal environment of the chamber sterile. The relative humidity inside the chamber remains high; therefore, correct osmolarity is maintained and the medium is protected from desiccation.

Centrifuges

There are different types of centrifuges based on speed. A low-speed centrifuge is needed for most of the cell cultures. The separated beads of cells are disrupted simply by a gentle breaking action. Commonly cells are centrifuged at 20° C. The motor evolves heat that raises the temperature. Therefore, the use of low temperature for centrifugation is preferred so that the cells should not be exposed to high temperatures.

Inverted Microscope

The use of an inverted microscope is important to observe cell cultures in situ. Because the cells are found on the bottom of the tissue culture vessel (e.g Petri plates). The culture medium remains above the growing cells in plates. If such plates are put over the stage of an ordinary microscope, the growing cells at the bottom cannot be observed. Therefore, the inverted microscope is used for such a purpose.

Culture Room

All types of cultured plant tissues are incubated under the conditions of well-controlled temperature, humidity, illumination and air circulation. The culture room should have light and temperature control systems. Generally, the temperature is maintained at 25 ± 2°C and 20 98% relative humidity and uniform air ventilation. Generally, cultures darkness each for a period of 12 hours.

Data Collection (Observation)

The cultures are monitored at regular intervals in the culture room for the growth and development of cultured tissues. Observation is also made under the aseptic area in laminar airflow.

Isolation of Animal Material(Tissue)

During the experimental work, attempts should be made that animal materials are not contaminated. When glassware and media are sterilized, animal materials should be handled. Before that, a balanced salt solution (BSS) is required. This solution consists of 1000 units of penicillin and 0.5 mg of streptomycin washed in BSS aseptically to avoid contamination. The tissue to be cultured should be properly sterilized with 70 percent ethanol, and removed surgically under aseptic conditions. Attempts should be made to avoid contamination. Thus, the tissue isolated is either stored in freeze or used immediately.

Establishment of Cell Cultures

There are many types of animal cells that can grow in such as tumor cells, pigmented melanoma cells, neuroblastoma cells, steroid-producing adrenal cells, growth hormone prolactin-secreting cells, teratoma cells capable of differentiating in cold trial conditions pigmented or cartilage cells, etc. On the basis of the purpose of the experiment, the immortal cell line can be developed from cultured tissues. Healthy animal tissues are cultured on artificial nutrient media that proliferate and differentiate into able of dividing into a mixture of different types of cells.

Cell Lines

The primary cell culture cannot long time because the cell utilizes all nutrients of the medium. Therefore, sub remains viable for culturing needs to be done on another fresh medium. Hence, the primary culture is removed adherent cells are dissociated enzymatically medium and passed into fresh culture flask or by repeated pipetting. Then the cells are diluted with fresh. This results in multiple copies of proliferating cells. During the course of repeated sub-culture and selection, the cell line gets evolved and properly established consisting of rapidly proliferating cells. Thus the unaltered form of cell fine (only for a limited number of generations) is called continuous cell line which propagates in a single type of cell with a negligible amount of non- logarithmic ways. This mixture of cells is in single-cell suspension which may be used as a primary culture or starter culture. The primary culture (in the form of single-cell suspension is subcultured by transferring into culture dishes/flasks containing special growth nutrients t optimal growth conditions Consequently, some cells attach to the surface and proliferate to yield a single cell line, in spite of being damaged in suspension. Therefore, while subculturing the suspension should be diluted with a fresh medium at a certain ratio and transferred into a flask (Anon, 1988). Primary Cell Cultures: The first step in establishing cells in culture is to dissociate organs into a single cell suspension. It is done by mechanical or enzyme methods. The cells are transferred into special glass or plastic containers containing a culture medium. Under these conditions, maintenance of growth of such cells is called primary cell culture

Hybridoma Technology

Blood consists of two major components, various kinds of cells and the fluid called serum consists of a heterogeneous population of antibodies. Different antibodies are different types of B-lymphocytes. Hence, serum consists of different types of antibodies which are often called polyclonal antibodies. The antibodies bind to specific domains of antigens (called antigenic produced by dif determinants or epitopes) and neutralize them.

Monoclonal Antibodies Production of Monoclonal Antibodies (MoAb)

The mouse MOAB has revolutionized the field of biology and more specifically immunology. The mice MoAb have been used in human patients with varying levels of success in leukemia, lymphoma, melanoma and colorectal cancer. Clinical trials have indicated several limiting factors such as: (i) heterogeneity of tumor cells (not all malignant cells carry relevant antigen), circulating free antigens (they bind Fab on antibody molecules and thus block Moab from binding to the target cells), (iii) antigenic modulation (antigen modulated off the cell surface as a consequence of the binding of MoAbs to the cancer cell as in leukemia. This problem can be overcome by applying introduced to hamster kidney cells.

Applications of Monoclonal Antibodies

The monoclonal antibodies are used in the four main ways i.e. disease diagnosis, disease treat- nt, passive immunization and detection and purification of biomolecules. However, their exploitation for various uses is based on two features: (i) exacting the specificity of antibody binding, and (ii) the presence of a similar structure of all antibody molecules.

Disease Diagnosis

Using MoAbs, medical conditions and diseases of sufferers can be diagnosed. One of the approaches is the 'antibody- sandwich' strategy which is also called ELISA (enzyme-linked immunosorbent assay). The circulatory antigens in the blood can be assayed using ELISA or radioimmunoassay (RIA) technique. By using ELISA, you can test for HIV, hepatitis, typhoid, herpes, etc. For each test, a separate ELISA kit is used; see Figure 1.1.

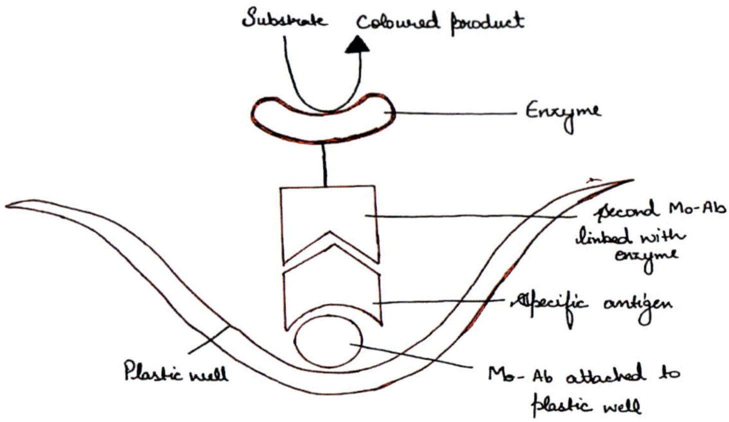

Figure 1.1: ELISA testing.

2. BASIC CELL CULTURE, CELL LINES, GROWTH CYCLE AND CULTURE MEDIA

Introduction

Cell culture is the process by which prokaryotic, eukaryotic or plant cells are grown under controlled conditions. But in practice, it refers to the culturing of cells derived from animal cells. Cell culture was first successfully undertaken by Ross Harrison in 1907. Roux in 1885 for the first time maintained embryonic chick cells in a cell culture. Historical events in the development of cell culture

- 1878: Claude Bernard proposed that the physiological systems of an organism can be maintained in a living system after the death of an organism.
- 1885: Roux maintained embryonic chick cells in a saline culture.
- 1897: Loeb demonstrated the survival of cells isolated from blood and connective tissue in serum and plasma.
- 1903: Jolly observed cell division of salamander leucocytes *in vitro*.
- 1907: Harrison cultivated frog nerve cells in a lymph clot held by the 'hanging drop' method and observed the growth of nerve fibers *in vitro* for several weeks. He was considered by some as the father of cell culture.

- 1910: Burrows succeeded in the long-term cultivation of chicken embryo cells in plasma clots. He made detailed observations of mitosis.
- 1911: Lewis and Lewis made the first liquid media consisting of seawater, serum, embryo extract, salts and peptones. They observed limited monolayer growth.
- 1913: Carrel introduced strict aseptic techniques so that cells could be cultured for long periods.
- 1916: Rous and Jones introduced proteolytic enzyme trypsin for the subculture of adherent cells.
- 1923: Carrel and Baker developed 'Carrel' or T-flask as the first specifically designed cell culture vessel. They employed the microscopic evaluation of cells in culture.
- 1927: Carrel and Rivera produced the first viral vaccine - Vaccinia.
- 1933: Gey developed the roller tube technique.
- The 1940s: The use of the antibiotics penicillin and streptomycin in culture medium decreased the problem of contamination in cell culture.
- 1948: Earle isolated mouse L fibroblasts which formed clones from single cells. Fischer developed a chemically defined medium, CMRL 1066.
- 1952: Gey established a continuous cell line from a human cervical carcinoma known as HeLa (Helen Lane) cells. Dulbecco developed a plaque assay for animal viruses using confluent monolayers of cultured cells.
- 1954: Abercrombie observed contact inhibition: motility of diploid cells in monolayer culture ceases when contact is made with adjacent cells.
- 1955: Eagle studied the nutrient requirements of selected cells in culture and established the first widely used chemically defined medium.
- 1961: Hayflick and Moorhead isolated human fibroblasts (WI-38) and showed that they have a finite lifespan in culture.
- 1964: Littlefield introduced the HAT medium for cell selection.
- 1965: Ham introduced the first serum-free medium which was able to support the growth of some cells.

- 1965: Harris and Watkins were able to fuse human and mouse cells by the use of a virus.
- 1975: Kohler and Milstein produced the first hybridoma capable of secreting a monoclonal antibody.
- 1978: Sato established the basis for the development of serum-free media from cocktails of hormones and growth factors.
- 1982: Human insulin became the first recombinant protein to be licensed as a therapeutic agent.
- 1985: Human growth hormone produced from recombinant bacteria was accepted for therapeutic use.
- 1986: Lymphoblastoid γIFN licensed.
- 1987: Tissue-type plasminogen activator (tPA) from recombinant animal cells became commercially available.
- 1989: Recombinant erythropoietin in the trial.
- 1990: Recombinant products in the clinical trial (HBsAG, factor VIII, HIVgp120, CD4, GM-CSF, EGF, mAbs, IL-2).

Major Development's in Cell Culture Technology

The first development was the use of antibiotics which inhibits the growth of contaminants. The second was the use of trypsin to remove adherent cells to subculture further from the culture vessel. The third was the use of a chemically defined culture medium.

Why is Cell Culture used for?

Areas where cell culture technology is currently playing a major role.

- Model systems for Studying basic cell biology, interactions between disease-causing agents and cells, effects of drugs on cells, process and triggering of aging & nutritional studies

Toxicity Testing

- Study the effects of new drugs.

Cancer Research

- Study the function of various chemicals, viruses & radiation to convert normal cultured cells to cancerous cells.

Virology

Cultivation of viruses for vaccine production also used to study their infectious cycle.

Genetic Engineering

Production of commercial proteins, large-scale production of viruses for use in vaccine production e.g. polio, rabies, chickenpox, hepatitis B & measles.

Gene Therapy

Cells having a functional gene can be replaced by cells that are having non-functional genes.

Tissue Culture

In vitro cultivation of organs, tissues & cells at defined temperature using an incubator & supplemented with a medium containing cell nutrients & growth factors is collectively known as tissue culture. Different types of cell grown in culture includes connective tissue elements such as fibroblasts, skeletal tissue, cardiac, epithelial tissue (liver, breast, skin, kidney) and many different types of tumor cells.

Primary Culture

Cells when surgically or enzymatically removed from an organism and placed in a suitable culture environment will attach and grow is called as primary culture. Primary cells have a finite life span. Primary culture contains a very heterogeneous population of cells. Subculturing of primary cells leads to the generation of cell lines. Cell lines have a limited life span, they passage several times before they become senescent. Cells such as macrophages and neurons do not divide *in vitro* so can be used as primary cultures. The lineage of cells originating from the primary culture is called a cell strain.

Continuous Cell Lines

Most cell lines grow for a limited number of generations after which they cease. Cell lines either occur spontaneously or are induced virally or chemically transformed into Continuous cell lines. Characteristics of continuous cell lines

- Smaller, more rounded, less adherent with a higher nucleus / cytoplasm ratio.
- Fast growth and have an aneuploid chromosome number.
- Reduced serum and anchorage dependence and grow more in suspension conditions.
- Ability to grow up to higher cell density.
- Different phenotypes from donor tissue.
- Stop expressing tissue-specific genes.

Types of Cells

- On the basis of morphology (shape & appearance) or on their functional characteristics. They are divided into three.
- Epithelial like- cells attached to a substrate and appears flattened and polygonal in shape.
- Lymphoblast like- cells do not attach remain in suspension with a spherical shape.
- Fibroblast like- cells attached to a substrate appears elongated and bipolar.

Culture Media

- Choice of media depends on the type of cell being cultured.
- Commonly used Media are MEM, GMEM, EMEM, DMEM, etc.
- Media is supplemented with antibiotics viz. penicillin, streptomycin, etc.
- Prepared media is filtered and incubated at 4°C.

Why Sub-Culturing

- Once the available substrate surface is covered by cells (a confluent culture) growth slows & ceases.
- Cells to be kept in healthy & in a growing state have to be sub-cultured or passaged.
- It's the passage of cells when they reach 80-90% confluency in flask/dishes/plates.

- Enzymes such as trypsin, dispase, collagenase in combination with EDTA breaks the cellular glue that attached the cells to the surface.

Culturing of Cells

- Cells are cultured as anchorage-dependent or independent
- Cell lines derived from normal tissues are considered as anchorage-dependent grows only on a suitable substrate e.g. tissue cells
- Suspension cells are anchorage-independent e.g. blood cells
- Transformed cell lines either grow as a monolayer or as suspension

Adherent Cells

- Cells that are anchorage-dependent.
- Cells are washed with PBS (free of ca & mg) solution.
- Add enough trypsin/EDTA to cover the monolayer.
- Incubate the plate at 37°C for 1-2 min
- Tap the vessel from the sides to dislodge the cells.
- Add a complete medium to dissociate and dislodge the cells with the help of a pipette which is remained to be adherent
- Add complete medium depending on the subculture.
- requirement either to 75 cm or 175 cm flask.

Suspension Cells

- It easier to passage as no need to detach them.
- As the suspension cells reach confluency.
- Aseptically remove 1/3rd of the medium.
- Replaced with the same amount of pre-warmed medium.

Transfection Methods

- Calcium phosphate precipitation.
- DEAE-dextran (dimethylaminoethyl-dextran).
- Lipid-mediated lipofection.
- Electroporation.

- Retroviral infection.
- Microinjection.

Cell Toxicity

- Cytotoxicity causes inhibition of cell growth.
- The observed effect on the morphological alteration in the cell layer or cell shape.
- Characteristics of abnormal morphology are the giant cells, multinucleated cells, a granular bumpy appearance, vacuoles in the cytoplasm or nucleus.
- Cytotoxicity is determined by substituting materials such as medium, serum, supplements flasks, etc. at a time.

Working with Cryopreserved Cells

- Vial from liquid nitrogen is placed into 37 C water bath, agitate vial continuously until the medium is thawed.
- Centrifuge the vial for 10 mins at 1000 rpm at RT, wipe the top of the vial with 70% ethanol and discard the supernatant.
- Resuspend the cell pellet in 1 ml of complete medium with 20% FBS and transfer to properly labeled culture plates containing the appropriate amount of medium.
- Check the cultures after 24 hrs to ensure that they are attached to the plate.
- Change medium as the color changes, use 20% FBS until the cells are established.

Freezing Cells for Storage

- Remove the growth medium, wash the cells with PBS and remove the PBS by aspiration.
- Dislodge the cells by trypsin-versene.
- Dilute the cells with a growth medium.
- Transfer the cell suspension to a 15 ml conical tube, centrifuge at 200g for 5 mins at RT and remove the growth medium by aspiration.
- Resuspend the cells in 1-2ml of freezing medium.

- Transfer the cells to cryovials, incubate the cryovials at -80 C overnight.
- The next day transfers the cryovials to Liquid nitrogen.

Cell Viability

- Cell viability is determined by staining the cells with trypan blue.
- As trypan blue dye is permeable to non-viable cells or death cells whereas it is impermeable to this dye.
- Stain the cells with trypan dye and load to a hemocytometer and calculate % of viable cells

$$\% \text{ of viable cells} = \frac{\text{Number of unstained cells x 100}}{\text{total number of cells}}$$

Common Cell Lines

Human cell lines

- **-MCF-7 breast cancer**
- **HL 60 Leukemia**
- **HEK-293 Human embryonic kidney**
- **HeLa Henrietta lacks**
- **Primate cell lines**
- **Vero: African green monkey kidney epithelial cells**
- **Cos-7 African green monkey kidney cells**
- **And others such as CHO from hamster, sf9 & sf21 from insect cells**

Contaminant's of Cell Culture

- Cell culture contaminants of two types.
- Chemical-difficult to detect caused by endotoxins, plasticizers, metal ions or traces of disinfectants that are invisible.
- Biological-cause visible effects on the culture are mycoplasma, yeast, bacteria or fungus or also from cross-contamination of cells from other cell lines.

Effects of Biological Contamination's

- They compete for nutrients with host cells.
- Secreted acidic or alkaline by-products cause the growth of the host cells.

- Degraded arginine & purine inhibits the synthesis of histone and nucleic acid.
- They also produce H_2O_2 which is directly toxic to cells.

Detection of Contaminants

In general indicators of contamination are turbid culture media, change in growth rates, abnormally high pH, poor attachment, multi-nucleated cells, graining cellular appearance, vacuolization, inclusion bodies and cell lysis. Yeast, bacteria & fungi usually show visible effects on the culture (changes in medium turbidity or pH). Mycoplasma is detected by direct DNA staining with intercalating fluorescent substances e.g. Hoechst 33258. Mycoplasma is also detected by enzyme immunoassay by specific antisera or monoclonal abs or by PCR amplification of mycoplasmal RNA. The best and the oldest way to eliminate contamination is to discard the infected cell lines directly.

Basic Equipment used in Cell Culture

- Laminar cabinet-Vertical is preferable.
- Incubation facilities- Temperature of 25-30° C for insect & 37 ° C for mammalian cells, CO2 2-5% & 95% air at 99% relative humidity. To prevent cell death incubators are set to cut out at approx. 38.5°C.
- Refrigerators- Liquid media kept at 4° C, enzymes (e.g. trypsin) & media components (e.g. glutamine & serum) at -20°C.
- Microscope- An inverted microscope with 10x to 100x magnification.
- Tissue culture ware- Culture plastic ware treated with polystyrene.

Rules for Working with Cell Culture

- Never use contaminated material within a sterile area.
- Use the correct sequence when working with more than one cell line.
- Diploid cells (Primary cultures, lines for the production of vaccines,etc.).
- Diploid cells (Laboratory lines).
- Continous, slow-growing line.
- Continous, rapidly growing lines.

- Lines that may be contaminated.
- Virus-producing lines.

Basic Aseptic Conditions

- If working on the bench use a Bunsen flame to heat the air surrounding the Bunsen.
- Swab all bottle tops & necks with 70% ethanol.
- Flame all bottlenecks & pipette by passing very quickly through the hottest part of the flame.
- Avoiding placing caps & pipettes down on the bench; practice holding bottle tops with the little finger.
- Work either left to right or vice versa, so that all material goes to one side, once finished.
- Clean up spills immediately & always leave the work place neat & tidy.

Safety Aspect in Cell Culture

- Possibly keep cultures free of antibiotics in order to be able to recognize the contamination.
- Never use the same media bottle for different cell lines. If caps are dropped or bottles touched unconditionally touched, replace them with new ones.
- Necks of glass bottles prefer heat at least for 60 secs at a temperature of 200°C.
- Switch on the laminar flow cabinet 20 mins prior to starting working.
- Cell cultures that are frequently used should be subcultured & stored as duplicate strains
- Use actively growing cells that are in their log phase of growth, which are 80-90% viable.
- Keep exposure to trypsin at a minimum.
- Handle the cells gently. Do not centrifuge cells at high speed or roughly re-suspend the cells.
- Feeding & subculturing the cells at more frequent intervals than used with serum-containing conditions may be necessary.

- A lower concentration of 10^4cells/ml initiate the subculture of rapidly growing cells & a higher concentration of 10^5cells/ml for slowing growing cells.

Growth Cycle

- Animal cell growths in culture have a characteristic growth pattern similar to bacteria.
- As cells grow and divide in a monolayer or in suspension, they usually follow a characteristic growth pattern composed of three phases:

1. **Lag Phase**
2. **Log Phase**
3. **Stationary phase**

Log Phase

- It is a period of zero growth when cells are first inoculated into the growth medium.
- The length of this phase depends on the type of cells and their metabolic states during inoculation.
- It is a period of adaptation during which the cells recover from trypsinization, reconstruct their cytoskeleton, secrete matrix to aid attachment and spread out on the substrate.

Log Phase

- The exponential growth phase is a period of continuous cell doubling.
- Animal cells normally exhibit a doubling time of between 15 and 25 hours.
- The length of the log phase depends on the seeding density, the growth rate of the cells and the density at which cell proliferation is inhibited by density.

Stationary Phase

- It is the period after growth when there is no change in the culture cell density.

- The phase occurs when the nutrients have been depleted or inhibitory metabolites have accumulated in the culture.
- All available growth surfaces are occupied and all the cells are in contact with surrounding cells.
- Further growth of cells can be obtained by subculturing the cells in a fresh medium.

Culture Media

The culture medium is a liquid or gel designed to support the growth of cells, tissue or organs. A typical cell culture medium is composed of a complement of amino acid, vitamins, inorganic salts, glucose and serum as a source of growth factors, hormones and attachment factors. There are different types of media used for animal cell culture that may be grouped into two broad categories: Natural media or artificial media.

Natural Media

It consists solely of naturally occurring biological fluids. It is very useful and convenient for a wide range of animal cell cultures. The major disadvantage of natural media is its poor reproducibility due to a lack of knowledge of the exact composition of these natural media.

Artificial Media

It contains partly or fully defined components that are prepared artificially by adding several nutrients (organic and inorganic). Artificial media supplemented with serum or with suitable formulations of organic compounds supports prolonged survival of the cell culture. It may be defined in different groups into the following four classes:

- Serum containing media
- Serum-free media
- Chemically defined media
- Protein free media

Subjective Questions

1. What is sterilization?
2. Explain the basic equipment in the cell and tissue culture laboratory?
3. Describe about the hazards and safety in cell culture laboratory?
4. What are the basics of animal cell culture?
5. What is media preparation?
6. What is the role of important components of culture media?
7. What is recombinant DNA?
8. Define major steps involved in the technology?
9. How does rDNA work?
10. Explain the applications of RNA technology?
11. What are model organisms?
12. What is the significance of model organisms?
13. Define types of molecular markers?
14. Describe the classification of molecular markers?
15. What are the principles of nucleic acid-based methods?
16. What is the need for cryopreservation?
17. Define the steps involved in cryopreservation?
18. What are the methods of cryopreservation?
19. What is cryopreservation as a strategy to establish somatic cell banking?
20. Explain the progress in somatic cell nuclear transfer?
21. Explain the methodology of animal cloning?
22. What is embryo developmental block and reprogramming in cloning?

Objective Questions

1. The foundation for the development of cell culture technique was laid by

(a) Roux

(b) Arnold

(c) Ross

(d) Harrison

Answer: A

2. **Who is regarded as the father of tissue culture**
   (a) Harrison
   (b) Arnold
   (c) Ross
   (d) Roux

Answer: A

3. **The limited replicative capacity of human cells in culture is called**
   (a) Hayflick effect
   (b) Brownian effect
   (c) Contact inhibition
   (d) None of the above

Answer: A

4. **HAT medium used for the selection of fused hybrid cells in hybridoma culture was introduced by**
   (a) Littlefield and Miller
   (b) Kohler and Milstein
   (c) Frish and Jentoft
   (d) Eagle and Karl

Answer: A

5. **Which of the following is HGPRT+ and survives in HAT medium**
   (a) B cells
   (b) Myeloma cells
   (c) Hybrid cells
   (d) Both a and c

Answer: D

6. **Use of trypsin to free cells from tissue matrix for cell culture was described by**
   (a) Jolly and Koch
   (b) Beebe and Ewing
   (c) Sims and Stillman
   (d) Gottlieb Maberland

Answer: C

7. **Cells used in feeder layer**
   (a) Should have the ability to divide
   (b) Have the ability to metabolize
   (c) These properties are obtained by exposing cells to irradiation
   (d) All of the above

Answer: D

8. **Cell culture technique became simpler only after the advent of**
   (a) Antibiotics
   (b) Trypsin
   (c) Cell culture media
   (d) All of the following

Answer: D

9. **Which of the following is true regarding the animal cell culture technique**
   (a) Lactic acid is the source of carbon
   (b) Cells have a high requirement of L-glutamine
   (c) Cholin is necessary for cell adhesion and cytoskeleton
   (d) All of the following

Answer: D

10. **Which of the following is the structural fiber in the cell culture system**
    (a) Collagen
    (b) Elastin
    (c) Fibronectin
    (d) Both a and b

Answer: D

11. **Optimum pH required for the growth of mammalian cells is**
    (a) 5.3-7.0
    (b) 6.5-7.0
    (c) 7.2-7.4
    (d) 8.1-8.9

Answer: C

12. The optimum temperature of any cell or organ is called

(a) Critical temperature

(b) Threshold temperature

(c) Ceiling temperature

(d) None of the above

Answer: C

13. For the culture of avian cells the optimum temperature requirement is

(a) 37°C

(b) 40°C

(c) 42°C

(d) 35°C

Answer: B

14. Which of the following is the oldest cell line

(a) Hela cell line

(b) Vero cell line

(c) CHO cell line

(d) BHK cell line

Answer: A

15. Which cell line is used for the production of recombinant sex hormones

(a) BHK cell line

(b) Vero cell line

(c) Hela cell line

(d) CHO cell line

Answer: D

16. The ratio of CO2: O2 used in cell culture system should be

(a) 1:5

(b) 1:13

(c) 1:19

(d) 1:25

Answer: C

17. Which of the following is the most commonly used cell fusing agent?

(a) PEG

(b) NaNO3

(c) Sendai virus

(d) Polyvinyl alcohol

Answer: A

18. Which of the following virus is used for cell fusion

(a) Sendai virus

(b) Herpes virus

(c) Myxovirus

(d) All of the following

Answer: D

19. Which of the following is an easy and rapid method to interpret the viability of cells in the culture system

(a) Trypan blue dye exclusion

(b) Neutral red assay

(c) Fluorescein dye assay

(d) All of the following

Answer: A

20. Vero cell line is obtained from Kidney cells of

(a) African green monkey

(b) Chinese hamster

(c) Calf

(d) Swine

Answer: A

CHAPTER 12

BIOINSTRUMENTATION AND TECHNIQUES

1. GENERAL PRINCIPLES OF BIOCHEMICAL INVESTIGATIONS AND PRINCIPLE METHODS OF SEPARATION OF BIOMOLECULES

Introduction

Cell biologists research the intricate relationship between structure and function at the molecular, subcellular, and cellular levels. However, a complex biological system such as a biochemical pathway can only be understood after each one of its components has been analyzed separately. Only if a biomolecule or cellular component is pure and biologically still active can it be characterized and its biological functions elucidated. Fractionation procedures purify proteins and other cell constituents. In a series of independent steps, the various properties of the protein of interest—solubility, charge, size, polarity, and specific binding affinity are utilized to fractionate it, or separate it progressively from other substances. Three key analytical and purification methods are chromatography, electrophoresis, and ultracentrifugation. Each one relies on certain physicochemical properties of biomolecules.

Chromatography

Chromatography is the separation of sample components based on differential affinity for a mobile versus a stationary phase. The mobile phase is a liquid or a gas that flows over or through the stationary phase, which consists of spherical particles packed into a column. When a mixture of proteins is introduced into the mobile phase and allowed to migrate through the column, separation occurs because proteins that have a greater attraction for the solid phase migrate more slowly than do proteins that are more attracted to the mobile phase.

Several different types of interactions between the stationary phase and the substances being separated are possible. If the retarding force is ionic in character, the separation technique is called ion exchange. Proteins of different ionic charges can be separated in this way. If substances absorb onto the stationary phase, this technique is called absorption chromatography. In gel filtration or molecular sieve chromatography, molecules are separated because of their differences in size and shape. Affinity chromatography exploits a protein's unique biochemical properties rather than the small differences in physicochemical properties between different proteins. It takes advantage of the ability of proteins to bind specific molecules tightly but noncovalently and depends on some knowledge of a particular protein's properties in the design of the affinity column.

Electrophoresis

Many important biological molecules such as proteins, deoxyribonucleic acid (DNA), and ribonucleic acid (RNA) exist in solution as cations (+) or anions (-). Under the influence of an electric field, these molecules migrate at a rate that depends on their net charge, size and shape, field strength, and the nature of the medium in which the molecules are moving. Electrophoresis in biology uses porous gels as the media. The sample mixture is loaded into a gel, the electric field is applied, and the molecules migrate through the gel **matrix.** Thus, separation is based on both the molecular sieve effect and on the electrophoretic mobility of the molecules. This method determines the size of biomolecules. It is used to separate proteins, and especially to separate DNA for identification, sequencing, or further manipulation.

Ultracentrifugation

Cells, organelles, or macromolecules in solution exposed to a centrifugal force will separate because they differ in mass, shape, or a combination of those factors. The instrument used for this process is a centrifuge. An ultracentrifuge generates centrifugal forces of 600,000 g and more. (G is the force of gravity on Earth.) It is an indispensable tool for the isolation of proteins, DNA, and subcellular particles.

2. CENTRIFUGATION TECHNIQUES-BASIC PRINCIPLES, ANALYTICAL AND PREPARATIVE CENTRIFUGATION AND THEIR APPLICATIONS

Introduction

Centrifugation is a process used to "separate or concentrate" materials suspended in a liquid medium. The theoretical basis of this technique is the effect of gravity on particles (including macromolecules) in suspension. The two particles of "different masses will settle in a tube at different rates in response to gravity". It is a process that involves the use of centrifugal force for the sedimentation of heterogeneous mixtures with a centrifuge used in industry and in laboratory settings. This process is used to separate two immiscible liquids. Denser components of the mixture migrate away from the axis of the centrifuge, while less-dense components of the mixture migrate towards the axis.

Principle

It is a device for "separating particles from a solution according to their size shape, density, the viscosity of the medium and rotor speed. The density is higher than that of the solvent sink sediment and the particles that are lighter then it floats to the top". The greater the difference in density, the faster they move. If there is no difference in density i.e. isopycnic conditions, the particles stay steady.

- **Routine centrifuge rotors**
- **Calculation of *g*-force**
- Differential centrifugation
- Density gradient theory

Centrifuge Rotor

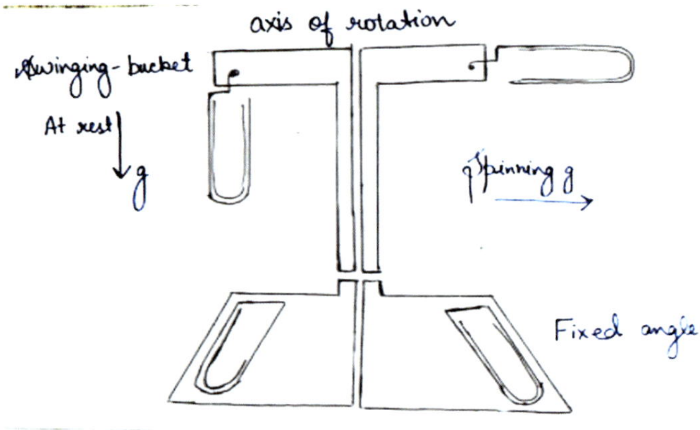

Geometry of Rotors

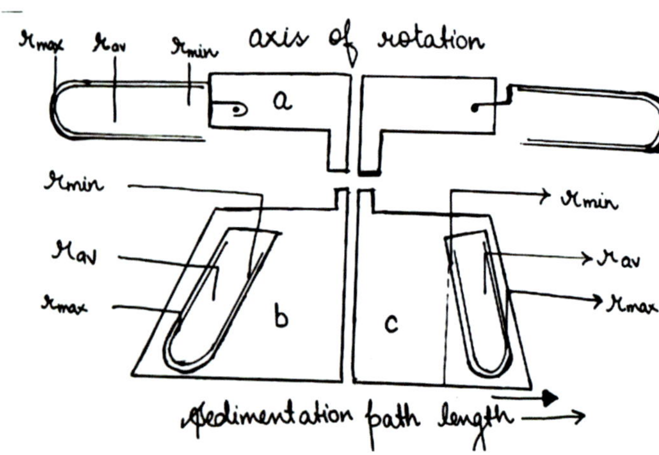

k'-factor of Rotors

The *k*'-factor is a measure of the time taken for a particle to sediment through a sucrose gradient. The most efficient motors which operate at a high RCF and have a low sedimentation path length, therefore, have the lowest *k*'-factors. The centrifugation times (*t*) and *k*'-factors for two different rotors (1 and 2) are related by:

$$t_1 = \frac{k_1 t_2}{k_2}$$

Calculation of *RCF* and *Q*

$$RCF = 11.18\, x\, r \left(\frac{Q}{1000} \right)^2$$

$$Q = 299 \sqrt{\frac{RCF}{r}}$$

- RCF in swinging-bucket and fixed-angle rotors at 40,000 rpm.
- Beckman SW41 swinging-bucket (13 ml)
- g_{min} = 119,850g; g_{av} = 196,770g;
- g_{max} = 273,690g
- Beckman 70.1Ti fixed-angle rotor (13 ml)
- g_{min} = 72,450g; g_{av} = 109,120g;
- g_{max} = 146,680g

Velocity of Sedimentation of a Particle

$$v = \frac{d^2(\rho_p - \rho_l)}{18\mu}\, g$$

v = velocity of sedimentation d = diameter of particle
ρ_p = density of particle ρ_l = density of liquid
g = centrifugal force μ = viscosity of liquid

Differential Centrifugation

The density of a liquid is uniform. The density of a liquid is << Density of particles. The viscosity of the liquid is low. Consequence: The rate of particle sedimentation depends mainly on its size and the applied *g*-force.

Size of Major Cell Organelles

• Nucleus	4-12 µm
• Plasma membrane sheets	3-20 µm
• Golgi tubules	1-2 µm
• Mitochondria	0.4-2.5 µm
• Lysosomes/peroxisomes	0.4-0.8 µm
• Microsomal vesicles	0.05-0.3µm

Differential Centrifugation of a Tissue Homogenate

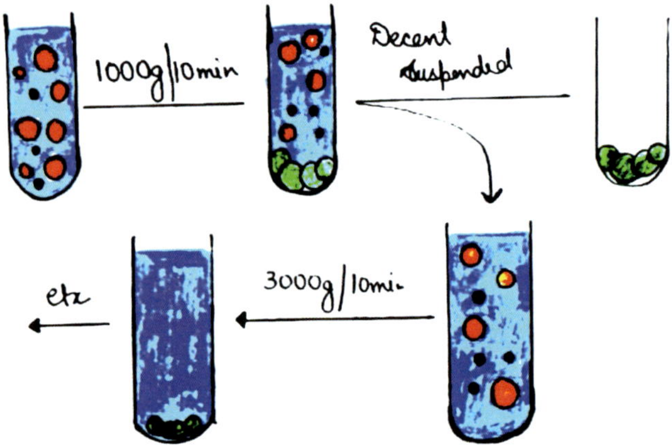

Differential centrifugation of a tissue homogenate

1. Homogenate – 1000g for 10 min
2. Supernatant from 1 – 3000g for 10 min
3. Supernatant from 2 – 15,000g for 15 min
4. Supernatant from 3 – 100,000g for 45 min

- Pellet 1 – nuclear
- Pellet 2 – "heavy" mitochondrial
- Pellet 3 – "light" mitochondrial
- Pellet 4 – microsomal

Differential Centrifugation (III) Expected Content of Pellets

- 1000g pellet: nuclei, plasma membrane sheets.
- 3000g pellet: large mitochondria, Golgi tubules.
- 15,000g pellet: small mitochondria, lysosomes, peroxisomes.
- 100,000g pellet: microsomes.

Differential Centrifugation

It has poor resolution and recovery because of particle size heterogeneity. The particles starting out at r_{min} have the furthest to travel but initially experience the lowest RCF. Smaller particles close to r_{max} have only a short distance to travel and experience the highest RCF.

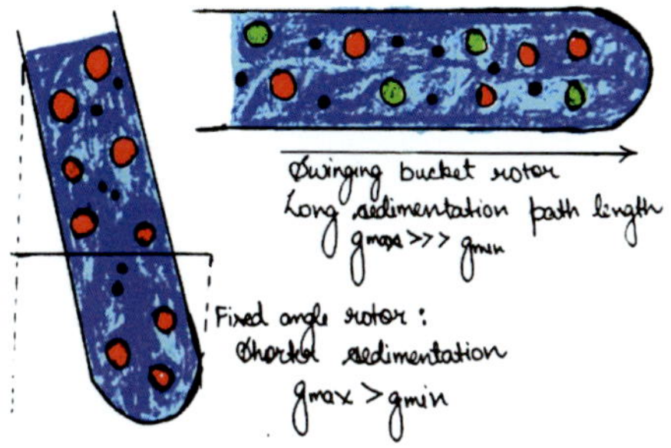

The rate of sedimentation can be modulated by particle density. The nuclei have an unusually rapid sedimentation rate because of their size and high density. Golgi tubules do not sediment at 3000g, in spite of their size: they have an unusually low sedimentation rate because of their very low density: $(\rho_p - \rho_l)$ becomes rate limiting.

Density Gradient Centrifugation

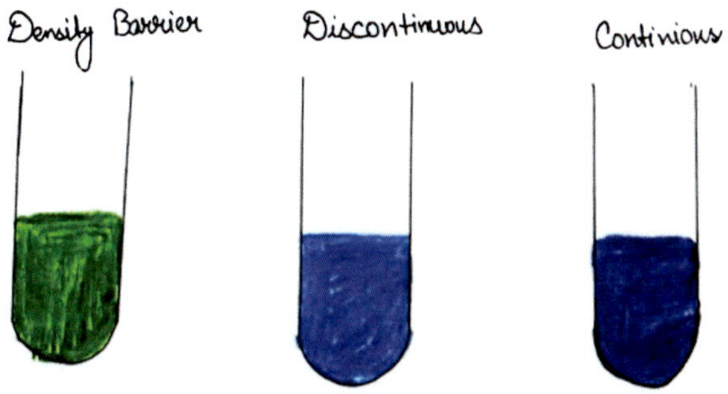

A particle will sediment through a solution if the particle density > solution density. If the particle density < solution density, the particle will float through the solution. When particle density = solution density the particle stops sedimenting or floating. The formats for the separation of particles are according to their density.

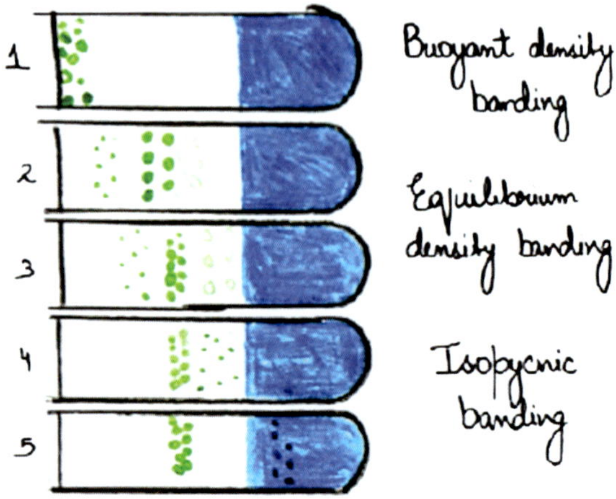

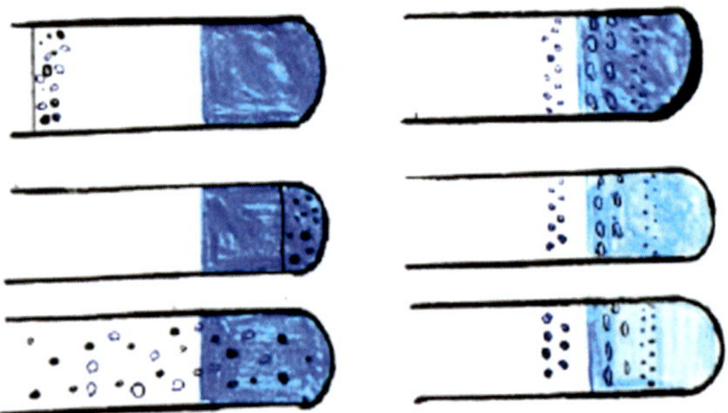

When density of particle < density of liquid V is -ve

Resolution of Density Gradients

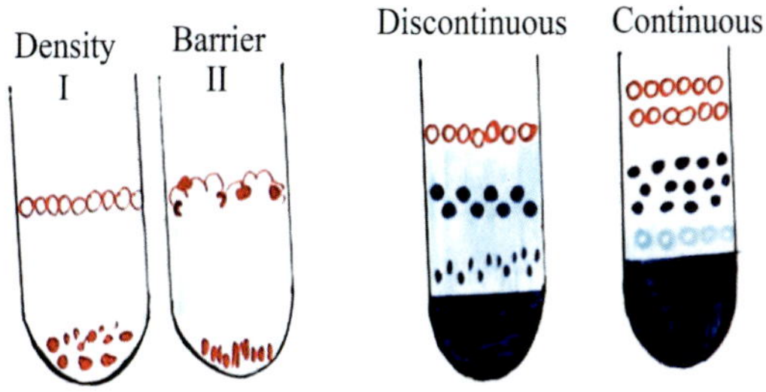

Separation of Particles According to Size

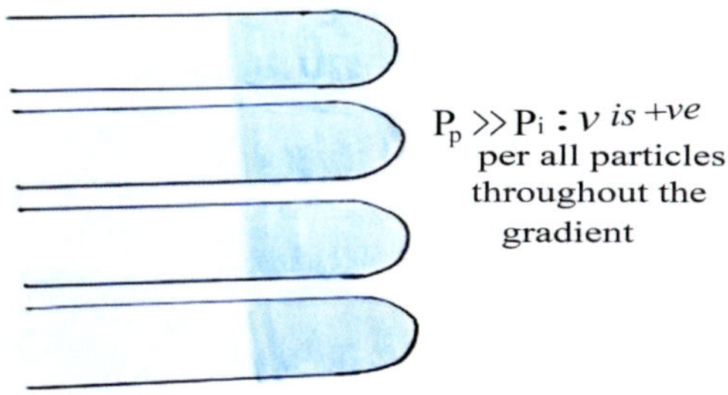

Application of Centrifugation

Centrifugation can be employed to separate a mixture of two different miscible liquids. This "technique can also be used in order to study and analyze macromolecules and their hydrodynamic properties. Mammalian cells can be purified with the help of a special type of centrifuge. Centrifugation is known to have a vital role in the fractionation of many subcellular organelles. Furthermore, centrifugation is also useful in the fractionation of membrane fractions and membranes. Centrifugation also has applications in the fractionation of membrane vesicles. Chalk can be separated from water with the help of a centrifuge. Skimmed milk is a form of milk that has a lower amount of dissolved fats. Skimmed milk can be obtained from regular milk with the help of the process of centrifugation. Here, the centrifuge serves to separate the fat from the milk, leaving the required skimmed milk behind. Cyclonic separation is an important process that has vital applications in the separation of particles from air flows. Another important application of this technique is in the stabilization and clarification of wine. This technique, in combination with other purification techniques, is extremely helpful while separating proteins. Other techniques that are used include salting-out techniques such as ammonium sulfate precipitation. Centrifuges are widely used in the field of forensic chemistry. In this field, the technique is employed for the separation of blood components from blood samples. Furthermore, the technique is also employed in certain laboratories for the separation of urine components from urine samples". Differential centrifugation, a distinct type of centrifugation, is known to have applications in the identification of organelles.

3. SPECTROPHOTOMETRY: UV-VISIBLE SPECTROPHOTOMETER AND FLUORIMETRY

Introduction

Photometry refers to the science of measurement light in terms of the "light's brightness that the human eye can observe. Photometers examine how light interacts with reflective materials as the basic principle. The arrangement of photometers allows for photometry to be applied in different fields of study. Some instruments are used to observe how light is absorbed and/or reflects wavelengths. Other instruments measure light by converting light into an electrical current and then measuring the electrical current produced by the light". Photometry is often used in the study of liquids and solutions in chemistry. Photometers can help measure the masses of organic or inorganic materials in a solution or liquid.

Colorimetry

The technique in which the concentration is measured by its ability to produce or change the color of a solution is called colorimetry.

- Changes the solution's ability to absorb light.

Spectrophotometry

The technique is used to light measure through chemical concentrations. A colorimetric is a method where an instrument is used to determine the amount of analyte in a sample by the sample's ability or inability to absorb light at a certain wavelength.

Measurement of ozone (O_3) above the South Pole:

- O_3 provides protection from ultraviolet radiation
- Seasonal depletion due to chlorofluorocarbons

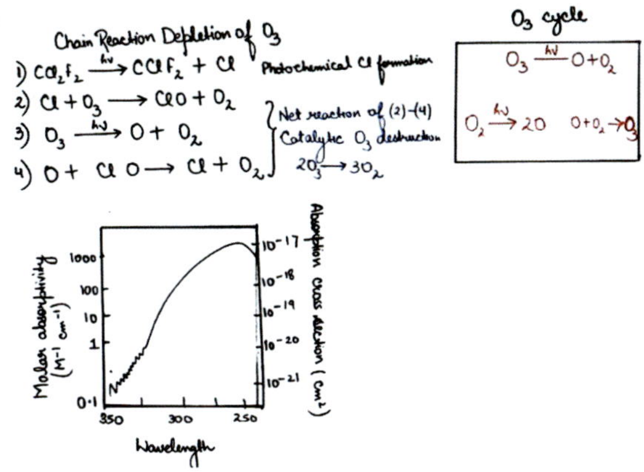

Properties of Light

Particles and Waves

Light waves are consisting of perpendicular, oscillating electric and magnetic fields. Parameters used to describe light.

- Amplitude (A): height of the wave's electric vector.
- Wavelength (l): distance (nm, cm, m) from peak to peak

Frequency (n): number of complete oscillations that the waves make each second

Hertz (Hz): unit of frequency, second^{-1} (s^{-1})

1 megahertz (MHz) = $10^6 s^{-1} = 10^6$Hz

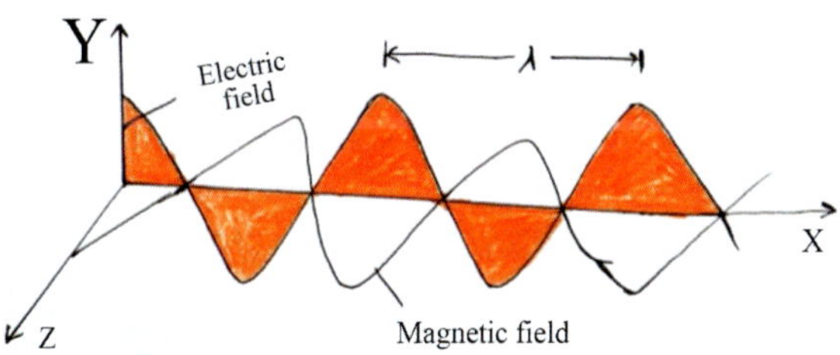

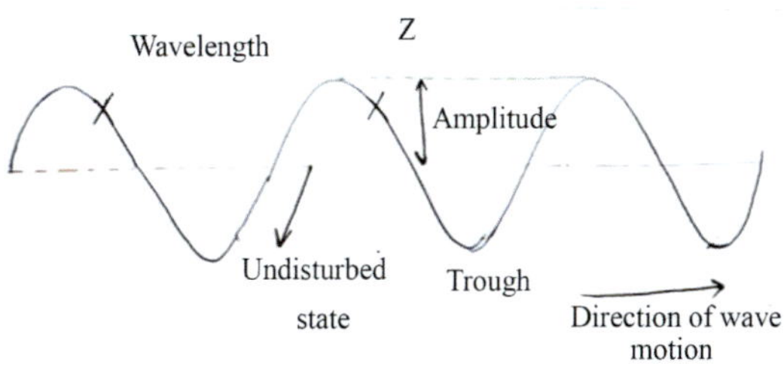

Properties of Light

Energy (E): the energy of one particle of light (photon) is proportional to its frequency.

E = hv

where: E = photon energy (Joules)

n = frequency (sec^{-1})

h = Planck's constant ($6.626x10^{-34}$J-s)

The frequency(n) is increases as also energy (E) of light increases.

Relationship between frequency and wavelength

λv = c = V

V = c/λ

where: c = speed of light ($3.0x10^{8}$ m/s in vacuum))

n = frequency (sec^{-1})

l = wavelength (m)

Relationship between energy and wavelength

E = hc/λ = hcv

where: = (1/l) = wavenumber

As frequency (l) is decreases, the energy (E) of light increases.

Types of Light: The Electromagnetic Spectrum

Again, the energy (E) of light increases as frequency (n) increases or wavelength (l) decreases.

Absorption of Light

1.) Colors of Visible Light

Many types of chemicals absorb various forms of light. The color of light absorbed and observed passing through the compound are complimentary.

Wavelength of maximum absorption (nm)	Color absorbed	Color observed
380—420	Violet	Green-yellow
420—440	Violet-blue	Yellow
440—470	Blue	Orange
470—500	Blue-green	Red
500—520	Green	Puple
520—550	Yellow-green	Violet
550—580	Yellow	Violet-blue
580—620	Orange	Blue
620—680	Red	Blue-green
680—780	Red	Green

2.) Ground and Excited State

When a chemical absorbs light, it goes from a "low energy state (ground state) to a higher energy state (excited state). Only photons with energies exactly equal to the energy difference between the two-electron states will be absorbed. Since different chemicals have different electron shells that are filled, they will each absorb their own particular type of light. Different electron ground states and excited states".

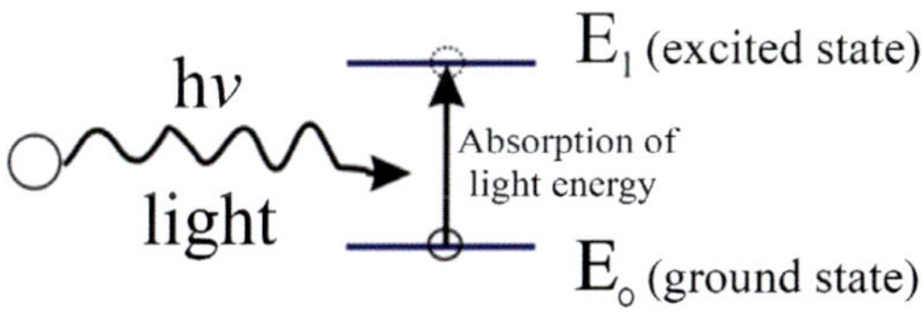

The energy required for a photon to give this transition:

$DE = E_1 - E_o$

3.) Beer's Law

The relative amount of a certain wavelength of light absorbed (*A*) that passes through a sample is dependent as the distance the light must pass through the sample (cell path length - b), the amount of absorbing chemicals in the sample (analyte concentration – c), the ability of the sample to absorb light (molar absorptivity - e), the relative amount of light making it through the sample (P/P_o) is known as the transmittance (T).

$T = P/P_0$

Percent transmittance $\%T = 100 * P/P_0$

T has a range of 0 to 1, %T has a range of 0 to 100%

Absorbance (A) is the relative amount of light absorbed by the sample and is related to transmittance (T). absorbance is sometimes called optical density (OD).

$$A = -\log\left[\frac{P}{P_o}\right] = -\log(T) = -\log(\%T/100)$$

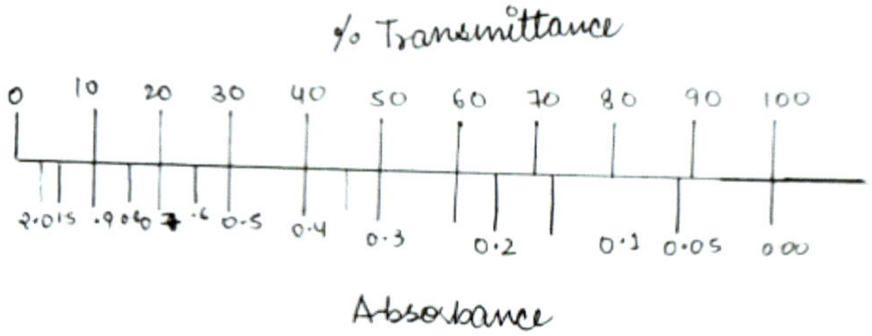

A has a range of 0 to infinity

Absorbance is useful since it is directly related to the analyte concentration, cell path length and molar absorptivity. This relationship is known as Beer's Law.

$$A = \varepsilon bc$$

where: A = absorbance (no units)
e = molar absorptivity (L/mol cm)
b = cell pathlength (cm)
c = concentration of analyte (mol/L)

Beer' law allows compounds to be quantified by their ability to absorb light, Related directly to concentration (c)

Different chemicals have different energy levels.

- The different ground vs. excited electron states.
- It will have different abilities to absorb light at any given wavelength.

The absorption spectrum is the plot of "absorbance (or e) vs. wavelength" for a compound. The greater the absorbance of a compound at a given wavelength (high e), the easier it will be to detect at low concentrations. By choosing different wavelengths of light (1_A vs. 1_B) as different compounds can be measured.

Spectrophotometer

An instrument used to make absorbance or transmittance measurements is known as a spectrophotometer.

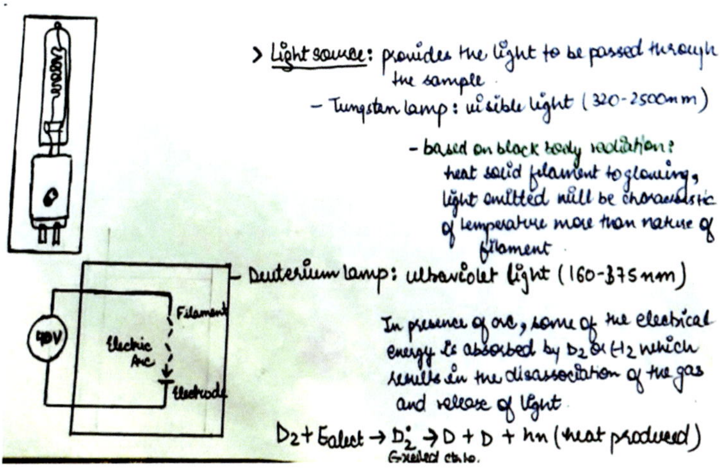

Wavelength selector (monochromator): used to select a given wavelength of light from the light source.

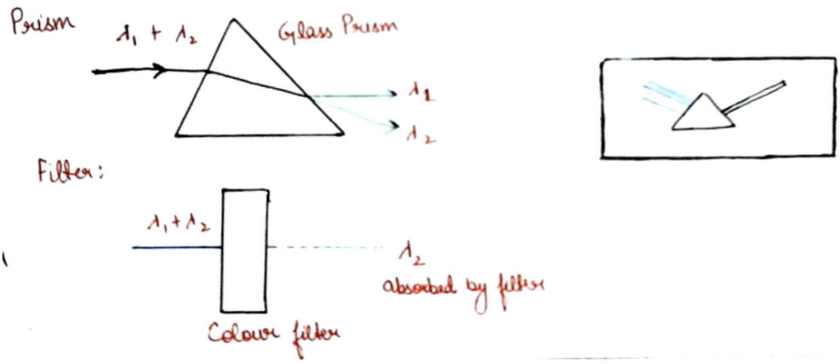

Reflection or Diffraction Grating

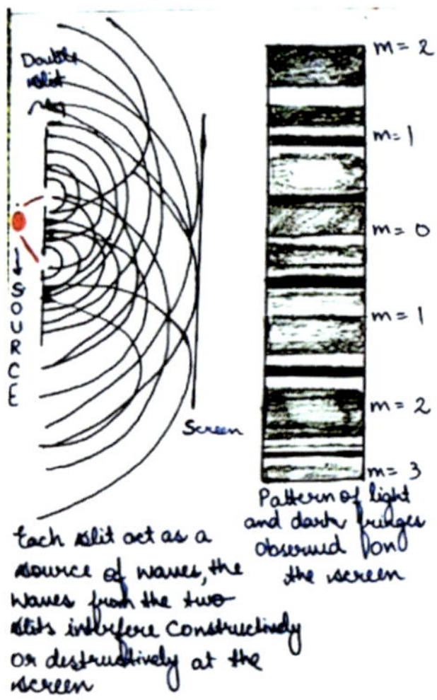

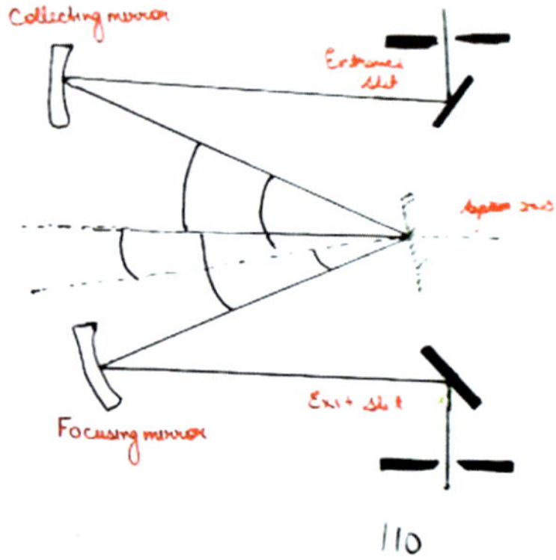

- **Sample Cell**: sample container of fixed length (b).
- It is usually a round or square cuvet.
- It is made from material that does not absorb light in the wavelength range of interest.
  - Glass – visible region
  - Quartz – ultraviolet
  - NaCl, KBr – Infrared region
- Light Detector: measures the amount of light passing through the sample.
- It usually works by converting the light signal into an electrical signal.

Process

Light hits photoemissive cathode and e^- is emitted. "An emitted e^- is attracted to electrode #1 (dynode 1), which is 90V more positive. It causes several more e^- to be emitted. This e^- is attracted to dynode 2, which is 90V more positive than dynode 1, emitting more e^-. The process continues until e^- is collected at anode after amplification at 9 dynodes. The overall voltage between anode and cathode is 900V. One photon produces $10^6 – 10^7$ electrons". The current is amplified and measured.

Types of Spectrophotometers

Single-Beam Instrument: Sample and blank are alternatively measured in the same sample chamber.

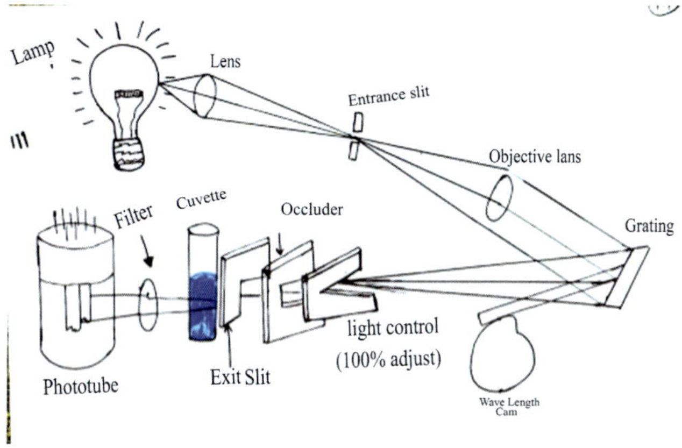

Double-beam Instrument

It continuously compares the sample and blank. It automatically corrects for changes in electronic signal or light intensity of the source.

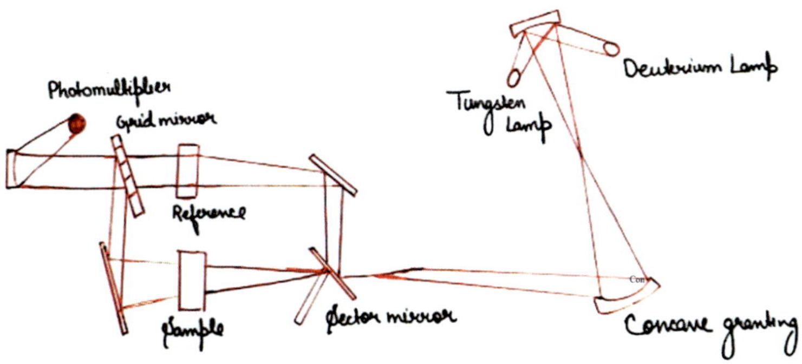

Chemical Analysis

1.) **Calibration**

- To measure the absorbance of a sample, it is necessary to measure P_o and P ratio
  - P_o – the amount of light passing through the system with no sample present
  - P – the intensity of light when the sample is present

– P_o is measured with a blank cuvet
– Cuvet contains all components in the sample solution except the analyte of interest

P is measured by placing the sample in the cuvetthen exactly measure an unknown concentration to obtain a calibration curve using a range of known concentrations for the analyte.

2.) **Limitations in Beer's Law**

The results in a non-linear calibration curve. Solute molecules influence one another because of their proximity at high concentrations.

- Molar absorptivity changes
- Affect on equilibrium, (HA and A^- have difference absorption)

Analyte properties change in different solvents. Errors are found in the reproducible positioning of cuvet. It is also problems with dirt & fingerprints. The instrument has electrical noise. Keep A in a range of 0.1 – 1.5 absorbance units (80 -3%T)

Precautions in Quantitative Absorbance Measurements

Choose a wavelength at an absorption maximum. Minimizes deviations from Beer's law, which assumes e- is constant. Pick peak in the absorption spectrum where the analyte is the only compound absorbing light. Choose a wavelength where the analyte has the largest difference in its absorbance relative to other sample components (www.scribd.com., www.slideshare.com., www.coursehero.com., www. biocyclopedia.com., www. molecular-plant-biotechnology.info., www.slideplayer.com., www. documents.mx., www.docplayer.net., www.bmb.psu.edu.,www.archive. org.,www.mafiadoc.com., www.powershow.com., www.bioscience.jbpub. com., www.outhorstream.com., www.docslide.com., www.msluay.com., www.scienceprofonlue.com., www.csus.edu., www.krishikosh.egrouth. ac.in., www.tankonyvtor.hu., www.silo.pub., www.studyblue.com., www. Preservearticles.com., www.biotechuniverse.blogspot.com).

4. CHROMATOGRAPHIC TECHNIQUES

Introduction

Mikhail Tswett invented chromatography in 1901 during his research on plant pigments. He used the technique to separate various plant pigments such as chlorophylls, xanthophylls and carotenoids.

Original Chromatography Experiment

- **Start:** A glass column is filled with powdered limestone ($CaCO_3$).
- **End:** A series of colored bands was seen to form, corresponding to the different pigments in the original plant extract. These bands were later determined to be chlorophylls, xanthophylls and carotenoids.

Chromatography

Tswett named this new technique chromatography (Greek = chroma "color" and graphein "writing") based on the fact that it separated the components of a solution by color.

Types of Chromatography

Tswett's technique is based on liquid chromatography. There are now several common chromatographic methods. These include

- Paper Chromatography
- Thin Layer Chromatography (TLC)
- Liquid Chromatography (LC)
- High-Pressure Liquid Chromatography (HPLC)
- Ion Chromatography
- Gas Chromatography (GC)

Paper and Thin Layer Chromatography

The solvent moves up paper by capillary action, carrying mixture components at different rates.

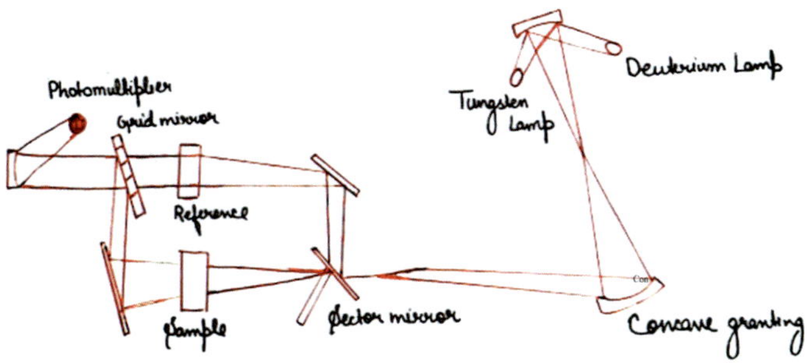

In all "chromatographic separations and the sample is transported in a mobile phase. The mobile phase can be a gas, a liquid, or a supercritical fluid. The mobile phase is then forced through a stationary phase held in a column or on a solid surface. The stationary phase needs to be something that does not react with the mobile phase or the sample. The sample has the opportunity to interact with the stationary phase as it moves past it. Samples that interact greatly then appear to move more slowly". Samples that interact weakly then appear to move more quickly. Because of this difference in rates, the samples can then be separated into their components.

Chromatography

It is based on a "physical equilibrium that results when a solute is transferred between the mobile and a stationary phase. In a chromatography column, flowing gas or liquid continuously replaces the saturated mobile phase and results in the movement of A through the column. As material travels through the column, it assumes a Gaussian concentration profile as it distributes between the stationary packing phase and the flowing mobile gas or liquid carrier phase".

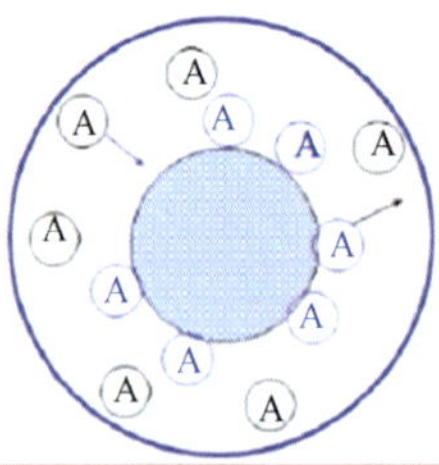

Cross Section of Equilibrium in a column.
"A" are adsorbed to the stationary phase.
"A" are traveling in the mobile phase.

K = distribution coefficient *or* partition ratio

$$K = \frac{C_S}{C_M}$$

Where C_S is the molar concentration of the solute in the stationary phase and C_M is the molar concentration in the mobile phase.

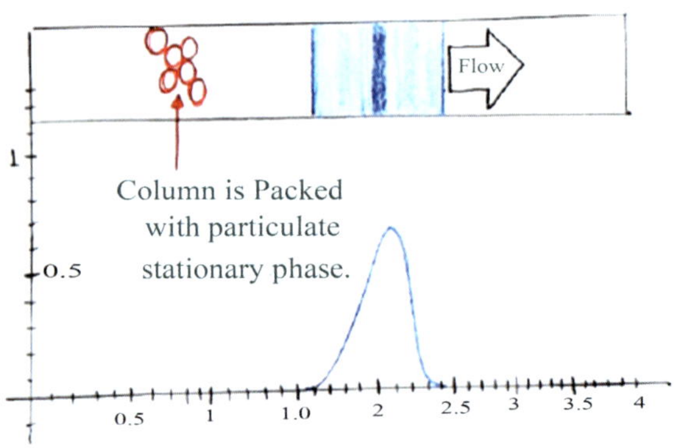

In a mixture, "each component has a different distribution coefficient, and thus spends a different amount of time absorbed on the solid packing phase vs. being carried along with the flowing gas. More volatile materials are carried through the column more rapidly than less volatile materials, which results in the separation".

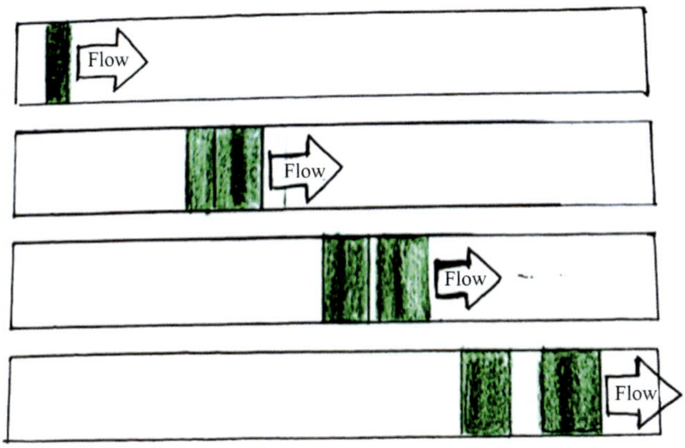

If a detector is used to determine when the components elute from the column, a series of Gaussian peaks are obtained, one for each component in the mixture that was separated by the column.

Note: The first two components were not completely separated.

Peaks in general tend to become shorter and wider with time.

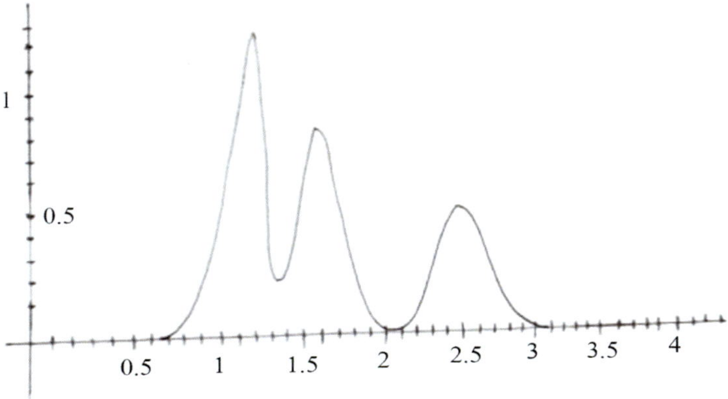

Theoretical plate is a term coined by Martin & Synge. It is based on a study in which they imagined that chromatographic columns were analogous to distillation columns and made up or numerous discrete but connected narrow layers or plates. The movement of the solute down the column then could be treated as a stepwise transfer. Theoretical plates (N) measure how efficiently a column can separate a mixture into its components. This efficiency is based on the retention time of the components and the width of the peaks.

N = Number of theoretical plates (a measure of efficiency)

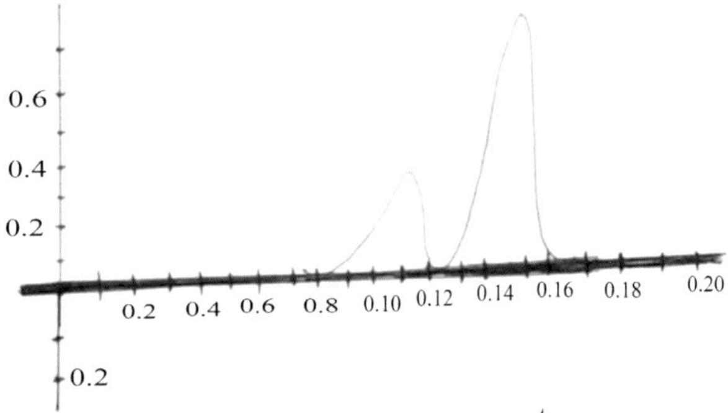

$$N = 16\left(\frac{t_R}{W_b}\right)^2$$

t_R is the retention time and it is measured from the injection peak (or zero) to the intersection of the tangents. W_b is the width of the base of the triangle; it is measured at the intersection of the tangents with the baseline. When the retention time, t_R, is held constant, the column that produces peaks with narrower bases, w_b, will be more efficient have a greater N value. Likewise, a column that produces wider peaks will be less efficient and have a smaller N value. This is because a smaller denominator, w_b, will yield a larger overall number and a larger denominator will yield a smaller number.

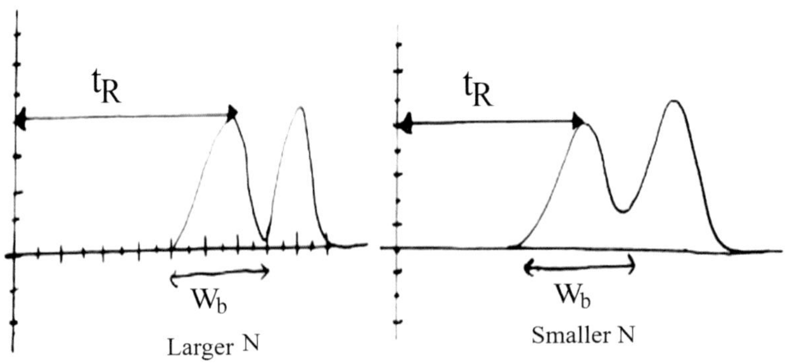

Gas chromatography

It is good for volatile samples (up to about 250 °C). It requires 0.1-1.0 microliter of liquid or 1-10 ml vapor. It can detect <1 ppm with certain detectors. It can be easily automated for injection and data analysis.

Components of a Gas Chromatography

- **Gas supply:** (usually N_2 or He).
- **Sample Injector:** (syringe / septum).
- **Column:** 1/8" or 1/4" x 6-50' tubing packed with small uniform size, inert support coated with a thin film of nonvolatile liquid
- **Detector:** TC - thermal conductivity and FID - flame ionization detector

5. ELECTROPHORESIS

Electrophoresis

Electrophoresis is a method whereby "charged molecules in solution, chiefly proteins and nucleic acids, migrate in response to an electrical field. Their rate of migration through the electrical field depends on the strength of the field, on the net charge, size, and shape of the molecules, and also on the ionic strength, viscosity, and temperature of the medium in which the molecules are moving. As an analytical tool, electrophoresis is simple, rapid and highly sensitive. It can be used analytically to study the properties of a single charged species or mixtures of molecules. It can also be used preparatively as a separating technique. Electrophoresis is usually done with gels formed in tubes, slabs, or on a flatbed. In many electrophoresis units, the gel is mounted between two buffer chambers containing separate electrodes, so that the only electrical connection between the two chambers is through the gel". In most electrophoresis units, the gel is mounted between two buffer chambers containing separate electrodes so that the only electrical connection between the two chambers is through the gel.

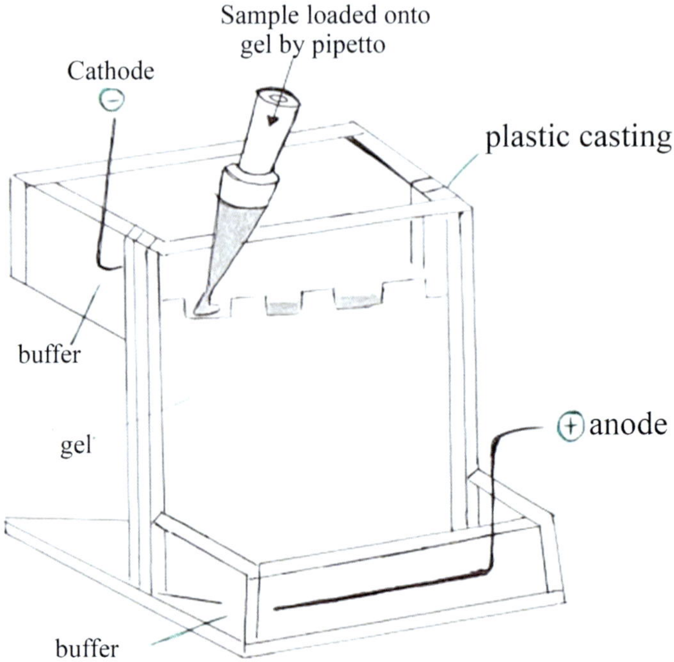

Interrelation of resistance, voltage, current and power and two basic electrical equations are important in electrophoresis.

- The first is Ohm's Law, I = E/R
- The second is P = EI
- This can also be expressed as $P = I^2R$

In electrophoresis, one electrical parameter, either current, voltage, or power, is always held constant.

Consequences

Under constant current conditions (velocity is directly proportional to current), the velocity of the molecules is maintained, but heat is generated. Under constant voltage conditions, the velocity slows, but no additional heat is generated during the course of the run. Under constant power conditions, the velocity slows but heating is kept constant. The net charge is determined by the pH of the medium. Proteins are amphoteric compounds, that is, they contain both acidic and basic residues. Each protein has its own characteristic charge properties depending on the number and kinds of amino acids carrying amino or carboxyl groups. Nucleic acids, unlike proteins, are not amphoteric. They remain negative at any pH used for electrophoresis.

During Polymerization

- Exothermic reaction
- Gel irregularities
- Pore size

During Electrophoresis

- Denaturation of proteins
- Smile effect
- Temperature regulation of buffers

It inhibits convection and diffusion, which would otherwise impede the separation of molecules. It allows a permanent record of results by staining after a run. It can provide additional separation via molecular sieving.

Agarose and Polyacrylamide

Although "agarose and polyacrylamide differ greatly in their physical and chemical structures, they both make porous gels. A porous gel acts as a sieve by retarding or, in some cases, by completely obstructing the movement of macromolecules while allowing smaller molecules to migrate freely. By preparing a gel with a restrictive pore size, the operator can take advantage of molecular size differences among proteins. Because the pores of an agarose gel are large, agarose is used to separate macromolecules such as nucleic acids, large proteins and protein complexes. Polyacrylamide, which makes a small pore gel, is used to separate most proteins and small oligonucleotides. Both are relatively electrically neutral. Agarose is a highly purified uncharged polysaccharide derived from agar. Agarose dissolves when added to a boiling liquid. It remains in a liquid state until the temperature is lowered to about 40° C at which it gels. The pore size may be predetermined by adjusting the concentration of agarose in the gel". Agarose gels are fragile, however, they are actually hydrocolloids, and they are held together by the formation of weak hydrogen and hydrophobic bonds.

Polyacrylamide gels are tougher than agarose gels. Acrylamide monomers polymerize into long chains that are covalently linked by a crosslinker. Polyacrylamide is chemically complex, as is the production and use of the gel.

Considerations with PAGE

- Preparing and Pouring Gels
- Determine pore size
  - Adjust the total percentage of acrylamide
  - Vary the amount of crosslinker
- Remove oxygen from the mixture
- Initiate polymerization
  - Chemical method
  - Photochemical method
- Analysis of a gel
  - Staining or autoradiography followed by densitometry.
  - Blotting to a membrane, either by capillarity or by electrophoresis, for nucleic acid hybridization, autoradiography or immunodetection.

SDS Gel Electrophoresis

In SDS gel electrophoresis, "migration is determined not by the intrinsic electrical charge of polypeptides but by molecular weight. Sodium dodecyl sulfate (SDS) is an anionic detergent that denatures secondary and non–disulfide–linked tertiary structures by wrapping around the polypeptide backbone. So, SDS confers a net negative charge to the polypeptide in proportion to its length. When treated with SDS and a reducing agent, the polypeptides become rods of negative charges with equal charge densities" or charge per unit length.

Continuous and Discontinuous Buffer Systems

A continuous system has only a "single separating gel and uses the same buffer in the tanks and the gel. In a discontinuous system, a nonrestrictive large pore gel, called a stacking gel, is layered on top of a separating gel. The resolution obtainable in a discontinuous system is much greater than that obtainable in a continuous one. However, the continuous system is a little easier to set up. It can determine of molecular weights of proteins by SDS-PAGE. Run a gel with standard proteins of known molecular weights along with the polypeptide to be characterized. A linear relationship exists between the $\log_{10}$ of the molecular weight of a polypeptide and its R_f. Rf = ratio of the distance migrated by the molecule to that migrated by a marker dye-front. The R_f of the polypeptide to be characterized is determined in the same way, and the log_{10} of its molecular weight is read directly from the standard curve".

Blotting

Blotting is used to transfer proteins or nucleic acids from a slab gel to a membrane such as nitrocellulose, nylon, DEAE, or CM paper. The transfer of the sample can be done by capillary or southern blotting for nucleic acids (Southern, 1975) or by electrophoresis for proteins or nucleic acids.

Isoelectric Point

There is a pH at which there is no net charge on a protein and this is the isoelectric point (pI). Above its isoelectric point, a protein has a net negative charge and migrates toward the anode in an electrical field. Below its isoelectric point, the protein is positive and migrates toward the cathode.

Isoelectric Focusing

Isoelectric focusing is a method in which "proteins are separated in a pH gradient according to their isoelectric points. Focusing occurs in two stages; first, the pH gradient is formed. In the second stage, the proteins begin their migrations toward the anode if their net charge is negative or toward the cathode if their net charge is positive. When a protein reaches its isoelectric point (pI) in the pH gradient" it carries a net charge of zero and will stop migrating.

Two-dimensional Gel Electrophoresis

Two-dimensional gel electrophoresis is widely used to separate complex mixtures of proteins into many more components than is possible in conventional one-dimensional electrophoresis. Each dimension separates proteins according to different properties.

6. RADIOISOTOPE TECHNIQUES

Introduction

Detection and measurement of different types of radioisotopes used in biology, incorporation of radioisotopes in biological tissues and cells, molecular imaging of radioactive material, safety guidelines. An atom is a fundamental piece of matter. An atom itself is made up of three tiny kinds of particles called subatomic particles: protons, neutrons and electrons. The protons and neutrons make up the center of the atom called the nucleus. The electrons fly around above the nucleus in a small cloud. Protons are positively charged while electrons are negatively charged subatomic particles. The number of protons present in the nucleus is known as the atomic number (Z). Neutrons are uncharged particles with a mass approximately equal to that of a proton. The sum of the protons (z)and neutrons(N) in a given nucleus is the mass number (A) A=Z+N). Atoms of a given element with different mass numbers (i.e. different numbers of neutrons) are called isotopes. Different isotopes of the same element have the same number of protons in their atomic nuclei but differing numbers of neutrons. A nuclear species is represented by a subscript number for the atomic number, and a superscript number for the mass number (atomic weight) followed by the symbol of the element. $^{16}{}_{8}O$ (8= atomic number of the oxygen atom, 16= atomic mass or mass number of the oxygen atom) The number of isotopes of a given element varies: there are three isotopes of hydrogen ($^{1}{}_{1}H$, $^{2}{}_{1}H$, $^{3}{}_{1}H$), seven of carbon ($^{10}{}_{6}C$ and $^{16}{}_{6}C$), etc.

The ratio of neutron and proton determines the stability of isotopes. For stable isotopes, the N/Z ratio will be 1.1-1.55. Isotopes for elements with low atomic numbers tend to have an equal number of neutrons and protons in the nucleus. Isotopes for elements with high atomic numbers require more neutrons. Unstable isotopes called radioisotopes and they tend to be stable by radioactive decay

Radioisotopes are radioactive isotopes of an element. They can also be defined as atoms that contain an unstable combination of neutrons and protons, or excess energy in their nucleus. The radioactive isotopes or radioisotopes are isotopes of an element having an unstable nucleus that decays (emitting alpha, beta, gamma rays) until stability is reached. The stable end product is a nonradioactive isotope of another element. More than 3000 radioisotopes are known, of which only about 84 are seen in nature.

Alpha, Beta and Gamma Radiation

Alpha radiation is a heavy charged particle and more toxic than other forms of radiation. It is nonpenetrating. Beta radiation is light charged particle and its toxicity is the same as electromagnetic radiation per unit of energy. Its penetration potential varies with the source. Gamma radiation is an electromagnetic radiation and it is highly penetrating. Atoms with an unstable nucleus regain stability by shedding excess particles and energy in the form of radiation. The process of shedding the radiation is called radioactive decay. The radioactive decay process for each radioisotope is unique and is measured with a time period called a half-life. One half-life is the time it takes for half of the unstable atoms to undergo radioactive decay. Unit of radioactivity. In SI system Becquerel (Bq) is the unit of radioactivity. It is defined as one disintegration per second. Curie (Ci) is another most popular frequently used unit of radioactivity. This is defined as the quantity of radioactive material in which the number of nuclear disintegrations per second is the same as that in 1gm of radium, namely 3.7x 1010 (37 GBq) (http://www.mgcub.ac.in/pdf/material/20200429185940c0c2312fbb.pdf).

Application

- Some radioisotopes used in nuclear medicine have short half-lives, which means they decay quickly and are suitable for diagnostic purposes; others with longer half-lives take more time to decay, which makes them suitable for therapeutic purposes.
- Industry uses radioisotopes in a variety of ways to improve productivity and gain information that cannot be obtained in any other way.
- Radioisotopes are commonly used in industrial radiography, which uses a gamma source to conduct stress testing or check the integrity of welds. A common example is to test aeroplane jet engine turbines for structural integrity.
- Radioisotopes are also used by industry for gauging (to measure levels of liquid inside containers, for example) or to measure the thickness of materials.
- Radioisotopes are also widely used in scientific research and are employed in a range of applications, from tracing the flow of contaminants in biological systems to determining metabolic processes in small Australian animals.
- They are also used on behalf of international nuclear safeguards agencies to detect clandestine nuclear activities from the distinctive radioisotopes produced by weapons programs.

7. MOLECULAR BIOLOGY TECHNIQUES

Introduction

Southern blotting techniques

- A method, developed genes in a DNA restriction fragment is called the Southern blotting technique.
- Southern blots can easily provide a physical map of restriction sites within a gene located normally and reveal the number of copies of the gene in the genome, and the degree of similarity of the gene when compared with the other complementary genes.
- A molecular biologist, E.M. Southern (1975) reported this approach for analyzing the related on a chromosome.
- The procedure starts with the digestion of the DNA population by one or many restriction enzymes Consequently, DNA fragments of unequal length are produced.
- This preparation is passed through agarose gel electrophoresis which results in the separation of DNA molecules based on their size.
- DNA restriction fragments present in the gel are denatured by alkali treatment. The gel is then put on top of the buffer saturated filter paper.
- The upper surface of the gel is covered with a nitrocellulose filter and overlaid with dry filter paper. The dry filter paper draws the buffer through the gel.
- The buffer contains single-stranded DNA. Nitrocellulose filter binds DNA fragments strongly after baking at 80 ° C, DNA fragments are permanent and fixed when come in contact of nitrocellulose filter.
- Then the filter is placed in a solution containing radio-labeled RNA or denatured DNA probe of known sequences.
- These are complementary in sequence to the blot-transferred DNA. The radiolabelled nucleic acid probe hybridizes the DNA on the nitrocellulose filter.
- The filter is thoroughly washed to remove the probe.

- The hybridized regions are autoradiographically detected by placing the nitrocellulose filter in contact with a photographic film.
- The images show the hybridized DNA molecules.
- Thus, the sequences of DNA were recognized following the sequences of the nucleic acid probe.

Northern Blotting Technique

- The southern blotting technique could not be directly applied to the blotting transfer of mRNA separated by gel electrophoresis, because RNA was found not to bind with nitrocellulose filter. Alwine et al (1979) devised a technique in which RNA bands are blot transferred from the gel to a chemically reactive paper.
- An aminobenzyloxymethyl cellulose paper prepared from Whatman filter paper No. 540 after a series of uncomplicated reactions, is diazotized and rendered into the reactive paper and, therefore, becomes available hybridization with radiolabeled DNA probes.
- The hybridized bands are found by autoradiography Thus, Alwine's method extends that of the Southern method and for this reason, it has been granted the name, Northern blotting.
- "There is not this northern or western like Southern." These blot transfers are reusable because of the firm covalent bonding of RNA to the reactive paper.
- Effective in binding the DNA as well. Small fragments of DNA can most effectively be transferred to the diazotized paper derivative to nitrocellulose.
- These techniques were more advanced and demonstrated that mRNA bands can also be blotted directly on nitrocellulose paper and appropriate conditions.

Western Blotting

- Towbin et al. (1979) developed the western blotting technique to find out the newly encoded protein by a transformed cell.
- Its working principle lies on the antigen-antibody reaction: hence is an immunodetection technique. In this method, radiolabelled nucleic acid probes are not used.

- This technique follows the following steps: (Extraction of protein from transformed cells. Separation of protein by using SDS-PAGE (sodium dodecyl sulfate polyacrylamide electrophoresis) where SDS acts as a solvent for electrophoresis.
- Transfer of electrophoresed gel in a buffer at low temperature (40 ° C) for half an hour blotting of proteins on nitrocellulose filter paper of nitrocellulose filter, Whatman filter and coarse filter in transfer buffer.
- Soaking of nitrocellulose filter, Whatman filter and coarse filter in transfer buffer.
- Placing of Whatman filter paper on a cathode plate followed by the stack of coarse filter Whatman filter, electrophoresed gel, nitrocellulose filter, Whatman filter paper, coarse filter stack, Whatman filter and anode plate transferred.
- Putting the complete set up in a transfer tank containing sufficient transfer buffer.
- Application of an electric field (30 V overnight for 5 hours) to cause the migration of proteins from the gel to nitrocellulose filter and binding on its surface. (The nitrocellulose filter has an exact image of the pattern of proteins as present in the gel. This type of blotting is called western blotting.)
- Hybridization of proteins by using radiolabelled antibodies (I^{123}-antibodies) of known structure, isolated from the rabbit.
- Washing of nitrocellulose filter with a wash solution (Tris-buffered saline + tween 20) to facilitate the removal of unhybridized antibodies.
- Detection of hybridized sequences by autoradiography. The dots of the diagram show the presence of the desired protein.

Biochemical Markers

- There are proteins produced by gene expression. These proteins can be isolated and identified by electrophoresis and staining. Isozymes are the different molecular forms of the same enzyme that catalyze the same reaction and proteins. They are revealed on electrophoresis through a colored reaction associated with enzymatic activity. They are the products of various alleles of one or several genes. Monomer and dimer isozymes are most often used because analysis of their segregation is easier. Isozymes are generally codominant.

Molecular Markers

- The molecular markers are DNA sequences that are readily detected and whose inheritance can be easily monitored. The use of molecular markers is based on naturally occurring DNA polymorphism, which forms the basis for designing strategies to exploit for applied purposes. A marker must be polymorphic, that chromosome carrying the mutant gene can be distinguished from the chromosome with the normal gene by markers it also carries. List of molecular markers such as:

Table 7.1: Molecular markers

Acronym	**Nomenclature**
RFLP	Restriction Fragment Length Polymorphism
VNTR	Variable Number Tandem Repeat
SSCP	Single-stranded Conformational Polymorphism
STS	Sequence Tagged Sites
STMS	Sequence Tagged Microsatellite sites
RAPD	Random Amplified Polymorphic DNA
AP-PCR	Arbitrarily Primed Polymerase Chain Reaction
DAF	DNA Amplification Finger Printing
SSR	Simple Sequenced Repeats
CAPS	Celeaved Amplified Polymorphic Sequence
MAAP	Multiple Arbitrary Amplicon Prefilling
SCAR	Sequenced Characterized Amplified Region
SNP	Single Nucleotide Polymorphism
SAMPL	Selective Amplification of Microsatellite Polymorphic Loci
ISSR	Inter-Simple Sequence Repeat
ALFP	Amplified Fragment Length Polymorphism
S-SAP	Sequence specific Amplified Polymorphism
RBIP	Retrotransposon-Based Insertional
IRAP	Inter-Retrotranspososon Amplified Polymorphism
REMAP	Retrotransposon-Microsatellite Amplified Polymorphism

The molecular markers have some desirable properties:

- It must be polymorphic as it is the polymorphism that is measured for genetic diversity studies.
- **Codominant inheritance:** The different forms of a marker should be detectable in a diploid organism to allow discrimination of homo- and heterozygous.
- The marker should be evenly and frequently distributed throughout the genome.
- It should be easy, fast cheap to detect.
- It should be reproducible.
- It should allow a high exchange of data between laboratories. Unfortunately, there is no single molecular marker, which meets all these requirements.

Non-PCR based Approaches

- Those are based on DNA-DNA hybridization between a DNA or RNA probe and total genomic DNA, e.g., RFLP (Restriction Fragment Length Polymorphism).

PCR based Approaches

- Those based on the PCR amplification of genomic DNA fragments: Random amplified polymorphic DNA (PARD), microsatellites or simples sequences repeat polymorphic (SSRP), amplified fragment length polymorphism (AFLP), arbitrarily primed PCR (AP-PCR), SCAR, SNP, etc.
- However, some assays combine features from both (e.g., retrotransposon-based insertional polymorphism-RBIP).

Targetted PCR and Sequencing:

Those based on sequencing: sequenced tagged sites (STS), sequence characterized amplified region (SCARs) sequence tagged microsatellites (STMs), cleaved amplified polymorphic sequences (CAPS) , etc.

NON-PCR based Approaches

Restriction Fragment Length Polymorphism (RFLP)

Procedure

- Isolate the DNA.
- Cut the DNA into smaller fragments using restriction enzymes (s).
- Separation of DNA fragments by gel electrophoresis.
- Transfer the DNA fragments to nylon or nitrocellulose membrane filter.
- Visualize specific DNA fragments using labeled probes (Incubation with a probe and allow for hybridization).
- Southern Hybridization (Autoradiography)
- Analysis of results (RFLP pattern with positive bands) **Figure 7.1.**

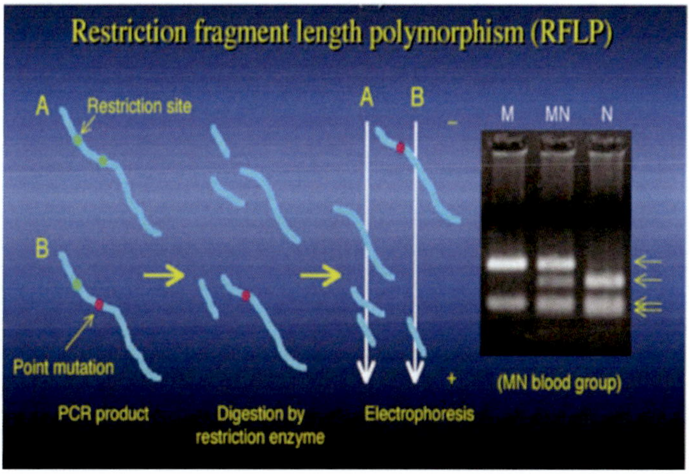

Figure 7.1: Restriction fragment length polymorphism.

Uses of RFLP

- It permits direct identification of a genotype or cultivar in any tissue at any developmental stage in an environment independent manner.

- RFLPs are codominant markers enabling heterozygotes to be distinguished from homozygous.
- It has discriminating power that can be at the species/population (single locus probes).
- The method is simples as no sequence-specific information is required.
- Indirect selection using qualitative traits.
- Indirect selection using quantitative trait loci.

PCR Based Approaches

Random Amplified Polymorphic DNA (RAPD) mArkers

Procedure

- Isolation of DNA.
- Keep the tubes in PCR thermocycler (Taq DNA polymerase, primer, dNTPs and buffer).
- Denature the DNA (94°C, 1 min).
- DNA strands separated.
- Annealing of primer (36°, 2 min).
- Primer annealed to template DNA strands (DNA synthesis 72°C, 1.5 min).
- Complementary strand synthesis (35 to 45 cycles).
- Amplified products are separated by gel electrophoresis.

Application of RAPD

- Construction of genetic maps.
- Mapping of traits.
- Analysis of the genetic structure of populations.
- Fingerprinting of individuals.
- Targeting markers to specific regions of the genome.
- RAPDs have been utilized for the identification of somatic hybrids.

- RAPDs have been recommended as a useful system for the evaluation and characterization of genetic resources.
- Some success has already been achieved in the characterization of collections from the genera Gliricidia and Thenobroma.

Limitations of RAPD

RAPD polymorphisms are inherited as dominant-recessive characters. This causes a loss of information relative to markers that show codominance.

RAPD primers are relatively short, a mismatch of even a single nucleotide can often prevent the primer from annealing, hence, there is a loss of band.

The production of non-parental bands in the offspring of known pedigrees warrants its use with caution and extreme care.

RAPD is sensitive to changes in PCR conditions, resulting in changes to some of the amplified fragments.

Amplified Fragment Length Polymorphism (AFLP).

Procedure

High-quality genomic DNA is cut with restriction enzymes (generally by two enzymes) and double-stranded (ds) oligonucleotide adaptors are ligated to the ends of the DNA fragments.

Selective amplification of sets of restriction fragments is usually carried out with ^{32}P labeled primers designed according to the sequence of adaptors plus 1-3 additional nucleotides. Only fragments containing the restriction site sequence plus the additional nucleotides will be amplified.

Gel analysis of the amplified fragments. The amplification products are separated on highly resolving sequencing gels and visualized using autoradiography.

Fluorescent or silver staining techniques can be used to visualize the products in cases where radiolabeled nucleotides are not used in the PCR.

Advantages of AFLP

This technique is extremely sensitive.

It has high reproducibility, rendering it superior to RAPD.

It has wide-scale applicability, proving extremely proficient in revealing diversity.

It discriminates heterozygotes from homozygotes when a gel scanner is used.

It permits the detection of restriction fragments in any background or complexity, including pooled DNA samples and cloned DNA segments.

It is not a single fingerprinting technique but it can be used for mapping also.

Disadvantages

It is highly expensive and requires more DNA that is needed in RAPD (1 µg per reaction).

It is technically more demanding than RAPDs as it requires the experience of sequencing gels. AFLPs are expensive to generate as silver staining, fluorescent dye or radioactivity detects the

Subjective Questions

1. What is the Spectral Method of Analysis?
2. What are UV Visible Spectrometers?
3. Describe the absorption instrumentation?
4. What are the single & double beam instruments?
5. Define instrumentation?
6. What are infrared spectrometers?
7. What is atomic absorption spectroscopy?
8. What are flame emission photometers?
9. Explain the chromatography basics?
10. What is the operation of chromatography?
11. Describe the chromatographic behavior of solutes?
12. Explain the column efficiency and column resolution?
13. What is gas chromatography?
14. Explain the detection system of gas chromatography?
15. What is gas-solid chromatography?
16. Explain liquid chromatography?
17. What is high-pressure liquid chromatography?
18. Describe the detection system of liquid chromatography?

19. What are amino-acid analyzers?
20. Explain the instrumentation of x-ray spectroscopy?
21. What are the X-Ray diffractometers?
22. Define X-Ray absorption meter?
23. What is X-Ray spectrometry?
24. Explain the X-Ray fluorescent spectrometer?
25. What is the total reflection X-Ray fluorescence spectrometer?

REFERENCES

1. Abollino, O., Giacomino, A., Malandrino, M., Mentasti, E., Aceto, M., Barberis, R., (2006). Assessment of metal availability in a contaminated soil by sequential extraction. Water Air Soil Pollution, 173(1-4), 315-338.
2. Adopted from Introduction to Bioinformatics Lopresti BioS 10, November 2009.
3. Agarwal, V. K, (2000). Molecular Biology. New Delhi: S. Chand.
4. Agarwal, V. K. and P. S. Verma, (2000). Concepts of Molecular Biology. New Delhi: S. Chand.
5. Ahmadpour, P., Ahmadpour, F., Mahmud, T., Abdu, A., Soleimani, M., Tayefeh, F.H, (2014). Phytoremediation of heavy metals: a green technology. *In:* Asrari, E. (Ed), heavy metal contamination of water and soil: Analysis, Assessment, and Remediation strategies. CRC Press, Taylor and Francis, 249.
6. Aldrich, J., and Vaker, R, (1970). Plant Dis. Reptr. 54, 446-448.
7. Alexander, M, (1977). Introduction to Soil Microbiology, John Wiley & Sons, New York.
8. Alexander, Renee R., and Joan M. Griffiths. *Basic Biochemical Methods,* 2nd ed. New York: Wiley-Liss, 1993.
9. Ali, H., Khan, E., Sajad, M.A, (2013). Phytoremediation of heavy metals-concepts and applications. Chemosphere, 91(7), 869-881.

10. Amirato, P.V, (1983). In Handbook of Plant Cell Culture. Vol. I, Techniques for Propagation and Breeding (Eds. D.A., Evans, W.R., Sharp, P.V., Amminrato and Y., Yamada), PP. 82-123, Macmillan., New York.
11. Anand, A., et al, (2004). Physiol. Biol. 10, 233-242.
12. Aneja, K.R, (2014). Experiments in Microbiology, Plant Patholoy and Biotechnology. Fourth Edition. New Age International (P) Limited Publisher.
13. Aneja, K.R, (2014) Laboratory Manual of Microbiology and Biotechnology. Scientific International PVT. LTD Medtech,
14. Anonymous (1980 a) Nature (London) 286,110.
15. Anonymous (1980 b). NBRI Newsletter 7, 28.
16. Atkinson, G.F, (1961). Mushrooms, 2nd ed. Hofner Publishing Co., New York.
17. Baker, R., and Nash, S.M, (1965). Phytopath. 55, 1381-1382.
18. Baker, K.F., and Snyder W.C, (1965). Ecology of Soil Borne Plant Pathogens-Prelude to Biological Control, John Murray, London, P. 571.
19. Bollag, Daniel M., and Stuart J. Edelstein. *Protein Methods.* New York: Wiley-Liss, 1991. Scopes, Robert K. *Protein Purification: Principles and Practice,* 3rd ed. New York: Springer Verlag, 1994.
20. Banerjee, P.K, (2011). Problems on genetics molecular genetics and evolutionary Genetics.
21. Bhattacharya, D., and R. Choudury, (2010). Statistcs theory and Practices.
22. Barabasz, A., Wilkowska, A., Tracz, K., Ruszczyñska, A., Bulska, E., Mills, R.F, (2013). Expression of *HvHMA2*in tobacco modifies Zn-Fe-Cd homeostasis. Journal of Plant Physiology, 170, 1176-1186.
23. Barell, B.G., Air, G.M., and Hutchindon, C.A, (1976). Nature. 264, 34-41.
24. Becking, J.H, (1982). In Advances in Agricultural Microbiology, eds. N.S. Subba Rao), pp. 89-110, Oxford & IBH Publish, Co., New Delhi.
25. Bergersen, F.J, (1971). Ann. Rev. Plant Physiol. 22, 121-140.
26. Betsch, David F, (2006). "DNA Fingerprinting in Human Health and Society." Retrieved: 12th April 2006 http://www.accessexcellence.org/RC/AB/BA/DNA_Fingerprinting_Basics.html

27. Carlson, P.S, (1973). Science 180, 1366-1368.

28. Bhaskaran, S, (1987). In Biotechnology in Agriculture, (eds. Natesh, S., Chopra, V.L., and Ramachandan, S.,) PP. 101-108, Oxford &IBH Publ. Co., New Delhi.

29. Biello, David, (2006). Tobacco Plant Transformed into Plague Vaccine Factory. Retrieved 10 Jan. 2006 http://www.sciam.com/article.cfm?chanID=sa016&articleID=000E3140-EB2E-13C2-AB2E83414B7F0000

30. Binns, A.N, (1981). In Propagation of Higher Plants Through Tissue Culture: Emerging Technoloies and Strategies, (eds. Constantin, M.J., Henke, R.R., Hughes, K.W., and Conger, B.V.,) Enviorn Exp. Bot. 21, 325-332.

31. Biotechnology: A Dutch Perspectives (1981). A report by the Netherlands Study center for Technology Trends, (ed. J.H.F. Van Apeldoorn).

32. Boffey, S.A, (1987). In Biotechnology: The Biotechnology Principles (eds. Trevan, M.D., Boffey, S., Goulding, K.H., and Stanbury, P.) pp 11-152. Open University Press: Milton Keynes, England.

33. Bouwer, E.J., and Zehnder, AJ. B, (1993). Trends Biotechnology. 11, 287-318.

34. Bouwer, E.J. et al, (1998), In Biological Treatment of Hazardous Wastes (eds. Lewandowaski, G.A., and Deflippi, L.J.) pp 237-270, John Wiley & Sons.

35. Brooks, R., Lee, J., Reeves, R.D., Jaffre, T, (1977). Detection nickeliferous rocks by analysis of herbarium specimens of indicator plants. Journal of Geochemical Exploration, 7, 49-57.

36. Bull, A.T., Holt, G., and Lilly, M.D, (1983). Biotechnology: International Trends and Perspectives, p. 84, Oxford & IBH Publ. Co. PVT. Ltd., New Delhi.

37. Carlson, P.S, (1983). Cell and Tissue Culture Techniques for Cereal Crop Improvement, pp. 407-418, New York.

38. Chaudhary, K., Khan, S, (2016). Physiochemical characterization of Fluoride (F) contaminated soil and its microbe-assited bioremediation by *Prosopis juliflora*. J. Plant Biology and Soil Health, 3(2), 8.

39. Chaudhary, K., Sumira, J., Khan, S, (2016). Heavy metal ATPase (*HMA2, HMA3 & HMA4*) genes in hyperaccumulation mechanism of heavy metals. Plant metal interaction (Emerging Remediation Techniques), 545-552.

40. Chaudhary, K., Agarwal S., Khan, S, (2018). Role of phytochelatins (PCs), metallothionines (MTs) and heavy metal ATPase (HMA) genes in heavy metal tolerance. R. Prasad (ed.), mycoremediation and environmental sustainability, fungal biology, https://doi.org/10.1007/978-3-319-77386-5_2 39-60 Springer.
41. Chaudhary, K., Khan, S, (2018). Role of plant growth promoting bacteria (PGPB) for bioremediation of heavy metals: an overview. (biostimulation remediation technologies for groundwater contaminants-IGI-GLOBAL).
42. Chawla, H.S, (2009). Introduction to Plant Biotechnology. Third edition Oxford & IBH Publishing Co. Pvt. Ltd (A unit of CBS Publishers and Distributors Pvt Ltd).
43. Christov, L., and Van Driessel, B, (2003). Indian J. Biotech. 2, 444-450.
44. Clark, D.S, (1962). Biotehnol Bioeng. 4, 1721.
45. Cohen, S.N., Chang, A.C.Y., Boyer, H.W., and Helling, R.B, (1973). Proc. Nat. Acad. Sci. (U.S.A) 70, 3240-3244.
46. Collins, H.A., and Watts, M, (1983). In Handbook of Plant Cell Culture: Techniques for Propagation and Breeding, Vol. I (eds. Evans, D.A., Sharp, W.T., Ammirato, P.V., and Yamada, Y.) pp. 729-747. Macmillan, New York.
47. Crueger, W., and Crueger, A, (1984). Biotechnology-A Text Book of Industrial Microbiology. Sinaver Associates Inc., Sunderland, USA, pp 23, 207-216.
48. Cummings, Michael R., and Williams S. Klug, (2004). Concepts of Genetics. New Delhi: Pearson Education.
49. Dodds, J.H., and Roberts, L.W, (1985). Experiments in Plant Tissue Culture, 2nd Ed. Cambridge University Press, Cambridge.
50. Dubey, R.C, (2009). A textbook of Biotechnology. S. Chand and Company LTD.
51. Ferber, Dan, (2004). “The New Science of Cell Hacking.” *Popular Science*, June pp: 43-44.
52. Frazier, W. C. and D. C. Westhoff, (1993). Food Microbiology. New Delhi: Tata McGraw-Hill.
53. Fisher, R.A. Statistical methods for Research Workers.

54. Gautam, N., Verma, P.K., Verma, S., Tripathi, R.D., Trivedi, P.K., Adhikari, B., Chakrabarty, D, (2012). Genome wide identification of rice class I metallothionein gene: tissue expression patterns and induction in response to heavy metal stress. Functional and Integrative Genomics, 12, 635-647.
55. Garg, N., Garg, K.L., Mukerjii, K.G, (2013) Laboratory Manual of Food Microbiology. I.K. International Publishing House Pvt. Ltd.
56. Gibbs, W, (2001). "Cybernetic Cells." Scientific American, Aug. pp: 54-57
57. Greipsson, S, (2011). Phytoremediation, National Education Knowledge, 3(10), 7.
58. Gritfith, A., Gelbart, W.M., Miller, S.H., and Lewontin, R.C., Modern genetic analysis.
59. Gu, C.S., Liu, L.Q., Deng, Y.M., Zhu, X.D., Huang, S.Z., Lu, X.Q, (2015). The heterologus expression of the Iris lacteal var. chinensis type 2 metallothionenin IIMT2b, gene enhances copper tolerance in *Arabidopsis thaliana*. The Bulletin of Environmental Contamination and Toxicology, 94, 247-253.
60. Guha, S., Maheswari, S.C, (1964). Nature (London), 204, 297.
61. Guha, S., Maheswari, S.C, (1966). Nature 212, 97-98.
62. Gupta, C.P., Sharma, A., Dubey, R.C., and Maheshwari, D.K, (1999). Cytobios. 99, 183-189.
63. Gupta, C.P., Dubey, R.C., Kang, S.C., Maheshwari, D.K., Current Science. 81, 91-94.
64. Haberlandt, G, (1902). Sber Akad. Wiss. Wien. III: 69-92.
65. Hanson, Earl D, (1983) Recombinant DNA Research and the Human Prospect. Washington, DC. American Chemical Society.
66. Henshaw, G.G, (1982). In plant Tissue Culture, pp. 789-792, Tokyo.
67. Huang, G.Y., and Wang, Y.S, (2010). Expression and characterization analysis of type 2 metallothionein from grey mangrove species (*Avicennia marina*) in responses to metal stress. Aquatic Toxicology, 99, 86-92.
68. ICGEB, (1998). Activity Report, 1998.
69. Khoshoo, T.N, (1988). Enviornmental Concern and Strategies. II Ed. PP. 253-404. Ashish Publ. House, New Delhi.
70. King, P.J, (1980). Int. Rev. Cytol. Supplement. 11A, 25-54.

71. Kumar, H.D, (2003). Modern Concepts of Biotechnology. New Delhi: Vikash Publishing House.
72. Lee, S, (2014). Artifical induction of cadmium tolerance and its further enhancement via heterologus co-expression of SpHMT1 and AtPCS1 in the yeast cells. Journal of Korean Society Applied Biological Chemistry. 57, 307-310.
73. Lopez, D. A., R. M. Williams., and Michlke, K, (1994). *Enzymes: The Foundation of Life*. München: The Neville Press.
74. Lund, Pete. "What is Bioinformatics?" Bioscience Course Manual, The University of Birmingham. pp: 35.
75. Ma, Y., Rajkumar, M., Luo, Y., Freitas, H, (2011). Inoculation of endophytic bacteria on host and non-host plants-effects on plant growth and Ni uptake. Journal of Hazard Materials, 195, 230-237.
76. Maestri, E., Marmiroli, M., Visioli, G., Marmiroli, N, (2010). Metal tolerance and hyperaccumulation: costs and trade-offs between traits and environment. Environmental and Experimental Botany, 68(1) 1-13.
77. Maheswari, S.C., Rasid, A., and Tyagi, A.K, (1983). IAPTC Newsletter 41, 2-9.
78. Mandels, M., and Stenberg, D, (1976). J. Ferment. Technol. 54, 267-286.
79. Mandal, P.K, (1988). Gene Therapy. Sci. Reptr. 25, 156-157.
80. Mason, H.S., and Arntzen, C.J, (1995). Trends Biotechnol. 13, 388-392.
81. Maxam, A.M., and Gilbert, W, (1977). Proc. Natl. Acad. Sci. (U.S.A) 74, 560-564.
82. Memon, A.R., Schroder, P, (2009). Implications of metal accumulations mechanisms to phytoremediation. Environmental Science and Pollution Research, 16(2), 162-175.
83. Miller, C.O., Skoog, F., Saltz, M. and Strong, F.M., (1955). J. Am. Chem. Soc. 77, 1992.
84. "Molecular Biotechnology in Life." Author not mentioned. Retrieved: 5th April 2006 http://www.goldenessays.com/free_essays/4/technology/molecular-biotechnology-in-life.shtml.
85. Morgan, T.H., (1901). Regerantion, Macmillian: London.
86. Nitsch, C, (1974). C.R. Acad. Sci. (Paris), 278, 1031-1034.
87. Murashige, T, (1974). Ann. Rev. Plant Physiol 25,135-166.

88. Peters, P, (1993). Biotechnology- a guide to genetic engineering. Wmc. Brown Publishers, Dubuwue, P. 253.
89. Prakash, J., and Giles, K.L, (1987). Int. Rev. Cytol. 107, 273-292.
90. Prescott, L.M., Harley, J.P., and Kein, D.A, (2003). Microbiology. McGraw-Hill High Education, New York.
91. Purohit, S, (2005) Agricultural Biotechnology. India: Agrobios.
92. Purohit, S, (2004) Biotechnology: Fundamental and Applications. India: Agrobios.
93. Rajkumar, M., Ae, N., Prasad, M.N.V., Freitas, H, (2010). Potential of siderophore-producing bacteria for improve heavy metal phytoextraction. Trends Biotechnology, 28(3), 142-149.
94. Reinert, J., and Bajaj, Y.P.S, (1977). In Applied and Fundamental Apects of Plant Cell, Tissue and organ ulture (eds. Reinert, J., and Bajaj, Y.P.S) pp. 251-267, Springer-Verlag Berlin.
95. Roberts, R.J, (1983). Nuceilc Acid Res. 11, 135-168.
96. Robson, R.L., (1979). Biochem. J. 181:569-575.
97. Ruschel, A.P., and Vose, P.B, (1983). In Hand Book of Plant Cell Culture-Techniques for Propagation and Breeding, Vol. 1 (eds. Evans, D.A; Sharp, W.R; Ammirato, pp. and Yamada, Y.) pp 828-851, Macmillian Publ. Co., New York.
98. Russel, P.J, (1998). Genetics (5th edition).
99. Sagar, G.V. (2008). MCQs in Biochemistry. New Age International (P) Limited Publishers New Delhi ISBN (13) 978-81-224-2627-4.
100. Sagar, G.V. (2008). MCQs in Microbiology. New Age International (P) Limited Publishers New Delhi ISBN (13): 978-81-224-2931-2
101. Sasson, A, (1984). Biotechnology: Challenges and Promises, UNESCO, paris.
102. Saxena, H.C, (2016). Elementary statistics.
103. Schell, J., and Van Montagu, M, (1983). Biotechnology 1, 175-180.
104. Schesinger, H, (1999). DNA Conductor. Popular Science, Aug. 47.
105. Schlegel, H.S, (1986). Genera Microbiology, 6th Ed. (Tranlated by M. Kogut) Camcridge University Press, London.
106. Sharp, W.R., Sondahl, M.R., Caldas, L.S., and Maraffa, S.B, (1980). Hort. Rev. 2, 268-310.

107. Shrof, Joannie M. "Miracle Vaccines." *US News & World Report*, 23 Nov. 1998. pp: 157
108. Singh, Mahabir, and Sandeep A, (2004). "Applied Genetics." MTBiology Today, April 27-29.
109. Skoog, F, (1944). Am. J. Bot. 31, 19-24.
110. Smith, P.H, (1966). Dev. Industrial Microbiol. 7, 156-159.
111. Southern, E.M, (1975). J. Mol. Biol. 98, 503-517.
112. Snustad, D. Peter, and Simmons, M.J, (2005). Principle of genetics (4th edition).
113. Srivastava, S.K., et al, (1995). Indian J. Exp. Biol. 33, 962-966.
114. Strick-Berger, M.W, (2012). Genetics (3rd edition).
115. Standfield, W.D., Schaums, (1969). Outline Theory & Problems is Genetics (2rd edition).
116. Swaminathan, M.S, (1980). Second Annual Day Lecture, NBRI, Lucknow.
117. Tangahu, B.V., Sheikh Abdullah, S.R., Basri, H., Idris, M., Anuar, N., Mukhlisin, M, (2011). A review on heavy metals (As, Pb, and Hg) uptake by plants through phytoremediation. International Journal of Chemical Engineering, 31, 939161.
118. Thomas, E and Davey, M.R, (1975). From Single Cell to plants, Wyked=ham: London.
119. White, P. R, (1934). Plant Physiol. 9, 585-600.
120. Willams, J.G, (1981). In Genetic Engineering, Vol. I, (ed. Williamson, R.,), PP. 1-59, AcademicPress London.
121. Wiseman, A, (1975). Handbook of Enzyme Technology, Halsted Press, New York.
122. Wikipedia Online Encyclopedia, Retrieved 20 Feb. 2006 http://www.wikipedia.org/wiki/Biotechnology.
123. Wikipedia Online Encyclopedia, Retrieved 20 Feb. 2006 http://www.wikipedia.org/wiki/genetic_engineering.
124. Wikipedia Online Encyclopedia, Retrieved 20 Feb. 2006 http://www.wikipedia.org/wiki/genetically_modified_food.
125. Wikipedia Online Encyclopedia, Retrieved 22 Feb. 2006 http://www.wikipedia.org/wiki/bioinformatics

126. Yamada, Y., aand Fujita, Y, (1983). In Handbook of Plant Cell Culture-Technique for Propagation and Breeding. (eds. Evans, D.A; Sharp, W.R., Ammirato and Yamada, Y) pp. 717-728. Macmillian Publ. Co., New York.

127. Zar, Jerrold, H, (1999). Biostatistical Analysis (4th edition).

INDEX

H

I

K

L

M

N

O

Kolibri 360 GmbH
Ferdinand-Jühlke-Straße 7
99095 Erfurt, Deutschland
produktsicherheit@kolibri360.de